From the Maluku to Molecules

Oliver Kayser

From the Maluku to Molecules

How Natural Substances Write History

Oliver Kayser
Fakultät Bio- und Chemieingenieurwesen
Technische Universität Dortmund
Dortmund, Germany

ISBN 978-3-662-69922-5   ISBN 978-3-662-69923-2 (eBook)
https://doi.org/10.1007/978-3-662-69923-2

Translation from the German language edition: "Von den Molukken zu Molekülen" by Oliver Kayser, © Der/die Herausgeber bzw. der/die Autor(en), exklusiv lizenziert an Springer-Verlag GmbH, DE, ein Teil von Springer Nature 2024. Published by Springer Berlin Heidelberg. All Rights Reserved.

This book is a translation of the original German edition "Von den Molukken zu Molekülen" by Oliver Kayser, published by Springer-Verlag GmbH, DE in 2024. The translation was done with the help of an artificial intelligence machine translation tool. A subsequent human revision was done primarily in terms of content, so that the book will read stylistically differently from a conventional translation. Springer Nature works continuously to further the development of tools for the production of books and on the related technologies to support the authors.

© The Editor(s) (if applicable) and The Author(s), under exclusive license to Springer-Verlag GmbH, DE, part of Springer Nature 2024

This work is subject to copyright. All rights are solely and exclusively licensed by the Publisher, whether the whole or part of the material is concerned, specifically the rights of translation, reprinting, reuse of illustrations, recitation, broadcasting, reproduction on microfilms or in any other physical way, and transmission or information storage and retrieval, electronic adaptation, computer software, or by similar or dissimilar methodology now known or hereafter developed.
The use of general descriptive names, registered names, trademarks, service marks, etc. in this publication does not imply, even in the absence of a specific statement, that such names are exempt from the relevant protective laws and regulations and therefore free for general use.
The publisher, the authors and the editors are safe to assume that the advice and information in this book are believed to be true and accurate at the date of publication. Neither the publisher nor the authors or the editors give a warranty, expressed or implied, with respect to the material contained herein or for any errors or omissions that may have been made. The publisher remains neutral with regard to jurisdictional claims in published maps and institutional affiliations.

Cover Illustration: © Epic Photos/Generated with AI/Stock.adobe.com

This Springer imprint is published by the registered company Springer-Verlag GmbH, DE, part of Springer Nature.
The registered company address is: Heidelberger Platz 3, 14197 Berlin, Germany

If disposing of this product, please recycle the paper.

Preface

From the Moluccas to the molecules? What kind of strange book title is that, my granddaughter asks me. I admit, the title is indeed peculiar and at second glance it reveals that the Moluccas are not just an archipelago in Indonesia between the large islands of Sulawesi in the west and New Guinea in the east, but have actually written a piece of natural substance and pharmacy history. The Moluccas also bear the name "Spice Islands", because from the 16th century first the Portuguese and from 1602 the Dutch with the Dutch East India Company traded for over 250 years. Above all, the brisk trade in nutmeg and cloves, which were grown on the islands, made the Dutch very rich. In the heyday of Dutch trade, nutmegs were weighed in gold, so coveted were they in Europe. Modern phytochemistry has been able to prove that the nuts contain alkaloids similar to the psychoactive substance ecstasy, which is why in the hippie era of the wild 1960s and 1970s in the USA, the powder of today's cheap supermarket nutmeg was taken when there was no LSD at hand. This will be discussed later when it comes to the question of whether so-called drugs do not also have the potential for innovative active ingredients (Fig. 1).

The Moluccas have also written world history and changed our political world to an extent that we cannot imagine. Perhaps it would be just as natural for us today that Dutch, not English, is spoken in the USA, if the British had not exchanged the small island of Run for the much larger island of Manhattan on April 18, 1667. This island was the easternmost of the Moluccas and the Dutch realized that they had no chance in the war against the naval power England. Trade would bring in more than expensive

Fig. 1 Oliver Kayser

wars against the rival. In the Peace of Breda (1667), they demonstrated great negotiating skills. Thus, Nieuw Amsterdam, which the Dutch had bought from the indigenous people of North America for the equivalent of 24 dollars in 1624 and was managed by Petrus Stuyvesant until 1667, became New York, because the British crown awarded it to the Duke of York.

Dear readers, you can already see how exciting the history of natural substances can be, and you will read further below that medicinal plants and later pharmaceuticals have influenced our civilization and all of our lives more than we are aware of today. I look forward to telling you some stories to show that chemistry and pharmacy are more than the work of nerds and quirky scientists in the lab. This is the right keyword to write a little about me, the author.

Allow me to introduce myself briefly. My name is Oliver Kayser, I am a pharmacist and natural product biotechnologist at the Technical University of Dortmund. For over 30 years, I have been dealing with medicinal plants, their extraction, and the pharmacological testing of natural substances. For about 15 years, I have been investigating the biotechnological production of plant natural substances and want to know if plant biosynthesis pathways can be replicated in baker's yeast. Why I am so passionate about natural substances and plants, I cannot explain. But I have always been fascinated by the enormous variety of chemical structures in nature and during my studies, I enjoyed reading the stories and anecdotes behind the first medical applications and medicinal plants. If one reads the stories and sometimes curious reports in the individual chapters with a smile, it should become clear that the history of natural substances and medicinal plants is also a history of toxicology, that is, the study of poisons, because out of ignorance about dosage and effect, not a few people have poisoned and killed themselves.

This fateful proximity, that a substance can be poison or remedy, I found great as a student and it aroused my interest to go into science and not to work as a pharmacist in a pharmacy. After my pharmacy studies in the beautiful Münster in Westphalia, I accepted the invitation of my mentor and later doctoral father Prof. Kolodziej to do my doctorate with him in Berlin. The topic was unusual and aroused my interest, as it was about the characterization of the ingredients of a South African medicinal plant, which today is very popular as the drug Umckaloabo® during the cold season. I had no idea during my doctoral thesis that this herbal medicine would experience such a boom. The changing history behind the plant was very exciting and unknown to me like many others treated with it. I researched in London how the plant came to England and Europe. Over 100-year-old documents on phytochemistry, effect, and a court case that I could view fascinated me and led to a chapter in this book. The court case Umckaloabo® and the suspicion of biopiracy by a company reappear in different colors again and again with medicinal plants, and it is exciting to learn that natural substances have also written legal history.

I was able to conduct a portion of my scientific studies at the Robert Koch Institute in Berlin under my current friend Dr. Kiderlen, who trained me excellently in the art of cell culture. For this, I am still very grateful to him today.

My own research began in the Netherlands at the University of Groningen. What comes to mind when one conducts research in the Netherlands? Of course, cannabis. In 2004, I had the idea to investigate whether tetrahydrocannabinol (THC), the psychoactive substance in the flowers of the cannabis plant, could also be genetically engineered in yeasts. At the beginning of the new millennium, scientists were very enthusiastic about the combination of molecular biology and genetics that formed synthetic biology. It seemed as simple as it was tempting to combine DNA as the genetic blueprint of any organism like Lego bricks and create new biosynthesis pathways. That's what I wanted to do, and the THC biosynthesis was to become my playground. It has remained so even after my move to TU Dortmund until today. It is incredible how complex nature is and how much there is still to explore, even if one believes that everything should be much simpler through genetics and modern bioinformatics.

But my love for natural substances and biochemistry has remained. I find it incredible what researchers discover about how natural substances work in our body and which biochemical networks they activate. This knowledge is the basis of modern drug research, even though many active ingredients today are synthetic substances or therapeutic proteins like insulin or interferons. A scientific study from the USA shows that about 40% of today's drug substances (excluding proteins) have their chemical origin in natural substances from plants or microorganisms. This is a very high number that will not decrease in the future, because nature is far ahead of us in chemical research in its creativity.

I would like to make a small note. For the sake of readability, I have not consistently chosen to use gender-neutral language.

You will alternately find the male and female designation. I have also taken the liberty of using the male designation in historical periods when it is highly likely that no female scientists were active in the working group at the mentioned location. The conversations between scientists described in some chapters will certainly not have taken place in the same wording. For dramatic reasons, I have taken the liberty of putting these words into the mouths of the wise men and women. Original quotes are always set in italics. This book is aimed at the interested layperson without chemical or medical prior knowledge. All scientists may forgive me if the correct botanical or chemical designation according to IUPAC is missing, as it is not my intention to write a specialist book for the circle of experts. Dear colleagues, you will surely find this in your textbook at the back of your bookshelf. The technical terms that cannot be paraphrased, I have explained in a glossary.

What distinguishes this book from a scientific work? In narrative form, I try to present the history of natural product chemistry as vividly and understandably as possible. Certainly, one or the other will miss their personal favorite natural substances. I'm sorry for that, but there are too many exciting natural substances and stories that can't be squeezed into just under 700 pages between two book covers. But there is hope that they may still make it to the stage in a second edition or another book and the curtain will rise.

Do not be afraid of chemical formulas and technical jargon, as I try to avoid them. In a profile, you will always find a chemical structural formula and some information about chemistry. Rest assured, even without reading these profiles, you will understand everything. The most important questions that are answered have little to do with chemistry and everyone is more interested in how natural substances have significantly influenced our lives to this day. What does cocaine have to do with grateful local anesthetics, why did a great love lead to the invention of the latex glove, and what does quinine have to do with tea in Sri Lanka?

I wish you a lot of fun reading this book and I am also looking forward to your criticism, suggestions, and of course, any hints if I have forgotten or misinterpreted something. Please send me an email at m2m@posteo.de.

Dortmund
in January 2024

Oliver Kayser

Acknowledgment

For you, dear Susa, with great thanks for your love and the great effort in reviewing with all the many hints to make the book readable for ordinary people.

Dear Franz-Josef, wherever you are now and reading this book, I thank you for your conversations about pharmacology and I know, if the world were to end tomorrow, you would still be planting cannabis today.

Preliminary Remark

The conversations that I have had my scientists conduct probably did not take place in this way. As a stylistic device and to give the reader an impression of how scientists, in my experience, think and decide, I have allowed myself to invent some dialogues from the literature. If it is such dialogues, they are presented in the usual upright script. If they are quotes, they are always set in italics.

I have not consistently used the female and male form, as it does not seem suitable to me in some places for the flow of reading. I also refrained from using the female address in the historical context when there were demonstrably no female members in a work group, institute, or organization. This is regrettable from today's perspective, but it corresponds to the historical facts.

Due to the mention of many medications, active pharmaceutical ingredients, and pharmaceutical companies, one might assume that the author has a connection to a pharmaceutical company. I have no conflict of interest and am not paid directly or indirectly by the pharmaceutical industry in third-party projects. My doctoral thesis (1992–1997) was on the chemical analysis and pharmacological testing of Umckaloabo, but there was never any funding or personal grant.

Contents

Morphine—the Gift of Morpheus	1
Opium Makes International Politics	13
Heroin	19
Synthetic Opioids	24
Dealers in white coats—How Oxycodone Makes America's Middle Class Addicted	32
Cocaine—from Drug to Local Anesthetic	37
The Path of Cocaine to Local Anesthetics	55
Aspirin®—the Most Used Drug in the History of Mankind®	71
The Invention and Worldwide Success of Aspirin®	89
Not Only the German Spirit Should be Healed by Aspirin®	92
On the Trail of the Mechanism of Action of Acetylsalicylic Acid	101
New Painkillers Follow Aspirin®	109
Paracetamol	110
Ibuprofen	114
Steroids and the Pill	119
Margaret Sanger—the Women's Rights Activist and Her Fight for the Contraceptive Pill	126
Margaret Sanger and Natural Product Chemistry	129

Female Sexual Hormones	132
The Path to the Contraceptive Pill	132
Adolf Windaus, the Lord of the Four Rings	140
The Birth of the Contraceptive Pill and its Midwives	150
About Black and Bright Heads	163
Does the Country Need New Men?	190
Male Sexual Hormones	191
The Discovery of Testosterone	194
Gossypol—the Pill for Men?	199
Glucocorticoids	201
Microorganisms in white coats	225
Cortisol as a Miracle Drug	230
Vitamin D? Then rather Cod Liver Oil!	232
Berlin Air and Light Against Rickets	235
Caffeine and Viagra	**245**
Caffeine, Theobromine, and Theophylline	245
About Chocolate Men in the Snow	247
Maté Tea—the Drink of the Gods	250
Coffee and Caffeine—a Popular Drug?	251
Caffeine and Theophylline—a Strong Coffee Against Asthma	255
Sildenafil—from Waterfalls and Hard Facts	260
Quinine	**269**
Kina Kina—the Miraculous Fever Tree	271
How Tonic Water and Gin Became Pretty Best Friends	274
Learning from Quinine does not Mean Learning to Win—the Lost Battle Against Malaria	287
The First Will Remain the Last	295
Coumarins—Rat Poison for the President	**297**
What the Hell—Blood, Blood, Damned Blood	299
Natural Substances as Starting Materials for Drug Synthesis	**307**
Of Sharks and Christmas Biscuits	307
Scopolamine—About Truths and Veracities	312
Scopolamine—from the Witch's Kitchen to the Pharmaceutical Lab	314
Scopolamine—from Drug to Psychotherapy	319

Scopolamine in Therapeutic Systems	321
Tiotropium and How Nuclear Power Became Something Good	323
Podophyllotoxin—Medicinal Weed	327
From Rubber to Medical Latex	332
The Invention of the Disposable Glove	335
The Alternatives to Natural Rubber	340
From Parisians and Fromsers	342
Modern Drug Research	355
Leo Sternbach and the Benzodiazepines	360
Paul A.J. Janssen—the Self-made Pharma Entrepreneur	363
Fashionable Diseases and Lifestyle Drugs	367
The Birth of the Pharmaceutical Industry	369
The State Intervenes—Drug Approval	376
Of Rats, Mice, and Humans—Animal Testing in Drug Research	378
How Should We Deal With Animal Testing?	386
Oncomouse—Probably the Most Famous Mouse in Science	388
Is Man Just a Mouse Too?	395
Cell Cultures—the Immortality of Henrietta Lacks	398
The Chimpanzee as a Pharmacist—Even Animals Know What's Good for Them	410
Healing Earth from Pharmaceutical Mining	413
Biopiracy—the Curse of Colonization	416
Kew Garden—the Marigote Bay of Biopirates?	417
The King of Biopirates	423
Biopiracy Today	424
Hoodia—the Thirst Quencher from the Dessert	426
Healing Curry	426
The Potion of the Gods from the Home Garden	426
Umcka What? The Strange Story of a Geranium	427
Are Drugs Only Bad?	437
Amphetamines—Drug or Medicine for the Fidgety Phil?	443
Ephedrine	445
Adrenaline	459
Salbutamol	468
Mothers Little Helpers and Methylphenidate	469
LSD—Lucy in the Sky of Diamonds	479

Cannabinoids   489
  Cannabis for the Treatment of Epilepsy   492
  Cannabis for the Treatment of Glaucoma   494
  The Declaration of War on Cannabis   496
  The Cannabis Moss of the Maori   501
  Cannabis—a Bright Future?   502

Healing Poisons   505
From Bee Stings and Frog Poisons   507
The Viagra of the Renaissance   509
Ergot—From Plague to Mr. Hofmann's Problem Child   513
Snakes—from Venom to Blood Pressure Reducer and the Birth of Computer Chemistry   524
Paclitaxel—an Academic Success Story   532
With Genetic Engineering for the Protection of Forests   538
Tetrodotoxin—a Devilish Delight for the Palate   540

Glossary   545

References   549

About the Author

Oliver Kayser is a pharmacist and internationally recognized university professor and natural product researcher at TU Dortmund. He is the author of many scientific publications and textbooks. His great interest lies in medicinal plants and natural substances that promoted the birth of modern medicinal substances. The focus of his work is the research of the biosynthesis of medically significant natural substances in plants, which today serve as lead structures for new active substances in cancer medicine, neurological diseases, and as antibiotics.

List of Figures

Morphine—the Gift of Morpheus

Fig. 1	Friedrich Sertürner (1783–1841) (public domain, Wikipedia)1	2
Fig. 2	Definition of the natural substance class of alkaloids by Carl F. W. Meißner (public domain, Wikipedia)1	7
Fig. 3	Gottfried Wilhelm Leibniz (1646–1716) (public domain, Wikipedia)1	10
Fig. 4	Antoine Baumé (1728–1804) (public domain, Wikipedia)1	12
Fig. 5	Opium poppy, *Papaver somniferum*, Otto W. Thomé, Flora of Germany, Austria and Switzerland 1885 (public domain, Wikipedia)1	14
Fig. 6	Packaging and trade of opium in 19th century China (Permission Wellcome Foundation, London, UK)1	15
Fig. 7	Felix Hoffmann (1868–1946) (public domain, Wikipedia)2	20
Fig. 8	Charles R. A. Wright (1844–1894) (public domain, Wikipedia)2	20
Fig. 9	Hans Walter Kosterlitz (1903–1996) 35 (Permission Royal Society of Chemistry, London, UK)2	27
Fig. 10	Otto Krayer (1899–1982) (public domain, National Library of Medicine, Bethesda, USA)2	29
Fig. 11	Synthetic Opioids (Work of the author)	31

Cocaine—from Drug to Local Anesthetic

Fig. 1	Coca plant (*Erythroxylum coca*) (public domain, Wikipedia)1	38
Fig. 2	Friedrich Wöhler (1800–1882) (public domain, Wikipedia)1	40
Fig. 3	Paolo Mantegazza (1831–1910) (public domain, Wikipedia)1	41

Fig. 4	Vin Marini, advertisement with Pope Leo XIII (public domain, Wikipedia)1	42
Fig. 5	Justus Carl Hasskarl (1811–1894) (public domain, Wikipedia)1	45
Fig. 6	Sigmund Freud, around 1905 (1856–1939) (public domain, Wikipedia)1	48
Fig. 7	Anton von Störck (1731–1803) (public domain, Wikipedia)1	52
Fig. 8	Carl Koller (1857–1944) (public domain, Wikipedia)1	56
Fig. 9	On Coca, Book 1885, Sigmund Freud (public domain, Wikipedia)1	57
Fig. 10	Richard Willstätter (1872–1942) (public domain, Wikipedia)2	59
Fig. 11	German Pharmacopoeia, Pharmacopoea Germanica, 1872 (public domain, Wikipedia)1	61
Fig. 12	Paul Ehrlich (1854–1915), 1915 (public domain, Wikipedia)1	63
Fig. 13	Hans von Euler-Chelpin (1873–1964) (public domain, Wikipedia)	66
Fig. 14	Holger Erdtman (1902–1989) (Permission, University of Stockholm, SE)	66
Fig. 15	Local anesthetics derived from cocaine (year of discovery in brackets) (Work of the author)	68

Aspirin®—the Most Used Drug in the History of Mankind®

Fig. 1	*Salix alba* (public domain, Wikipedia)1	72
Fig. 2	Cesare Bertagnini (1827–1857) (public domain, Wikipedia)1	75
Fig. 3	Raffaele Piria (1814–1865) (public domain, Wikipedia)1	76
Fig. 4	Friedrich Bayer (1825–1880), around 1863 (public domain, Wikipedia)1	78
Fig. 5	Friedrich Carl Duisberg (1861–1935), 1909, painted by Max Liebermann (public domain, Wikipedia)1	80
Fig. 6	Arthur Eichengrün (1867–1949), around 1900 (public domain, Wikipedia)2	84
Fig. 7	Heinrich Dreser (1860–1924), 2nd from right sitting, in the Pharmacological Laboratory at Bayer, circa 1897 (public domain, Wikipedia)1	84
Fig. 8	Bottle with Aspirin powder, circa 1899 (Permission Bayer Archive)1	93
Fig. 9	First page of the New York World from August 15, 1915 (public domain, Wikipedia)1	100
Fig. 10	John Robert Vane (1927–2004) (Permission Wellcome Foundation, London, UK)2	103
Fig. 11	Non-Opioid Painkillers*[Selection](in brackets year of chemical or pharmacological discovery)* (Work of the author)	109
Fig. 12	Stewart Adams (1923–2019) (Permission, Boots UK)	114

Steroids and the Pill

Fig. 1	Margaret Sanger (1879–1966) (Permission, University of Claifornia Library, Los Angeles, USA)2	122
Fig. 2	Self-built chair of the Zurich testicle bathers, around 1980, left top view and right view (Permission, Der Muger, 27.04.2017)	124
Fig. 3	Representation of the child-rich "Schmotzerin" (Barbara Stratzmann, 1448–1503) in the Cyriakus Church Bönnigheim (Permission Reinhard Kaarsch, Wikipedia)	125
Fig. 4	Katharine Dexter McCormick, suffragette, April 22, 1913 (public domain, Wikipedia)2	131
Fig. 5	Ludwig Haberlandt (1885–1932) (Permission, Museum for Contraception and Abortion, Vienna, AT)	134
Fig. 6	Steroids and steroid-active natural substances *(Selection)* *(Work of the author)*	136
Fig. 7	L. Haberlandt, Publication: The hormonal sterilization of the female organism, 1931 (Permission, Museum for Contraception and Abortion, Vienna, AT)	138
Fig. 8	Adolf Windaus (1876–1959) (Release, University of Freiburg, DE)2	141
Fig. 9	Adolf Butenandt (1903–1995)	143
Fig. 10	Heinrich Wieland (1877–1957)	144
Fig. 11	Edward A. Doisy (1893–1986)	146
Fig. 12	Edgar Allen (1892–1943), around 1933	147
Fig. 13	Hans Herloff Inhoffen (1906–1992)	152
Fig. 14	Walter J. V. Schoeller (1880–1965), 1935	153
Fig. 15	Schering main laboratory, Berlin-Wedding, 1937	154
Fig. 16	Opium-smoking Chinese in a Hamburg opium den	156
Fig. 17	Hans Herloff Inhoffen (l) and Huang Minlon (r), Schering AG, Berlin, around 1936	159
Fig. 18	Arthur Serini (1897–1945), around 1940	161
Fig. 19	Estradiol pilot plant at Schering AG, Berlin, 1935. In the background is laboratory assistant Johannes Schultze, who carried out the ethinylations for Inhoffen	162
Fig. 20	Russell E. Marker (1902–1995), right image person on the right	163
Fig. 21	Louis Fieser (1899–1977), 1965	167
Fig. 22	*Dioscorea mexicana* 1837 (public domain, Wikipedia)	169
Fig. 23	George Rosenkranz (r) and Luis E. Miramonte (l), 2001	176
Fig. 24	Carl Djerassi (1923–2015), 2004	178
Fig. 25	*Xenopus laevis,* African clawed frog	184

Fig. 26	Alan M. Turing (1912–1954)	193
Fig. 27	Bernhard Zondek (1891–1966)	196
Fig. 28	*Gossypium hirsutum* (public domain, Wikipedia)	199
Fig. 29	Tadeus Reichstein (1897–1996), circa 1950 (l) and 1970 (r)	203
Fig. 30	Leopold Ružička (1887–1976)	203
Fig. 31	Edward C. Kendall (1886–1972)	205
Fig. 32	Fritz Verzár (1886–1979)	210
Fig. 33	Hermine Raths (1906–1984) (r) and Marguerite Steiger (1909–1990) (m), 1972	217
Fig. 34	John F. Kennedy (1917–1963)	219
Fig. 35	First page of the US patent for acetylsalicylic acid by Bayer AG, 1900	224
Fig. 36	Frederick W. Heyl (public domain, www.upjohn.net)2	226
Fig. 37	Alejandro Zaffaroni (1923–2014)	228
Fig. 38	Children suffering from rickets. In: De Rachitide sive morbo puerili …, Leiden, 1672	234
Fig. 39	UV irradiation of children of factory workers in the BASF "nurse ambulance" for rickets prophylaxis, 1937	236
Fig. 40	X-ray of the right hand of the boy Arthur H. before irradiation (27.01.1919) and after three months of irradiation on 09.04.1919	237
Fig. 41	Elmer McCollum (1879–1967)	238
Fig. 42	Edward Mellanby (1884–1955)	239
Fig. 43	Harry Steenbock (1886–1967)	241

Caffeine and Viagra

Fig. 1	*Coffea arabica*	246
Fig. 2	*Camellia sinensis*	247
Fig. 3	*Theobroma cacao*	249
Fig. 4	Friedlieb Ferdinand Runge (1794–1867)	252
Fig. 5	Caffeine, physiological Messenger substances and derived Potency drugs	261

Quinine

| Fig. 1 | Tu Youyou (*1930) (r) with her tutor Lou Zhicen (1920–1995) (l) at the Chinese Academy of Medical Sciences, 1951 (public domain, Wikipedia) | 270 |
| Fig. 2 | *Cinchona officinalis L. (public domain, Wikipedia)* | 272 |

Fig. 3	The Aurich government reports on seven cases of malaria in Northern Germany, August 20, 1923 (Federal Archives, Koblenz, DE)	274
Fig. 4	William Jackson Hooker (1785–1865) (public domain, Wikipedia)1	275
Fig. 5	Clements Markham (1830–1916) (public domain, Wikipedia)1	276
Fig. 6	Richard Spruce (1817–1893) (public domain, Wikipedia)1	277
Fig. 7	Charles Ledger (1818–1905) (Permission of the Wellcome Foundation, London, UK)1	278
Fig. 8	Manuel Incra Mamani (?–1871) (public domain, Wikipedia)1	279
Fig. 9	Pierre-Joseph Pelletier (1788–1842) (public domain, Wikipedia)1	281
Fig. 10	Joseph Bienaimé Caventou (1795–1877) (public domain, Wikipedia)1	282
Fig. 11	Franz Junghuhn (1809–1864) (public domain, Wikipedia)1	284
Fig. 12	Pieter van Leersum (1854–1920) (public domain, Wikipedia)1	286
Fig. 13	William Henry Perkin (1838–1907) (public domain, Wikipedia)1	288
Fig. 14	Quinine and derived antimalarial drugs (selection) (work of the author)	290

Coumarins—Rat Poison for the President

Fig. 1	*Melilotus officinalis* (commonly known as, wikipedia)	300
Fig. 2	Alfred Winterstein (1899–1960) (gemeinfei, wikipedia)1	305

Natural Substances as Starting Materials for Drug Synthesis

Fig. 1	Squalene (work of the author)	308
Fig. 2	*Illicium verum Hook* (public domain, wikipedia)	310
Fig. 3	*Illicium anisatum L.* (public domain, wikipedia)	311
Fig. 4	Anisatin (work of the author)	312
Fig. 5	*Hyoscyamus niger L.* (public domain, wikipedia)	315
Fig. 6	Scopolamine (work of the author)	315
Fig. 7	Witches' Sabbath, Luis R. Falero, 1880 (public domain, wikipedia)	316
Fig. 8	*Cicuta virosa L.* (public domain, wikipedia)	318
Fig. 9	Cicutoxin (author's work)	319
Fig. 10	Karl Damian von Schroff (1802–1887) (public domain, wikipedia)	321
Fig. 11	Butylscopolamine (author's work)	323
Fig. 12	Tiotropium (author's work)	323

Fig. 13	*Duboisia leichhardtii* (public domain, wikipedia)	324
Fig. 14	Chemical worker, plant manager and master in the alkaloid operation at Boehringer Ingelheim (Permission Boehringer Ingelheim, Ingelheim, DE)	325
Fig. 15	*Podophyllum peltatum L.* (public domain, wikipedia)	329
Fig. 16	*Anthriscus sylvestris L.* (HOFFM.) (public domain, wikipedia)	330
Fig. 17	*Hevea brasiliensis* (public domain, wikipedia)	333
Fig. 18	William Stewart Halsted (1852–1922) (Release, Johns Hopkins University, USA)	337
Fig. 19	W. S. Halsted (1852–1922) in the operating room at Johns Hopkins Hospital, around 1900 (Release, Johns Hopkins University, USA)	339
Fig. 20	*Taraxacum kok-saghyz* (public domain, wikipedia)	341
Fig. 21	Julius Fromm (1883–1945) (public domain, wikipedia)	343
Fig. 22	The girls of the Haller Revue in their Quadriga figure at the Admiralspalast on Friedrichstraße hold up a condom (permission bpk Berlin, DE)	345
Fig. 23	Arsphenamin or Arsenobenzol (Salvarsan®) (work of the author)	346
Fig. 24	Paul Ehrlich (l) and Hata Sahachirō (r) (public domain, wikipedia)	347
Fig. 25	Julius Fromm's condom factory, Berlin-Köpenick, circa 1922	350
Fig. 26	Casanova (l) testing the tightness of a condom (Permission Landesarchiv Berlin, Photographer Henri Fromm)	352

Modern Drug Research

Fig. 1	Aspartame (work of the author)	359
Fig. 2	Leo Sternbach (1908–2005), 1991 (Permission Roche, Basel, CH)	360
Fig. 3	Finasteride (work of the author)	368
Fig. 4	Friedrich Koch (1786–1865) (Permission CK Koch Winery, Oppenheim, DE)	374
Fig. 5	Merck's Pharmaceutical-Chemical Novelties Cabinet, English, 1889 (Permission Merck AG, Darmstadt, DE)	375
Fig. 6	Development of a product, phases of drug approval, around 1920 (public domain, wikipedia)1	378
Fig. 7	Milton J. Greenman (1866–1937)	383
Fig. 8	Ralph Brinster (*1932) (public domain, wikipedia)	390
Fig. 9	Beatrice Mintz (1921–2022) (Permission from Smithsonian Institute Archives, Washington DC, USA)	392
Fig. 10	George Otto (1899–1970) and Margaret Gey, circa 1950 (Permission Johns Hopkins Hospital, Baltimore, USA)	401

Fig. 11	G. O. Gey's lab at Johns Hopkins Hospital, circa 1950 (Permission Johns Hopkins Hospital, Baltimore, USA)	403
Fig. 12	Roller bottles in G. O. Gey's lab, circa 1950 (Permission Johns Hopkins Hospital, Baltimore, USA)	404
Fig. 13	Adolf Just (1859–1936) (Permission Luvos-Heilerde, Friedrichsdorf, DE)	414
Fig. 14	Historical Luvos Healing Earth Production in Blankenburg, Harz (Permission Luvos-Heilerde, Friedrichsdorf, DE)	415
Fig. 15	Adolf Engler (1844–1930) (public domain, wikipedia)2	420
Fig. 16	*Pelargonium sidoides* DC (public domain, Wikipedia)	428

Are Drugs Only Bad?

Fig. 1	Henry Hyde Slater (1823–1871) (public domain, Wikipedia)1	444
Fig. 2	L-(-)-Ephedrine (work of the author)	445
Fig. 3	Nagayoshi Nagai (1844–1929) (public domain, wikipedia)2	445
Fig. 4	*Ephedra distachya* C.A.MEY (public domain, wikipedia)1	446
Fig. 5	Nagai couple, right Therese Nagai née Schumacher, 1896 (public domain, copyright expired) 1	448
Fig. 6	Drawing of ephedrine by Nagayoshi Nagai in his lab book (public domain, copyright expired) 1	452
Fig. 7	L-Adrenaline (work of the author)	453
Fig. 8	Avicenna (980–1037) (public domain, wikipedia) 1	455
Fig. 9	The first page with invocation of Allah from a manuscript of Avicenna's Canon from the 16th century (public domain, wikipedia) 1	456
Fig. 10	Takamine Jōkichi (1854–1922) (public domain, wikipedia)2	461
Fig. 11	R-Salbutamol (work of the author)	468
Fig. 12	Albert Hofmann (1906–2008), 1993 (Permission Novartis, Basel, CH)	479
Fig. 13	*Claviceps purpurea* (public domain, wikipedia)1	480
Fig. 14	*Turbina corymbosa* (public domain, wikipedia)1	486
Fig. 15	*Cannabis sativa* L. (public domain, wikipedia)1	490
Fig. 16	Raphael Mechoulam (1930–2023) (public domain, wikipedia)2	491
Fig. 17	Harry Jacob Anslinger (1892–1975) (public domain, thanks to Pennstate University Library and Archive)2	496

Healing Poisons

Fig. 1	Batrachotoxin (work of the author)	507
Fig. 2	Epibatidine (work of the author)	508
Fig. 3	*Lytta vesicatoria*, Spanish Fly (public domain, wikipedia)	510

Fig. 4	Cantharidin (work of the author)	510
Fig. 5	*Sorbus aucuparia* L. (public domain, wikipedia)1	511
Fig. 6	Psoralen (work of the author)	512
Fig. 7	Henry Hallet Dale (1875–1968) (public domain, wikipedia, creative commons 4.0) 2	518
Fig. 8	Miguel A. Ondetti (1930–2014) (public domain, wikipedia)2	528
Fig. 9	Monroe E. Wall (1916–2002, l) and Mansukh Wani (1925–2020, r) (Permission Research Triangle Institute, RTI)	534
Fig. 10	Paclitaxel (work of the author)	535
Fig. 11	*Taxus baccata* (public domain, wikipedia)1	538

List of Tables

Morphine—the Gift of Morpheus

| Table 1 | Heroin sales of Bayer AG from 1898 to 1925 by countries (Bayer Archive 15 D5.4 and 166/8)[13] | 23 |

Modern Drug Research

Table 1	Drugs found by chance or intuition	357
Table 2	Chemical or pharmaceutical companies founded by pharmacists in Germany (*not emerged from a pharmacy laboratory)	370
Table 3	Discovery of important alkaloids	372
Table 4	Drugs derived from natural substances	373

Are Drugs Only Bad?

| Table 1 | Famous personalities and their rarely or often highly valued plant drugs, amphetamine, and alcohol | 473 |

Morphine—the Gift of Morpheus

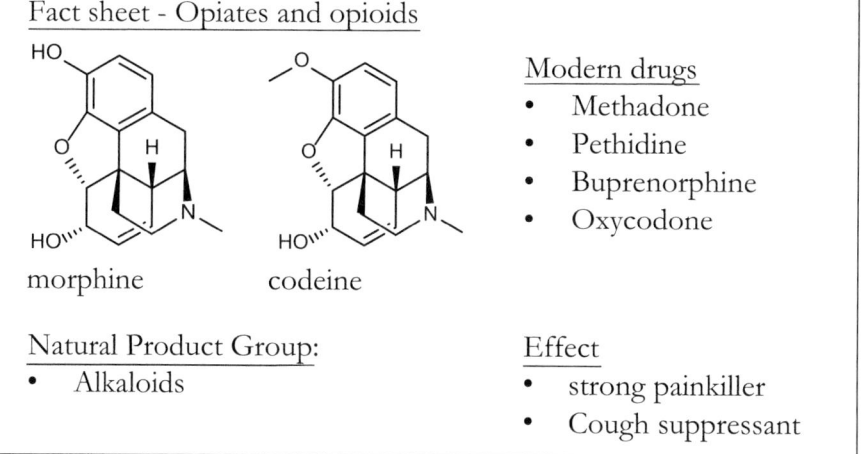

The history of morphine and its pain-relieving effect has the makings of a true scientific thriller. Morphine and Heroin have brought much quality of life to humanity in the treatment of diseases that cause severe pain, such as cancer or accidents with severe injuries. However, the history of this extraordinary natural substance has also brought much suffering when we think of drugs like heroin and Fentanyl that are produced with criminal intent and sold illegally in all societies of the world. This exciting contrast makes it interesting to take a closer look at what morphine is, where it comes from, and what we have been able to learn from this natural substance in medicine and pharmacy.

When did it all begin? Of course, a very long time ago with the ancient Egyptians and later with the Greeks. But let's leave them aside for now and travel with a time machine to a city that today hardly anyone would associate with pharmaceutical research, let alone with morphine—Paderborn. In 1804, Paderborn was a tranquil city in Westphalia, part of the Kingdom of Prussia, and known for its theological university, which was founded as early as 777. It was a turbulent time at the beginning of the 19th century when Napoleon was waging war in Europe and was to take the city without a fight two years later. It was during this time that Friedrich Sertürner (Fig. 1) worked as a pharmacy assistantin the Adler-Apotheke, after starting his apprenticeship in 1799 and passing his exam in 1803. He was born in 1783 as the son of the Austrian land surveyor Josephus Simon Sertürner near Paderborn. What prompted his father to enter the service of Prince Friedrich Wilhelm of Paderborn and Hildesheim and move to Paderborn is not known. The Sertürner family led a pleasant life, with little evidence of science and pharmacy. Was the son bored or what else ignited his scientific fire?

Fig. 1 Friedrich Sertürner (1783–1841) (public domain, Wikipedia)1

> **Chronologie**
>
> **1804** – First morphine isolation by Bernard Courtois
> **1804** – Discovery of morphine by F. Sertürner
> **1817** – Doctorate F. Sertürner in Jena on morphine
> **1818** – Publication on the medical effects of morphine
> **1820** – Technical production in France and the German states
> **1839-1842** und **1856-1860** – Opium wars in China
> **1873** – Heroin synthesis by C.R.A. Wright
> **1923** – Structural elucidation of morphine
> **1937** – Methadone synthesis by M. Bockmühl and G. Ehrhart
> **1937** – Pethidine synthesis by O. Schaumann
> **1960** – Fentanyl synthesis by P. Janssen
> **1950** – First total synthesis of morphine
> **1958** – Haloperidol synthesis by Paul Janssen
> **1967** – Tilidine synthesis
> **1975** – Discovery of enkephalins by Kostlitz

Interest in chemistry would not have been unusual at this time, as the Steve Jobs of their time, who successfully experimented in their own pharmacy labs, like Friedrich J. Merck in the Engel-Apotheke in Darmstadt or Paul C. Beiersdorf in the Merkur-Apotheke in Hamburg, were just getting started. In the 19th century, Germany was the Silicon Valley of today in California, where later Krupp, Thyssen, Mercedes and Benz or Siemens turned ideas and tinkering in the backyard into global corporations. This was not the case for Friedrich Sertürner, who after his pharmacy exam in 1809 acquired a second pharmacy in Einbeck, which belonged to Westphalia and was incorporated into the Kingdom of Prussia after the Congress of Vienna in 1817. But his son Viktor found less pleasure in the pharmacy. He had entrepreneurial genes and in 1873, as a pharmacist, he founded his own chemical factory in Hameln.[1]

It is fascinating to learn why Sertürner, in the deepest province, far away from the nearest university where chemistry or pharmacy was taught, was involved in the study of opium. He had neither a university degree nor a solid knowledge of the theory of extraction and isolation of natural substances. Perhaps this lack of knowledge was also an advantage. Uninfluenced by the prevailing doctrine that plant ingredients can only be organic acids, he avoided tunnel vision and approached his work with an open mind. Perhaps this was his luck, because before him, other chemists like Antoine Beaumé and Charles Louis Derosne in France had tried to isolate morphine. They failed to discover this natural base due to the false doctrine that misled them. Of course, Sertürner was not the only one in the world of natural substances, and the chemistry of natural substances was very much alive, especially in France. In Paris lived and worked Sertürner's competitors, of whom he might have only heard. Charles Louis Derosne (1780–1846) is worth mentioning, who, as a pharmacist and owner of the highly respected pharmacy in Rue St. Honoré in Paris, dealt with opium and morphine. It was common for chemists to conduct their research not in a university laboratory, but in their own pharmacy laboratories. Derosne extracted opium and believed he had found morphine. He called it narcotine and sent a report on his work to the prestigious Société de la Pharmacie in Paris in 1803. In it, he wrote that he had developed evidence for a crystalline salt of opium. He noticed that this salt showed poor basic properties and tried to explain this with impurities from potash. Well, there was a lot of ifs and buts. It seems that he had found an alkaloid, but whether it was really morphine can be doubted.

Today, however, we must not be arrogant and believe that only alchemists were at work back then. This was by no means the case. From history lessons, we know that with the Enlightenment and the French Revolution, the natural sciences and medicine were led out of the darkness of alchemy. The scientists of that time recognized that there must be natural laws, and they were very keen to introduce rational principles into their research. The research of that time cannot and should not be compared with our current high-performance research, in which our knowledge doubles every two to a maximum of five years. It is remarkable that 200 years ago it was hardly possible to identify the chemical structure of simple molecules without knowledge and without instrumental equipment. Until the introduction of modern techniques, a structure was only considered proven when it was fully synthesized. The exact elucidation of the chemical structure without synthesis, using only analytical methods, was not only achieved with morphine, but with other natural substances that I would like to introduce to you in this book, only many decades, yes, centuries later. For morphine, the exact structure was only clarified in 1923 by J. Masson Gulland (1898–1947) and Robert Robinson (1886–1975). It took another 27 years until

Marshall D. Gates (1915–2003) and Gilg Tschudi succeeded in a chemical synthesis.

How is modern structure elucidation conducted? Since the 1960s, two devices have made their way into the laboratories of chemists, revolutionizing structure elucidation and making it possible to examine almost any molecule in the milligram range. One technique is nuclear magnetic resonance spectroscopy or in English Nuclear Magnetic Resonance Spectroscopy (NMR), the other is mass spectrometry (MS). In the following, I will always abbreviate the Nuclear Magnetic Resonance Spectroscopy with NMR, as this abbreviation has become established in chemistry and in the German language and is constantly used in laboratory jargon. You will now close the book in boredom if I explain the physical basics of NMR to you. But take away just one piece of information that can be very useful to you. In NMR, which you can also find as magnetic resonance imaging in medicine, the analytics are based on the change in the magnetic field of hydrogen and carbon. Almost all molecules in a biological system like our body consist of these two elements. Hydrogen and carbon have atomic nuclei. These consist of neutrons and positrons, around which the electrons orbit. All of this follows no real order, but a magnetic field can align them for a moment or relax them, as they say in chemistry. When falling back into the disordered state, this relaxation occurs differently, depending on which atoms are nearby. Who stands next to whom is revealed by different resonance frequencies when falling back. The atoms of a molecule can then be arranged with a lot of experience and a little mathematics to give a chemical structural formula, as you see in all profiles.

In contrast to nuclear resonance spectroscopy, mass spectrometry is destructive. While we can reuse the sample multiple times in NMR spectroscopy, unfortunately, this is not possible with mass spectrometry. Why is that? In mass spectrometry, the substance is bombarded with charged electrons. When these hit the substance to be examined, they explode, as you know from Star Wars movies. From the whole, many small parts of the molecule are obtained, which carry a charge. This is important because only charged molecule fragments can be detected in the detector. Now the puzzle game begins. Again, experience, knowledge, and often intuition are necessary to discover the original structure. I would like to explain this puzzle game to you using an example. Imagine you are holding a blue cup in your right hand and a red cup in your left hand and you drop both on the floor. The cups break and you now find numerous blue and red shards. The energy that was necessary to break the two cups are the electrons mentioned above.

Your eye is the detector and your task is to reconstruct the red and the blue cup from the many shards. The mass spectrometer does nothing else.

It is not surprising that many assumptions and claims about the discovery of natural substances from the 19th and 20th centuries were not accurate. Even Sertürner was often wrong in his 57 described opium experiments. Thus, he discovered the meconic acid, which he mistook for morphine, but after administering it to his dog, who showed no signs of consciousness disturbance, he discarded the substance and continued his research. The breakthrough came when he made the opium extract alkaline and shook it against ethanol. From this alcoholic phase, he obtained crystals, which he again gave to his dog. The young dog of just under two years became drowsy and vomited after some time. This was a sign of overdose. Nevertheless, Sertürner gave him another 6 grains after another hour. A grain is an old unit and corresponds to 360 mg. The opium was incorporated into a kind of gum slime, but the dog immediately vomited again. He began to twitch and convulse, which lasted for several hours. Sertürner was sure that *"this body is the actual numbing basic substance of opium"*. Sertürner had found morphine. But what about these acids and bases? Let's take a closer look. Morphine, like many other substances that will be described later, is an alkaloid. When Sertürner was researching in Paderborn, this term did not yet exist. If you had asked him what alkaloids are, he would have just shrugged his shoulders. The term alkaloids was only introduced in 1819 as a coined word and is supposed to indicate that these are nitrogen-containing natural substances that react alkaline, i.e., basic. Basic means that the pH value on a scale from 1 to 14 with a value of over 7 is in the basic range (Fig. 2). Now you might ask: What is the pH value? In science, this is understood to mean the concentration of hydrogen ions (in the acidic range) or of hydroxide ions (in the basic range) in an aqueous solution. The first attempt to define the pH value was made by Søren Peter Lauritz in 1909 at the Carlsberg Laboratories in Copenhagen and had to deal with national sensitivities as to whether there should be a French, German or Danish unit. From 1924, a neutral unit was established, which has a real scientific basis and is still valid today.

The definition of alkaloids, tentatively established by the Halle pharmacist Carl Friedrich Wilhelm Meißner (1792–1853), was not clear even then and has undergone many interpretations and changes to this day. Starting from the assumption that alkaloids could only be found in plants, the individual classes were defined according to their occurrence in plant families. Thus, there were the *Catharanthus* alkaloids, the *Coca* alkaloids, the ergot alkaloids, and many more. Chemists and botanists were faced with problems when similar alkaloids appeared in different plant families, which had the same

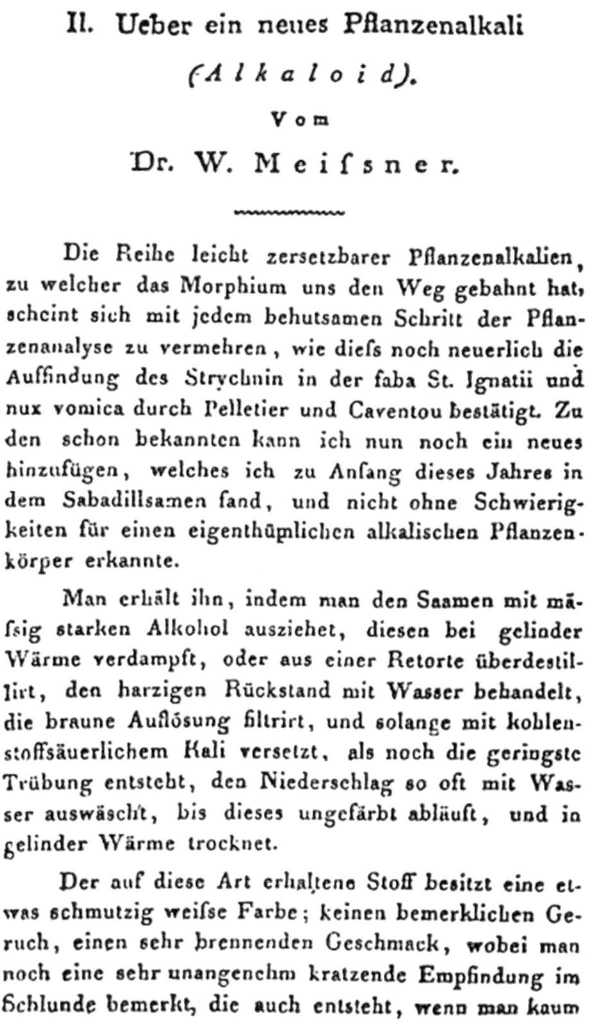

II. Ueber ein neues Pflanzenalkali
(A l k a l o i d).
Vom
Dr. W. Meifsner.

Die Reihe leicht zersetzbarer Pflanzenalkalien, zu welcher das Morphium uns den Weg gebahnt hat, scheint sich mit jedem behutsamen Schritt der Pflanzenanalyse zu vermehren, wie diefs noch neuerlich die Auffindung des Strychnin in der faba St. Ignatii und nux vomica durch Pelletier und Caventou bestätigt. Zu den schon bekannten kann ich nun noch ein neues hinzufügen, welches ich zu Anfang dieses Jahres in dem Sabadillsamen fand, und nicht ohne Schwierigkeiten für einen eigenthümlichen alkalischen Pflanzenkörper erkannte.

Man erhält ihn, indem man den Saamen mit mäfsig starkem Alkohol auszieht, diesen bei gelinder Wärme verdampft, oder aus einer Retorte überdestillirt, den harzigen Rückstand mit Wasser behandelt, die braune Auflösung filtrirt, und solange mit kohlenstoffsäuerlichem Kali versetzt, als noch die geringste Trübung entsteht, den Niederschlag so oft mit Wasser auswäscht, bis dieses ungefärbt abläuft, und in gelinder Wärme trocknet.

Der auf diese Art erhaltene Stoff besitzt eine etwas schmutzig weifse Farbe; keinen bemerklichen Geruch, einen sehr brennenden Geschmack, wobei man noch eine sehr unangenehm kratzende Empfindung im Schlunde bemerkt, die auch entsteht, wenn man kaum

Fig. 2 Definition of the natural substance class of alkaloids by Carl F. W. Meißner (public domain, Wikipedia)1

chemical structure. Suddenly there were tropane alkaloids in the coca plant and atropa alkaloids in the angel's trumpets, which belong to the nightshade family. Nightly pondering to structure the chaos now led to chemical classifications such as the division into tropane alkaloids, quinoline alkaloids, nicotine alkaloids, and so on. This made sense until it was discovered that there are mixed types and in recent decades alkaloids have also been isolated from animals such as insects. This was far too confusing. Today there is an intelligent solution, which I believe will endure. Alkaloids are named after the amino acid that is at the beginning of the biosynthesis. Now there are the ornithine alkaloids, to which cocaine belongs, the phenylalanine/tyrosine alkaloids, to which morphine belongs, and all biochemists are happy.[2]

Before we return to the publication of Friedrich Sertürner's scientific results, let me briefly explain how the publication of scientific data works today. I always emphasize to my doctoral students that a scientific work is only completed with the first of the publication. Publishing or releasing is an essential part of proving proper work and the correctness of one's own results in science. Even though we read a lot today about falsified or plagiarized doctoral theses and publications, I hope that this is only the case with a vanishing minority in the scientific community. If not, we as scientists not only lose our reputation in society. We also harm ourselves because it is no longer possible to reproduce and repeat experiments. Every publication in the natural sciences is preceded by an intensive period of experimental research in the laboratory. It is the doctoral students who implement their ideas in experiments and test series in the laboratory. A doctoral thesis can take four to five years in the experimental natural sciences, as many experiments are carried out and most are discarded. About 95% of the experiments are not successful and take a lot of time. They are not only carried out once, but are repeated for confirmation and validated with new substances, cell lines or references for security. When the doctoral student and supervisor are satisfied, the figures are written in a text that presents the hypothesis briefly, presents the results and discusses everything in the context of the current literature. A strenuous process that costs both a lot of nerves again, sometimes makes tears flow and usually ends with a good bottle of red wine. When the supervisor and doctoral student agree, the manuscript manuscript is sent to a publisher who commissions the work for review. These reviewers are colleagues who remain anonymous and either tear the manuscript apart or are full of praise. No, they usually lie somewhere in between and try to help fairly. With the request for responses and changes, the manuscript with the comments is sent back to us. We are happy to do this, and the new, annotated version goes back to the reviewers. The manuscript is accepted when the reviewers have probably also drunk a bottle of red wine and are satisfied. The publisher sets the text and gives it an appealing layout. The authors look for errors one last time in a proof copy and the manuscript can go to print. Depending on the field, this review process takes between three and six months. If it is successful, the author can find his publication in the scientific databases on the Internet. Even if the own manuscript is good, it may be rejected by very well-known and large journals. The rejection rate is very high and as an editor of a scientific journal, the author knows this all too well. Of the over 1300 submitted manuscripts that he gets to read in a year, only a few can be published and the rejection rate is unfortunately around 85%. Scientific publications are very important in our scientific world, because only in this way can it be ensured that the professional audience quickly learns what has been newly discovered in their field. Publishing the results on your own website, which must first be found, is not enough. Who guarantees that this website with the research results will still exist in 20 or 50 years? We all know how quickly websites change and disappear. With the deletion, the knowledge also goes and in the worst case is lost forever. In addition to these perhaps vain motives, publications are the currency in the tough third-party funding business. No money without publication, because who gives money voluntarily if there is no guarantee of good research. The funders are also vain, because they expect to be mentioned as friendly donors in every acknowledgment of a publication.

The publication of scientific results looked quite different in the times of Friedrich Sertürner and other researchers until the beginning of the 19th century. Scientific journals, as we know them, did not exist. The journals were not specialized and accepted all publications of a field, be it surgery, pharmacy, or botany. Many journals were only regionally distributed, and those that were available nationwide reached the rest of the world only with a considerable delay of one to two years. The language of science was Latin and only in the 18th century was published in German, French, and English. In addition to journals, knowledge was also published in books, and some researchers found it very glorious to see all their work published in one volume at the end of their scientific career. The current pace of at least five publications per chair per year was unthinkable at that time. How available knowledge was thought of before our modern times, I would like to explain using a few examples. Konrad Gesner (1516–1565) was a polymath who studied Latin, Greek, medicine, and natural sciences in Basel. He was eager to summarize the knowledge of his then-known world in a book, which he published after ten years of work in his work "Bibliotheca universalis sive catalogus omnium scriptorum locupletissimus in tribus linguis Latina, Graeca et Hebraica". What he did not know: His work covered only 15–20% of the actual knowledge of the world at his time. Another polymath who realized that the knowledge of the world could no longer be squeezed between two book covers was Gottfried Wilhelm Leibniz (1646–1716, Fig. 3). The world was in upheaval, Europeans sailed to all continents and brought back plants, animals, minerals, or drawings. Gottfried Wilhelm Leibniz, who did not invent the Geman butter buscuit with his name as inprint the butter biscuit, realized that there now needed to be places of collected knowledge, namely

Fig. 3 Gottfried Wilhelm Leibniz (1646–1716) (public domain, Wikipedia)1

libraries. He reorganized the ducal library in Wolfenbüttel, which was very well equipped for its time with 30,000 volumes. For comparison: The Library of Congress in Washington D. C., followed by the British Library in London, is the largest library in the world and today houses 170 million media.

The first known scientific journal is the Journal des Sçavans, which was published in Paris in 1695. The aim was to publish scientific results from all areas of natural sciences, be it chemistry, biology, or physics. Interestingly, it was characteristic of this journal to print obituaries of deceased researchers. Groundbreaking and still indispensable today was also the printing of experiment descriptions by researchers. Until the Second World War, there were only a small number of specialist journals. In a biology journal, everything was published that had to do with the areas of biochemistry, botany, zoology, or behavioral biology. Today, this has fundamentally changed and each discipline has its own specialist journal. Only in the large and renowned journals like Science or Nature are cross-disciplinary publications published that can be of interest to all disciplines. The language of science today is almost always English. Every researcher who values himself and wants to make his research results accessible to a broad public writes and publishes his work in a journal that publishes in English.

Johann Trommdorff's Pharmacie was also published in German, and in 1805 the publication of a preliminary report on the isolation of morphine by the pharmacy assistant Sertürner was accepted. Today, this publication would be referred to as a non-peer-reviewed preprint, a pre-publication, to keep the competition at bay. The actual publication followed a year later under the title "Principium Somniferum". If you look at the digitized work on the internet, you will be surprised to find that Sertürner almost apologetically points out that he was not aware of the publication by his competitor Derosne. This is credible in the time of the Napoleonic turmoil and the very slow spread of knowledge. But Sertürner seems to have doubted his work. But self-doubt also makes a good scientist. Further publications by him followed in the years 1811 (Journal der Pharmacie) and 1817 (Annalen der Physik), in which he brought his work from the laboratory of the Rats-Apotheke in Einbeck to the world. In the latter publication, he described a self-experiment with three young men. They took 100 mg of morphine and found that it was probably too much of a good thing. The intake of vinegar as a supposed antidote and an emetic did not improve the intoxication, as the morphine was already in the bloodstream despite the urge to vomit.

How did the Parisian competitors react when they learned of Mr. Sertürner and his work from 1811? The gentlemen were shocked and felt cheated of their laurels. "We were the first!", cried Antoine Beaumé (Fig. 4) and Charles Louis Derosne in Paris. They were stunned that a third party in the provinces and in a city whose name they had never heard dared to claim that he had isolated morphine, which they called narcotine. It is not surprising that the vain scientific dispute over who discovered morphine broke out. The dispute was brought to the institute with the request to bring about a clarification. All parties passionately argued over who deserved the honor of being the intellectual and experimental father of morphine. This dispute lasted almost 15 years and was only settled with the awarding of the Monthyon Prize to Sertürner in 1831.

If a natural substance with this enormous potential were discovered today, it could rightly be called a blockbuster or "One-Billion-Dollar-Drug", as the Americans like to do. Its discoverer would be rich and companies would be vying for licenses and patents. Why didn't Sertürner become rich? Why did he have to continue to struggle as a country pharmacist with trade freedom and unclear ownership of his pharmacy? Quite simply, because there was no patent law in Germany at that time. A patent is a document issued by

Fig. 4 Antoine Baumé (1728–1804) (public domain, Wikipedia)1

a state or today also by a supranational institution like the EU that grants the holder a monopoly. Sertürner would have been able to market morphine alone. What does it mean to own a patent? It is an agreement between the state and the patent holder that he or she can do whatever he or she wants with the patent for 20 years. But nothing is given away. In return, the patent holder must disclose all information on how he or she came to the invention. This includes detailed descriptions of the technical process, the chemical structure, the pharmacological tests, the synthesis, and so on. A lot of knowledge that is now also available to the competition, which, however, is not allowed to copy or market the product for 20 years. If things go well, it's almost like owning a money printing machine.

Patents are viewed very critically today, and some have the impression that the term of 20 years is much too long and that companies are only exploiting the market with high prices. We will shed light on this area of tension in the last chapter, but at this point it should be noted that the development of drugs has become very expensive today, with up to 900 million EUR per active ingredient.

The first patents were granted in Great Britain (1652) and in the USA (1790). Friedrich Sertürner lived in the wrong country and at the wrong time, because in the German Empire, patent law only came into force in 1877 at the urging of Werner von Siemens to the German Kaiser. So it was a matter of course that third parties used Sertürner's knowledge without paying license fees, which would not be possible today. We know that as early as 1826 a certain E. Merck in the Engel-Apotheke in Darmstadt extracted morphine from opium and that pharmacists and druggists also tried their hand at extraction in France and other countries. No one thought of respecting intellectual property, as we know it as a patent, and paying Sertürner license fees. A self-evident taking, as we perhaps know it today from many reports about Chinese companies, prevailed.

Before we turn to another important section in the history of opium and morphine: What did Sertürner do with his "world fame" as the discoverer of morphine? In 1821, he was able to buy the Rats-Apotheke in Hameln and worked there until his death in 1841 at the age of 58.[1] He also wrote chemical textbooks and published a scientific journal, which is largely unknown today. Sertürner died from the effects of a gout disease, which can be very painful. Because of severe pain, he probably took morphine, which promised relief. After his death, his body was transferred to Einbeck and buried there.

Opium Makes International Politics

Anyone who has ever taken a long-distance trip to Asia will have noticed that many Asian airports, especially in Southeast Asia, display a very drastic warning on the welcome sign: Drug smugglers are threatened with the death penalty. It is not uncommon for the media to report that tourists are executed for illegal drug importation. Where does this extreme penalty, which is applied so rigorously, come from? To understand this, one has to go back about 170 years in time. China was a large empire around 1850 and was of interest to the emerging European colonial powers such as England, France, and the emerging German Empire. However, in the mid-19th century, the Chinese Empire was still a developing country. The great era of the Ming Dynasty was almost history and the country was characterized by rural, peasant structures. Industrialization as in Europe was largely unknown.

The social problems were great and England saw a great opportunity to trade and make money against the backdrop of a weakened nation.

Ancient China before the 8th century did not know opium consumption, at least there are no text sources known that describe habitual use in society.[4] How did opium poppy (*Papaver somniferum* L.) and Opium get to China? It is assumed that opium poppy (Fig. 5) and opium were brought to China by land from the north and by sea. On the land route, it was monks, camel drivers, and traders from Afghanistan, today's Uzbekistan, and Kyrgyzstan. By sea, it was Muslim traders and missionaries who certainly brought seeds in the poppy capsules. Cultivation in fields in the Sichuan province has been known since this time. Opium was valued, as the emperors Song Taizu (927–976) in 973 and Renzong in 1057 had medical works written

Fig. 5 Opium poppy, *Papaver somniferum*, Otto W. Thomé, Flora of Germany, Austria and Switzerland 1885 (public domain, Wikipedia)1

in which the healing effect of opium was described. By the end of the 13th century, opium consumption had spread so far that it became a problem. The domestic opium production for consumption did not seem to be sufficient, as large quantities also reached China via Arab traders (Fig. 6). Of course, the rulers, especially Emperor Ming Taizu (1368–1398), were not pleased that his subjects indulged too much in opium consumption.[5] However, for fear of a popular uprising, he refrained from banning opium. It remained a tolerance until the well-balanced trade relations between Arabs and Chinese over centuries were abruptly disturbed by the Portuguese in 1498, who rounded the Cape of Good Hope and were seen in the Chinese Sea. Their strong military superiority quickly led to the establishment of Portuguese seaports along the African coast to India, which initiated a lively trade with China. The country was flooded with opium and resisted.

It is not surprising that obstacles to foreign trade were erected from the mid-16th century onwards. The result was that the supply of opium became scarce and prices on the market soared. Higher prices made domestic cultivation attractive again and the land was no longer used for growing rice as a food source. It is easy to imagine that the prices for rice also rose and it became scarce. This is very familiar to us today when we think of South America and Afghanistan, where coca and poppy displace the areas for conventional agriculture. But it got even worse in China. With the discovery

Fig. 6 Packaging and trade of opium in 19th century China (Permission Wellcome Foundation, London, UK)1

and conquest of South America, the tobacco plant also became known to Europeans. Tobacco contains the stimulating alkaloid nicotine, which the Indios liked to smoke. The Spaniards adopted this habit and brought tobacco as smoking goods first to Europe and then with the other colonial powers to the whole world. With a delay of about 100 years, tobacco arrived in China in 1620 with the Portuguese and was very successfully cultivated in the Shima region around Zhangzhou. Too successful for the Portuguese, as they turned Brazilian tobacco into expensive junk. As in all societies, smoking enjoyed great popularity, which is probably due to the high addictive potential is. Tobacco shops sprang up in China from the mid-17th century like mushrooms. The last Ming Emperor Huaizhun banned smoking in 1644, but no one adhered to it. Interestingly, this did not happen for health policy reasons, as we know them from modern democracies, no, it was a question of morality. The principles of Confucian teaching were shaken and the nobility as a role model succumbed to vice. It didn't help. Smoking developed into a health-threatening popular sport over the next 100 years.

Was there a connection between tobacco- and opium smoking? Certainly, because opium smoking had been known since the Ming period and from the 17th century it increasingly mixed with tobacco smoking. Opium was cut with tobacco. The Portuguese and Dutch probably saw no difference whether tobacco or opium was smoked and expanded their business purposefully. All coastal cities were supplied, from where opium smoking gradually spread to the mainland. The tobacco ban by the last Ming emperor of 1644 threatened anyone with death who traded with the "barbarians", i.e. the Europeans. How did the opium pipe come into the world? With the Dutch, who liked to smoke tobacco in a pipe in Europe. Because of the high price for tobacco, opium was first cut and later the tobacco was completely left out of the pipe, which they had learned from the Dutch, and opium was smoked pure.

In the 18th century, opium addiction developed into a national drama. Reports came from the provinces that the army was no longer operational because the soldiers had no more morale and fighting spirit. Opium dens sprang up all over the country. The rich Chinese had a private smoking room, the poor smoked in shabby basement rooms. A ban on opium dens followed in 1729, but it was not until Mao Zedong and the communist leadership that the ban was consistently enforced. The numbers spoke for a real drug problem in China. While 60 tons were officially imported in 1767, it was already 1390 tons in 1835 and the number rose to 6500 tons in 1880. The actual amount smuggled is still unknown today, it must have been many times higher. In 1940, 40 million opium addicts were known and Chinese society was falling apart. The corruption in the administration flourished, income structures collapsed and the loss of authority of the government was great.[4]

When we talk about the Opium Wars, we must consider another aspect. At this time, the British sold opium to China in order to get a product in return for which there was a high demand at home. It was tea. However, the Chinese did not want to trade with the British, as English products were not in demand due to their simplicity and foreignness. Although the Chinese middle class was relatively wealthy compared to European societies, consumer culture was not very pronounced. People were satisfied with what they had in China.

Although tea contains stimulating alkaloids such as caffeine, the dried leaves of the tea bush *Camellia sinensis* cannot be classified as a drug. Tea is considered mildly stimulating, but it was more appreciated for its taste as a beverage. You have just read the Latin name and perhaps you noticed the word *sinensis*, which means Chinese in German. Tea was cultivated in China and it gained great popularity in England. Until the mid-17th century, tea was unknown in England, but then the hot drink became known through the East India Company and demand increased sharply. The nobility enjoyed drinking tea and it was a status symbol. Queen Anne (1665–1714) promoted its popularity by replacing the usual warm beer for breakfast with a cup of tea. Suddenly, no one wanted to do without tea, the king saw a new source of income and imposed high import duties. Tea also made world politics and led to the independence of the USA. Surely, you are also well aware of the Boston Tea Party from your school lessons, which ultimately triggered the American Revolution.

Let's go back to China. The English imports from China were paid in silver. The English quickly ran a trade deficit and the silver reserves became smaller over the years as the ships sailed towards China. Attempts were made to exchange other goods such as wool, but the Chinese rejected them. Other goods also failed because there was still no demand in China. Only opium was accepted, but it was not available in England and the East India Company, which conducted the trade, did not want to get involved in the opium trade. The triangular trade was reinvented. Silver in exchange for tea in China through the Company and opium from India via private trading ships, which was a British colony and where the East India Company had the monopoly on opium. In this way, silver flowed into the pockets of the British via India and the Chinese precious metal balance slowly but surely became negative. It got even worse, because despite the Chinese ban on opium for recreational use in 1796, as we would say today with embarrassment, smuggling flourished. The smugglers were pursued and if they were caught, they had to wear the Kang, a huge wooden shackle around their neck, and received 100 strokes with the bamboo cane. The opium trade in Asia was conducted with hard gloves and is more reminiscent of piracy. The Chinese government had little power, as the monopoly on the opium trade initially lay with the Portuguese, before the British drove them out in India. After the Battle of Palashi, a mosquito-infested village between Kolkata and Murshidabad in India, the British took over the rule of the Portuguese in India and thus also the opium monopoly.

In 1838, the high official Huang Jiaoze submitted a memorandum to the emperor, in which he stated that *"opium is the cause of all the suffering of his people"*, and he urgently asked Emperor Dao Guang to impose harsher penalties up to the death penalty to curb the evil. The emperor was aware of these problems from his court and was not averse to the request, as even the crown prince had succumbed to opium addiction. Commissioner Lin, who was considered a hardliner at the court in Beijing, was sent to the provinces in 1840 to enforce the opium prohibition. He was the right man for this task. In the province of Canton, he asked the population to hand in their opium pipes and cleaned up the corrupt coast guard, who made a handsome profit from the opium trade. For a good price, they let the ships of small Chinese middlemen pass, who took over the cargo from private Indian ships sailing under British control off the coast. Officially, the British East India Company was never present. Raids were carried out on the mainland and Lin particularly targeted the Hong merchants. He met with the clan chiefs of the twelve large families and made it unmistakably clear to them that the opium trade had to be stopped within three days, otherwise he would execute the heads of the two most suspicious families and confiscate their wealth (Fig. 6). He was clever, because he did not say who he meant, and reluctantly the powerful families stopped the trade. All these measures now hit the white devils, as the British and other European nations were called.[6] They tried to bribe Lin, but he remained firm.

The British, particularly the trader David Sassoon, who possessed monopoly rights from the British government, made millions from this business. But hard times came for the East India Company with the new anti-opium policy. The workers stayed away, many European traders had passport difficulties, they were no longer allowed to move freely in the country and also had to hand over existing opium stocks, about 1230 tons, without compensation. The business was on the ground. David Sassoon turned to Queen Victoria and the first Opium War began. Whether one can speak of an Opium War is questionable, because it was more of a Tea War, as opium was only a pretext to balance the poor trade balance of the British. The aim of the first Opium War was to enforce a compensation of 2 million pounds demanded by the traders from the Chinese for the destruction of the opium and its loss of profit. The overpowering British won, and in the Peace of Nanking it was stipulated that opium was henceforth legal in parts of China and a compensation of 7 million pounds had to be paid for the 20,000 boxes of opium confiscated at the time by Lin Zexu. But Her Majesty also did not skimp and demanded 21 million pounds as compensation for the war effort and had the island of Hong Kong transferred as a British leasehold. After the Second Opium War, Hong Kong became a British colony and a small mainland strip, the "New Territories", today the district and administrative district Kowloon with surrounding areas, were leased for a period of 99 years in 1898. It is interesting that a plant and a natural substance could write so much contemporary history.

The British distrusted the peace and uncertain markets in China. They smuggled tea plants from China to India to grow them there. In this book, you will read about other cases of biopiracy and find that plant smuggling and cartel formation have driven entire national economies into ruin. China was not so severely affected, but the tea as we know it definitely does not come from India. The British East India Company brought plants to the provinces of Assam and Sikkim, thereby changing the economic map of the subcontinent. Demand in England and its colonies such as Australia, South Africa and New Zealand increased, the first cultivation in the region of Nabua expanded to other areas like the Nilgiri-Hills, Darjeeling and Ceylon. In 1859, only 3% of the tea came from India, but imports to England rose to 90% by 1890, with a growth rate of more than 837% in the first 75 years of tea trade. China had become uninteresting and English tea producers like Lipton, Twining shape the tea trade to this day.[7]

Heroin

Another milestone in the history of morphine was the development of heroin by the Bayer company, which was sold under the same brand name in 1898. The chemist Felix Hoffmann (Fig. 7), who worked at the Bayer factories in Wuppertal-Elberfeld, took up the described work of the English chemist Charles Romley Alder Wright (1844–1894, Fig. 8). He had been very intensively involved with the conversion of alkaloids with acetic anhydride. In his short but industrious life, he synthesized hundreds of new opiates, including heroin, which he diamorphine called. Acetic anhydride

Fig. 7 Felix Hoffmann (1868–1946) (public domain, Wikipedia)2

Fig. 8 Charles R. A. Wright (1844–1894) (public domain, Wikipedia)2

is formed when two molecules of acetic acid combine. This anhydride is quite something, as it carries out very energy-rich acetylations with other substances. If you look at the chemical formula of morphine, you will find that it contains two hydroxyl groups that like to react with acetic anhydride. These reactions are extremely popular in chemistry, as they make substances very lipophilic, i.e., fat-loving.

Wright was a lecturer in chemistry at the medical faculty of St. Mary's Hospital in London. His interest was in the structural elucidation of morphine, which he hoped to accelerate through acetylation reactions. In 1874, he converted morphine, but also codeine, isolated a white substance, and calculated the sum formulas, from which he could conclude the double acetylation. These experiments are undoubtedly the birth of heroin. *"If you boil morphine […] for several hours with […] glacial acetic acid, a large part is converted into a substance that is related to morphine in the same way as diacetylcodeine is to codeine"*, Wright reports in his publication, which can be seen as the birth certificate of heroin.[8] He initially called heroin first diacetylmorphine, but did not recognize the medical benefit of his work. The later name "heroin", derived from the Greek word Heros, was taken by Friedrich Bayer & Co. because their employees, who were allowed to take heroin experimentally, reported feeling heroic after taking it. The name fit the German imperial era, in which everything had to be heroic and superior.

In addition to Wright, other researchers such as the alkaloid researcher Grimaud in 1881, Otto Hesse in 1882 and Wilhelm Danckwortt on were very active in this field and reported independently of each other about acetylations. For the chemistry professor Otto Hesse, it was mistrust, as he could not explain Wright's synthesis and repeated his work. What he understood better than Wright was the interpretation of the number of hydroxyl groups of morphine. In Justus Liebig's Annals of Chemistry from 1884, he describes again the acetylation, but also the saponification of the acetyl residues, *"because if the diacetylmorphine is heated in alcoholic solution with a little potassium hydroxide, it is quickly decomposed into acetic acid and morphine"*.

The strong effect and the potential to cause strong dependence were seen early on. However, pharmacologists had no knowledge of the mechanism of action and wanted to learn more about opium and morphine through initial animal experiments. In 1895, the pharmacologist von Mering, in collaboration with the company E. Merck in Darmstadt, sought new morphine derivatives, but even before him, there was high interest in these wonderful substances. In England, F. M. Pierce was already conducting animal experiments on frogs, dogs, and rabbits in 1874, injecting them with diacetylmorphine under the skin. In the 1880s, D. B. Dott and Ralph Stockmann

(1861–1946) took up the work and published the article "Pharmacology of Morphine and its Derivates" in 1890, in which they reported very clearly about respiratory depression and seizures when the animals received too high a dose.[9] Felix Hoffmann and the Bayer pharmacologist Heinrich Dreser were aware of these publications and nevertheless included the synthesis and retesting in animal experiments in the Bayer research program. Interestingly, Dreser later reported on the known effects in publications, but did not mention the original authors Dott and Stockmann.[10] The impression was created that Dreser had discovered and tested diacetylmorphine for the first time and alone. Even the well-known German pharmacologist Erich Harnack (1852–1915) was surprised and commented with the words: *"Dreser does not mention [...] the work of Dott and Stockman at all. [...] Even the circulars recommending heroin from the Elberfelder Farbenfabriken (1898/99) give the impression, through their literature registers, that Dreser was the first and only one to have conducted animal experiments with diacetylmorphine."*[11]

The synthesis of diacetylmorphine at Bayer was undertaken by Felix Hoffmann, who would later become known as the father of Aspirin®. The chemist was born in 1868 in Ludwigsburg and studied chemistry and pharmacy in Munich. After his dissertation "On some derivatives of dihydroanthracene and decahydroquinoline" (1893) and a short stint at the Bavarian State Laboratories, he got a job at Bayer and moved from the Isar to the Rhine. In the chemical laboratories, he encountered the "acetylation mania" of Rosenthaler, as his then colleague Fritz Hofmann (1866–1956) ironically noted in 1920. Every era has its fashions—even in science. Acetylation was certainly one of them, but it must also be acknowledged that with Heroin® and Aspirin®, two acetylation blockbusters came onto the market.

Heroin® entered clinical trials, and cough patients received Heroin® over a period of six months. The Bayer company doctor Dr. Theobald Floret, who accompanied the study, was very confident when he described the study results in 1898 in the Therapeutic Monthly: *"The heroin prescribed by me [...] proved to be an extraordinarily useful, prompt and reliable means of combating cough and cough irritation, as well as chest pain primarily in inflammations [...] Angina, [...], Bronchitis. [...]. One of such patients told me that no medicine had done him as much good as my powder."* As if he could not believe it or had to convince himself, Floret took Heroin® himself and could only confirm the miraculous effect. He also denied any habituation: *"[...] Thus I was enabled to carry out my activity for hours without being bothered by coughing anymore. [...] The preparation was otherwise very well tolerated. [...] A habituation to it does not seem to occur [...]."* [12] Convinced of having a very good medication in hand, he did not hesitate to prescribe Heroin®

to children who had whooping cough had. With these results, Bayer was to bring another successful product to the market. In the USA and Europe, Heroin®, as it was now officially called, came into the pharmacies and lined up on the shelves next to Aspirin®, Salophen® and Lycetol®. The USA was the main sales market, followed by Russia and England. Also in Germany, the Bayer product was very popular, as the company's sales up to the ban show (Table 1). It was the drug of choice for whooping cough in pediatrics, as a painkiller in gynecology and the over-the-counter drug next to cocaine and cannabis on the pharmacy shelf.

Was there no one who warned of the dangers and recognized that a substance was being administered here that was highly addictive? Yes, there were these critics, but there were no effective laws prohibiting the sale of dangerous drugs. The aforementioned pharmacologist Harnack from Halle raised his voice and warned against heroinism: *"After everything we know so far about heroin from animals and humans, I am of the opinion that the handover of the drug to practice was premature, and that a very dangerous poison has been given into the hands of the unsuspecting practitioner, about which not enough caution can be urged."*[11] The addictive potential was clearly demonstrable and as we will see later in the example of the first opiate crisis in the USA, there were thousands of addicts in Europe and the USA. It was probably the first worldwide opiate crisis at the beginning of the 20th century, as many patients quickly became addicted out of habit. The Berlin doctor G. Strube wrote in 1898: *"The patients generally liked the drug, those accustomed to morphine were satisfied with the exchange, many, for whom it was given under the correct indication, asked for it again when I left it out or replaced it with codeine"*, as he reported about his 50 patients in the Berlin Urban Hospital.[14]

After the problematic experiences with heroin and morphine, it was an urgent development to find new opiate-like painkillers that should solve three problems. First, the reduction of the addiction potential, second, the

Table 1 Heroin sales of Bayer AG from 1898 to 1925 by countries (Bayer Archive 15 D5.4 and 166/8)[13]

Country	Sales (in %)
United States of America (USA)	58.4
Russia	7.1
Germany	6.8
England	5.1
Italy	3.1
Austria	2.3
Spain	1.5

risk of respiratory depression, and third, the short half-life with oral administration. After the chemical structure of morphine was clarified in 1923, researchers tried to find substructures that were responsible for the effect and possibly for the side effects. From today's perspective a daring venture, because in the 1920s and 1930s of the last century, it was not even known that there are opiate receptors and endogenous opiates, so-called endorphins. It was a bit like trial and error. The animal model test was the only way to separate the good candidates from the pharmacological waste. From the large number of synthetic opiates, I would like to discuss some interesting representatives that, unlike heroin®, have found their way into the clinic as drugs.

Synthetic Opioids

Heroin® is not a synthetic opiate, as it was only slightly modified at two points. Synthetic opioids are a complex group of chemical structures. The difference between opiate and opioid is that opiates are the natural substances of the opium poppy and opioids are chemical or synthetic opiates. As in the lectures of Pharmaceutical Chemistry, one could describe all substances in this book according to their substance class, but that makes no sense here. We prefer to proceed chronologically, starting with the older opiates and working our way up to the present.

The first real synthetic opioid was **Pethidine,** which was synthesized in July 1937 by the Austrian doctor Otto Schaumann (1891–1977) and Otto Eisleb (1887–1948) at the I. G. Farben in Frankfurt-Hoechst. Schaumann's career is remarkable, as he was one of the few doctors who took up a doctoral study in chemistry at the Technical University of Munich in 1921. After completing his doctorate, he joined the company Kalle, which later merged with I. G. Farbenindustrie AG. After the Second World War, he accepted the call to the chair of pharmacognosy at the University Innsbruck, which he held until his retirement in 1962. After his successful retirement, he was still so interested in research that he accepted the offer to lead the pharmacology and later the entire medical research of the company Boehringer Mannheim. His most important discovery was the aforementioned Pethidine, which was found by a "guided accident".

The examination of the effect and toxicity of pethidine was carried out early after synthesis. All substances were routinely tested on mice. The test is simple and the experimenter injects the substance to be tested into the mice via the tail vein and observes the behavior of the animals. Apart from the very rare immediate death, mice behave noticeably when they are unwell. They become hyperactive, run through the cage and no longer instinctively stay on the walls. They lick themselves more often or become apathetic when the active ingredient also reaches the brain. A particular behavior was observed by the German pharmacologist Straub in 1911 during his mouse experiments in Freiburg after administering morphine. The mice dragged their tails in an S-shape over their backs, as if they were driving bumper cars. He found this strange, and this reaction subsequently served as forensic evidence of opiates. When Otto Schaumann went in search of antispasmodics in 1940 and injected one of the synthesized substances, pethidine, into a mouse, he saw it arch its tail upwards. This typical mouse tail phenomenon made him sit up and take notice. Was his pethidine perhaps a painkiller? To prove this, Schaumann conducted a second experiment, the so-called "Hot-Plate-Assay". In this experiment, a mouse is placed on a hot plate. This plate can be a glass plate under which hot water flows, or a metal plate that is electrically heated. Because we have thermoreceptors in the upper layers of skin that are supposed to protect us from burns or scalds, heat above 55 °C triggers a pain stimulus. These thermoreceptors, for the discovery of which the Nobel Prize was awarded in 2021, are connected to neurons of the pain pathway that signal to the brain: Attention, pain in the paw! The normal reaction of the mouse is to lick its paw to cool it. To determine the pain stimulus, the researcher measures the time from when the animal is placed on the plate to the first lick of the paw, until the animals jump up or run away. In most cases, this time span is between 40 and 60 seconds.

The second important synthetic opiate was also discovered in 1937 but by Max Bockmühl (1882–1949) and his assistant Gustav Ehrhart (1894–1971). In the lab, it still bore the number VA 10820 and became known under the name **Methadone**. Methadone was a child of the laboratories of I. G. Farben during the Nazi era, but it was not pharmacologically tested until 1942. Due to the lost war, all patents had to be surrendered and the US pharmaceutical company Eli Lilly took over methadone, which it brought to the market as Dolophine® on. The Americans allowed any manufacturer to produce and distribute methadone without license fees. This was certainly the reason why many methadone drugs came onto the market. The opioid is a typical active ingredient of its time, as drug research until the late 1960s was characterized by the minimization of the morphine structure towards simple structures that should no longer pose a risk of addiction. A look at the structure in the structural formulas below does not suggest any similarity to morphine at first glance, but one must imagine the structure three-dimensionally in the binding pocket of the receptor.

Neither for Pethidine nor for methadone or other opioids (Fig. 11), the goal of strong analgesic effect without dependency potential was achieved. The number of dependent patients and addicts increased rapidly after the Second World War. From the 1970s onwards, methadone was dispensed as a substitute for heroin, as its half-life in the body was much longer at four hours and addicts were well covered with two doses over the day and night. The very short half-life of heroin of about 20 minutes in the body is the reason why drug addicts come into withdrawal so quickly and need the next shot.

The structure of morphine was further minimized and in 1967, with Tilidin, an opioid came onto the market that resembled the structure of the natural opiate. The drug with the name Valoron® has developed into a fashion drug to this day. In 1978, it was withdrawn from the market because there were too many cases of patients with symptoms of opiate addiction. Tilidin was placed under the Narcotics Act and was no longer available on the open market without a prescription. In the same year, the pharmaceutical company Gödecke brought Valoron N® onto the market. The special thing about the new drug was that it contained the opiate blocker Naloxone in a ratio of 50 to 4 with Tilidin to prevent misuse. Resourceful drug addicts, who had a solid basic knowledge of chemistry and physics, found out that Valoron® on toast destroys the inhibitor. In internet forums, you can find posts advising to drizzle Valoron® on a slice of toast and put the slice in the toaster. The glowing wires reach a temperature of 500 °C, which is sufficient to separate Naloxone (465 °C) from Tilidin (96 °C) due to its boiling point. However, whether the Valoron toast works is doubtful. Although Naloxone has a very high boiling point and evaporates more difficultly than Tilidin, whether it is a complete removal at the mentioned mixing ratio? Unfortunately, there are only a few chat posts on the internet that refer to subjective experiences and do not provide an answer.

In 1975, opiate research made a major breakthrough. The physiological and chemical messenger for the opiate receptors was found. The researchers were able to show that the activation of the opiate receptors is not caused by small molecules like adrenaline or acetylcholine, but by short protein chains, so-called peptides. These small peptides were called neuropeptides because they are released as messengers by nerve cells. Hans Walter Kosterlitz (1903–1996) took mouse brains and isolated molecules whose effect was inhibited by Naloxone. After long nights, two peptides were found that consisted of five amino acids and only differed by the amino acid at the end. These were Tyrosine-Glycine-Glycine-Phenylalanine-Methionine, called Met-Enkephalin, and Leu-Enkephalin, which contained Leucine instead of Methionine. Today, about 100 neuropeptides with an effect on the opiate receptor are known, which are divided into further groups, such as the endorphins, the Dynorphins or Deltophins. They were not only discovered in humans, but also in the giant leaf frog *(Phyllomedusa bicolor)*.

Hans Walter Kosterlitz (1903–1996, Fig. 9) was born in 1903 as the son of a doctor in Berlin and wanted to study mathematics. His interest was sparked by the acquaintance with Albert Einstein's stepdaughter and he liked the idea of dedicating his life to the natural sciences. His father considered this to be a breadless art and recommended at least the study of jurisprudence, which Kosterlitz found too dry. He complied with his father's wish and went to study law in Heidelberg. After six months, he already lost interest in studying law and enrolled in medicine first in Heidelberg, then in Freiburg and from 1925 in his birthplace Berlin. Under the strict wings of Peter Rona (1871–1945) and Leonor Michaelis (1875–1949), who made himself immortal with the mathematical description of enzymatic kinetics, he learned the essential tools of analytical, synthetic and physical chemistry necessary for biochemical and physiological research. Kosterlitz later wrote that he had all the freedoms in his doctorate. *"I was given a key to the lab"*

Fig. 9 Hans Walter Kosterlitz (1903–1996) 35 (Permission Royal Society of Chemistry, London, UK)2

and *"was instructed not to break anything"*, he remembered his mentor Peter Rona. He never left the lab before midnight, he lived for science.

In 1928, he received his medical license and two years later he graduated from the Charité. Kosterlitz was a doctor and researcher through and through. During the day, he stood by the bedside and took care of his patients as a young doctor, in the evenings he worked with the Swiss Wilhelm His (1863–1934), who was interested in the pathology of carbohydrate metabolism, in the laboratory. However, his work was not allowed to cost anything and had to take place outside of his working hours, as His emphasized to him. Research in the field of diabetes, as the pathology of carbohydrate metabolism was called, was highly topical, and real breakthroughs were expected. John J. R. Macleod (1876–1935), Frederick G. Banting (1891–1941) and Charles H. Best (1899–1978) had obtained insulin for the first time in 1921 at the Institute of Physiology in Toronto, Canada, and for this groundbreaking work, Macleod received the Nobel Prize for Medicine in 1923. The world was interested in this research, as diabetes was an incurable disease that could end terribly with diabetic coma and early death. Kosterlitz became an expert in the metabolism of galactose and other sugars.

At the end of the 1920s, it was a turbulent time in Berlin and in 1933 the National Socialists took power in Germany. Old family records showed that Hans-Werner Kosterlitz came from a bourgeois Jewish family. Although he did not actively practice his faith and felt rather alienated from Judaism, he was denaturalized by the new rulers in Berlin and emigrated to Aberdeen, Scotland in 1934, where he worked with John James R. Macleod. Macleod advised him to come, but did not want to give him any hope of a permanent position. Kosterlitz came because he saw no future in Nazi Germany. The newly founded established Diabetes Foundation in Great Britain gave him its first scholarship of 50 pounds. That was a start, the Medical Research Council (MRC) would later support him much more generously. The real shock for him was the unexpected death of his mentor Macleod a year after his arrival. He considered whether he should not better leave Aberdeen right away. He stayed and died in the city he fell in love with.[15]

After the war, he was concerned with the question of how opiates work. He knew from previous work that opiates have an effect on the intestine and he believed that there must be receptors responsible for the relaxation of the intestine. To learn more and practice experiments in the lab, Kosterlitz made a trip to Otto Krayer (1899–1982, Fig. 10) at the Pharmacological Institute of Harvard Medical School. One might suspect that his host Otto Krayer also emigrated to the USA because of his Jewish faith, but things

Fig. 10 Otto Krayer (1899–1982) (public domain, National Library of Medicine, Bethesda, USA)2

were a bit different. Otto Krayer was an upright man in the time of lies. The Jewish pharmacologist Philipp Ellinger (1887–1952) was removed from his chair of pharmacology in Düsseldorf by the Nazis in 1933 and Otto Krayer was supposed to follow. He declined and Wolfgang Heubert (1877–1957) remembered: "[Krayer came] *at noon in person to tell me that he had just presented his inner doubts to Ministerialrat Achelis about taking over for a man who, in his opinion, had been dismissed without just cause, whereupon he had been dismissed with the remark that he would then have to look for another one. Magnificent!*" He drew the consequence and left.

Krayer was considered by the British after Sir Henry Dale as *"the only German Gentlemen"*, because he had publicly refused to take up a position at the Medical Academy in Düsseldorf. The anger in Berlin and Düsseldorf was great, and Krayer was subjected to a professional ban that went so far that he was no longer allowed to enter a library. These were not good prospects for being able to work in medical research in the German Reich again. He decided to emigrate to London to the University College, but moved almost immediately after his arrival to Beirut, where he led the pharmacology at the American University of Beirut from 1934 to 1937. In 1939, he accepted the call to the chair of pharmacology at Harvard University, which he held until 1966.

Kosterlitz was unable to take concrete results from his guest stay at Harvard, but the fire was kindled and he became even more interested in the control of intestinal peristalsis in fat- and sugar-rich diets. The work was all neat and solid, but it lacked the kick, the special something he was looking for. Then he came across a publication that electrified him. As he was looking through a nearly forgotten publication by Paul Trendelenburg (1884–1931), the mentor of Otto Krayer, in the Archive for Experimental Pathology and Pharmacology, he read words like pendulum movement in guinea pigs, motor skills, cutting out the intestine, and morphine. "Stop!" he thought. Why morphine? What role could this substance play outside the brain? It's amazing how his brain worked and allowed this thought while he read the very long article of 126 pages. Take the article in hand, look for the word morphine and you will only find it on page 106. He probably hesitated like you and pondered. It couldn't be that this plant alkaloid is formed in the intestinal cells. There had to be another natural substance that the body produced. Despite or perhaps because of his 65 years and the beginning of retirement, he wanted to know exactly. Kosterlitz, the old veteran, had not forgotten his analytical tools from his Berlin days with Peter Rona and Leonor Michaelis (1875–1949) and went back to the lab.

Time was pressing, as Kosterlitz was only appointed professor in his mid-sixties. He bore with humility the fact that he was often overlooked for appointments in Aberdeen. He never gave up and finally he was appointed to the Chair of Clinical Pharmacology. The Dean of the University of Aberdeen was an impatient man and thought he was too old for new research and wanted Kosterlitz to finally leave. Then fresh dollars came from the USA. The American National Institute of Health (NIH) gave Kosterlitz a decent financial injection and paid for a lab for him to study narcotics. Kosterlitz was allowed to stay and dug into research. Just before the peak of his research and his discovery, competition from Sweden and the USA in the field of endorphins caught up. However, he benefited from a new technique that required a good hand from the experimenter, which the old fox possessed, to keep the competition at bay. Ultra-thin tissue sections could be made and with the help of probes he found tissue receptors and their activating substances, which he isolated. Two years later, he painstakingly purified the first enkephalins in the lab and published his work in 1975. The breakthrough was achieved. A flood of honors and awards came down on him. Although he was so close to the Nobel Prize, he did not receive it. His son John Michael did better. In 2016, as a professor of physics at Brown University, he received the great award for his work on the theory of various topological phases of matter. Mathematics is indeed worthwhile.[15]

Let's stay with intestinal peristalsis and opiates for a moment. Many of us take a medication on holiday trips, when things literally go down the drain, of which few know that it is an opioid and is derived from morphine. The active ingredient is called loperamide and the medication has become known under the trade name Imodium®. The drug substance discovered in 1969 in the laboratories of Janssen Pharmaceuticals, which no one wants to do without in the treatment of diarrhea, acts on the opiate receptors in the intestine. It relaxes the smooth muscle, which normally transports the contents to the exit by longitudinal and cross-sectional contraction. The medication is well tolerated as it only works locally in the intestine and does not cause respiratory depression. Cough suppression or pain relief are also not known. For the chemist, this is surprising, as it has the same substructure of the piperidine ring as in Pethidine shows (Fig. 11).

In addition to the exciting opiates, scientifically called agonists, the first inhibitor, with Nalorphine, was found in 1942, although the two discoverers Weilard and Erikson had no idea about the opiate receptors in our body and did not know what an inhibitor was. The great advantage of this and the coming antagonists, also called opiate antidotes, was that a remedy for

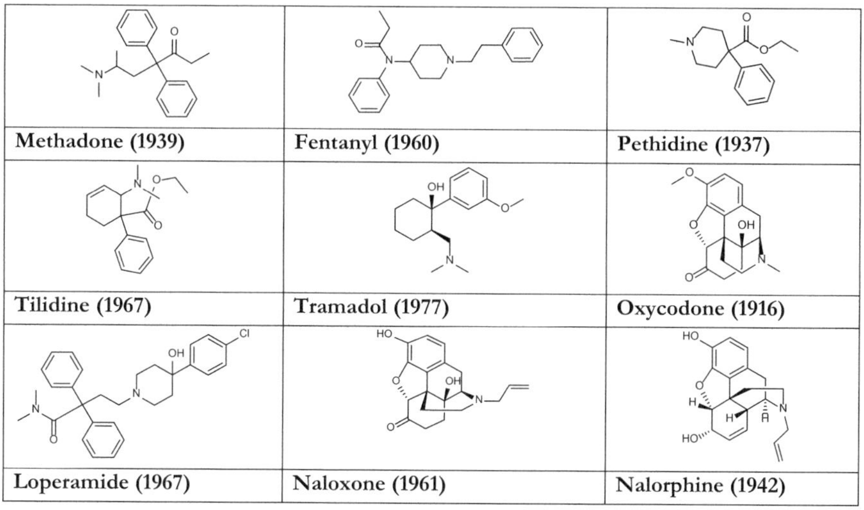

Methadone (1939)	Fentanyl (1960)	Pethidine (1937)
Tilidine (1967)	Tramadol (1977)	Oxycodone (1916)
Loperamide (1967)	Naloxone (1961)	Nalorphine (1942)

Fig. 11 Synthetic Opioids (Work of the author)

respiratory depression was finally found. Nalorphine (Fig. 11) as the first antidote showed but a problem in patients, because it triggered anxiety.

One can well imagine the situation where the patient is brought back from his morphine rush and begins to riot in the ambulance out of fear of the circumstances that he is not aware of and that surround him. With the further development to Naloxone in 1961, the first pure antagonist was found, which cancels the effect of the opiates on the opiate receptors and is considered safe to use.[16] Naloxone only became known during the second opiate crisis in the USA, as it brought more than 26,000 patients back to life after an overdose of heroin according to the American health authority CDC.

Dealers in white coats—How Oxycodone Makes America's Middle Class Addicted

It was a disgrace for President Donald J. Trump when he stepped in front of the press on October 27, 2017, and announced his fight against opioids, especially against Oxycodone, with which the US pharmaceutical company Purdue made billions in profits. Everyone had known the problem for 20 years, as greed and well-paid lobbying had first made parts of the American middle class addicted and then driven them into the illegal drug market. Oxycodone or Oxycontin®, as the product was called on the market, became the "Hillbilly Heroin" and has since dominated the illegal so-called recreational drug market. No other worldwide, newer drug epidemic has driven so many people into addiction, into death, into prisons and despair, or made orphans like this one.[17]

The pharmaceutical company Purdue launched a drug called Oxycontin® in 1996, which was supposedly a weak opioid that was well tolerated and not supposed to cause addiction. If addiction did occur, as many doctors reported to the company, it was a pseudo-addiction and Purdue recommended giving even more Oxycontin® to alleviate the symptoms of addiction. It's madness to want to drive out the devil with Beelzebub. Since the year 2000, all experts have been clear that a problem would become really big in American society. It mainly affected the impoverished white middle class in the "Rusty Belt", who were left sick and unemployed after the closure of the mines and steelworks. Many worked underground, their joints and backs ached. The doctors who visited the people often prescribed synthetic opioids like Oxycodone because the health insurance companies put pressure on them and wanted to keep costs low. Small towns in West Virginia like Kermit and Huntington, a few hundred miles west of the capital, became the Ground Zero of the opioid crisis, as the life expectancy dropped dramatically to the level of El Salvador. The loss of heavy industry led to one of the last places of all US states for quality of life according to the last US census. In some cities, there was no longer a working population. The older ones were sick and dependent, the younger ones failed the urine tests for drugs to get simple jobs like truck drivers or garbage collectors.

The US federal government was already aware of the full extent in 2017, as the US Drug Enforcement Agency DEA and the Ministry of Health revealed. Since the year 2000, 300,000 Americans have died from opioids, that's 175 people per day and more than in the entire Vietnam War. More than 12 million Americans, who are older than 12 years, regularly take opioids and one million of them consume heroin. It is astonishing that 85% of all opioids that are legally sold in the world are prescribed and consumed in the United States. The rate of babies born with drugs has increased by 500% in the last 20 years, and many children are given up for adoption because the parents find the drugs more important than themselves and their child. What was this drug, what was Oxycodone, that millions of people longed for? Let's take a look back to the year 1916.

The world was at war when Martin Freund (1863–1920) and Edmund Speyer (1878–1942) introduced the new opioid of the company Merck at the University Frankfurt in 1916.[18] It was later called later Eukodal® and was used for coughs and severe pain. The Darmstadt company launched it on the market in 1919 and was successful. There was a chemical peculiarity, because Freund and Speyer did not use morphine as a starting material, they synthesized oxycodone from Thebaine, a biosynthetic precursor of plant morphine. Thebaine itself has no effect on the opioid receptors and is now only used as a synthesis precursor to produce to Buprenorphine, Naloxone and Naltrexone to produce. Buprenorphine is a painkiller with a very short duration of action, making it ideal for surgery. Naloxone is an opioid antagonist administered by emergency doctors in cases of overdose and has already been introduced. The opioid antagonist or the opioid antidote is often administered in combination with a strong opioid to prevent abusive overdosing and reduce the risk of constipation. The idea of the combination was also transferred to oxycodone, as this opioid quickly leads to abuse, which has been known since the 1920s. In professional journals of that time, there was talk of "Eukodalism", i.e., addiction to oxycodone and opioid dependence. It was said that Adolf Hitler was supplied with oxycodone by his personal physician Theo Morell, and the leader swore by this painkiller as much as by the amphetamines he received from his personal physician. In the years when Morell was the leader's personal physician, he administered about 800 injections and Eukodal® according to his own diary. This earned him the title "Reichsspritzenmeister" as Göring spitefully called him.[19]

Oxycodone became known as a strong and quickly addictive opioid and was placed on the list of painkillers to be controlled by the authorities of the Weimar Republic in 1929. For the right kick, as the addicts knew already in the 1930s, the tablets had to be crushed, dissolved, and injected into the veins, so that the opioid arrived quickly and strongly in the brain. The euphoric potential was twice as high as that of heroin® and the analgesic effect surpassed that of morphine. But the doctors also knew about the danger: Oxycodone was the queen among the opioids and one of the first designer drugs that was even legal. Klaus Mann (1906–1949), the son of the great writer Thomas Mann, tried this opioid: *"I don't take pure morphine. What I take is called Eukodal. Little sister Euka. We find it has nicer effects."* Despite the known addictive potential, the painkiller Eukodal® surprisingly remained in the drawers of pharmacies for a long time. Only in 1990 was it removed from the market and replaced by Oxygesic® sustained-release tablets, which could no longer be crushed and injected into the vein.

The opioid crisis of this millennium was not the first painkiller crisis in the USA. Although former President Donald J. Trump claimed in 2017, *"that the nation had never before experienced such a crisis"*, a look at the history books shows that almost exactly 100 years ago, a first opioid crisis shook the country, even if oxycodone was not the trigger. This crisis was accompanied by racism against blacks because of their cocaine use and against Chinese immigrants who brought opium smoking from their homeland to the USA. Most white Americans of the 19th century saw and tolerated opium smoking as a typical Chinese characteristic, but equated this behavior with prejudiced irrationality, weakness, and effeminacy. *"There is something in the character of these vast Asian populations"*, a newspaper report said in 1866, *"with their corrupt and effeminate manners and their decided preference for negative pleasures and a dreamy and contemplative life, which seems to draw them in a special way to the stimulus that opium provides"*. Harry Hubbel Kane was among the dedicated authors who pursued opium smoking and later wrote against it. He saw not only the general decay of morals and customs but also the poisoning of the "respectable class" and their children. To solve this problem, he called for a halt to the immigration of Chinese to the USA. Between 1860 and 1869, opium imports to the USA increased from 10 to 75 tons, and the number of active opium smokers was estimated at 100,000 to 150,000, including not only Chinese but all immigrant nationalities. The public debate about the opium ban was fought with hard gloves and terms like madness, debt, suicide, poison for hope and ambition, impotence, and destruction of physical and mental functions were not missing in the public debate. Already at the end of the 19th century, US politicians demanded laws to criminalize consumption and President Theodore Roosevelt (1858–1919) came under public pressure. In 1908, he appointed the doctor Hamilton Wright from Ohio as the first drug commissioner of the USA, who was to deal with the fight against the drug epidemic in the USA.[20] *"The habit has this nation in its grip to an astonishing extent"*, Wright said in a 1911 interview with the New York Times. *"Our prisons and hospitals are full of victims, it has robbed ten thousand businessmen of their moral sense and turned them into beasts who exploit their fellow men, […] it has become one of the most fruitful causes of misery and sin in the United States."* It is not surprising that the assessments of then and now are similar. Hamilton Wright wrote the law that banned opium smoking, and from 1909 every opium pipe between San Francisco and New York went out. At least that's what the authorities believed, but after the Second World War, 700 opium smokers were still arrested in New York, and the last opium den was smoked out in 1950.

Not only smoked opium, but also swallowed and injected morphine spread in the societal veins of the white American middle class. Even though the New York Times thought morphine was a problem of blacks, Italian immigrants, and tramps, and it hardly occurred in the working class, many doctors disagreed, who saw in morphine *"the relief of the pains of existence"* in a unleashed assembly line industry. What we know today as burn-out was diagnosed as Neurasthenia, a state of exhaustion. However, it was diagnosed less among workers than in the intellectual middle class and was widespread in the USA. As 100 years later, doctors prescribed their patients opiates like morphine and made them dependent. The New York Times turned against the practice of doctors who considered neurasthenia a fashionable disease and stated that the widespread morphine dependence is not a "fashionable disease" and *"is in reality a disease that is caused not by the fault of its victims, but their physical and mental defects mainly through the mediation of their doctors"*.[21]

The "Harrison Act" of 1914 was the preliminary legislative climax until the "Anslinger Tax Act" 30 years later and criminalized the non-medically justified consumption of narcotics and cocaine. Francis Burton Harrison (1873–1957) was a representative of the Democratic Party in the Congress of the United States of America and responsible for health policy. His law also had the support of the Republicans, who were bothered by the unrestrained drug use of their fellow citizens. Just a year later, the crackdown began with full force, and doctors who prescribed opiates and cocaine to drug addicts to avoid withdrawal symptoms had to expect punishment or imprisonment.

Of course, the first opioid crisis in the USA between 1870 and 1920 cannot be directly compared with the crisis of the 2000s, but some parallels are obvious.[22] Like today, there was lax legislation to the advantage of the pharmaceutical industry with little protection for the patients. It was also possible to prescribe opiates for diseases that were not therapy-appropriate, and many doctors and pharmacists did not have to fear professional consequences if they prescribed or dispensed strong painkillers very laxly or even negligently. However, there is a fundamental difference to today's crisis. In the first opiate crisis, the state was convinced that help comes before punishment. Even the tough Harrison Act allowed doctors to do everything that enabled the rehabilitation of their patients. This changed abruptly with a certain Mr. Anslinger, who saw the war on drugs also as a war on addicts. We will get to know Mr. Anslinger in the chapter on cannabis, but it can already be revealed here that he stigmatized addicts as characteristically bad and in his eyes they were per se criminals. A trauma that American society still carries within itself today. We will later read how this simple bureaucrat created one of the most powerful agencies and shaped drug policy on both a national and global level as we know it today.

Cocaine—from Drug to Local Anesthetic

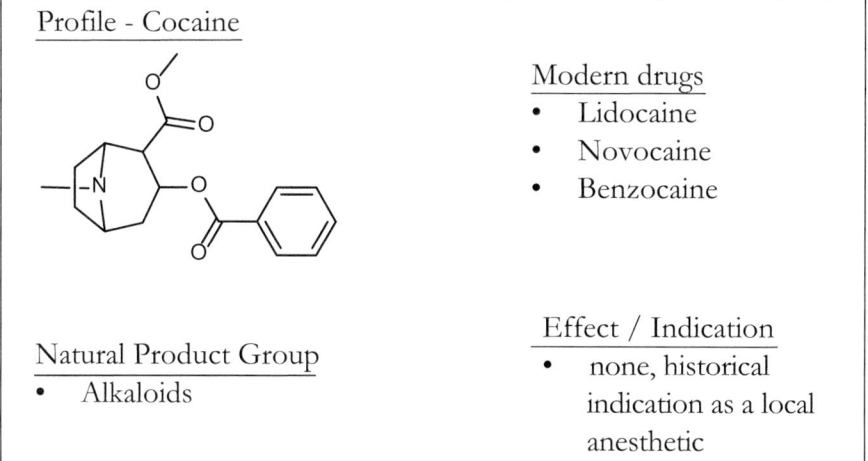

No less fascinating than the history of morphine and strong painkillers is that of cocaine. It is a pharmaceutical promise and a pharmacological nightmare at the same time. If it weren't for the local anesthetics derived from cocaine, we would fear going to the dentist. The other side of the coin is the pandemic cocaine flood, the flourishing of drug cartels and the misery of cocaine users in the metropolises of this world. Cocaine is an alkaloid that is extracted from the coca bush, which is native to the highlands of the eastern Andes of South America (Peru, Colombia, and Bolivia). The first description of the coca plant *(Erythroxylum coca)* (Fig. 1) by a European comes from Nicolas Bautista Monardes (1493–1588) in his trilogy "Historia

Fig. 1 Coca plant (*Erythroxylum coca*) (public domain, Wikipedia)1

medicinal de las cosas que se traen de nuestras Indias Occidentales", which was published in 1565 in Seville, Spain. It describes the use of the coca plant, which was a sacred ritual for the Indians, the embodiment of a god, and whose consumption was reserved for outstanding religious leaders. The Spanish conquerors broke this customary right at the Second Council of Lima in 1569 and allowed all Indians free access. It took another 200 years for the coca bush to reach Europe from South America in 1750 and attract the interest of science. Almost another hundred years would pass before the zoologist, geographer, and explorer Eduard Friedrich Poeppig (1798–1868) as a young scientist became interested in the healing properties of the plant. He was one of the most important researchers of America and visited the young American states of the USA, Cuba, Chile, and of course Peru for research purposes. His life was highly interesting, because like a Charles Darwin or Alexander von Humboldt he mostly explored unknown landscapes and indigenous peoples alone. Poeppig must have been a reserved, perhaps shy person. Despite his enormous collecting mania of 17,000 plants, several hundred stuffed animals, maps, and plant drawings, which can be seen today in museums and archives in Dresden and Leipzig, he published little. His publications were rarely translated into English or French, but into Spanish and Portuguese and partly also into Russian. Perhaps this is also the reason why his person and his work have received so little attention in Germany.

> **Chronology**
>
> **1836** – First description of the biological effects of the coca plant by Eduard Poeppig
> **1860** – First isolation of cocaine by Albert Nieman
> **1884** – Freud publishes his work "On Cocaine"
> **1895** – Carl Koller describes the local anesthetic etic effect on the eye
> **1894** – Willstätter clarifies the correct cocaine structure.
> **1890** – Eduard Ritsert discovers benzocaine as a local anesthetic

During his expedition from 1829–1831, he lived as a hermit in a simple hut in Pampayaco on the eastern slopes of the Andes. Just before Christmas, his hermit existence almost came to an early end when he was bitten by a poisonous snake. Poeppig believed that his last hour had come here in Peru. Near death, the Indians gave him a drink and leaves of a strange plant. They did not give up on him, even though he had already written his will. Fortunately, he recovered from the poisoning. During these two years in Peru, he got to know the coca bush. Like any good researcher of his time, after his recovery, he chewed coca leaves and studied the effects in a self-experiment. In 1836, he noted the effects after ingestion and described typical effects such as euphoria, suppression of hunger, and increased activity, which we all know well today. The bush seemed to have something special and aroused the interest of Friedrich Wöhler (1800–1882, Fig. 2) in Göttingen. He asked the Austrian Karl von Scherzer to bring back plant material from an expedition so that his doctoral student Albert Niemann (1834–1861) could isolate the miraculous substance. Karl von Scherzer(1821–1903) promised and brought back coca leaves from the Austrian Novara expedition 1857–1859, which the young Niemann immediately extracted.[23] In 1860, a year before his death, Niemann held the white crystals of the alkaloid, which he called cocaine, in his hands. Unfortunately, Albert Niemann fell very ill, the exact causes are not known. It is assumed that it was the late effects of chronic mustard gas poisoning. Before studying the coca plant and cocaine, he dedicated himself to mustard gas as a new chemical, which gained a very sad significance as a warfare agent Lost in the First World War. Niemann

Fig. 2 Friedrich Wöhler (1800–1882) (public domain, Wikipedia)1

underestimated the danger and must have come into contact with the chemical. After his death, the work was continued by Wilhelm Lossen (1838–1906), who succeeded in determining the molecular formula of cocaine in 1862.

Whether Albert Niemann was actually the first to discover cocaine is disputed in the professional world. Also in the ring were Friedrich Gaedcke (1828–1890), who found a substance in 1855 that he Erythroxylin called and also numbed the tip of the tongue. Also, the Italian Paolo Mantegazza (1831–1910, Fig. 3) raises his hand as a candidate, as he presented cocaine in pure form at the University of Pavia in 1858. It is only certain that Albert Niemann is the author of the published first description of cocaine, and that is the decisive factor.

In contrast to the aforementioned gentlemen, Paolo Mantegazza was neither a chemist nor a biologist. He studied medicine and conducted research as a neurologist and physiologist in Pisa, Pavia, and Milan. Coming from a wealthy bourgeois background, he afforded himself the luxury of practicing as a doctor in India, Argentina, and Paraguay. In South America,

Fig. 3 Paolo Mantegazza (1831–1910) (public domain, Wikipedia)1

he often visited his patients in the countryside and observed coca farmers who chewed coca leaves and pushed them into their cheeks. Becoming curious, he followed the farmers' example, chewed the leaves, and this changed his life. Paolo Mantegazza can certainly be described as a pioneer of drug research. In addition to his botanical descriptions of the coca bush, its effects on humans, and probably also of pure cocaine, his interest in psychotropic substances went far beyond. He published writings on the effects of alcohol, mate tea, opium, hashish and ayahuasca, which he classified according to their effects, as we still do today.

Coca extracts quickly became popular as a pleasure substance in Europe, but their potential as a medicinal substance was not yet recognized. The French pharmacist Angelo Mariani (1838–1914) introduced his Vin Mariani to the market, an alcoholic extract from the coca plant. This Bordeaux wine was very popular with the creative bourgeoisie like Émile Zola, Jules Verne, Henrik Ibsen and the upper ten thousand. Queen Victoria of England took coca wine for her menstrual discomfort and the two Popes

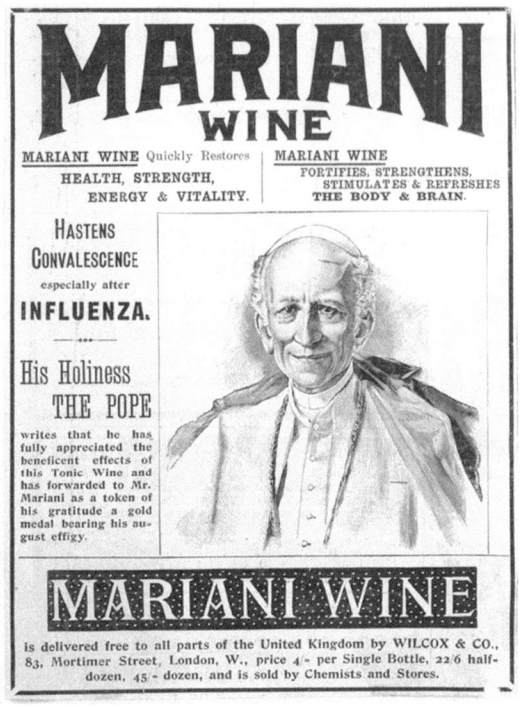

Fig. 4 Vin Marini, advertisement with Pope Leo XIII (public domain, Wikipedia)1

Leo XIII. and Pius X. for physical strengthening (Fig. 4). Pope Leo indirectly advertised the sinful drink by awarding a gold medal and featuring his head on advertising posters. One could describe the coca wine as functional food of the 19th century, as it was highly recommended against rickets, anemia and general weakness. Naturally, such a bestseller found imitators, especially in the USA, where the morphine-addicted pharmacist John Stith Pemberton (1831–1888) distributed "Pemberton's French Wine Coca". His noble drop aroused the wrath of a part of the American population. It was outrageous to bring a wine to the market. The stumbling block was not the cocaine, but the alcohol. The temperance movement in the USA saw the danger not coming from South America, but from the breweries and cider mills of their own country. Pemberton removed the controversial, godless alcohol and created a refreshment drink only with cocaine, which he called Coca-Cola. A product icon and a global mega-corporation in Atlanta, USA, were born. Anyone who believes that the first Coca-Cola bottles were a lemonade as we know it today should be told that Coca-Cola was also intended as Functional Food. Pemberton advertised his Coca-Cola as *"valuable brain food that can cure all possible nervous symptoms: nervous headaches, neuralgia, hysteria, and melancholy."* The main ingredient was cocaine, which kept customers loyal. It is not exactly known how much cocaine was in a bottle, but it can be assumed that the amount was sufficient to draw a line. The customers remained loyal and patiently waited for each new delivery of their tonic. When cocaine was banned in sodas in 1903, the end of Coca-Cola seemed to have come. The company became inventive and extracted the cocaine from the leaves before the extract was added to the drink. The old target group fell away, the corporation invested huge sums in advertising, especially targeting teenagers who wanted to identify with the risqué image. It worked, as we know, because Coca-Cola is still a global corporation today and represents the American Way of Life like no other product.

During the short period of its legality, the coca bush was not only cultivated in South America. Early on, the European colonial powers recognized that coca could also be grown far from the Andes if similar climatic conditions were present. The Netherlands had today's Indonesia with the capital Batavia, later Jakarta, as a colony. The capital is located in the northwest of the island of Java on the Java Sea. However, the location in a tropical country was unsuitable due to the hot and humid climate. It had to be a climate that, similar to Peru, showed a mountainous location with moderate temperatures. Bogor, today about an hour's drive southeast of Jakarta, offered itself to cultivate coca plants in a botanical garden with the name "Buitenzorg" (Dutch for Without Worries), in reference to the name of the Potsdam Palace Sanssouci. The pleasant high-altitude climate was highly appreciated by the Dutch, and Bogor and Buitenzorg quickly developed into the residence of the governors-general and their administration, who ruled Indonesia. Perhaps it was the Dutch love of flowers or the search for a piece of home far from Europe that prompted the Dutch East India Company to take care of the maintenance and preservation of a small botanical garden or royal park, which had been used for 500 years in Bogor for the cultivation and preservation of rare trees. King Sri Baduga Maharaja (1401–1521) is said to have lived a proud 120 years, which may have been due to the good climate in Bogor. He was ruler over the area of Bogor and established the garden for the cultivation of tropical plants. The Dutch East India Company(VOC), however, was accountable to its shareholders and had a different, rather pecuniary interest in the garden. In 1817, part of the garden was converted by decree into an agricultural research station, where, among other things, the cultivation of the cinchona tree *(Cinchona pubescens)* and rubber tree *(Hevea brasiliensis)* was researched. Both, like the coca plant, come from Peru and prefer to grow in mountain regions. The cinchona bark and its extracts were formerly used as a medicine against malaria and fever, as we will see in the chapter on the history of quinine. In the colonies of Africa and Asia, heavily affected by malaria, it was not only a blessing to be able to protect oneself from this infection, but also decisive for war. For most soldiers did not die from the enemy's guns, but from the bites of the infected Anopheles mosquito.

The stimulating and invigorating effect of cocaine promised high demand and good business in Europe and the USA. It was just too bad that the Peruvians and Bolivians were watching very closely who had the plants. Export was forbidden and the theft, which was carried out by the Dutch and British, was a typical case of biopiracy, as we call this kind of robbery today. For the British and Dutch, these legal subtleties were not important, they were more interested in who they could hire as a biopirate and where the plant could be grown in their Asian colonies. For the British, the mountains near Nilgiri in India were a closer choice. For this project, Kew Garden near London was to take the lead and we will read in a later chapter how this botanical garden became the imperial center of biopiracy.

For the Dutch, it was clear that the coca plant was being cultivated in Bogor. In December 1854, Justus Karl Hasskarl (1811–1894, Fig. 5) returned from an expedition deep into the Peruvian rainforest, where he had been sent under a false name by the Dutch government. His secret mission was to research cinchona trees for cultivation in Java. He had no success with the cinchona trees, as all the seedlings died on the way to Java and the seeds only produced trees with very low yield. Naturally, he had not overlooked the coca-chewing indigenous people. He reported back to his employer and recommended that the Dutch East India Company try cultivating the coca bush in Java. As the company did not expect large profits and local government officials feared that the Javanese might misuse the plant as a drug, the proposal was discussed very cautiously. A few years later, this opinion would change fundamentally. Cocaine found great popularity in Europe, and the first companies, such as Merck in Germany, began production. The first two coca plants arrived in Java in 1875 or 1876 and were purchased by the company Hermann Linden in Ghent, Belgium. These two mother plants were the basis for later cultivation in Bogor. A plantation of 300,000 plants in Tjikeumeuh developed near the Botanical Garden, after the discovery and representation of cocaine by Carl Koller led to an explosion in the demand for pure cocaine.

The directors of the Dutch East India Company, such as Rudolph Herman Christiaan Carel Scheffer (1844–1880), were rubbing their hands

Fig. 5 Justus Carl Hasskarl (1811–1894) (public domain, Wikipedia)1

together, thinking they had struck a green gold vein, but things were to turn out differently. The purchased plants of the genus *Erythroxylum coca,* variety *spruceanum*, contained little cocaine compared to the English varieties grown in India and on Ceylon (now Sri Lanka), but almost twice as many secondary alkaloids. This was bitter for the growers, and the disappointment was all the greater when 20 tons of the dried plant found no good market either on the London Stock Exchange (1890) or on the Amsterdam Stock Exchange (1891). Later, Java coca, as it was called by the traders, was no longer offered on the exchanges due to its poor quality. The area of the plantations shrank to 200 hectares. Coca plants from Peru formed a natural monopoly and often led to far overpriced prices, which could be easily enforced due to lack of competition.

German cocaine producers wanted to free themselves from Peruvian dependency and open up cheaper sources that offered them price and delivery security. This is the only way to explain the visit of Willie Merck to Buitenzorg in 1888, who wanted to learn about coca cultivation in Java. The solution to the trade problem and the rescue of Javanese coca cultivation was only brought about by the German chemical industry. Two teams of scientists were in a race to figure out how to produce cocaine from by-products through conversion. Two competing research groups were working with Carl Liebermann(1842–1914) and Fritz Giesel(1852–1927) on one hand, and Alfred Einhorn(1856–1917) and Otto Klein on the other hand, on the question. Liebermann was a professor of chemistry at the Technical University in Berlin-Charlottenburg and his colleague Giesel was employed at the quinine factory Buchler & Co. in Braunschweig at that time. Their competitors were also respected and well established: Alfred Einhorn was an academic advisor at the Polytechnic University of Aachen, today's RWTH Aachen, and Otto Klein was working at Boehringer Mannheim. In their 1888 published work, Einhorn and Klein described the splitting of esters by acids and the subsequent re-esterification to methyl and benzoic acid esters. This was the breakthrough for the originally considered inferior Java coca. With their methods, 15% cocaine could now be obtained instead of the original 2%, and the coca plants and their plantations in Java flourished again.

Why share the money with the Germans when you can benefit from the value chain yourself? A Swede and a Prussian founded the Nederlandsche Cocaine Fabriek, or NCF for short, in Amsterdam in 1900. Cocaine and later morphine, as well as synthetic substances Novocain, Ephedrine and Heroin were produced in Schinkelstraat in the city center, where the Hells Angels Amsterdam also had their address at one time. The founders Georg Boldemann and Otto Eberhard were not squeamish and used German patents. This was not punishable at the time and did not constitute a patent infringement, as there was no patent law in the Kingdom of the Netherlands and therefore no patent infringement. Boldemann was considered the financial expert and visionary of the NCF, while the Mecklenburger Eberhard took care of the technical matters. The NCF was successful, because with the outbreak of World War I, Merck KGaA in the German Reich lost its markets, the Netherlands were neutral and the NCF could deliver. It should be noted that Merck in Darmstadt had a global market share of 39% in cocaine at the beginning of the 20th century. In 1920, the NCF was able to deliver around 2000 kg of cocaine, but the infamous success story continued until shortly after World War II, when amphetamines were produced for the German Wehrmacht.

Most of the money was not made with the plant, but with the isolated alkaloid. The major companies that divided the world market among themselves came from the Netherlands, Germany, and the USA. In Germany, the companies E. Merck KGaA, Boehringer Mannheim, Knoll, Gehe and Riedel produced cocaine, in the USA Parke-Davis and Mallinckrodt. All produced the hydrochloride as a salt and sold it loose to intermediaries, such as the chemical and pharmaceutical wholesale trade. Before World War I, the two German companies Merck KGaA and Boehringer Mannheim accounted for 50% of world production and almost 70% of German production. For the company Merck KGaA, sales figures are also known from the books. These rose from just under 1 kg in 1883 to 1553 kg in 1888 and served 25% of the world market. The record amount of 5678 kg in 1914 collapsed to 235 kg with the outbreak of World War I in 1915. In 1913, the NCF came very close to its two competitors and overtook them after World War I. It is not surprising that the NCF already extracted 500 kg of pure cocaine from Java coca in 1913.

It is little known that soldiers in World War I were given cocaine on both sides of the front to alleviate their fear of coming out of the trenches and facing machine gun fire. After World War I, this led to many human tragedies, as tens of thousands of soldiers became addicted. Reports came from London that war veterans were robbing pharmacies, and in Berlin up to 10,000 cocaine addicts were registered in hospitals.

But back to the 60s and 70s of the 19th century. Cocaine was a legal chemical substance extracted from the imported coca plant. It can be assumed that the effects and especially the addictive potential of cocaine were known, even if this was not necessarily seen that way in society. Everyone could get it at the pharmacy, and in medicine it was believed to help with withdrawal from morphine and alcohol. A madness, because one drug was replaced by another. In the end, the so-called polytoxicomaniacs filled the graves, as they were now dependent on several drugs at the same time.

Today, the white powder must be seen as an unexplored fashion drug of its time. It was fashionable in a world that was in upheaval. With the invention of the automobile, the railway, the airplane, and the discovery of electricity, the world exploded from the mid-19th century onwards. Many buildings grew into the sky thanks to steel frame construction, at least that's what people believed when they saw the 91 m high Flat Iron Building in New York. Cars that drove 100 km/h were considered record-breaking sports cars. Industrialization brought consumption, and the work week was no longer characterized by 16-hour days, but for the first time there was free time for pure pleasure. Cocaine enhanced the euphoria of modernity and, along with alcohol, cigarettes, and sometimes morphine, was a sign of success and abundance in the emerging American and European society.

One of the most prominent cocaine users of this time was "Sniffing Siggi", better known by his name Sigmund Freud (1856–1939, Fig. 6), who wanted to try out this "magic substance", as he called it, out of scientific curiosity.[24] From the bridal letters to his fiancée and later wife Martha Bernays we know today about his motives and experiences with cocaine.[25] Sigmund Freud was an assistant doctor at the General Hospital in Vienna in 1898 and was looking for a brilliant idea for his research that would make him beautiful, rich and famous. This was not easy. As a Jew, he was not taken seriously and building a scientific career was more difficult then than it is today.

If you want to become a professor today, you must successfully complete your doctorate after your studies. It is expected that you will publish your scientific works for the doctorate in very good scientific journals. From the day you hold your doctoral certificate in your hands, it becomes tricky if you want to stay in science. Young scientists rarely stay at their alma mater, they move as so-called postdocs into the world to work scientifically in foreign laboratories again. If the enthusiasm remains high and you find a mentor who wants to keep you at the university, the signs for a scientific career are not bad. The postdoc employment contract is extended and the opportunity to work scientifically is given. Some also speak of the scientific

Fig. 6 Sigmund Freud, around 1905 (1856–1939) (public domain, Wikipedia)1

proletariat here, as chain contracts often determine income over many years. Now, according to the German microbiologist Paul Ehrlich, it's the 4 Gs in German like Geduld, Glück, Geld und Geschick (patience, luck, money and skill), that decide about the career. If everything goes well, after a few years follows the appointment to a professorship or with the habilitation another degree. This entitles the young scientist to carry the title "Privatdozent". This brings honor and a longer title on the business card, but no higher income or an immediate professorship. For many scientists, a multi-year odyssey of invitations and lectures begins now, which are jokingly referred to as "auditions" in university circles. Professorships are a scarce commodity in the scientific operation and it can take until the middle or end of the fourth decade of life with an appointment. It can often also be assumed that a helpful network of lectures, personal cooperations and good relationships with former superiors and mentors is certainly not a disadvantage to accelerate one's own career.

What did the appointment look like in Sigmund Freud's time? It is astonishing how great the parallels to the present day are and how many structures in the scientific community have remained to this day. Dr. Freud was in a difficult situation. He had no formal position that allowed him to fully focus on research. He was not a postdoc, and the love letters of the then 28-year-old from the years 1884 and 1885 express his wishes, hopes, but also his frustration that he did not quite know what he should research and how he could achieve a professorship. Money was tight, there was no health insurance with fee schedule for a secured income, and he needed money for the trip to Wandsbek near Hamburg, where his fiancée was waiting for him. He wanted to get married and the costs for the wedding, which was to take place soon, he also had to raise. But Sigmund Freud wanted to become a respected doctor and specialist: "*I do it*[the research, note by the editor], *because I want to make people aware that I exist and that I will deal with nervous diseases*". A look at his over 150 publications shows how flexible or rather fickle Sigmund Freud was in order to survive in the scientific community of his time. This was both a claim and an interest, because in addition to his very extensive medical studies, he also attended lectures in logic and philosophy. He read the psychological writings of Franz Brentano(1838–1917) and delved into special subjects such as internal medicine, physiology, dermatology, and neurology. In neurology, he conducted brain anatomical studies, developed a staining technique for brain sections, and dealt with the just-understood electricity and its application in the diagnosis and therapy of nervous diseases. To make matters worse, the instruments were not as readily available in the labs as they are today. No, he had to buy them with his own money. We find in some of his letters listings of income and expenses that well document his investments in scientific equipment and not just in cocaine.

Unhappy to be employed as a secondary doctor for neurology at the General Hospital (AKH) in Vienna, he found cocaine and his new field of research in 1884. It could have made him world-famous back then, but his scientific curiosity and his talkativeness brought his colleague Carl Koller the honor of going down in medical history as the discoverer of the local anesthetic. Perhaps that's just as well, because who knows if Sigmund Freud would then have had the desire and time to think about psychoanalysis, for which he received the Nobel Prize. But let's stay in the year 1884 for now.

Freud read about the effects of the coca bush and was interested in this cocaine from South America. A publication by the doctor Dr. Theodor Aschenbrandt in the German Medical Weekly prompted him to obtain cocaine from the Angel Pharmacy at the Viennese court: "Good day, very pleased, I am Mr. Freud and want to draw a line with cocaine." That's probably not how it was when he came into the pharmacy, because the literature about Mr. Freud attests him noble motives. Where did the cocaine actually come from? Were they then as now dark drug labs that met the demand? No, reputable pharmaceutical companies took over the extraction and distribution. A company from Darmstadt we have already met in the chapter about morphine. The company E. Merck, specialized in plant alkaloids, distributed cocaine since 1862 and sold it for 6 marks per gram when Freud acquired it in Vienna in 1884. That was a lot of money for the young secondary doctor, who had to spend almost 10% of his monthly salary on his addiction. Freud immediately negotiated a quantity discount with the pharmacist, because after his self-studies he was convinced of the effect, prescribed it to his patients and also to his mother-in-law. From our point of view naively, he gave his friend Fleischl, who suffered from severe pain, cocaine for morphine withdrawal. It did not go well, unbearable pain made further morphine injections necessary and he died early as a morphine and cocaine addict. For Freud, a world collapsed. Why did Freud believe in cocaine as a substitute drug for morphine addicts? It was due to another article titled "Coca in the Opium habit" that the American Bentley published in 1880 in the Therapeutic Gazette appearing in Detroit. What Sigmund Freud perhaps did not know or did not take seriously enough was the fact that the editor of this supposedly scientific journal was George S. Davis, who also ran the pharmaceutical company Parke, Davis & Co., the largest cocaine producer in the USA. This was a clear conflict of interest, and George Davis used the reputation of the journal for his advertising purposes by publishing articles that were pseudoscientific.

The death of his friend Fleischl made Sigmund Freud thoughtful, he regretted his naivety and carelessness towards cocaine and his recommendations. However, Freud did not give up and looked for further applications for the alkaloid, such as depression, nausea during pregnancy and sea travel, heartburn, and migraines. Neither patients nor relatives were thrilled by the effect and Martha sometimes reacted with mockery. But Sigmund did not let up and defended his "magic remedy" and the cocaine research in self-experimentation. He found that he became much more relaxed during lectures, his depressive moods disappeared, and he was less tired.

At this point, I have to interrupt and discuss the question, are self-experiments allowed today? We have already read in the first two chapters that many researchers did not shy away from trying out chemical substances themselves and the list is long and crazy.[26–28] Self-experiments with drugs or active pharmaceutical ingredients are as old as humanity. An early example is James Jurin (1684–1750), who conducted a self-experiment in 1740 to find out whether the intake of soap suds alleviates bladder stones. Samuel Bard (1742–1821) and the intake of opium in 1765, Samuel Hahnemann(1755–1843) founded homeopathy with a self-experiment in 1790, Carl Koller discovered the local anesthetic effect of cocaine on the eye in 1885, Albert Hofmann(1906–2008) took LSD in 1943 and Alexander Shulgin(1925–2014) tested ecstasy in a self-experiment in the 1970s and produced 200 more ecstasy derivatives in his private laboratory on his farm in California. Alexander Shulgin had humor when he referred to Ecstasy as "Low-Calorie" Martini. These are just a few researchers from the field of pharmacology. The list would be even longer if we were to list all infectious disease specialists and surgeons. We see that scientific curiosity knows no bounds when it comes to one's own health.

From today's perspective, the scientific approach of Anton von Störck (1731–1803, Fig. 7) was logical and pioneering. Von Störck was the personal physician of Maria Theresa (1717–1780), but also a studied philosopher and a very successful doctor in Vienna. He earned a good reputation as a doctor for the poor and his healing successes quickly spread in Vienna. He became the personal physician of Maria Theresa when he cured the empress of smallpox. When the empress was well, von Störck was active at the University of Vienna as a professor of medicine, with a particular interest in pharmacology. The term pharmacology at the time is of course incorrect, as the study of the effects of drugs was then equated with pharmacy. It would be more correct to assign the studies of the Protomedicus and the Rector magnificus to pharmaceutical biology. For he dealt with the effects of water hemlock *(Cicuta virosa)*, autumn crocus *(Colchicum autumnale)* and

Fig. 7 Anton von Störck (1731–1803) (public domain, Wikipedia)1

jimson weed *(Datura stramonium)*. What makes his work appear modern are the clinical phases, as we would call his step-by-step testing of plant extracts today. But let's let him speak for himself: *"However, I will undertake nothing that may be daring. I will first try everything on animals and myself before I give anything to a sick person."* Each plant or the extracts obtained from it were first tested on animals. If they showed no side effects or toxicity, he took them himself before giving them to his patients or even the empress. This order animal experiment before administration to subjects or patients is still valid today.

So, how do self-experiments fit into our well-regulated legal world? The concept of self-experimentation does not exist in either German or European law. It starts with the question of the purpose and meaning of a self-experiment. Do I undertake it as a therapeutic trial to alleviate or cure a disease I suffer from? Or is it a scientific experiment to, for example, understand the biological effect of a natural or chemical substance? With these exciting questions, the lawyer Maria Huber made it to the Olympus of sciences with her dissertation at the University of Vienna.[29] She states that it is undisputed that every person can do what he or she wants with his or her body if he or she is a consenting adult. This ability to consent presupposes that I, as a participant in a self-experiment, am of legal age and not mentally impaired. If I now need the help of third parties, who in particular monitor the self-experiment, call for medical help in case of problems or are involved in the implementation, it becomes complicated. The author of the dissertation therefore distinguishes three categories of self-experiments. In the first category, researchers and research subjects are united in one natural person. In the second category, this person receives help or support through the participation of third parties, and the third category describes the collective self-experiment when a group of researchers conducts a self-experiment.

These questions are not easy to answer because there is no law that prohibits self-harm, whether pure scientific curiosity or healing is the goal. A look in the Duden shows that a self-experiment *is a trial (for research purposes) undertaken on one's own body*. This means that the researcher, at the moment he or she conducts the self-experiment, is also the object of knowledge acquisition. Can or may then only trained scientists conduct self-experiments and the right is denied to other non-studied people? Or may only doctors conduct self-experiments with chemical substances and medicinal substances on themselves? No, in principle there is no prohibition for any of us. Self-experiments are also known among patients who, in their desperation of a therapy-exhausted treatment, grasp at the last straw. We can read exemplarily about the diverse and often questionable self-treatment with cannabis, which is fueled by many opinions in blogs. Whether cannabis is useful as medicine will be discussed later.

A self-experiment also occurs when a drug, which is not approved in its type or dosage for a disease, is taken by a person. What a drug, a medicinal substance, and an approval are will be explained in a later chapter. Approval is understood to be the permission of an authority that gives a pharmaceutical company the right to sell. At this point, it is important to understand that doctors may also prescribe drugs that do not comply with the approval. This so-called "Off-Label Use" is viewed very critically and in case of damage due to side effects, the pharmaceutical company bears no responsibility or liability. Examples are Nifedipine, a blood pressure reducer, used as a labor inhibitor, Trazodone as a sleep aid, or Retinoids for flat warts on the skin.

Back to Sigmund Freud and his self-experiments with cocaine. According to today's legal view, his self-experiment would not be legal, as he took an illegal, i.e., non-prescription and non-marketable substance. Possession and intake are punishable and even trying out of curiosity is not legally covered. Today, Freud would have to obtain a permit from the ethics committee Ethics Committee of the University of Vienna and with this in turn apply to a state authority for a permit for the acquisition and intake of cocaine. If it is serious research, the state control bodies usually have understanding and allow controlled studies. Whether they would have approved a self-experiment without prior animal and toxicity tests, I doubt. According to Ms. Huber, *"the scientific progress must always outweigh the disadvantages of the self-experiment"*.

But as so often in jurisprudence, one gets into a dispute, which is definitely not a self-experiment. Where does the self-experiment end? I think it is difficult to explain to the judge that one wanted to try heroin in a self-experiment. This is a clear violation of the Narcotics Act. Running naked through the city center when the unknown plant is psychotropic? More likely, but it's not unlikely to end up in front of the judge for causing public annoyance. Will she take it seriously when there are already 400 proceedings on the table? It can be stated that a self-experiment must be reconciled with the concept and essence of research. We must read from Ms. Huber that other laws complicate the question of legality. Thus, in addition to medical law, labor law is also decisive. If the employer expressly prescribes that self-experiments are to be refrained from, the self-experimenter must also refrain from these out of a duty of loyalty beyond working hours in order not to lose his or her performance capability. It is also interesting to note that a self-experiment that leads to an invention must be reported to the employer.

Self-experimentation is therefore not punishable, permitted, and falls under the freedom of research. However, its limits are found in the restriction of the physical integrity of third parties or by laws, as we have seen above. If the person conducting the self-experiment is harmed or even dies, the causing of death cannot be considered as suicide. Mrs. Huber refers to this as negligent self-harm or negligent self-killing. In summary, it can clearly be referred to as self-experimentation when it involves researchers who carry out a new, previously unexplored scientific question on their own body with full mental self-responsibility and this experiment does not violate the good manners clause.

What about other involved persons who may be asked to witness the self-experiment? The request can be made to help in an emergency or to call for medical help. Self-experimenters and involved third parties would be wise to draw up a contract about their actions and the expected assistance.

As a consequence of a not well-regulated legal situation, it can be said that almost no limits are given to self-experimenters by our current laws, disregarding third parties. Only a few restrictions, such as labor law or the fact that drugs, psychotropic substances or natural substances must be legally obtained according to the Medicinal Products Act or Narcotics Act, can make self-experimentation more difficult.

The Path of Cocaine to Local Anesthetics

Cocaine found its way as a model substance for local anesthetics through self-experiments and abuse. In the time of the k. u. k. double monarchy, in which Freud lived, snorting a line of cocaine was not known. Freud and all other cocaine users did not snort, they took the cocaine orally. This type of intake leads to a weaker effect, as a not insignificant part of the cocaine is destroyed by stomach acid. The reason why cocaine users snort today is certainly related to the rapid onset of action through the nose and the acid lability in the stomach.

Understandably, Freud noticed that the cocaine numbed his tongue tip when taken orally. During a cigarette break with colleagues in the courtyard of the AKH in Vienna, an assistant doctor with severe toothache toothache passed by the smoking group and Freud addressed the young colleague: *"I believe I can help, I have found that cocaine causes numbness of the tongue."* The colleagues were amazed and looked at him. Some shrugged their shoulders, but why shouldn't they believe colleague Freud? They followed the toothache sufferer and Freud into his lab, where he relieved the young assistant doctor of his pain with a few drops of cocaine solution. The group was thrilled, especially Carl Koller, who no longer felt the throbbing tooth. Koller immediately recognized the potential of cocaine. That same day, he took a frog pond frog from the institute aquarium of his mentor Professor Stricker, which was then a popular model for animal experiments. He dripped cocaine solution into the right eye of the croaking frog and waited a minute. The assistant doctor Gustav Gärtner later reported: *"…after about a minute, such insensitivity of the cornea had occurred that no reaction occurred to its touch, or even to its injury."* The dream of every ophthalmologist to operate on the eye without pain had been fulfilled.[30]

In a hurry, Carl Koller (1857–1944, Fig. 8) wrote his results on a piece of paper and gave it to the ophthalmologist Joseph Brettauer (1835–1905), who was briefly in Vienna and on his way to Heidelberg for an ophthalmological conference. Koller had no money for the trip to present his results and trusted that Brettauer would present the results and initiate the triumph of cocaine. For the first time, the sensitive eye could be operated on without pain. In Heidelberg was also Henry Drury Noyes, who immediately understood what he saw, and immediately reported to the Journal Medical Record

Fig. 8 Carl Koller (1857–1944) (public domain, Wikipedia)1

that he was an eyewitness to the cocaine experiment on the eye. In New York, ophthalmologists successfully repeated Koller's work, and cocaine was introduced into eye surgery [31].

Who was the actual discoverer of cocaine as a local anesthetic? Koller or Freud? Freud wrote an essay in 1884 "On Coca" (Fig. 9), in which he predicted a great future for his favorite substance as a local anesthetic.[32] Perhaps Freud did not take this entirely seriously, as he liked to talk and was more than surprised when he returned to Vienna well-rested from a vacation in Wandsbek in late summer 1884. He got off the train of the Empress Elisabeth Railway, walked past the newspaper stands at the Vienna Westbahnhof, and read the headlines about the sensational discoveries of his colleague Koller. He grabbed a newspaper from the stand and walked quickly past the Westend Café to the tram line to get to the AKH. In his love letters, we read that Freud was deeply hurt because he considered cocaine a child of his research. Now Koller had achieved what Freud longed for, the scientific recognition, the idea of the century, to start a medical career at the university. During his vacation, the bomb had dropped and Koller had become world-famous overnight. Freud later wrote that he was annoyed to have gone to Hamburg, but in later letters, he recognized Koller as the rightful discoverer of local anesthesia [33, 34] (Fig. 9).

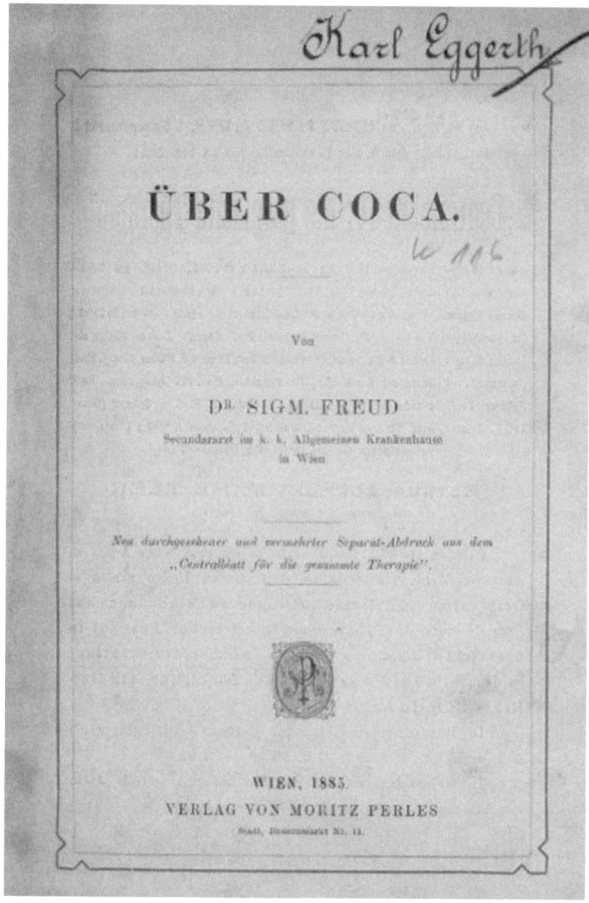

Fig. 9 On Coca, Book 1885, Sigmund Freud (public domain, Wikipedia)1

What happened to Carl Koller after his great discovery? During the time of strong anti-Semitism in Austria-Hungary, Koller was subjected to many hostilities. A serious incident also ended his career in Vienna and prompted him to emigrate to the USA, where he worked as an ophthalmologist at the Mount Sinai Hospital in New York and died in 1944.

What happened? In 1885, Koller turned against his colleague Zimmer, who was the assistant of the highly respected Professor Billroth, and removed the too tight tourniquet from a patient's finger, who was threatened with the loss of the organ. The hierarchical structures among the doctors in Vienna were almost militarily strict at that time. Zimmer shouted at Koller, one word led to another, fists flew in the hospital, and Koller's fist hit Zimmer's ear. Both were reserve officers and challenged each other. It was a matter of honor. At dawn on January 6, the brawlers met in the Josefstadt infantry barracks and shot at each other. Both injured each other and were admitted to the hospital where they worked. The police interrogated both duelists, a police investigation was initiated, but everything quickly fizzled out. Freud, who could have been resentful because of the publication of cocaine as a local anesthetic, supported Koller and stood by him. For Koller there was no future in Vienna anymore. He had to leave. Koller left the city in 1887 with a heavy heart and after stays in the Netherlands and Germany, he boarded the S. S. Saale to New York, where he lived from 1888 and practiced as an ophthalmologist. He died at the age of 84 and never saw Vienna again.[30, 35]

It is astonishing that the numbing effect of cocaine on the tongue was not recognized by many high-ranking scientists and that it took almost 25 years after the discovery of cocaine for its application to the eye, later in dentistry, and finally in surgery to become established. But there was no stopping it. Even Sigmund Freud's father underwent a glaucoma operation with cocaine as a local anesthetic, which Carl Koller personally performed. Spinal blocks were performed in surgery by W. S. Halsted in New York, dentistry adopted cocaine, and all possible applications were tried out. Unfortunately, the surgeon W. S. Halsted, whom we will get to know better in the chapter on the rubber glove, took a liking to cocaine and became addicted like many of his colleagues, whose addiction usually began with self-experimentation. The TV series "The Knick", which is about an imaginary hospital in New York at the end of the 19th century and in which a prominent cocaine-addicted doctor plays the role of Halsted, is recommended.

For a long time, there was a desire to find a local anesthetic that would allow the doctor to use it safely. The pharmacists of that time took up this demand and tried to separate the numbing effect from the euphoric effect. However, the development of new local anesthetics is not possible without knowledge of the structure of cocaine. Richard Willstätter (1872–1942, Fig. 10) took on this task in 1898 and published the structure of the alkaloid. He was the expert in the field, as he was the specialist in his habilitation with the title "Investigations in the Tropine Group". The cocaine structure, in which two rings with a nitrogen atom are, which you can see in the short profile, is such a tropine structure. Willstätter was offered a professorship at the University of Munich in 1899. To increase his chances, his mentor Adolf von Baeyer suggested that he should be baptized. As a Jew, he firmly declined and left Munich to accept a professorship in Zurich in 1905. The man was too good, because he received the Nobel Prize for

Fig. 10 Richard Willstätter (1872–1942) (public domain, Wikipedia)2

Chemistry in 1915 and was offered professorships in Berlin and Munich, which he accepted. Certainly, a highlight of his career was accepting the call as a professor and successor to Adolf von Baeyer, but he was not to be happy in Munich. Anti-Semitism increased with the Hitler coup and after quarrels at the university, he resigned his post and became a private lecturer, consultant for the chemical industry, and fellow of the Rockefeller Foundation. Willstätter believed too late that the Nazis would let him emigrate to Switzerland after they took power. The opposite was the case. He lost almost all his property, like the villa in Munich, his private library, which he had in seven rooms, and almost his life, when he wanted to flee illegally to Switzerland at Lake Constance. He was lucky that he was allowed to leave for neutral Switzerland before the outbreak of the Second World War, where he died of a heart attack after three years.

Almost simultaneously, chemists threw themselves at the now known cocaine structure and literally took it apart. Today, this research approach is referred to as structure-activity relationship analysis, to find out which parts of the chemical structure are relevant for the effect and which subunits could be important for the side effects or the metabolism.

Chronologically, it was Eduard Ritsert (1859–1946) who discovered Benzocaine in 1890 and brought the first local anesthetic to market in 1902 that did not chemically correspond to cocaine. Later we read that Alfred Einhorn (1856–1917) was the discoverer of the more effective Procaine in 1905, but it was first Ritsert who found Benzocaine in 1902 with a lot of work and intuition. But let's go back to the year 1890, when Ritsert, after several moves to Aachen, London and a doctorate in Berlin (1890), dedicated himself to the research of Benzocaine in the Rosen-Apotheke in Offenbach.

Eduard Ritsert (1859–1946) was one of six brothers who did not follow the commercial training of his brothers, but found pleasure in the natural sciences.[36] As the son of a manufacturer, he had a sheltered childhood that fostered curiosity and cosmopolitanism in him. He completed training as a pharmacy assistant and worked in various pharmacies until he decided to study pharmacy in Giessen in 1882. However, he did not follow the typical academic career with a doctorate and appointment to a chair at home. He went to Berlin, where he received his doctorate in 1890 on "Investigations on the Rancidity of Fats". At the Moabit Hospital, he found time to deal with the synthesis of benzocaine, a substance, *"which gave such beautiful crystals"*. This group of substances the para-aminophenols had already attracted him during his time as a pharmacy assistant, when he tried to separate acetanilide, an important drug at the time, cleanly from impurities. Whether he accidentally or deliberately became aware of the local anesthetic effect of benzocaine is not known. He tasted his substances with his tongue and felt that it became numb. Ritsert was surprised himself that his benzocaine had a local anesthetic effect without the dreaded side effects of cocaine. The substance, which he called Anaesthesin, was tested at his request by the Rostock pharmacologist Robert Kobert (1854–1918), who congratulated Ritsert on his breakthrough after presenting the very encouraging results. Benzocaine was to change the world of local anesthetics, but the Hoechster Farbwerke in Frankfurt, to whom Ritsert offered benzocaine, declined. For another ten years, the idea and the benzocaine remained in the drawer until Ritsert finally made his breakthrough. After marrying the pharmacist's daughter Elisabeth Schleußner (1865–1904) in 1891, he was doing so well financially that he dared to become a pharmaceutical entrepreneur and successfully marketed his drug Anaesthesin® himself in clinics. It was finally the alternative to cocaine that doctors had been longing for. As early as 1905, it was included in the German Pharmacopoeia – the accolade for outstanding drugs.

What are pharmacopoeias? Pharmacopoeias are formally legal texts that scientifically define the quality of drugs and medicinal plants and which pharmacies or pharmaceutical companies must adhere to if they are intended for use in humans or animals. Historically, pharmacopoeias are better known as recipe collections, in which since the Antiquity (Papyrus Ebers or Materia Medica) fewer chemical substances than rather medicinal plants were described. The inclusion of chemical substances only took place with the flowering of the chemical and pharmaceutical industry, which was able to synthesize drugs or cleanly isolate natural substances for the first time. An example is the Pharmacopoea Germanica (Fig. 11), called the German Pharmacopoeia from 1890 onwards, which was written in Latin. The term pharmacopoeia is not uniform and was formerly

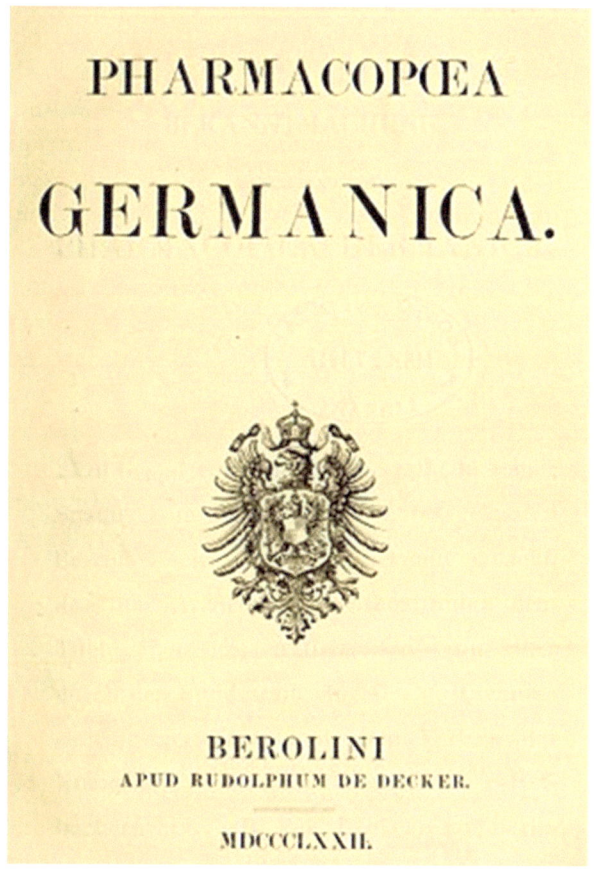

Fig. 11 German Pharmacopoeia, Pharmacopoea Germanica, 1872 (public domain, Wikipedia)1

referred to especially in the Anglo-Saxon language area as a Medical Book or Pharmacopoeia. The term is derived from the Greek word pharmakopoieĩn, which can be freely translated as "drug preparation". Today there are many national pharmacopoeias worldwide such as the German Pharmacopoeia (DAB), the US Pharmacopoeia (USP) or also the Japanese Pharmacopoeia(JP) in several editions. With the European Union, the first supranational pharmacopoeia was created, which like all other pharmacopoeias sets rules for the quality, testing, storage and designation of drugs. New editions appear at irregular intervals, in which new drugs are included, which are usually a bit older because they have fallen out of patent protection. But also outdated drugs are removed from the pharmacopoeia if there are better or more modern substances for the indication area. Each drug is described very systematically in its own chapter, the so-called Monograph. This includes the chemical and microbiological purity, the analysis methods to be used, the list of chemical by-products that may not be detectable or only within certain limits.

How did the discovery of benzocaine occur in detail? Was it the result of a structured search? No, as so often happens, everything came quite differently and unexpectedly. In Offenbach, an employee pulled him aside by the shirt sleeve because he had an analytical problem. The new substance phenacetin for reducing fever seemed impure to him, and he needed Ritsert's advice. The identity check was done by determining the melting point, which was not 120 °C, but 114 °C. This lowering of the melting point is always an indication of contamination in pharmacy. The same applies to salt as a spreading agent in winter, which causes the snow to melt. If you put salt on ice, the common melting point, the so-called eutectic point, is lowered. The two gentlemen suspected that a synthesis precursor contaminated the phenacetin or a by-product caused the lowering of the melting point. Discussions followed and both believed that a substance, which we call benzocaine today, was responsible for the poor quality.

Perhaps there was a lack of time or the Rosen Pharmacy was unsuitable for the synthesis, because for the moment the two left it at theoretical considerations. But Ritsert did not let it rest later when he went to the hospital in Berlin-Moabit in 1890. In his spare time, he set about synthesizing benzocaine, which he succeeded in doing extraordinarily well. Beautiful white crystals formed, indicating a high purity of the substance. As so often, the chemists and pharmacists of their time tried out the substances or at least tasted them. As Sigmund Freud described for cocaine, Ritsert noticed that his tongue tip became dull and numb. He became euphoric and dared to do an animal experiment. He dissolved the substance and dropped a few drops into the eye of a rabbit. The animal no longer reacted and was insensitive. How toxic this new substance was, was the next question to be answered. Since cell experiments could not be carried out around the turn of the century, this question was difficult to answer. Isolating and cultivating cells from tissues or tumors was practically impossible. Only after the Second World War and especially with the discovery of HeLa cells, cell-based in-vitro testing was to become routine in pharmacological laboratories.

Ritsert looked around to see who could help him in Berlin with the question of the toxicity of drugs. The Moabit Hospital was located in Turmstraße, a good three kilometers away was the Charité, where Paul Ehrlich (1854–1915, Fig. 12) lived, taught and researched. Paul Ehrlich is often equated with his great works in the field of serology, vaccine research and bacteriology, but during his time in Berlin he was also a cancer researcher. This digression into his Berlin time between 1882 and 1899 may seem strange, but he habilitated with the monograph "The Oxygen Requirement of the Organism. A Color Analytical Study", in which he described various dyes such as methylene blue that seemed suitable for tumor diagnosis. In his laboratory at the Charité he tried to grow tumors, which was not possible without animal experiments. It is assumed that Ritsert went to Ehrlich at the Charité and asked him to feed his new

Fig. 12 Paul Ehrlich (1854–1915), 1915 (public domain, Wikipedia)1

substance to laboratory mice. Ehrlich must have liked the idea of feeding benzocaine to his mice, and he wrote to Ritsert after completing these experiments: *"Mice eat incredible amounts of the new substance without it harming them."* Ritsert was thrilled and wrote to August Laubenheimer in Gießen that he had a very interesting and non-toxic local anesthetic.

His former teacher, now director of the Hoechst dye works, asked the Breslau pharmacologist Wilhelm Filehne (1844–1927) for an assessment of the economic exploitability. He was unsure and saw no market potential. He recognized the special properties of benzocaine as a local anesthetic, but the low water solubility was a knockout criterion for him. In his opinion, benzocaine would therefore not replace cocaine for a long time. A fatal misjudgment and the dye factories gratefully declined further development.

All the more, the question of the solubility of a drug occupies pharmacy today. Low water solubility is said to be when it is below 10 mg per liter. This is very little, but for comparison: Even marble dissolves better in water with 50 mg per liter. The question of the solubility of highly effective drugs has occupied Pharmaceutical Technology intensively in recent years. About 40% of all interesting substances are not further researched because they do not dissolve sufficiently in water or organic solvents to achieve sufficient therapeutic blood levels.

Eduard Ritsert would not accept "No" from the major pharmaceutical companies. He believed in his idea and in 1903 founded his own company for the development and distribution of benzocaine in pain ointments, which became very successful. The company has been family-owned for generations and is still active today. Currently, benzocaine is no longer marketed as a local anesthetic for injection, but it is of great importance as an active ingredient in lozenges for sore throats. Condoms are also coated with benzocaine to delay premature ejaculation.

Shortly after the discovery and commercialization of benzocaine, procaine followed in 1905. It immediately outperformed benzocaine and was celebrated by its discoverers as a *"glorious step"*. The discoverers were the chemists Alfred Einhorn (1856–1917), whom we already know from earlier cocaine chemistry, and Emil Uhlfelder (1871–1935). He synthesized the local anesthetic procaine in the Hoechst dye works, which came onto the market under the trade name Novocain®. The name spoke volumes and was meant to show doctors that it was something new (novo) that replaced cocaine (cain). Today we talk a lot about so-called blockbuster drugs, which are found in cancer research or are supposed to conquer new diseases like Alzheimer's or dementia. But back then, a simple structure like procaine with an indication like local anesthesia, which is rather insignificant today, was also part of it. Who would voluntarily go to the dentist if there was no local anesthesia? The scene in the tenth chapter of the novel "The Buddenbrooks" by Thomas Mann is very impressively described. Thomas Buddenbrook dies of pain and fainting after a botched tooth extraction, and the described scene shows how dramatic such a procedure was before the discovery of local anesthesia. All patients were glad that minor procedures could now be performed painlessly, everyone demanded it and the fear of pain was taken away.

The joy over procaine was quickly clouded when its high allergenic potential became apparent. Patients suffered anaphylactic shocks and, with the then still insufficient intensive care medicine, deaths were not avoided. Further syntheses and structural changes did not reduce this disadvantage and risk factor, for example, the tetracaine introduced in 1930 was ten times stronger, but also significantly more toxic. But until the discovery of a new type of local anesthetic, lidocaine, the path of knowledge took some turns and detours.

In the 1930s, the biologist Hans von Euler-Chelpin (1873–1964, Fig. 13) was researching in Stockholm on a topic that had nothing to do with local anesthetics. Euler-Chelpin, who received the Nobel Prize for alcoholic fermentation in 1929, was interested in the inheritance of chemical properties of metabolism in barley. He examined the genes and enzymes of the plant for chemical differences in metabolic mutants and discovered that it could be the indole alkaloid gramine involved. Euler assumed it was an alkaloid, but the structural elucidation was still in its infancy and apart from mass determination by elemental analysis, there were not many clues. If the structure was not too complicated, attempts were made to synthesize the conceived structure. This is exactly what Euler agreed with his assistant Holger Erdtman (1902–1989, Fig. 14). The young colleague was not thrilled with his boss's assignment and immediately delegated the task to his doctoral student Nils Löfgren (1913–1967), who worked as a chemist in the lab. Löfgren was now supposed to synthesize gramine as part of his doctoral thesis, which he seemingly succeeded in doing in 1935. To everyone's surprise, it turned out not to be the gramine that Euler had found in barley, but a chemically related substance, the isogramine. Chemists like to prefix their substances with "iso" when the basic structure is the same as the original, but some atoms, which are grouped together, are moved to another location in the molecule.

Erdtman and Löfgren met in the lab and looked at all the samples that lay as white powders on the table. We have often read in this book that isolated or synthesized substances are gladly tested, and the two began with the feast

Fig. 13 Hans von Euler-Chelpin (1873–1964) (public domain, Wikipedia)

Fig. 14 Holger Erdtman (1902–1989) (Permission, University of Stockholm, SE)

from the molecular kitchen. They were surprised to find that a chemical precursor left a numb feeling on the tip of the tongue. Löfgren probably recognized the medical potential first and persuaded Erdtman to look for new structures that were even more effective than this isogramine precursor. He grumbled, but let his doctoral student continue. After another seven years in the lab and more than 55 compounds, there was none that was better than the reference procaine. That was disappointing.

But Löfgren did not give up the work. In early 1943, as a doctoral assistant, he resumed work on amide local anesthetics. He had managed to recruit Bengt Lundquist (1922–1953) as an assistant, who unfortunately died prematurely of a brain hemorrhage. They continued the synthesis and tasting and found the compound LL30. The abbreviation stood for the first two letters of the names Löfgren and Lundquist. LL30 seemed to convince both of them and they asked the Karolinska Institute if there was interest in clinical development. The doctors at the Medical Institute tested the compound for toxicity and effect and found it convincing and pharmacologically safe.

The breakthrough was achieved by Nils Löfgren and Bengt Lundqvist with LL30, which was later called lidocaine, by giving up the typical ester bond and introducing an amide bond into the molecule.[37] Ester bonds are formed between an acid and an alcohol, as we have already seen with the benzoyl ester of cocaine. Amide compounds are similar, but here the bond is between an acid and an amine. These amide structures are very typical for proteins. With the transition from ester to amide, the strong allergenic effect of the new local anesthetics disappeared.

Löfgren was not only a scientist, but also a businessman who wanted to bring the idea of lidocaine as a local anesthetic from the academic world to the market. But who should he ask? Certainly not the German Bayer. Although Sweden was neutral during the Second World War, Löfgren was very critical of the Nazi occupation of Norway and Finland. He asked the Swedish Pharmacia, who declined. Lundquist also had good contacts with Astra AB, who invited both for a discussion at the company headquarters in Södertälje. With lidocaine in their pocket, the two scientists took the train to the tranquil Södertälje west of Stockholm. Then Lundquist remembered that they had not yet tried numbing the inferior alveolar nerve. When he arrived, he went to the guest toilet, injected himself with a lidocaine syringe, and shortly afterwards sat in the conference room with the numbing effect in his cheek. The talks were successful and Astra AB wanted to manufacture and distribute lidocaine under the brand name Xylocain®. The name Xylocain is a combination of the precursor Xylidine and the general suffix for local anesthetics "cain".

The collaboration between the two discoverers of lidocaine was perfect. Löfgren was an eccentric with a lot of assertiveness. Today he would be described as a "doer" who was passionate about the idea of lidocaine. Lundquist was a small, highly intelligent man who was always ready to try lidocaine and other derivatives on himself. He was the first to perform a spinal anesthesia on himself using a mirror. It is almost unbelievable that anyone would do this, as it required not only manual skills but also very good anatomical knowledge. Of course, the question of toxicity and side effects arose, and one might naturally ask whether the toxicity is decisive for the dentist when injecting one milliliter at a low dosage? With a usual one to two percent solution, this is certainly not so dramatic. But lidocaine was also often chosen for larger procedures in the early days of anesthesia, which must be viewed critically. The outstanding benefits and few side effects convinced the doctors, which is why all previously known local anesthetics such as benzocaine and procaine were almost completely displaced from the market. Lidocaine is now on the WHO list of essential medicines and has also gained medical significance as a treatment for cardiac arrhythmias since it was introduced in Sweden in 1948 by the anesthetist Torsten Gordh (1907–2010). Interestingly, lidocaine is abused as a cutting agent and drug substitute for cocaine.[38]

The last substance in the development of modern local anesthetics (Fig. 15) was bupivacaine. The Swedish company Bofors AB revisited lidocaine in the 1950s. The idea was to achieve a prolonged effect through a chemical change. This was achieved with Bupivacaine, which had a slow onset of action and a duration of action of eight to twelve hours. It was not suitable for the dentist because of the long duration of action, as no patient wants to go through the day with a swollen cheek for hours. However, the

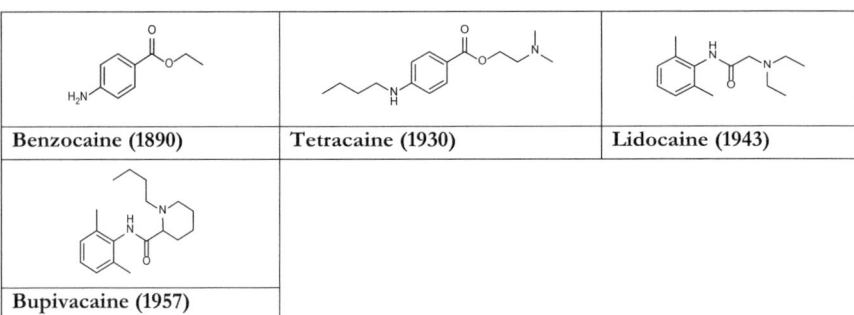

Fig. 15 Local anesthetics derived from cocaine (year of discovery in brackets) (Work of the author)

long duration of action was ideal for spinal and epidural anesthesia for moderate surgical procedures.

Despite the bad reputation of cocaine as a drug that causes a lot of suffering and misery, the discovery of local anesthesia was a blessing. Medical chemistry has managed to synthesize the addictive, pharmacological effect from the natural substance, so that today, more than 80 years after the discovery of lidocaine, we have a safe local anesthetic that no one would want to miss on the dentist's chair.

Aspirin®—the Most Used Drug in the History of Mankind®

Profile - Salicylic acid

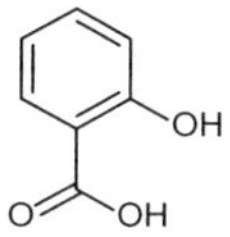

Salicylic acid

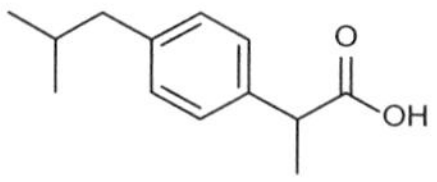

Ibuprofenic acid

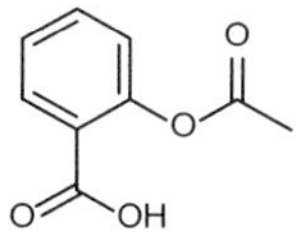

Acetylsalicylic acid

Natural Product Group

- Phenols

Modern drugs

- Acetylsalicylic acid
- Ibuprofen
- Benzocaine

Effect / Indication

- Painkillers
- Antirheumatic drug
- Anti-inflammatory
- Antipyretic

A few facts upfront: Acetylsalicylic acid is the most commonly used drug in human history. The annual production volume, at 44,000 tons, surpasses any other drug, with 26,000 tons alone in the USA. Aspirin was present during the first ascent of Mount Everest by Edmund Hillary in 1952 and was part of the carry-on luggage for the Apollo-11-Mission, which brought the drug to the moon. During the Argentine crisis between 1998 and 2002, aspirin tablets served as a barter item for the worthless Argentine Peso. The drug, which was discovered exactly 125 years ago in 2022, still causes disputes over who its inventor was. A dispute that has the makings of a thriller,

O. Kayser, *From the Maluku to Molecules*, https://doi.org/10.1007/978-3-662-69923-2_3

as vanity, greed, and racism played a not insignificant role. But more on that later.[39]

Like many drugs that we have discussed in the previous chapters, acetylsalicylic acid is a synthetic substance that finds its model in free salicylic acid. The name already reveals in which plant this acid was found. It is the willow, and from its Latin name for the genus *Salix* (Fig. 1), the chemical designation "salicylic acid" is derived. Its fever-reducing and anti-inflammatory effects were known in ancient Greece. The German pharmacologist Johann Andreas Buchner (1783–1852) believed in 1828 that he had found the fever-reducing substance in the willow. He isolated a yellow mass of *"bitter taste,"* as he later wrote after tasting it himself. He thought he had found the active ingredient with salicin, as he called the mixture, but he

Salix alba

Fig. 1 *Salix alba* (public domain, Wikipedia)[1]

was mistaken. In his pot was not the acid, but the salicyl alcohol, which was connected at one point with a sugar. With his discovery, he only confirmed the finding that two years before Buchner, the two Italians Luigi Valentino Brugnatelli (1761–1818) and Francesco Fontana (1794–1864) had made, who received an impure extract. They also believed they had found salicin, but Buchner was the first to fish the sharp crystal needles of salicin out of the mixture with tweezers and held pure salicin pinched in the tweezers in front of him. Henri Leroux was the first to present the first 25 g of salicin in pure form in 1830, and just three years later (1833), the pharmacist E. Merck in Darmstadt offered salicin as a substitute for the expensive quinine for fever reduction at half the price of the South American alkaloid[40].

In the 1810s, Europe was in a difficult economic situation as French Emperor Napoleon isolated England to ruin the island's economy. Europe was occupied by French troops, and he imposed a continental blockade from 1806 to 1813, which tried to prevent any trade with England. Of course, this blockade was very unpopular in most countries like Prussia, Denmark, and Russia and was circumvented by smuggling, but Napoleon's military victories forced the occupied countries to respect the blockade. England defended itself and in turn hindered trade by ships of the participating states, which also affected the Spaniards, who were also occupied by the French. Ships with cinchona bark from Peru no longer arrived. The shortage became a real problem, as malaria was rampant in Europe and cinchona bark was the only remedy for this parasitic infectious disease. The belief of doctors was that the cinchona bark was a means to combat fever, as it was not yet known that quinine killed the parasites, whose release after multiplication in the liver was the cause of the fever. Plasmodia as the cause of malaria were not detected until 1880 by Alphonse Laveran (1845–1922) in Algeria, and the true antiparasitic effect of quinine was not recognized until the beginning of the 20th century. In Napoleonic times and earlier, the medical advice was to combat the fever with an extract of the cinchona tree, but what to do when there was no more cinchona bark?

Many remembered the willows in Europe, which were also taken in folk medicine for the treatment of fever before quinine was available. The fever-reducing effect had been known since the ancient Greeks with Dioscorides (40–80) and some researchers believe to have read about the effect of willows on Sumerian tablets. Dioscorides (40–80) recommended extracts for joint pain and gout. Willow bark made a comeback and the people of the 19th century saw in the extracts of this tree an alternative and extracted them as tea. The tree became a bestseller, although it did not help against malaria, but the fever-reducing effect spread across the continent.

Large quantities were demanded and willows were cut down to extract from the raw material extracts that contained the as yet unknown salicin. The overexploitation of the willows went so far that the basket weavers in Germany felt threatened in their existence and the willows had to be legally protected.

The modern history of the willow and its effect begins with the English clergyman Edward Stone (1702–1768), who reported to the Royal Society of London about the astonishing effects. Edward Stone loved plants and was a hobby botanist who tasted all unknown plants on his walks in Chipping Norton. One morning, when he was suffering from fever, he tasted the bark of the silver willow and noticed the bitter taste. He immediately thought of the bitter quinine. Was it coincidence or intuition that he believed, if the willow tastes bitter, then it must work like quinine? Although Johann Andreas Buchner (1783–1852) isolated salicin in 1828, he did not come up with the idea to taste the found, slightly yellowish needles himself.

The discovery of the anti-inflammatory and analgesic effect of salicin is said to be an Italian story in the beginning, as the Italians Brugnatelli and Fontana are said to have isolated phenol from the plant in 1826, two years before Buchner. But here the literature must be wrong, because Brugnatelli had already been dead for many years in 1826. It was probably the chemist Cesare Bertagnini (1827–1857, Fig. 2), who died early at the age of 30 from tuberculosis, who tried out salicin and salicylic acid in a self-experiment with 6 g. He was also the first to report about tinnitus in case of overdose. Thus, the Tuscan chemist reported that on the first day *"no discomfort"* occurred, but on the second day *"a constant ringing in the ears and a kind of dizziness,"* which forced him to stop taking it.[41]

Bertagnini had access to a large amount of salicylic acid, because as an assistant at the side of the great Raffaele Piria (1814–1865, Fig. 3), who dealt successfully with the synthesis, a chemical way should be found to meet the high demand for salicin without deforestation of willows. Raffaele Piria, a Calabrian and the third Italian we get to know in this chapter, studied medicine and was a professor at the venerable University of Pisa from 1842. He liked chemistry, which he also studied at the Sorbonne with Jean-Baptiste Dumas after his medical studies. There he conducted his first research on salicin and salicylic acid, which until then had only been extracted from willows. He was probably also the first chemist to synthetically produce salicylic acid.

On the other side of the Alps, the chemist Hermann Kolbe (1818–1884), who found salicylic acid very intriguing, took a step further in 1859 with his technical synthesis. The Professor of Chemistry at the University of Marburg

Fig. 2 Cesare Bertagnini (1827–1857) (public domain, Wikipedia)1

had been the successor of Robert W. Bunsen (1811–1899) since 1852, who is known to many as the inventor of the Bunsen burner, although it is actually an invention of the Englishman Michael Faraday (1791–1867). Kolbe succeeded in the technical production of salicylic acid from phenol in the presence of carbon dioxide and sodium through a chemical process. The complete elucidation of the reaction mechanism of salicylic acid formation and a further improvement of the synthesis was achieved by Rudolf Schmitt (1830–1898), who received his doctorate under Kolbe and qualified as a professor in 1864 with the work "On some derivatives of salicylic acid". Kolbe recognized the economic potential of his technical synthesis of salicylic acid. The substance was popular and sold like hotcakes. So why not make money with the idea? In Dresden, he met Friedrich von Heyden, who directly asked Kolbe if he wanted to become a partner in the "Salicylic Acid Factory Dr. F. v. Heyden" in Radebeul near Dresden? The offer was tempting, Kolbe accepted and for the first time salicylic acid was produced on an industrial scale as a pharmaceutical. The chemical factory still produces under the name Arevipharma Medicines in Radebeul and is considered one of the historic sites of German chemistry.

Fig. 3 Raffaele Piria (1814–1865) (public domain, Wikipedia)1

Hermann Kolbe brought his knowledge about the Kolbe-Schmitt synthesis into the company and Friedrich von Heyden took care of the commercial matters of the factory. Finally, salicylic acid could be produced in unlimited quantities. In the past, natural plant sources such as the wintergreen *(Gaultheria procumbens)* or the real meadowsweet *(Spiraea ulmaria)* had to be laboriously collected, and every year too dry summers, too much rain, storms or insect plagues determined the available quantity. Now there were no more limits. Phenol was obtained almost unlimitedly by distillation of coal tar according to the old Runge method. It was a bothersome by-product of the gas plants and the emerging German chemical industry like Hoechst, BASF, Boehringer or the Friedrich Bayer & Co. color works gave away carbolic acid, as phenol was formerly called, at low prices. The price for salicylic acid fell from 10 to 1 thaler per 100 g after the extraction from wintergreen was replaced by the new Kolbe-Schmitt synthesis. Already in the first year after the foundation of the salicylic acid factory, 4 tons of salicylic acid were produced, four years later it was already 25 tons. Kolbe and von Heyden filed further patents in other European countries as well as the USA and thus became a threat to a competitor located 600 km further west on the Rhine.

Was there something new in the West? On the Wupper, the color factories of Friedrich Bayer (1825–1880) did not develop so splendidly in the 1880s. In the 1860s, the chemical industry emerged, which grew with tar dye chemistry. From coal tar, which was cheaply obtained in the Ruhr area, the company founded in 1863 by the dye dealer Friedrich Bayer in Barmen near Wuppertal isolated basic substances such as aniline and phenol, which were needed for the synthesis of dyes such as mauveine and fuchsine, the purple and magenta dyes. These were then important textile dyes, which today have no significance due to their questionable nature. The synthesis was known since the English chemist William Henry Perkin (1838–1907) accidentally found the purple mauveine during quinine synthesis. This dark purple dye was all the rage in England, and when the French Empress Eugénie, the wife of Napoleon III, and Queen Victoria wore skirts and blouses dyed with mauveine at official appearances in England, there was no stopping it. The press reported on the new clothes and everyone in England and Europe now also wanted to wear purple-magenta. William H. Perkin was a wealthy man at 35 and founded his textile dye company on the Grand Union Canal in London in 1857.

Friedrich Bayer (1825–1880, Fig. 4), originally named Friedrich Beyer, was a son of the silk weaver Peter Heinrich Friedrich. Silk weavers are weavers who process and dye silk and half-silk. He changed his name to "ay" because in the 1850s a fraudulent merchant in Leipzig, who bore his name, was causing trouble. To avoid damage to his reputation and business, he changed his name to the now well-known Bayer. The young Friedrich began his apprenticeship at the age of 14 with the cloth dyer Wesenfeld & Co in Barmen and learned the basics of the profession and the chemistry of dyes. Friedrich was intelligent and did not want to come home from hard work with dyed hands for the rest of his life, so at the age of 20 he started a trade in natural colors, which was very successful in Paris, London and Saint Petersburg.

In addition to Friedrich Bayer, Johann Friedrich Weskott (1821–1876) also recognized the great potential of dyes for the textile industry. Both lived in Barmen and were friends. Johann Friedrich Weskott already ran a dyeing business, which he took over from his father. The two Friedrichs saw the future in synthetic colors and in the very good complement of their personal skills. They founded the company Friedrich Bayer and Compagnie in Barmen in 1863, where the Wupper river provided a lot of fresh water, which they needed for bleaching textiles. Bayer took over the commercial

Fig. 4 Friedrich Bayer (1825–1880), around 1863 (public domain, Wikipedia)1

management, Weskott the technical management of the new company, which quickly flourished. Just one year after its foundation, the company had 50 employees. Today (2023), the global corporation employs 99,538 employees.

The factory Friedrich Bayer & Co. produced Mauveine and other dyes like Fuchsin, Aniline Blue or Alizarin. They were now in strong competition with other German chemical factories. The fiercest competitor was the Badische Anilin- und Soda-Fabrik, or BASF, founded in 1865 in Ludwigshafen. The business was tough: falling raw material prices with constant personnel costs, an unpredictable market in the 1870s and 1880s, and many new dye innovations made the business ruinous. Everyone knew that the race for low prices could not be won in the long run. To make matters worse, shortly after the end of the Franco-Prussian War in 1871, the founding crisis broke out, which was replaced by a stock market crash two years later (1873). How times between then and now are similar: The trigger of the stock market crash was the bursting of a speculative bubble due to risky company foundations and the associated overheating of the economy. There was too much money in the market, which flowed into the German economy due to the French reparations. The German economy boomed because of the fresh and cheap money. The concession obligation for stock corporations was abolished and in the two years after the war, 928 stock corporations and 61 banks were founded in Prussia alone. The country was in a stock market fever and speculated wildly.

In addition to the national problems, there were plenty of other problems to solve locally. Fuchsin could be sold well, but the increasing production polluted the Wupper. The residents, who used the river for drinking water extraction, were not pleased about the pink-colored water that flowed past their houses. Bayer had to make compensation payments, which were not insignificant and noticeably reduced the profit. As an anecdote, it should be mentioned that the Wupper was considered one of the richest rivers in Germany in terms of fish before industrialization. With the beginning of industrialization and the rise of the cloth industry, this biotope became a sewage canal.

Carl Rumpff (1839–1889), a partner and chairman of the supervisory board of the Bayer Corporation, which was converted in 1881, saw the economic hardship and sought new research areas for his company. He was aware that the colors had no future and something new had to be discovered. He wanted to enter into active ingredient research as a new economic sector, which—like biotechnology today—was a trend. New synthetic drugs were emerging and plant extracts were playing an increasingly minor role. For this change, Carl Rumpff sought new employees and hired 1883 alongside Martin Herzberg and Oscar Hinsberg, the later discoverers of the pain-relieving phenacetin, the local chemist Carl Duisberg (1861–1935, Fig. 5). Duisberg had just received his doctorate and had dealt with the new field of acetylation in his doctoral thesis. The young researcher committed himself fully to the company and irreversibly turned the helm to take course towards medicinal chemistry. However, this idea did not become a reality until 1891 with the construction of a new building and a separate pharmaceutical institute at Bayer.

Until this year, no chemist at the new Bayer AG had salicylic acid, let alone acetylsalicylic acid, on their strategic radar. And it almost happened that the miracle substance would have triggered a medical revolution without contributions from Bayer chemists. What happened? Charles Frédéric Gerhardt (1816–1856), son of a bank clerk in Strasbourg, studied chemistry in Karlsruhe, Leipzig, Dresden, and Giessen with Liebig. He later became a professor of chemistry at the University of Montpellier and his research area

Fig. 5 Friedrich Carl Duisberg (1861–1935), 1909, painted by Max Liebermann (public domain, Wikipedia)1

was acetylation. In his 1852 published work "Précis de Chimie Organique" he described the conversion of various substances with acetic anhydride and acetyl chloride.[42] Interestingly, he also mentions in his book the acetylation of salicylic acid to acetylsalicylic acid, although this was rather a side reaction, as the substances were impure at the time and the understanding of organic chemistry was limited, so many reactions turned out differently. Unexpectedly often comes, but Gerhardt was demonstrably the first chemist to synthesize acetylsalicylic acid.

Carl Duisberg was the outstanding chemist and entrepreneur in the history of Bayer AG. It is always worth learning more about this extraordinary chemist.[43] Duisberg was born in Heckinghauser Street in Barmen and came from humble circumstances. His father was a cloth maker and had conservative views about life in general and his son's education in particular. Carl was a quiet boy who did what his father asked of him until he discovered chemistry for himself. At the age of 14, he learned about the natural sciences at school and told his mother: "I want to become a chemist." But senior Duisberg had other plans and saw the pupil in the cloth factory. This science was useless and a study anyway too expensive. As a father, he could not afford to send young Carl to university. But his strong-willed mother Wilhelmine saw her son's talent, supported his wishes, and stood up to her husband: "The boy goes, you can stand on your head." His father grudgingly gave his permission, but Carl knew his father better and knew that this could be withdrawn just as quickly if money became tight. Carl accelerated and graduated from high school at the age of 16. At his father's request, he attended the vocational school for chemistry in Barmen, which he did not complete after twelve, but already after eight months. Finally, he enrolled as a chemistry student in Göttingen and studied like a man possessed. What others needed three years for, he achieved in one year and had all the certificates to register for the exam. There was only one problem that Carl had not thought of. He lacked the Great Latinum, which entitled him to study. A regulation of the Prussian Ministry of Culture required all high school students in 1879 to make up the Latinum. The need was great, the ingenuity even greater. After a year in Göttingen, he moved to Jena to the chair of Professor Johann Georg Anton Geuther (1833–1889). Jena, which belonged to Thuringia, waived the Latinum and he enrolled. Geuther wanted to bring his high-flyer back down to earth and ordered him to do lab work. Duisberg grumbled, but complied and used the time for his doctorate. On June 14, 1882, the then 20-year-old graduated on the chemistry of acetic acid esters. *"So I spent the best time of my life in Jena!"*, he looked back happily and contentedly on his doctorate, because he came into contact with philosophy and was involved in the Natural Science Association, where he made many acquaintances. This can explain why he chose geology and national economics as minor subjects for the defense of his dissertation, which were rather unusual for a chemist.

Duisberg was now a graduated chemist with financial problems. His father wrote him letters and did not hold back his opinion in Barmen that this budding career of his son would only lead to disaster and how expensive this study abroad already was. At the beginning of the 1880s, the job market for young chemists was very clear. The last effects of the founder's crash were still noticeable and he wrote many applications, which unfortunately remained unsuccessful. In his distress, he took a poorly paid position with his doctoral supervisor, which did not challenge him scientifically, offered too much for dying and too little for living. Geuther paid him 80 marks from private funds and provided free food and lodging at the institute.

He often applied unsuccessfully and it annoyed him that he was always asked where he had served. He had never served. The way out was to volunteer for a year with the First Bavarian Infantry Regiment. There was money, he had served and he was in Munich, where the great Adolf von

Baeyer taught, whom he could visit. But things did not go well in Munich, he returned after a year of service under his father's roof in Barmen and remained in search of work. He did not give up and did not want to spend the rest of his life in his father's business. He wrote many applications again and luck was on his side. He got an interview with Carl Rumpff, who hired him on a temporary basis. Carl Duisberg was to go as a delegated chemist to the University of Strasbourg in Alsace and work there on a third-party funded project. As if there were parallels to our time, Bayer AG Third-party funded projects awarded to universities, which were carried out by chemists outside the company. Today, my colleagues and I are very familiar with this funding because chemical and pharmaceutical companies like to outsource tricky research to universities.

Duisberg did not find the offer ideal. Bayer was too close to his parents' house. He wanted to leave, but what was he to do? There were no other offers, so he accepted. The insoluble task that was set for him in Strasbourg was the technical synthesis of indigo dye. Indigo dyes are the typical blue dyes used to dye denim fabrics. Although the blue Texas trousers, as the jeans were long called in Germany, did not yet exist in Europe, but in the USA the invention of jeans in 1872 by the German emigrant Levi Strauss (1829–1902) from Franconia and the Latvian Jacob Davis (1834–1908) was very popular among the gold diggers. The sewn canvas and later cotton, which was woven after Denim style, were dyed with indigo and provided with orange stitches and rivets. Indigo is a natural dye of the *Indigofera* species, which gets its blue color by oxidation in the air. The synthesis was very difficult at that time, but the king of dyes could already be clarified in 1878 by Adolf von Baeyer. The synthesis paths devised by Baeyer worked well in the lab, but they were not suitable for industrial production, which the companies BASF and Farbwerke vorm. Meister, Lucius & Brüning (later Farbwerke Hoechst) tried unsuccessfully. Duisberg also failed because the raw materials were too expensive or missing. Only the synthesis developed by Karl Heumann in 1890 made it possible to produce indigo dye at a low cost. Despite the failure, he was offered a permanent employment contract, because the indigo order was considered an insoluble probationary task. Carl Duisberg received an annual salary of 2100 Reichsmark and returned from Strasbourg to Barmen, where he worked as a chemist. This time the return was not so hard for him, because he had fallen in love with Johanna Seebohm (1864–1945), the niece of his boss, and might marry her.

Duisberg's talent as a chemist was quickly recognized by his boss. In a short time, the new employee succeeded in synthesizing three dyes, including the Congo red. The name raises questions about its origin, which are not clarified. Probably the Berlin company Agfa gave the name for marketing reasons. Because in Berlin at the same time the Congo Conference took place, which divided Africa among the European colonial powers. The word Congo was on everyone's lips and a Congo red could be marketed well. Through the marketing success, Agfa made a fortune with the red dye. But Bayer was doing badly and was under so much pressure due to collapsing sales markets in the color business that analysts at the Berlin Stock Exchange already saw the imminent end of the company. Carl Duisberg had to fix it with Congo red.

With a new synthesis path, he bypassed the patent of the former Bayer employee Paul Böttiger, who had registered a patent for the Congo red synthesis on his own after his dismissal. Things went well for Carl Duisberg and he attracted attention. The company management noticed him, because with 17 patents and 21 new dyes he was a welcome employee and he received private invitations from the chairman of the supervisory board Carl Rumpff. At one of the parties at the Rumpff house, he met the niece Johanna Seebohm. They got engaged in 1887, married a year later and he became the father of four children. The company flourished, he was granted power of attorney in 1888 and Carl Duisberg wanted to restructure "his company" like Carl Rumpff, because dyes were no longer the future. There should be two departments. In one, the old chemistry of dyes remained, but the second department should now deal with the chemistry and pharmacology of drugs.

In 1894, Felix Hoffmann (Fig. 7), and two years later Arthur Eichengrün (1867–1949) (Fig. 6) as a freshly graduated chemist, and in 1897 Heinrich Dreser (Fig. 7) as an older experienced pharmacologist were hired for Medical Chemistry. All three are considered the fathers of Aspirin, but their relationship with each other was complicated. The two young chemists probably reported directly to Carl Duisberg, which is why Hoffmann likely dealt with acetylation, which was in vogue in chemistry at the time. This did not have to be to his disadvantage, because we know from the chapter on morphine that the production of heroin would become a huge success for him personally and for Bayer AG. Perhaps heroin was his masterpiece, but the chemical conversion of salicylic acid to aspirin or better acetylsalicylic acid would make him and his colleagues world-famous. Both substances also made Bayer a global player, as we would call a multinational corporation today. To round off Duisberg's work at Bayer AG, it should be noted that he became director in 1900 and general director in 1912, who introduced the 9-hour working day at Bayer in 1905 as a *"patriarch with a social program"*. However, he is also remembered as the visionary of I. G. Farben, which was

Fig. 6 Arthur Eichengrün (1867–1949), around 1900 (public domain, Wikipedia)2

Fig. 7 Heinrich Dreser (1860–1924), 2nd from right sitting, in the Pharmacological Laboratory at Bayer, circa 1897 (public domain, Wikipedia)1

very problematic as a large monopoly of the German chemical industry during the time of National Socialism.

The work in the scientific laboratories (Fig. 7) was in great contrast to the hard work in the factories and mines, as we know it from history books. Not that performance was not demanded here, but a modern leadership style prevailed and Bayer showed its appreciation for the scientific minds in the labs in an unusual way. We have all seen documentaries on television about the corporate culture in American start-ups. All employees are very relaxed, there are flat hierarchies, everyone can have a coffee with the boss and yoga mats and table football are always available for relaxation. These Americans are cool with their ideas, but they are not the inventors of the relaxed working style. One can safely say that things at Bayer in Barmen 120 years ago were not like at Google, but Carl Duisberg knew how a good working atmosphere promotes creativity. Everyone valued mutual respect, exchange of ideas, a good collegial atmosphere and flat hierarchies. He had a very modern leadership style and gave his employees a lot of autonomy in implementing the tasks he assigned. Even casual clothing and no draped skirts, corsets, mutton sleeves for the ladies or tailcoat and frock coat for the men, which were common in the stiff German Empire, were worn in the Bayer laboratories. The jacket made its appearance and only a few wore a white lab coat.

The three who were entrusted with the development of acetylsalicylic acid could hardly be more different. Dreser, as head of the pharmacological department, was a loner who appreciated himself and his work rather than integrating into the team. Hoffmann as an outstanding chemist, about whom we know little about his private life, who lived and worked very secluded. The third in the group was Eichengrün, who was considered charismatic and sought dialogue with the others. Hoffmann was the youngest of the three, but he was the first to be hired and therefore had some privileges. On the whole, the atmosphere between the two chemists and the pharmacologist was spring-like warm at the beginning, only to become winter-like frosty at the end. But let's take a closer look at the three psychograms.

Felix Hoffmann (1868–1946) was born in 1868 in Ludwigsburg and initially worked as a pharmacy assistant in Geneva, Hamburg and Neuville before he began studying chemistry and pharmacy in Munich in 1889. He studied quickly, passed the state examination as a pharmacist in 1891 and received his doctorate just two years later in 1893 with the topic "On some derivatives of dihydroanthracene". After his doctorate, he did not stay long in Munich. Although he worked there at the State Laboratory in Munich, Adolf von Baeyer recommended him to contact Carl Duisberg, who hired him directly in his newly founded pharmaceutical-scientific laboratory. Hoffmann stayed with Bayer AG all his life and moved to Lausanne after his retirement in 1926, where he died in 1946. Hoffmann remained unmarried and had no children. Little is known about his private life compared to the other Aspirin discoverers.

Arthur Eichengrün (1867–1949) was hired by Bayer in 1896 and was to become one of the most innovative chemists, working directly with Dreser and providing active ingredient candidates.[44] Eichengrün was the son of a Jewish cloth merchant and studied chemistry at the Polytechnic Institute in Aachen, today's RWTH Aachen. His dissertation "On Methoxy-dioxy-dihydrocarbostyril", which was accepted in 1890, dealt with a cocaine-like local anesthetic. One of the main producers and traders of cocaine, in addition to the well-known company Merck, was the family business C. H. Boehringer in Ingelheim, where he was involved in the purification of cocaine until 1896. Upstream, Bayer AG was interested in him and made him an attractive offer. He moved to Wuppertal, where he took over the management of the newly founded pharmacological laboratory. This was certainly unusual for a young chemist and perhaps too demanding, as after only half a year the new colleague, a Dr. Dreser, was to take over this task. But Eichengrün remained loyal to Dreser, who tested the new substances. He stayed in chemical research and was so successful with two discoveries and patents that he left the company again in 1908. The man had an entrepreneurial spirit. He moved from the Rhine to the Spree and founded the Cellon Laboratory Dr. A. Eichengrün in Berlin-Charlottenburg on Tegeler Weg, which he managed until the end of his entrepreneurial career. Cellon is a chemical polymer made of cellulose acetate and camphor, for which Eichengrün received a patent in 1909. Cellon is a precursor of cellophane and was valued in the industry for its rigidity and strength. Thus, Cellon could be used for covering airplane wings or zeppelin hulls: Cellon was found in some windshields and men needed thin plates to stiffen the collar. For all these applications, Arthur Eichengrün held a patent monopoly and was able to market it personally.

With additional money from further Bayer inventions for Protargol for the external treatment of gonorrhea and Aspirin®, Eichengrün lived comfortably with his family in Berlin-Charlottenburg. Things were going well for him and with success came even more money. A yacht, cars, art, books, and a chic apartment were naturally part of his life. He enjoyed life, was generous, and loved to invite friends. Eichengrün was not averse to the pleasures of life and also appreciated the female sex, to which he felt attracted. He was married three times and his third wife was the nanny who took care of the education of his six children.

The 1920s had their ups and downs, and the Eichengrün company experienced them like a roller coaster ride through the world economic crisis. Eichengrün repeatedly brought the company out of the valley, and he also liked to go to the laboratory to supervise projects directly. But then came the biggest challenge of his life. In Germany, National Socialism was on the rise and Eichengrün, a Jew who did not see himself as religious, saw the changing world without suspecting the chaos he would be plunged into. The enemy literally lived next door. In his elegant house, a new neighbor, a big Nazi, had recently moved in. He greeted him friendly and tapped his fingers on the cap that the fat man wore as part of his oversized uniform. When a rumor in a leading newspaper raised the question of why the fat man was living in a house with a Jew, it became too much for Hermann Göring. He forced Eichengrün to move out.

But Eichengrün also had no luck with his second National Socialist neighbor, as he shared the same good taste with him. As a sad anecdote, it must be mentioned that Eichengrün was also a neighbor of the Führer on the Obersalzberg. Until the Aryanization of the Mitterwurf house in 1932, the Eichengrün family spent their summers there. From this year on, the SS forced Eichengrün to give up his holiday home. It was only a matter of time before the regime expropriated his company in 1939, more precisely, put it into the hands of a Volksdeutscher. His good reputation as a chemist allowed him to survive under the Nazis for a while, but from 1944 he spent 14 months in the concentration camp Theresienstadt. He also survived this ordeal and settled down in Bavaria, where he died at the age of 82 in Bad Wiessee near Lake Tegernsee.

Eichengrün also critically examined the still young pharmaceutical industry. As early as mid-1898, he criticized the pharmaceutical companies' all too reckless handling of untested drugs and did not shy away from criticizing his own company.[45, 46] He criticized the handling of the many new drugs that came onto the market in an article in the journal Angewandte Chemie: *"Unfortunately, however, such remedies, which serve new indications, belong to new areas, are presented from new perspectives or possess new properties, are far in the minority, and they almost disappear in the large number of unnecessary and in part even inappropriate preparations, which are often brought onto the market without sufficient, often even without any clinical testing."*[47] Eichengrün's article does not deny his incomprehension, reporting that a Berlin company brought ten new drugs onto the market in one day without clinical recommendation or indication. Against this background, it is interesting to know that it was not until the Drug Act of 1961 in the Federal Republic of Germany that there was a legal basis that made clinical trials mandatory and put a stop to the arbitrary sale of drugs. Eichengrün was far ahead of his time and proposed a model for the clinical testing of drugs in various phases in a memorable essay from 1899 from today's perspective. It was a system that we now take for granted. Eichengrün also had in mind the special testing of highly effective drugs, which he absolutely wanted to place under the constant control of the medical profession.

The third in the group was Heinrich Dreser (1860–1924). He was the oldest and had already had a remarkable career when he joined Bayer AG.[48] After studying medicine and receiving his doctorate in Heidelberg, he went on to further scientific stations in Breslau, Strasbourg, Tübingen, Bonn, and Göttingen. In Göttingen, he was an associate professor with a low salary and poor laboratory equipment. Probably dissatisfied with this situation, Dreser decided to accept the offer from Bayer AG and move to Wuppertal. Carl Duisberg recognized the potential of his new employee and promoted him by expanding the pharmacological laboratory. On April 1, 1897, he took over the management of the Pharmacological Laboratory as the successor to Arthur Eichengrün, who as a chemist was not very happy with the pharmacological management. Dreser was considered difficult by his colleagues because he was very conscientious in his methods and was considered a perfectionist. In the obituary by Ernst Lomnitz, who knew Dreser personally, the pharmacologist is described as an *"original, edgy, prone to sarcasm, unsociable and unsociable personality."*[48] This very conscientious way of working must have been inherited from his father, who as a professor of physics had a pronounced understanding of numbers and a love of methodology. The publication "Pharmacological about Aspirin" is recommended reading, which impressively shows how Dreser describes the dissolution behavior in simulated gastric juice using integral calculus.[49] The methodical approach in animal experiments in Heinrich Dreser's pharmacology earned him great respect, and the statistically correct handling of results from animal experiments is still valid today.

The reasons for his departure from the services of Bayer AG after 17 years are not fully known. Hans Horst Meyer, also in the services of Bayer AG, remembers that Dreser one day appeared unannounced in his office. *"For reasons unknown to me, he turned to Liège, where an experimental therapeutic laboratory was open for collaboration,"* Meyer said in his obituary in the Archives for Experimental Pathology and Pharmacology.[48] In Belgian Liège, Dreser had bad luck, because after a few months the First World War broke out. He had to flee with his wife and leave his personal belongings behind. Dreser was a wealthy man, because from the royalties he received from Bayer

AG, he could live more than well. This money enabled him at the end of his life to found a biochemical-pharmacological institute in Düsseldorf. Here he received an honorary professorship at the Institute for Biochemistry of the Medical Academy Düsseldorf. In 1924 he went to Zurich to work there with Max Cloëtta (1868–1940) in the field of inhalation anesthetics, but his death a few weeks after his arrival prevented him from doing so. He died of kidney failure.

What do we know about his private life? Little. We know that he played the violin, viola, and cello excellently, spoke English, French, and Italian in addition to German as his mother tongue. He was married twice, but had no children. In conversations, he was sarcastic, ironic, and had little joy in dealing with people, except for scientific instruction. Meyer remembers a person who was bitter and closed in conversation. Meyer's obituary certainly does not read like a reckoning, but is full of praise when it comes to scientific achievement. The quiet and strange impression that Dreser was not satisfied with himself and his environment remains, however. The obituary ends with a description of his social competence, which today is colloquially compared with that of a scientific nerd (English scientific nerd): *"Warm participation in others was not his thing, so that his relationship even with the big-hearted and lively feeling W. v. Schröder during their many years of coexistence in Schmiedeberg's Institute remained cool and completely impersonal. Dreser was hardly to be persuaded to attend meetings of colleagues; if he did, he kept as far back as possible, external recognition seemed completely indifferent to him and in this respect he was free from all vanity. But the setback, the ongoing difficulties, which he experienced in his scientific career, and which only ended with his entry into the Elberfelder Farbenfabriken, probably contributed to his basic mood."*[48]

The Invention and Worldwide Success of Aspirin®

Aspirin® was a great success for Bayer AG, which received the trademark rights in 1900 and conquered the American market. Neither for heroin nor for acetylsalicylic acid did Bayer AG receive the patent rights for the synthesis, as Hugo von Gilm (1831–1906) had already received a patent for the synthesis in 1860. Although it was not about academic recognition or the Nobel Prize, the question of who was the discoverer of Bayer Aspirin® would become a scientific crime story.

Although the synthesis was not pure, Hugo von Gilm is considered the father of acetylsalicylic acid, but Felix Hoffmann carried out the technical synthesis with Eichengrün to a highly pure product. The exact course of the synthesis, who did what and how the pharmacological testing was organized, remains confused to this day. Envy and vanity led to a dispute over the intellectual authorship of Aspirin®. After the Second World War, a discussion broke out among historians about the intellectual and actual authorship. A laboratory journal by Hoffmann shows that he was already dealing with acetylations and methylations early on. The laboratory journal from 1897 describes the acetylation of salicylic acid on 10.08.1897, the acetylation of morphine on 21.08.1897, and the conversion to acetylquinine on 09.12.1899. However, Eichengrün claimed in a publication shortly before his death in 1949 that he had had the idea for the acetylation. As a "privileged full Jew", he was not taken seriously. Hoffmann was only the executor of his idea.

What did Dreser think? He saw himself as the sole inventor. In his 1899 publication on the pharmacology of Aspirin® (acetylsalicylic acid), he does not mention either Hoffmann or Eichengrün.[49] He was the researcher who had recognized acetylsalicylic acid, even though he did not believe in the substance and had rejected the animal experiment because of the very likely cardiac arrhythmias. This is quite astonishing, as Eichengrün went directly into Dreser's office with the acetylsalicylic acid in his hand. "Oh, you know," Dreser could have replied, "Dear colleague, you are misjudging something, you don't understand that as a chemist. I believe this is a heart-damaging agent. I will not test it." Eichengrün must have looked astonished. They had made salicylic acid safer and Dreser didn't want to test it? Eichengrün was surprised and did not let go. He secretly packed a package and posted it to send the acetylsalicylic acid to Berlin. In the next few days, he wanted to take it himself and find out if Dreser was right. Eichengrün took acetylsalicylic acid in the following days and noticed—nothing! The substance was safe, no effect on the heart, a drop of iron chloride solution in the urine and the degraded salicylate turned blue-green. Dreser could not be right, Eichengrün looked at the white powder on the table. He wanted to wait for what the colleagues from Berlin wrote, then he wanted to confront Dreser with the results. The letter came, and it was very favorable. There were no indications of a toxic effect. The patients had tolerated the acetylsalicylic acid well and had spoken positively.

Eichengrün was now clear that acetylsalicylic acid had the potential of a money printing machine. Eichengrün went to Duisberg's office with the letter from the Berlin doctors and wanted to discuss further marketing with him. "Sit down, my dear, what brings you to me? I heard that there was a small dispute between you and the esteemed colleague Dreser." "Well, I was very surprised that colleague Dreser did not want to test the acetylsalicylic acid. So I took the liberty of sending a sample for testing to Berlin. Look here, the Berlin doctors are very satisfied." He pushed the letter across the desk. His boss immediately recognized the potential and wanted to bring acetylsalicylic acid to the market under the brand name Aspirin® as soon as yesterday. "Hm, my dear Eichengrün, we still need a nice name. What should we call our new child?" Eichengrün shrugged. Chemists, merchants, and nowadays entire marketing departments love to find catchy names for their products. They came up with "Aspirin". The found trade name can be decrypted as follows: The "A" stands for acetylsalicylic acid, "spir" for the salicylic acid-containing meadowsweet, and the "in" served at that time as an ending to indicate that it was a drug. The name of an industry icon was born.

So who was the father of Aspirin®? Michael de Ridder, the pharmacy historian Walter Sneader and the chemistry historian Elisabeth Vaupel have once again combed through the literature and laboratory journals from the Bayer archive.[13, 50] They come to the conclusion that it was a team effort and none of the three researchers made the intellectual invention alone. The fact that Eichengrün had the idea for acetylation is very credible, even if it is not in written form. The fact that Dreser published his publication without mentioning the names of the two chemists may have been normal for the time, as it was a pharmacological work. But from today's ethical point of view, it must be stated where the substances come from, and the names of the synthesis chemists must follow. Perhaps the question is too academic today, because then as now, new drugs can only be found and developed in a team. Eichengrün would certainly not deny today that Hoffmann had his share in it. Eichengrün's merit is probably to be seen on another level, that he prevailed against the stubborn Dreser, who despite the positive Berlin results wrote a memo that he thought nothing of acetylsalicylic acid. *"This is the usual Berlin bragging. The substance has no value"*, Dreser wrote quickly. But the importance of Eichengrün for Aspirin® was already known at the beginning of the new century. The Daily Mail of 1920 and the Chemical Trade Journal of 1929 attributed a prominent role to Eichengrün.

Then came the time of National Socialism, which took injustices for granted. The National Socialists did not want an Eichengrün, who as a Jew invented an industry icon. His name and his work were to be erased. For Eichengrün with his pronounced sense of justice, the disappointment must have been great when he visited the Hall of Honor of the German Museum in Munich in 1938. Whispering and murmuring greeted him as he entered the hall. A group of the Hitler Youth stood around the display cases and admired the exhibited objects while they were engrossed in the inscriptions. He made his way through the group of Hitler boys in their shorts. "Look, yes, that's cool. I want to become a chemist too," it whispered to him. He stood between the uniforms and looked at the white crystals in the showcase. *Aspirin, invented by Heinrich Dreser and Felix Hoffmann, Bayer AG.* He moved on to the other display cases. In a rear showcase he found acetyl cellulose, Cellit, his invention. Where was his name here? Nowhere, an invention without an inventor. Eichengrün left the museum, disappointed and betrayed.

The Bayer company had applied for a patent for the synthesis in the German Reich, which was rejected. At first, the Imperial Patent Office was very benevolent, but the examiner changed his mind and was of the opinion that only new processes and products could be patented. He had found a Mr. Gerhardt and a Mr. Kolbe, who had already shown everything. Duisberg, Hoffmann and Eichengrün had suspected this. The confidence to receive a patent dwindled, as most countries would follow the German view. A change in strategy was necessary and Bayer AG first secured the trademark rights for the trade name, which was granted in March 1899. But that was not enough, because Duisberg was clear that any competitor could immediately bring his Aspirin® to the market, even if it was not called that. He needed exclusivity. The idea was to get further patents in the USA and in England as well as in the British colonies, which were not subject to German law. There was the largest market in the world and was just good enough for a success-spoiled Duisberg as a great organizer in Bayer AG. It was his task to make a global corporation out of Aspirin® and his company. From a tar dye factory in a tranquil part of the German Reich, a giant was to emerge. He succeeded in this with further new drugs such as Suramin (Germanin®) and diacetylmorphine (Heroin®), which followed and expanded the position of Bayer AG. The big hit, however, was Aspirin®. Today we acknowledge that he managed to forge a global corporation from the small chemical factory for coal tar, which he joined as a 23-year-old employee, which he left at the age of 74.

Not Only the German Spirit Should be Healed by Aspirin®

Duisberg ramped up production and sent out some samples, which we would today call clinical samples, to leading doctors in the German Reich and Europe. Duisberg pursued two goals with Aspirin®. On the one hand, he wanted to distinguish himself from the rest of the competition with a high-quality product, which mainly offered contaminated acetylsalicylic acid. In addition, he wanted to keep the competition in check, which was so audacious as to counterfeit his Aspirin® and foist an inferior product on customers, damaging the good reputation of his company (Fig. 8). He had a brilliant idea, which we still know today as a milestone in the history of marketing. He invented the Bayer cross, which he had pressed into each tablet. Incidentally, he also invented the machine-made tablet. If Aspirin® was administered like many other medications as a powder in a bag, the patient now received a tablet that was handy, had a fixed dose, and could be conveniently taken along. Every patient saw the Bayer cross when taking it and was sure that it was the original.

Now he needed the US and UK patent, as a large market was expected in the USA. Duisberg received his patent in the USA on December 22, 1899, with the number 22,088, even before the trademark rights were granted to Bayer in the German Reich. On June 27, 1900, the patent was registered in the USA under the name of Felix Hoffmann. This was initially no problem, but the US Patent Office would later recognize a problem when it became clear that a "herb" had filed the application and not Henry Edward Newton, who, as a Briton, acted in the name and on behalf of Bayer AG.

Fig. 8 Bottle with Aspirin powder, circa 1899 (Permission Bayer Archive)1

As early as 1909, Bayer AG generated a third of its sales with Aspirin® in the USA. The strategists in Leverkusen decided not to transport the drug from the German Reich to the USA anymore, but to produce it directly on site in Rensselaer on the Hudson River in the state of New York, near the capital Albany, where a Bayer factory had been standing for a long time. Reasons were the high import duties, which could now be circumvented, and the immense damages that Bayer had suffered from the smuggling of acetylsalicylic acid from Europe via Canada and Mexico. The decision was made and the rest of the River Aniline and Color Works Company in Rensselaer was bought up and expanded into a Bayer site. Here, *"America's Home of Aspirin"* was to be created. There was a good network to the decision-makers in the USA and many German immigrants were hired as workers. The gold mine could now be exploited and a marketing plan developed when a telegram with the headline "London Calling" arrived in Leverkusen and the Patent Office in England caused trouble.

It was quiet in the courtroom on the morning of May 2, 1905, in London. The weather was unusually warm for the city and overall it seemed to be rather dry this year. His Lordship Justice Joyce scratched his head, the wig he wore as a judge itched. He would have to give it to the valet for cleaning. Had vermin infested it again? He looked at the paper in front of him. The sun dazzled him a little. Aspirin? Bayer patent? He looked up and in front of him sat the representatives of these German chemical companies, whispering softly. To the left the gentlemen of Friedrich Bayer AG, to the right the lawyers of the Chemical Factory of Heyden. Never heard of them, they were from Dresden, and there was supposed to be the Frauenkirche, of which he had heard a lot. His Lordship pursed his lips, a lot of paper on the tables on both sides. He looked to the right and saw the book he had fetched from the library—"Principles of Organic Chemistry". He didn't understand much about chemistry, but you never know. Now he saw the other gentlemen at the back of the door. The experts, who were waiting nervously. It would be a long day. He signaled to Baron George Moulton KC, the representative of the plaintiffs against Bayer, who immediately jumped up. The oak floor creaked. The Baron wanted to teach Bayer a lesson and roared about property rights, the history of medicine, and the high standing of this court, which in the name of the Crown finally wanted to speak justice and fairness: *"My Lord, this is an infringement of patent 27,088!"*

So there sat two German companies in London, fighting for the sole right to sell acetylsalicylic acid. A lot was at stake for Bayer AG, because owning a British patent had enormous value. It not only gave a monopoly on the island, but also in the colonies of the Commonwealth such as Canada,

Australia, and South Africa. It was about a lot of money. What did this upstart from Dresden dare to take on the venerable Bayer house? Everyone knew that the invention of acetylsalicylic acid and all medical knowledge came from Bayer in Elberfeld, according to the protocols of the time. Infamous, what this Saxon chemical soup kitchen allowed itself! The fight was hard, but everyone knew it was an old rivalry. That the patent would not be in good shape in the end was clear to Duisberg since the delivery of the complaint. But wasn't it Felix Hoffmann who had first produced acetylsalicylic acid of such high purity? Didn't that count? They had to conduct this trial, because the Chemical Works of Heyden had not only been producing salicylic acid since 1859, but also acetylsalicylic acid in the German Reich since 1901, as Bayer AG had been denied the patent. Now von Heyden wanted to displace the top dog and the chances were good.

The Bayer lawyers bought two pounds of acetylsalicylic acid from Heyden in England and had it chemically analyzed. The analysis protocol showed that the substance was almost identical. This was now the occasion for the claim by the Leverkusen lawyers that the product was a forgery, because the high quality of the purchased sample could only be explained by the fact that Bayer had developed the synthesis and here was a product that could only have been developed with a stolen synthesis. This was vehemently denied by the opposing lawyers, causing outrage. The trial lasted eight days, with cross-examinations, wild insults, and appeals from the judge to both sides to please moderate themselves. He learned a lot about chemistry and the book he had borrowed became his constant companion in the search for the truth. Terrible, he thought, all the chemical structural formulas, the German professional journals, and the poor expert English. His Lordship became tired. But one thing was clear to him: Aspirin® was extremely profitable, and that was the reason for the gentlemen's war. He listened to the experts James Dewar and Adolf Liebmann, who represented the plaintiffs. He found their arguments convincing. Then the statements of Frankland, Armstrong, and Rosenheim, who represented the other side. Yes, in principle, he also agreed with these people. He wanted to make a fair decision. This took time and he adjourned the session until July 8, 1905.

On that morning, the adversaries were gathered in the courtroom of the honorable Judge Lord Joyce to hear his verdict. The judge began with a solid summary of the case and shone with an unheard-of knowledge in the field of chemistry and synthesis. He noticed that the Bayer lawyers were not very attentive. They suspected that the judge before them knew too much about chemistry and would have liked to disappear into the ground, for they saw the bad end coming. *"Faulty," "misleading," "the subject obscured by chance, error, or intent,"* they heard from the judge's mouth. The faces of the Bayer lawyers turned pale and the joy of the plaintiffs was clearly noticeable. It was a total defeat for Bayer AG, sealed by the sentence *"In my opinion, it was not a new body or compound and I hold the patent in question in this case to be invalid."* [In my opinion, it is not a new active ingredient or a new compound, and I declare the patent in question in this case to be invalid.] The lawyers from Heyden jumped up and hugged each other. They had won and pocketed fat fees, which the losers also had to pay. Who's going to tell Duisberg now, the Bayer lawyers thought. Yes, how best to tell the boss?

He took it stoically and focused even more on the US business, where there was further trouble with counterfeit Aspirin®. Heyden first had to set up a distribution network in America. This could take time and was not even Duisberg's biggest problem. Bayer had meanwhile become a victim of its own success, because Aspirin® was being counterfeited in every backyard kitchen and sold as a supposed Bayer original. Duisberg wanted to set an example and sued Edward A. Kuehmsted in Chicago. Duisberg wanted to drive Kuehmsted into financial ruin and thus warn all other counterfeiters that Bayer AG would crack down hard. Kuehmsted defended himself and, encouraged by the London verdict, went to court. He had done nothing illegal, because in England it had been proven that the patent was wrongly based on false information, he argued. This was a very high risk. Because there remained the uncertainty whether the US judge would also overturn the patent as in England. But if they let the bunglers in Chicago have their way, then even more chemists would quickly get the idea to counterfeit Aspirin®. Duisberg decided to go to court. But the strategy was to drag out the legal dispute as long as possible, because in the meantime Bayer could sell the Aspirin® alone, Kuehmsted was not allowed to sell it and everyone else preferred to wait. Duisberg's logic was simple. On February 27, 1917, the patent expired and until then he did not want to give up the business. That was brave, but Duisberg was right. The trial lasted over five years, and this time the judge ruled in favor of Bayer AG.

But it was to get even worse. The First World War was the abrupt end of Bayer's activities in the USA. In 1917, the Bayer AG was effectively expropriated and all assets were seized as enemy property. For 5.3 million USD, the US company Sterling Products acquired the name, the famous Bayer cross as a trademark, and also the brand name Aspirin®. How did this drastic measure by the Americans come about, who only declared war on the German Empire late on April 6, 1917? The sentiment towards the Germans was by no means anti-German in the years 1914 to 1916. Aspirin®, as a German product, was very popular. This changed when a German submarine sank the passenger ship RMS Lusitania on May 7, 1915, taking 128 American women and their children to the bottom off the Irish coast. The mood on the other side of the Atlantic fluctuated and suddenly turned. The Americans were on the side of the British, and England's efforts to directly influence the American government to stop the trade route between the Empire and the USA ended the Cold War and made the Americans a party to the war. The Cold War had its origins in the USA early on. The sinking of American ships and British merchant ships by the Germans was an issue. The Germans and Brits fought each other mercilessly in the Atlantic. Great Britain was still a naval power, and the superior strength at sea was exploited. German merchant ships to and from the USA were consistently pursued and sunk wherever possible. But the war weighed heavily on the economies of both countries. The British had to do without many products of German chemistry because of the embargo. When Bayer stopped supplying Aspirin® to Great Britain, the domestic chemical industry could not supply the required acetylsalicylic acid. The new acetylsalicylic acid tablets from domestic production were not even popular at the front. A field doctor wrote from the front in France to his British colleague, *"the tablets are chalky, cause nausea and are hard to swallow"*. Indeed, the high impurities were the reason for the increased side effects, as later analyses revealed. To make matters worse, the chemical-pharmaceutical industry, as in the German Empire, had to switch from civilian products such as fertilizers to explosives and other war goods. Phenol became increasingly important. Phenol had been needed for the production of acetylsalicylic acid since the time of Kolbe-Schmitt. This could be very elegantly converted into salicylic acid. Now phenol was the required starting material for the explosive trinitrophenol, which was filled into shells. The question was: victory or headache?

The Americans did not want to do without German chemical products, as their own chemical industry was completely inferior to the German one on the Rhine at the beginning of the war. Aspirin® was just one example of the dependency and the need for greater independence and adaptation. Loyalty to the British was strong. Both countries wanted to make trade as difficult as possible for the Germans. Carl Duisberg saw this and wanted to do everything to take advantage of the last three profitable years of his monopoly, take the profits and strengthen the brand before the patent expired. He changed the legal form of the Bayer factory in Rensselaer and transferred the trademark and patent rights to an American company, the Synthetic Patents Company, which de facto belonged to the German Bayer AG. A trick that was quickly exposed. The trade and transport of salicylic acid to the USA was hardly possible anymore, and new partners in the USA had to be found for the delivery of Phenol. Since phenol was a scarce and sought-after commodity, this was no easy task. On the one hand, it was important for the war industry, but on the other hand, it was also a basic material for the emerging American electrical industry for the production of non-conductive plastics such as Bakelite, which were needed as insulators. And then there was the pharmaceutical industry that wanted to produce Aspirin®. An urgent problem had to be solved, and in desperation, they called on the former employee Hugo Schweitzer, who had been working as a chemist and consultant for his old employer in the USA since 1890 and was trying to manage the Synthetic Patents Company.

Did Carl Duisberg know that Schweitzer was leading a double life as an industrial spy? Schweitzer was a chemist, had earned his doctorate in coal tar chemistry in Freiburg, moved to the USA, and became head of the pharmaceutical department of Bayer AG in New York. Schweitzer convinced Duisberg of the plan to build a factory in Rensselaer, north of New York. In the spring of 1915, Hugo Schweitzer was received by Heinrich Albert (1874–1960), who was sent to the German embassy in Washington D.C. as the Interior Minister during the war. Heinrich Albert was also chairman of the Emperor's Secret War Council and was looking for agents in the USA who could help him. Heinrich Albert knew that Schweitzer, although he had adopted American citizenship, was a staunch German with the right mindset and great loyalty. The man was perfect, and over a cup of tea, they discussed how the USA could be kept out of World War I. Schweitzer was very taken with this conversation. Especially since Albert hinted that it would be in the interest of the Empire to use phenol not for the US Army, but better for other purposes. Could he not consider these possibilities in the interest of the Empire? The Kaiser would thank him. Not only him, but also Carl Duisberg, and a great plan was born, which went down in history as the "Great Phenol Plot" (The Great Phenol Conspiracy).[51]

Schweitzer developed a certain independence in relation to his local Bayer activities. In a time when there was no email, no fax, and no telephone connection between Rensselaer and Leverkusen, there were many freedoms and little control for management in North America. Perhaps this was also the reason for the alleged offensive involvement of Bayer AG in the great phenol conspiracy, as was later claimed, because they were not informed or could not intervene. In the procurement and purchase of phenol, which was needed for the synthesis of the explosive picric acid, Schweitzer was not successful. His attempt at business initiation was a failure, even before it really got started. And it got even worse, as the price for phenol skyrocketed. Every morning, Schweitzer saw the fever curve of phenol in the stock market news, against which only a good dose of Aspirin® could have helped. Because the supply chain simply no longer existed, Bayer seriously considered closing the plant in Rensselaer. But then it was reported in the newspaper that Thomas Alva Edison (1847–1931) was in need and required phenol. The great inventor with over 1000 inventions, the Elon Musk or Steve Jobs of the penultimate century, this man wanted to build his own factory for the supply of phenol. Even so much phenol that he didn't know what to do with the surplus. Schweitzer was happy to help him. He arranged a meeting with George Simon, a representative of the former arch-enemy Chemical Factory of Heyden. They worked out a secret battle plan and it looked like this: Simon was to buy phenol unnoticed with the money that Albert got from the German embassy, through the company Chemical Exchange Association. The company was to conclude a contract with Bayer Rensselaer for the delivery of the phenol in the form of salicylic acid. The coup was successful. A few weeks later, Schweitzer hosted a private dinner at the Astor Hotel in New York, to which Heinrich Albert was also invited. Schweitzer earned handsomely through a commission, the Chemical Factory of Heyden was in business like Bayer, and the Interior Minister was glad that now between 1 to 2 million kg of phenol no longer flowed into the American war industry—a Win-Win-Win-Business.

But this business did not last long. On August 14, 1915, all efforts, all secrets, and finally the big business from the German perspective were ruined by a stupid mistake. During a subway ride in New York, Albert had forgotten his briefcase in the car. There was no honest finder, because Karl Burke, who took the bag, was a member of the Secret Service and had been ordered to observe Heinrich Albert and the German ambassador. Albert noticed the loss when he saw the New York sun again. He ran back, but it was too late. The evidence in the form of documents and notes that the German Empire had violated the neutrality laws passed by the US Congress in 1915 were in the bag, and the bag was gone. The Americans knew and suspected that some diplomats were moving conspiratorially in the country. Stealing a diplomat's documents was illegal, and the US government was unable to prosecute Albert criminally, but the public damage was done. The documents were leaked to the newspapers, and the anti-German newspaper New York World, which belonged to a certain Joseph Pulitzer, ran the story under the headline *"How Germany has worked in US to shape opinion, block the allies and get munitions for herself, told in Secret Agents' letters"* (Fig. 9). Thomas Edison then sold the remaining phenol to the US military and was rightly more than annoyed. Carl Duisberg on the other side of the Atlantic

Fig. 9 First page of the New York World from August 15, 1915 (public domain, Wikipedia)1

was no less outraged and tried to keep Bayer AG as far away as possible and to shift the blame onto Schweitzer and the German embassy. It didn't help. Bayer could no longer gain ground in public. Perhaps it was this phenol conspiracy in connection with the intention to enter the war against the German Empire that led to the execution order for the de facto dissolution of the company in the USA. Carl Duisberg was powerless and could only watch helplessly from his office in Leverkusen on the Rhine.

After the Second World War, the Leverkusen company tried in vain to regain its name and rights. It would take 76 years until this wish came true in 1995. However, the company remained in the lucrative North American market and called itself Mobay. In 1978, Mobay bought the company Miles, whose product Alka-Seltzer® many will be familiar with. Irony of the takeover: The product also contains Aspirin®. How is this possible? Anyone who has ever been in an American drugstore knows that Americans use the name Aspirin as a synonym for acetylsalicylic acid. This of course has something to do with the unpronounceable name for Americans. But also with the fact that the only brand name in the USA had long lost its legal protection. It is now used quite naturally by many American pharmaceutical companies.

How the old German Bayer came to its name in the USA may have something to do with a side business of a competitor. The company Sterling was taken over by Kodak in the 1980s. Kodak was known in the photo business and not as a pharmaceutical company. Perhaps out of ignorance of the pharmaceutical market, they decided to sell the acquired self-medication business to the British pharmaceutical group SmithKline Beecham (SKB). In the 1990s, the name was transferred again, and Bayer AG had the opportunity to buy back the name for 1 billion USD. Some speculated whether the purchase of SKB by Kodak was only intended to sell the name to Bayer at a high price? Only a rogue would think evil of it.

So far, we have dealt with the interesting entanglements and confusions about a drug that helps many people with mild pain and rheumatism. Acetylsalicylic acid was taken billions of times until the time after the Second World War, but nobody knew what the biochemical mechanism of action was. A circumstance that we no longer have to deal with in modern drug research.

On the Trail of the Mechanism of Action of Acetylsalicylic Acid

It started with a bang and black smoke in the kitchen of a house in Birmingham in 1940. "Boy, what happened? Are you okay? Do we need to call the fire brigade?" the father of 13-year-old John Vane shouted. "Nothing, Dad, nothing," he coughed back and stumbled out of the kitchen. "I just tried out my Bunsen burner." The father took it calmly, because what should he expect from a boy of this age who was given a chemistry set? Certainly not that he would put it unpacked in the cupboard. The war had only broken out a few weeks ago, the Germans were dropping bombs on England, and now he was supposed to be angry with his boy who had caused this small explosion compared to the Wehrmacht? He grimaced a little, but took the little one in his arms and felt if everything was still in its place. Two arms, two legs, a head and no blood. So far, so good. He looked into the kitchen, where the smoke was gathering under the ceiling and slowly dissipating. He had just painted it with a little paint, which was hard to get in these war days. The boy was now not allowed to experiment in the kitchen and possibly burn down the house. Mr. Vane senior was a craftsman and preferred to build his offspring a small wooden shed between the house and the air-raid shelter, where he could experiment. Here it was allowed to stink and bang, but the house was safe.

John wanted to become a chemist, and this spectacular start into chemistry rarely describes the budding career of a future Nobel laureate. Young Vane attended King Edward VI High School in Birmingham, but the theoretical instruction and lack of practice quickly tired young John Vane, and his interest in chemistry waned. The end of his school days was approaching. Vane knew he would study anything but chemistry, even though he was one of the best in his class. He found everything boring, the teachers gave him instructions like in a recipe that he followed, and the only joy he felt was when he got more of the product than was indicated in the recipe. One midday at the end of the school year, his chemistry teacher Maurice Stacey stood before him. "John, come over here for a moment. I need to ask you something." Had he been up to mischief? Had the last experiment gone wrong? "Yes, Sir, what can I do for you?" "John, what do you plan to do with your life when you finish school?" "Sir, I don't know yet, but anything but chemistry." Maurice Stacey paused and wondered if he had misheard. "Well, you know, I received a letter from a professor in Oxford. He's looking for a good student who wants to study pharmacology." Pharmacology? He had never heard of it, but Oxford, of course, was the opportunity to get into the natural sciences and do something big. "Mr. Stacey, of course, I've always wanted that. I'm looking forward to going to Oxford."

John Vane (1927–2004, Fig. 10) went to Professor Harold Burn (1892–1981) in Oxford in 1946. He learned some basics of biology that he had missed during his chemistry studies, and now stood at the professor's door. Unlike the rather boring library he briefly visited, his new mentor was a true inspiration. The newly founded School of Pharmacology by Burn was to become one of the best addresses in England and Europe. Burn was enthusiastic and motivated his students to be curious and approach pharmacology through experiments. This was music to the ears of the young student Vane, who liked working in the lab. Vane became a pharmacologist who loved his work and found his calling in the subject. After his studies, he went to Sheffield, New Haven, Connecticut, and became an assistant professor at Yale University. In 1955, he returned to London and accepted the call to teach Experimental Pharmacology. There he also met Harry Collier (1912–1983)[52]. He was the son of a Brazilian civil engineer and an Englishwoman. Harry moved to England with his mother at an early age because his father was too much on the move and the marriage broke up.

Fig. 10 John Robert Vane (1927–2004) (Permission Wellcome Foundation, London, UK)2

The boy was intelligent and learned quickly. He was as knowledgeable in art as he was in science. He wrote scripts for radio and small documentaries, but his first love remained biochemistry. He received his doctorate in zoology in 1938, but was already appointed lecturer in comparative physiology at the University of Manchester in 1937. The salary was rather modest and not enough, which is why he tried his luck in the industry at de (ICI). In 1941, he was seconded by ICI to the Liverpool School of Tropical Medicine, where he stayed for 15 years. Collier changed again and came to the pharmaceutical company Parke-Davis, which gave him all the freedom to research what he found exciting, and here a rapid career began.

Physiology shaped him and aroused his interest in the emerging field of pharmacology, which tries to explore the biochemical mechanisms of action of drugs. His interest was broad, and he dealt with allergic bronchospasms as well as inflammation and pain. In the 1940s and 1950s, there was great interest in the search for new analogues of morphine and acetylsalicylic acid, which would later also lead to the discovery of ibuprofen. However, the big problem remained understanding the biochemical effect and developing experimental approaches. How did these substances work? Little was known about this. Dreser had hypothesized 50 years ago that Aspirin® worked centrally in the brain, but was it that simple? Collier could imagine researching everything, but it should not be acetylsalicylic acid. He was mistaken.

Collier turned to short peptides that affected smooth muscle contraction. These peptides are called bradykinins and he had discovered them in 1948 in his old homeland, Brazil. They were actually a byproduct of his research when he was studying snake venoms, which a large part of the Brazilian population suffered from. After the bite and the uptake of the venom, the body releases substances that cause dilation of the vessels and a drop in blood pressure. A phenomenon that years later was the basis for the development of a blood pressure lowering drug from snake venom. But we will read more about this later.

Mauricio Rocha e Silva (1910–1983) and his team were quite surprised in Brazil when they saw the difference in their patients after a snake bite. Short-chain proteins, which were identified as bradykinins or better as peptide or tissue hormones, were isolated as triggers and contributed to the sensation of pain, which was read in London and found very exciting. Collier noticed that people who were injured by snakes, where bradykinins played a role, took Aspirin® and felt better. He became curious. Did Aspirin® also have an effect on the peripheral organs and muscles? It was as always confusing. Collier set about the tangle of ideas and hypotheses and tried to find the right thread. In 1958, Collier came up with the idea of conducting an animal experiment in which bradykinin was injected into the lungs of guinea pigs. The results confirmed the prevailing opinion that bronchospasm must occur. The guinea pig, which was not doing well, showed this clearly. In a second experiment, he gave the guinea pigs acetylsalicylic acid before and after the bradykinin administration. A look into the cage confirmed to him that the guinea pigs that had received acetylsalicylic acid before the administration were normal and active, while the other group behaved anxiously and shyly and sought out the dark spots in the cage. Now his curiosity was greatly aroused. The acid kept the airways open and it must have neutralized the bradykinin. How could that be, he wondered.

Was Dreser right 70 years ago that acetylsalicylic acid works centrally in the brain? He severed the vagus nerve in guinea pigs, which leads from the brain to the lung. This nerve is the tenth cranial nerve in humans and, as the largest nerve of the parasympathetic system, is involved in the regulation of almost all internal organs, including the lung. Collier repeated the experiments and saw that no changes occurred. The group of guinea pigs that had previously received acetylsalicylic acid was lively. The other group did not feel well at all. Since the nerve had been severed in both groups, a central effect could be ruled out. Dreser was wrong! It was a breakthrough. Because 50 years after the introduction of Aspirin®, hardly anything was known about the mechanism of action. He reported to his boss, but neither he nor Parke-Davis saw the breakthrough, because no money could be made with these results. However, he let him continue and Collier continued his research. His son Joseph Collier, who later conducted research as Professor of Clinical Pharmacology at St. George's Hospital Medical School in London, said in an interview about his father: *"I'm not sure why they decided that way. He was an industrial scientist and not an academic researcher. Maybe they thought that if one of their people would do pure science, it would bring prestige to the company."*

With the methods available in the 1960s, elucidating the mechanism of action was a real Herculean task. Analytics was still in its infancy, and biochemical characterization of the receptors and their cloning was not even considered. Collier got help, because a young woman who had just earned her doctorate in pharmacology in London introduced herself. Dr. Priscilla Piper (1939–1995) was hired to assist him, and she proved to be a talented researcher.[53] It was planned that she should complete part of her training in the industry, and so she came to Parke-Davis, where she was to stay for several years. Collier made her his first assistant in research and together they conducted hundreds of experiments on guinea pigs, rabbits, and rats. It turned out that the task of bradykinin action was not easy to solve. The animal models were not uniform, the obtained lung tissue died off or was difficult to cultivate. They needed help and turned to Vane in London. Collier and Vane had known each other for a long time and were good friends. They met at the conferences of the Pharmacological Society, sometimes they went on vacation together with their families to the south of France. Harry's son Joseph spent two days a week in Vane's lab, which certainly influenced his decision to accept a professorship in clinical pharmacology. Vane's lab had excellent equipment and the questions posed for the detection of chemicals in fluids could be answered. One of the techniques established in Vane's lab was the Cascade Superfusion Bioassay, which worked as follows: A first piece of lung from a guinea pig is stimulated with a stimulant, so that tissue hormones are released, which are then transferred to a second piece of lung from another species (e.g., rabbit or rat). If an irritation can also be detected here, there must be something in the supernatant that the guinea pig cells have released. How beautiful, how simple, how incomprehensible. Vane himself urgently needed help and other researchers to support him. Then his old friend Collier called at the right time and Vane offered Priscilla Piper to teach her the techniques, as they were at their wits' end in his lab. Piper moved into the new lab at the Royal College of Surgeons, made the discovery of her life, and stayed there until her retirement.

She quickly learned the techniques of dissecting guinea pigs, cultivating tissue samples, and triggering or suppressing biochemical reactions with chemicals. Whether it was coincidence or good observation is unclear, but during one experiment they had a 30-second view of an aorta that should not have twitched. Was it something new? Was the guinea pig different or had no one noticed before? They isolated a new substance that no one knew, and named it Rabbit Aorta Contracting Substance (RCS), which directly leads to a laugh, because they were at the Royal College of Surgeons. Piper, who had been frustrated with acetylsalicylic acid for years, came up with the idea of injecting her new all-purpose weapon into the lungs of guinea pigs. Perhaps out of habit, and Vane, who knew little about her work at Parke-Davis, had no objections. She repeated the experiments, administered acetylsalicylic acid before and after the provocation, and found that it inhibited the release of RCS. Their and Vane's first Nature publication on this work titled "The release of rabbit aorta contracting substance (RCS) from chopped lung and its antagonism by anti-inflammatory drugs" was a recognized success, but it drove Collier up the wall, because he lost his best employee to Vane and saw his own research lose importance because he lacked the laboratory methods.

How could this happen? Vane was getting closer to the question of the effect of acetylsalicylic acid. He suspected that it might be the prostaglandins known since the 1920s, which the Swede Ulf von Euler (1905–1983) believed to have isolated from prostate secretions in 1935. However, it turned out that prostaglandins occur in all organs, but their physiological function remained unclear. Collier had previously tested prostaglandins in the guinea pig model, but there was no remarkable effect. From today's perspective, it could have been an impure extract or a wrong prostaglandin, and Collier soon gave up. Vane had it figured out. As he pondered the data for another publication on a quiet April weekend in 1971, it dawned on him that he had not given the prostaglandins enough credit. Could it be that prostaglandins and acetylsalicylic acid do have something to do with each other? Was the RCS actually a prostaglandin? He sat with a cup of tea and thought. So far, he had considered RCS as a substance that prevents the effect, but could it be that RCS is not formed at all? What if Aspirin® blocks prostaglandin biosynthesis? It struck him. Yes, that could be it. No, that had to be it.

Vane was in the office early Monday morning and called Piper to him. "Priscilla, I know how aspirin works. Everything became clear to me yesterday." Her look was tired. She rarely saw her boss in the office so early. "I need to go to the lab and do some work. Let's see if I'm right." "Boss, should I help, do you need help?" He hesitated, he had not been to the lab for a long time and let his assistants cut tissue. No, he wanted to do it himself. He wanted to be sure that everything was done properly and that he could rely on the data. He was a bit shaky and the first cuts through the guinea pig's lung were a disaster, but the routine returned. Like riding a bike, you never forget, he thought, transferred the tissue samples into the test tubes and performed the stress test. As always with and without acetylsalicylic acid and in the sample with the active ingredient he could not detect any prostaglandins. It was the chain reaction of inflammation that he had broken. The acetylsalicylic acid was the bolt that slid between the falling dominoes and stopped the fatal biochemical chain reaction. If the cell was injured, no more prostaglandin could be formed. The tissue was spared. He didn't know exactly what acetylsalicylic acid inhibited, but he knew it was so. He had solved the puzzle that Collier had been racking his brains about for ten years. But how was he going to tell his old friend that he had found the solution? It would be a heavy blow for him. He invited him to dinner to tell him about the results. Very modestly and kindly, Vane described the situation at the time: *"We went out for dinner. Harry congratulated me wholeheartedly, even though I knew he was very excited. I think he would have liked to have been the discoverer. In the end, it's luck, patience or coincidence. Whatever, but luck plays an important part."*

Collier was disappointed, but kept his anger in check, because he knew he couldn't stop Vane from continuing his work. Knowing that something big was underway, Vane swore all employees to the project. On June 23, 1971, he and Piper published their results in the prestigious journal Nature under the title "Inhibition of prostaglandin synthesis as a mechanism of action for aspirin-like drugs". He wanted to make this research watertight, to show everyone that he, John Vane, had found out how simple painkillers work. Like a firework, he published further articles on the mode of action of other light painkillers such as paracetamol and indomethacin in 1971 and 1972. A search in medical databases revealed that he authored or co-authored 17 publications in these two years. By now it was clear that acetylsalicylic acid inhibits the enzyme cyclooxygenase and thus prevents the formation of prostaglandins.

The question that preoccupied Vane was: What is this RCS? It would take another six intense years before Vane and Piper knew that in addition to prostaglandins, there are also thromboxanes, which are responsible for blood clotting. They discovered even more about thromboxanes, namely that they are also responsible for the constriction of vessels, as the two had seen and not understood in their guinea pig model for more than ten years. Thromboxan A2R, as RCS was later called, gave aspirin® another groundbreaking significance. Now the physiological process of blood clotting was understood, because the aggregation of platelets is triggered by thromboxanes. Billions of platelets, called thrombocytes, circulate in our blood, constantly waiting for a signal to clump together when an injury occurs. The vast majority of blood circulates in our veins and arteries and is broken down again after a short life in the bloodstream. When injuries occur, arachidonic acid is released, which attracts the platelets to stop the bleeding. Millions of people have taken aspirin® in low doses to prevent heart attacks by slightly inhibiting platelet aggregation, even though data from recent studies have revised the presumed benefit. In 1982, Vane, Sune Bergström (1916–2004) and Bengt Samuelsson (1934) received a call from Stockholm informing them that they would receive the Nobel Prize in Medicine. The discovery of prostaglandins and other related eicosanoids, as the collective term for prostaglandins, leukotrienes and thromboxanes is called, was cited as the reason. In an obituary for John Vane in 2004, Der Spiegel on 22.11.2004 called him the discoverer of the effect of aspirin, for which the Nobel Prize had been awarded 12 years earlier. This was not formally and substantively correct. The awarding of the Nobel Prize marked the end of almost 70 years of research, about which the American professional journal Medical World News claimed in 1970 that *"Prostaglandins could become as important for medicine in the seventies and eighties as penicillin and steroid hormones were in the fifties and sixties"*.

The success story of aspirin® is enormous and pharmacological research is far from over. In conclusion, some figures should illustrate that we are probably dealing with the most successful drug of all time with aspirin®. It is the most widely used drug in human history and as a brand like Coca-Cola or Apple, it is indispensable from our culture. It has made medical history, it was part of espionage, it has received a Nobel Prize, and it was the first drug ever to come on the market in tablet form. Hardly any other drug has been researched as intensively as aspirin®. However, Bayer would probably not get approval for aspirin today because the ratio of benefit to risk is too low from the authorities' point of view. But we should not be surprised if new indications for this oldie are found in the future.

New Painkillers Follow Aspirin®

Aspirin® was a great success, but the problems with side effects remained, more precisely, they increased in public perception. Because the more people took Aspirin®, the more reports of stomach bleeding there were. The term stomach bleeding is however incorrect, as these were not massive open wounds, but microbleeds in the stomach lining. This side effect was marketed very aggressively, and competitors to Aspirin® used the image of a blood-filled swimming pool to illustrate how much blood the British lost annually through the intake of Aspirin®. It was macabre, but the image was part of a marketing strategy to establish its product Panadol® against Aspirin®. Panadol® contained paracetamol and was the biggest competitor to acetylsalicylic acid before ibuprofen. The success was moderate. People stuck with Aspirin®, but the miracle drug slowly but surely lost its importance. The search was on for new painkillers that, like Aspirin®, but not like the opiates, should relieve pain (Fig. 11). Unaware of the mechanism of action, which was only discovered for opiates and Aspirin after the Second World War, the new drugs similar to acetylsalicylic acid were classified as non-opioid analgesics. A selection of the most important drugs today is shown in the structural formulas in Fig. 11. The development of Indomethacin, Diclofenac (Voltaren®) and Metamizole (Novalgin®) and Nefopam (Silentan®) as classic and modern non-steroidal drugs is indeed exciting, but it will not be pursued further in this book, as they do not originate from natural substances.

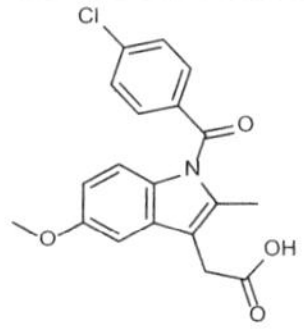 Indomethacin (1960)	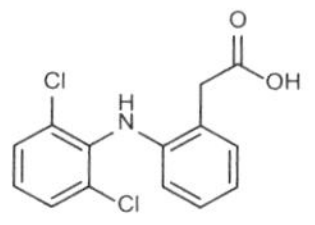 Diclofenac (2007)	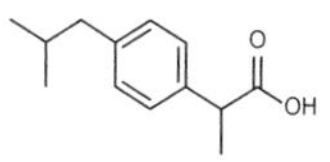 Ibuprofen (1961)
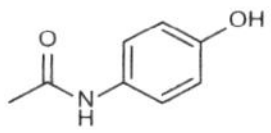 Paracetamol (1878)	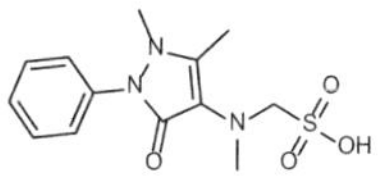 Metamizole (1922)	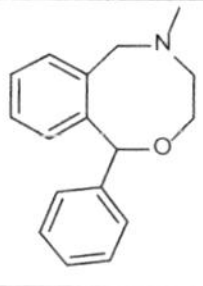 Nefopam

Fig. 11 Non-Opioid Painkillers*[Selection](in brackets year of chemical or pharmacological discovery) (Work of the author)*

Paracetamol

What was Panadol®, which should no longer fill a swimming pool with blood? The new painkiller was N-Acetyl-*para*-aminophenol, which is now called acetaminophen or more commonly paracetamol. It was not new, as its discovery in 1878 was much older than that of acetylsalicylic acid. The chemical precursors were acetanilide and phenacetin, simply synthesized from aminophenol, a waste product of tar chemistry, which was considered useless. The American chemist Harmon Northrop Morse (1848–1920) managed to synthesize paracetamol, but it was not marketed as a fever and pain relieving drug until many years after his death. Paracetamol, along with acetylsalicylic acid and ibuprofen, is one of the most successful over-the-counter painkillers in human history, which will be discussed in the following chapter. This is surprising, because at the time of its discovery in 1878, the active ingredient was classified as too dangerous by the pharmacologist Joseph von Mering (1849–1908) because paracetamol, like acetanilide, can trigger methemoglobinemia. This side effect means that a derivative of the red blood pigment increases in the red blood cells (erythrocytes), which is no longer able to transport oxygen. The cause is the oxidation of iron in hemoglobin, which is the binding partner for oxygen. Josef von Mering was a doctor with special achievements in the field of sleep and pain medication. The now obsolete sleeping pill barbital, which was introduced under the trade name Veronal®, he developed together with the chemist Emil Fischer. Barbiturates were considered the first choice for the treatment of sleep disorders until the discovery of benzodiazepines. Today they are no longer available, Pentobarbital is used in veterinary medicine to euthanize animals. Sadly, pentobarbital also gained notoriety as a lethal poison in the lethal injection, used to carry out the death penalty in the USA.

The pharmacological assessment of methemoglobinemia by von Mering seems strange from today's perspective. Because the supposed alternative was the phenacetin, which only differed by an acetylation. Bayer started selling phenacetin in 1888, believing that this substance was safe. Whether it was purely supposed scientific reasons or a feared patent infringement by the competition also needs to be considered. The company's acetanilide was removed from the market early on, leaving phenacetin as an alternative to acetylsalicylic acid. Phenacetin was used in drugs to relieve rheumatism, migraine and neuralgia.

But phenacetin had even more to offer, because with the increasing consumption of coffee in modern society, its euphoric effect also became known. Especially in the watch industry, this coffee with a chemical shot was taken by workers to increase performance. The intake of phenacetin and even more the abuse of caffeine was difficult to control and its carcinogenic effect and the phenacetin kidney, a permanent and sometimes fatal kidney damage, often showed. It is astonishing that a drug like phenacetin would never have been approved under today's legal requirements, but was only banned in 1983. The authorities were aware of the side effects and did not consider a withdrawal necessary despite concerns from the medical profession. The drug remained on the market for more than 100 years.

The assumption that methemoglobinemia is seen in the chemical structure of paracetamol was wrong. In 1948, Bernard B. Brodie (1907–1989) and Julius Axelrod (1912–2004) showed that paracetamol had been wrongly suspected by von Mering, as he had probably received paracetamol contaminated with phenacetin from the chemists for his study. Pure paracetamol showed no methemoglobinemia, and phenacetin alone was responsible for the methemoglobinemia. While studying his 1895 publication "Contributions to the Knowledge of Antipyretics and Analgesics," von Mering suspected something was wrong when he observed his dogs that had been administered phenacetin. He wrote: *"Allow me to make a brief remark at the end, which may have practical significance. In my numerous experiments on dogs, I noticed that the same substance in the same dose can have a different strong effect. I do not think I am wrong if I attribute this behavior to a different extensive decomposition in the organism."* His problem was that the effect was different despite the same dose and was not the same in all cases. Von Mering attributed this to the non-uniform absorption in the intestine in the dogs. He believed that the state of filling of the stomach and its pH value were important, and he was certainly partly right. He suspected that phenacetin changes before it is absorbed. Von Mering was a clever observer, but the cause lay in the metabolism after the phenacetin was taken up. A substance like phenacetin, which is not active itself but only becomes active through metabolism in the body, we now call a prodrug. Paracetamol was already discovered in 1899 as a breakdown product of phenacetin in human urine, but no one was simply interested in it.

Paracetamol found its way into medicine. After Brodie and Axelrod had shown that paracetamol could be a new old painkiller, it was marketed by McNeil Laboratories in the USA under the trade name Tylenol® Children's Elixir. A year later, Panadol® followed in the UK and in 1959 it came onto the market in Germany under the name Ben-u-Ron®. Today, paracetamol is the second most commonly used non-opioid painkiller after acetylsalicylic acid. It is offered by many manufacturers as a generic, as the patent protection expired decades ago. Despite all the controversies about its benefits, paracetamol has been on the World Health Organization's (WHO) Essential Drug List since 1977. It is not surprising that the mechanism of action, like that of Aspirin®, was not known. That paracetamol also inhibits cyclooxygenases was only made known by John Vane's investigations into acetylsalicylic acid, who tested paracetamol in parallel in his experiments. There was no Nobel Prize for the discovery of how paracetamol works. The substance was anyway part of the large acetylsalicylic acid study and was considered too insignificant in scientific perception.

Paracetamol is hardly soluble in water, which poses a problem when babies or toddlers need to take a syrup for fever. The little ones can't swallow a tablet. But the syrup with a sweet or peppermint flavor is gladly taken. Because of the insolubility, the syrups are made with propylene glycol, which is a cheap solvent. But even these costs some manufacturers in developing countries wanted to save and replaced the propylene glycol with the antifreeze diethylene glycol or ethylene glycol with fatal consequences. In 1990, 40 deaths were reported from Nigeria, which could be attributed to poisoning by the diethylene glycol contained in the syrups. It was even worse in Bangladesh, where between 1990 and 1992, 339 people died after taking paracetamol syrup, which was also adulterated with diethylene glycol. The greed was too great and the manufacturers tried another forgery in 1998. This time, 24 children died as a result.

Paracetamol was also a case of production sabotage in 1982, with 15 poisoned preparations and seven deaths. Everything began on September 29, 1982, when a student in Chicago was given a 500 mg Tylenol® tablet by her father. Shortly after she took the tablet, she fainted and died. A few hours later, a Polish immigrant who had taken the day off from his job as a postal worker to travel with his children met the same fate. He too was suffering from a sore throat, took some Tylenol® tablets, had convulsions, and died shortly thereafter. His wife and brother also took Tylenol® tablets on the same day and died within a few hours. The doctors who were called in assumed it was an epidemic. Two paramedics, who had also been at the deceased student's, remembered that she had taken Tylenol®. The blood was tested, and the toxicologists found that it contained not only paracetamol but also a lethal dose of potassium cyanide, which turns into highly toxic prussic acid in the stomach. The mayor of Chicago called on the citizens to return all Tylenol® preparations. The company Johnson & Johnson promised to replace the products and offered a reward of $100,000 for the capture of the perpetrator or perpetrators.

Although the police could not reconstruct the course of events beyond doubt, they quickly assumed it was a deliberate poisoning. The poisoning triggered a wave of activities. Many Poles living in Chicago sent Tylenol® to their homeland, which was still part of the Eastern Bloc at the time. China banned the import of Tylenol®, and in Chicago, nurseries where potassium cyanide could be purchased were searched. The identification of the fingerprints on the many Tylenol® packages naturally yielded nothing, and panic broke out because people now believed their toothpaste tasted strange, the eye drops were a bit different. A resident of Chicago called the police and claimed his dead squirrels had been poisoned with potassium cyanide.

These discussions were also held at the level of professionals, and in the business world, the question arose of a possible conspiracy between competitors. Memories were revived of the year 1975, when Johnson & Johnson and its partner McNeil Laboratories had invested $85 million in advertising to boost Tylenol® sales to $400 million. This was a huge sum and a clear lead over Anacin®, the product of the German competitor from Leverkusen, at a market volume of $1 billion. Americans consumed 400 million paracetamol tablets that year, in stark contrast to 90 million acetylsalicylic acid tablets. Was a market shifting here? No, on the contrary, the pharmaceutical industry showed remarkable solidarity. The competitors promised to refrain from advertising that would now have cast their product or acetylsalicylic acid in a better light. Instead, Johnson & Johnson destroyed 22 million capsules and tablets with a market value of $88 million. In a litigious country like the USA, lawyers took Johnson & Johnson to court and filed a class action lawsuit. They demanded $600 million because the manufacturing process had caused the contamination. However, this was not confirmed by the FDA, and the FDA sided with Johnson & Johnson. The lawsuit was dismissed. Nevertheless, the product sabotage was severe, and Johnson & Johnson lost its paracetamol market share from 37% to 7%.

Despite intensive search, the police had no success in apprehending the perpetrator. This is surprising, as there was never any talk of a demand for money or extortion at Johnson & Johnson. No hot lead to the perpetrator showed itself to the police. But there were annoying criminal copycats. They mixed herbicides into orange juice, pethidine into brownies, or Excedrin® tablets with mercuric chloride. The search extended across the entire country. Typical suspects from databases were summoned, lie detector tests were conducted, all to no avail. The only good thing about this product sabotage was that the FDA issued guidelines to provide medicines with safety features. Safety features that we are well familiar with today are the well-known plastic rings on bottles that break, or blister packs and no longer individual tablets in a tube. We patients can now immediately see if a medication has already been opened.

Ibuprofen

Although many patients consider Ibuprofen a painkiller, and they are right, the idea of Ibuprofen was quite different, namely not to be a painkiller, but a synthetic steroid. We go back to the 1950s and find ourselves between two gigantic developments in the history of pharmacy. One is that of painkillers, the other is the no less exciting story of steroids, which I will tell later. Stewart Adams (1923–2019, Fig. 12) and John Nicholson (1925–1983) were chemists at the British Boots Pure Drug Company, who had set themselves the goal of finding a drug for rheumatism that should not have the side effects of a corticosteroid. It was to become the super-aspirin, as it was called in the corridors at Boots.[54] In the following years, they synthesized and tested 600 active ingredients, one of which particularly aroused their interest—Ibuprofen.

Stewart Adams came from modest circumstances. His father was an employee at the railway and money was always tight. Whether it was due to money or rather difficulties at school, at 16 he dropped out of school

Fig. 12 Stewart Adams (1923–2019) (Permission, Boots UK)

and started an apprenticeship at Boots. His aunt was a magistrate and her nephew worked in a senior position at Boots. So the kind lady arranged for young Stewart Adams to get an apprenticeship in 1939, which he started with joy, according to his own account. Boots is today one of the major retail chains on the island as well as in some countries in Europe and Asia. Until the 1980s, the company operated its own pharmaceutical research, which is little known. Jesse Boots (1850–1931), the son of the founder John Boots (1815–1860), sold the company in 1920 to the United Drug Company in the United States. The business did not go well for the Americans, and the emerging Great Depression, which plunged the USA into its worst economic recession prompted Jesse Boots to buy back the company to Great Britain. John Boots (1889–1959), the grandson of his namesake grandfather, then successfully continued it with a research department for the discovery of Ibuprofen.

In the 1940s, no formal study was required to become a pharmacist in Great Britain. Formally, he was also not a pharmacist, but Adams wanted to know and learn more. He stayed the required three years in the apprenticeship and then went to university. He did not want to stand as a salesman or pharmacist in a pharmacy all his life. He liked pharmacy and the natural sciences so much that he enrolled at the University College of the University of Nottingham with a scholarship of 40 pounds per year to get his bachelor's degree in pharmacy. On weekends and during semester breaks, he earned some extra money at Boots. But that was not the only challenge, as he studied during the Second World War and had to expect to be drafted every day.

Not only did he seem to like the study, but also Boots, because in 1945 he returned to his old employer and was given the task of dealing with the new wonder substance Penicillin to deal with, which as an antibiotic was to change the world. His task was to monitor the production in Nottingham, which unusually took place by fermenting the brush fungus *Penicillium* in milk bottles. He did not like this and after 18 months he asked for a new task. He switched to a project that became his later doctoral thesis, and here he met a young assistant whom he married. But let's stick to science. Boots and the Pharmaceutical Society granted him a scholarship of 900 pounds for the preparation of his doctoral thesis. His promotion was about heparin and Adams discovered that histamine forms when he heated heparin in solution. The freshly graduated doctor returned to Boots in 1952 and was immediately involved in a project that was to change his life fundamentally. Boots had wondered if there were not drugs that could better treat rheumatism. The administration of corticosteroids plus Aspirin® was then the standard therapy. To the chagrin of many patients, the side effects were considerable: Allergic reactions and stomach problems were frequently reported. Developing a non-steroidal drug that relieves pain and acts like a corticoid was Boots' idea and Adams' task.

The research began in the kitchen of an apartment in Nottingham, which was not bombed by the Germans and can still be visited today.[55, 56] Not only by today's standards did the kitchen not meet the requirements for a laboratory. He accepted the kitchen lab as it was. Adams was under time pressure, as the same idea was being pursued in the USA and the Americans had already synthesized hundreds of new substances. He now had to set up a laboratory and establish scientific methods. The new lab was to follow six years later. The most important task remained the establishment of a new animal model. There was no established model, but everything would later depend on an animal study, he was sure of that. Adams found an idea during a literature study in an old publication. It showed an erythema on the skin of a guinea pig, which did not exactly fit rheumatism, but it was a start. His team now considered how to improve chemical structures. Various structural formulas were drawn on paper, revolving more or less around acetylsalicylic acid or simple salicylic acid. This was not really innovative, and this was confirmed by the pharmacological tests, which showed little or no toxicity, but also hardly any anti-inflammatory effect.

Perhaps it was this inadequacy of the guinea pig experiment that gave Adams the idea to test the chemical substances on himself after they had been classified as harmless in a 30-day feeding trial on rats.[57] What came, was taken. He had no inhibitions and was also the first to take Ibuprofen in a dose of 600 mg. Later, Adams told that after a night of heavy drinking he had no aspirin at home. He had a lecture to give, and his headache was a torment. He took a teaspoon of his synthesized Ibuprofen and the lecture came easily to his lips.

In August 1958, John Nicholson joined the team to support the research department. The team presented him with the not so nice results of the 200 compounds that had been synthesized so far, and it was hoped that Nicholson would bring in new and good ideas. They came, and he suggested investigating phenylpropionic acid as he thought he recognized similarities with the structure of morphine. The idea was brilliant, but based on a false assumption. Because neither the mechanism of action of acetylsalicylic acid nor that of morphine was known, and we know today that they differ fundamentally. Nicholson knew that in the neighboring chemical departments some of these propionic acids were being synthesized as candidates for herbicides, and he tested them in the model with the guinea pigs developed by

Adams. The success was modest, but it was a start, as the slight anti-inflammatory effect gave hope.

The team got to work. 600 more substances were synthesized, including BTS8402. However, clinical tests showed no analgesic or antipyretic effect in humans. The anti-inflammatory effect, however, was significant. Structure-activity analyses were carried out and further derivatives with a free acid were developed. BTS10335, BTS10499 and Ibufenac were tested as representatives for many new substances. The first two substances were immediately ruled out due to allergic reactions, only Ibufenac was to be further investigated. However, after approval in 1966, high liver toxicity was shown in patients. Ibufenac was withdrawn from the market. Was that it? Was all the research in vain? Should the project be filed away after four clinical studies with four active ingredients? No, Nicholson did not want that under any circumstances. He and Adams once again looked over all the data and structures that they had patented in 1962. They looked each other in the eye, and one of them pointed to a structure that was not particularly active, but had proven to be well tolerated in animal testing—Ibuprofen. *"There were many assumptions in our work"*, Adams commented later, *"but Ibuprofen was now to become number five, the last bullet in the revolver"*. The shot was a hit, a direct hit. No toxicity to the liver, it was three times as strong as aspirin, there were no allergic reactions. Boots wanted to bring Ibuprofen to market as quickly as possible, as the patent term was getting shorter and shorter. For Boots, Ibuprofen (Brufen®) was the blockbuster and saved the business with the high investments. Adams later remembered that the approval in the UK and the USA was considered difficult and this was his most exciting time. Approval was granted in 1969 in the UK and a year later in the USA. Only in 1983 was Ibuprofen available over the counter in pharmacies in England. In Germany, Ibuprofen was available without a prescription from 1989 in a dose of 200 mg and from 1998 in a dose of 400 mg. Today, 20,000 tons are produced worldwide and it is a bestseller among the free drugs in the pharmacy.

It surprises no one when he or she reads that Ibuprofen and many of the subsequent nonsteroidal analgesics, which are referred to in technical language as Me-too substances, also inhibit the Cyclooxygenases I and II. Ibuprofen is also known for unwanted effects such as stomach discomfort or heartburn. Nevertheless, the drug can be classified as safe and is often recommended as a substitute for acetylsalicylic acid. What is to be made of the history of Ibuprofen? Was it a lucky coincidence that everything turned out for the best? Perhaps, because in response to this question, Adams quoted

Louis Pasteur, who was his role model, with *"Fortune favors only the prepared mind"*. But it was more than that, it was his persistence, his more than commendable ambition, to research in an unknown world. Because the groundbreaking works of John Vane were not to come until 1972. He advised his young colleagues: *"Later as a senior scientist, I always encouraged my younger colleagues to take a walk in the woods and to explore basic concepts, as long as they come out further along the road towards the development of a new drug."* [Later as a senior scientist, I always encouraged my younger colleagues to take a walk in the woods and explore basic concepts, as long as they progress on the path to developing a new drug.] Adams remained with Boots for his entire scientific life, as he saw his employer as a *"very ethical company"*. Adams was also an upright person and researcher, as he never claimed to have earned even a "penny" from Ibuprofen. His achievement was recognized in the scientific world, because in addition to the honorary doctorate from the University of Nottingham, the Royal Society of Chemistry honored his achievement upon his death with the words that his invention *"continues to help many billions of people around the world."*

Steroids and the Pill

Brief profile – Steroids

Structural formula:

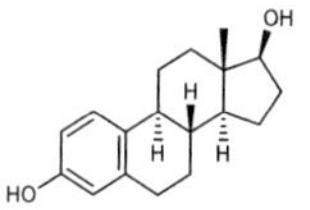

Estrogen

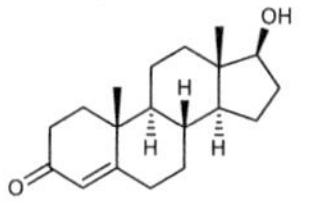

Testosterone

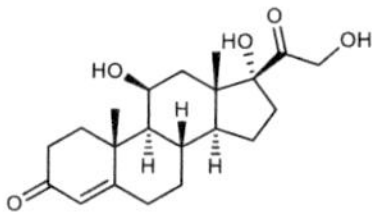

Cortisol

Modern drugs:

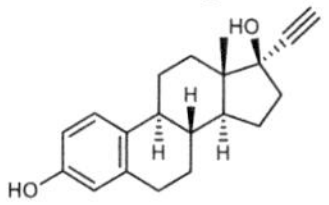

Ethinylestradiol (oestrogen)

Noethisterone (progestogen)

Natural Product Group:

- Steroids

Effect:

- Anticontraceptive
- Anti-inflammatory
- Anabolic agent

O. Kayser, *From the Maluku to Molecules*, https://doi.org/10.1007/978-3-662-69923-2_4

Be fruitful and multiply (Genesis 1, 22; 9,7), so it is written in the Bible, *and the Lord created man and woman of different sex.* The cultural history of sexuality certainly does not begin with Adam and Eve and the expulsion from paradise, but religion was formative for the Jewish-Christian cultures in Europe. Sexuality was lived, even if the rules of practice were dictated by the church and some rulers. Already in the Bible and in almost all times, sexuality played a role, which was usually labeled with a moral reproach depending on the culture. Among Christians, it had a bad reputation after the fall from grace and expulsion from paradise, whereas sexual ethics in Judaism, among Muslims and Buddhists are not inherently negative. Intercourse before and outside of marriage was ostracized in most cultures and was sometimes punished draconically. If these affairs also resulted in children, this was very often a social catastrophe for the mother. The father could often escape the nurturing fatherhood by paying alimony, as reported from the Middle Ages is. Free love, as propagated by the 1960s generation, was virtually unthinkable in Europe until modern times. In this respect, however, our Neanderthal ancestors, who were probably non-religious, were ahead of us. Anthropologists from the University of Liverpool suspect, based on finger length, that monogamy and social consideration were rather unknown. The ratio of both fingers to each other depends largely on the amount of male sexual hormones that the fetus was exposed to in the womb during pregnancy. If the testosterone is too high, the index finger is too short in relation to the ring finger.[58] Was there no love among Neanderthals? Certainly, but the authors also suspect that sex among Neanderthals was driven by testosterone.

Pregnancy and sexuality were private areas into which religion and politics liked to interfere. Nothing was personal, and in the name of God or the emperor, rules were established that, guided by their own interests, shaped almost all societies not only in the Western world. Neither in family planning nor in the question of desire for children did science play a role. Especially for the woman, who was often left alone in history with the carrying and care of a child, control over her own desire for children was an important concern, which was constantly disregarded by men in a society dominated by them. With the beginning of modern times, against all religious commandments and all societal norms, pregnancy was not necessarily accepted as God-given and both sexes became cunning in order to break the supposed commandment of lust equals sex equals pregnancy. How normal it was, how often to sleep with each other, was actually a question of culture and morality. Thus, a couple already had excessive sex in the USA of the second half of the 19th century if they indulged in love more than once a month.[59]

The desire for contraception was not just a desire for promiscuous sex, as we might believe from the hippie communes of the 1960s, but also a strong social aspiration in a rural and later industrial society marked by poverty and poor health. High infant mortality, sudden infant death, death of mothers in childbirth, and horrendous health conditions in the working quarters of all major cities, as we know them today only from the slums of developing countries, did not contribute to an honest and planned desire for children. It was precisely the distress of women and families in the 20th century that led to demands for family planning, which were particularly voiced by a woman in the USA who we would like to mention as representative of many women's rights activists and activists of the movement for birth control and forced sterilization—Margaret Sanger (1879–1966). In particular, the justification of sterilizing women against their will and the advocacy of racial hygiene, without referring to a specific ethnicity, her opinion of treating beggars, criminals, prostitutes, and drug addicts as certain social problem cases, as she called them, with appropriate contraception, also brought her criticism. A brief internet search reveals that she was defamed as a modern abortion witch, an alleged Ku Klux Klan member, and even accused of being a fascist who would organize a genocide. In the USA, she is still controversial today, as Margaret Sanger (Fig. 1) advocated for forced sterilization and eugenics, but was also an important pioneer and courageous women's rights activist who not only fought against social injustices, but also actively supported the research of the birth control pill with large sums from her foundation.[60] But let's take a step back into the history of contraception before we delve into the life of Margaret Sanger. I would like to briefly touch on the history of contraception before the modern *"chemical temporary sterilization"* was discovered, as Ludwig Haberlandt (1885–1932), its discoverer, described it in a very masculine and functional way.

Chronology of estrogens and progestins

1905 – Proof of the function of the ovaries for the female cycle by Kauer (...)
1921 – Explanation of the physiology of the female cycle by Haberlandt
1929 – Discovery of estrogen by Doisy and Allen
1929 – Isolation of the estrone by Butenandt
1935 – Diosgenin degradation to progesterone by Marker
1937 – First ethinyl synthesis by Hohlweg and Inhoffen
1948 – Estradiol total synthesis by Anner and Miescher
1949 – Market launch of Progynon C
1953 – Norethisterone synthesis by Carl Djerassi
1961 – Market launch of Anovlar®
1965 – Market launch of Ovosiston®

Fig. 1 Margaret Sanger (1879–1966) (Permission, University of Claifornia Library, Los Angeles, USA)2

The question of how sex can be enjoyable without immediately producing children has occupied men and women since ancient Greece and Egypt. In an ancient Egyptian text, which can be understood as a medical book of ancient Egypt, the woman is recommended in the Papyrus Ebers to make a tampon from acacia leaves, pomegranate seeds, and crocodile excrement, which is smeared with honey. The role of crocodile excrement remains more than questionable. Acacia leaves and pomegranates contain polyphenols, which destroy proteins that are abundant in sperm. The same principle applies to surface-active and spermicidal surfactants like Triton-X-100 or Triton N, which were sold in the 1960s. The honey tampons were supposed to seal the cervix and slow down or stop the migration of sperm. The recommendation to breastfeed for as long as possible was also reported, as this would hormonally maintain the woman's infertility. This can neither be scientifically proven nor was the method successful in practiced reality. The use of plant extracts remained the first choice in the upcoming Arabic and

Greek medicine, but the successes must have been meager. In a publication, Kumar and colleagues mention 577 plants that are allegedly used successfully worldwide in folk medicine. The biological effects of these extracts and natural substances are often unclear. These can be phytoestrogens or alkaloids that change the hormone balance, reduce uterine blood flow, lead to an abortion, or strengthen the belief in ritual actions to be temporarily infertile.[61]

In the Middle Ages, it was the man's skill to withdraw his member from the woman's vagina at the right moment, a pleasure that was often sacrificed. There was also the temporary thermo-sterilization of the man, as I heard on a hike in Japan to hot springs in a volcanic area from the story of my Japanese companion. Japanese men often bathed in the very hot water of volcanic springs from the Middle Ages to the early present. It is known that temperatures from 55 °C deactivate proteins and the justified assumption was that at similar temperatures in men, fried eggs are more likely to be present and he is no longer capable of procreation for the coming evening due to dead sperm. Unfortunately, there are no written reports on the effectiveness of this contraceptive method. A similar practice of thermal sterilization is said to have been carried out by African men when they sat in the hot desert sand. From Europe, anecdotal reference is made to the Zurich testicle bathing. Eleven friends, but not footballers, from the left-autonomous scene in Zurich came up with the idea in 1984 that equality between men and women should also be consistently implemented in contraception. Their anti-sexist theory was that sperm could be thermally sterilized using baths. The eleven experimented with testicle bathing chairs (Fig. 2) and immersion heaters, which maintained exactly 45 °C water temperature to kill the sperm, but without injuring either skin or member. *"We did this at home. Each for himself. Only twice did we bathe together. Carrying the chair around was too cumbersome, and besides, the testicle bathing took up an hour and a half every evening,"* says Beat Schegg as an old leftist. The matter was approached very professionally. Special underpants were designed, testicle weights were devised, and the sperm count was checked under the microscope in the bath. While the unbathed showed 80 to 120 million sperm per milliliter, this actually dropped to almost zero after a few weeks.[62] In fact, urologists pointed out in 2007 the problem of poor sperm quality in men who spend too long in the whirlpool or have hot laptops lying over their best piece for too long.[63] I am curious when the first publication will deal with sperm quality and seat heating in modern cars.

Back in Christian Europe of the Middle Ages, the question of deliberate contraception in a "moral agreement", as marriage was called, was probably

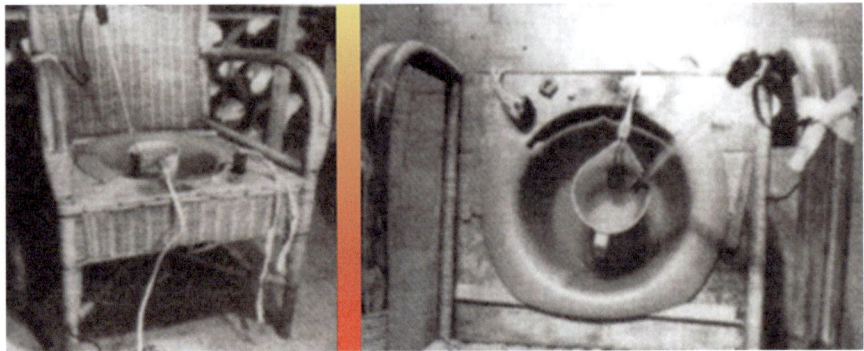

Fig. 2 Self-built chair of the Zurich testicle bathers, around 1980, left top view and right view (Permission, Der Muger, 27.04.2017)

rarely asked. If a man and woman were a "right couple" according to the religious commandments of the Middle Ages, high child mortality was the regulation of the number of offspring. Coupled with a low life expectancy in the Middle Ages, contraception was not really an issue back then. Women could not freely choose their role as mothers and were victims of their own fertility and, in extreme cases, pregnant all their lives.

An absurd story from the Middle Ages is that of Barbara Stratzmann (1448–1503) born Schmotzerin from Bönnigheim, who is said to have given birth to a total of 53 children, all of whom died early (Fig. 3). On a late Gothic painted wooden panel, the boys are nicely depicted on the right side of the father and the girls on the left with Barbara. In total, Barbara is said to have given birth to 18 singletons, five pairs of twins, quadruplets four times, and sextuplets and septuplets once each. According to her information, however, 19 children were stillborn. Of all the many children, the oldest was just eight years old. A lot of work for the notary Friedrich Deumling from Wimpfen, who documented the *"true historia"* in writing.[64] The high number of 53 children of Barbara Stratzmann is impressive, even though the number of children in families of the Middle Ages and modern times was generally high. It is known that Empress Maria Theresa gave birth to 16 children, Albrecht Dürer's (1471–1528) mother had 18 children, and Catherine of Siena (1347–1380) was the 25th child of the cloth dyer Benincasa. The sister of the German Emperor Henry VI. gave her second husband, Margrave Leopold III. of Babenberg, 30 children in 30 years of marriage. The superwoman of the world with 69 children, who holds the world record according to the Guinness Book, is said to be the wife of the

Fig. 3 Representation of the child-rich "Schmotzerin" (Barbara Stratzmann, 1448–1503) in the Cyriakus Church Bönnigheim (Permission Reinhard Kaarsch, Wikipedia)

Russian farmer Fyodor Vasilyev (1707–1782). However, there is no unequivocal confirmation.

The high child- and infant mortality was found in all social classes, including the nobility, who could afford better medical care. Excavations at medieval cemeteries in Västerhus, Sweden, and Espenfeld in Thuringia showed in the skeletons that in Sweden about half of the children did not survive the seventh and about a third did not survive the first year of life. In Thuringia, the death rates were lower at about a third for thirteen-year-olds and just under a tenth for one-year-olds. Perhaps the harsh climate in Sweden and poorer nutrition were the reason for the higher mortality.[65]

Mortality was particularly high in multiple pregnancies and premature births led to a very high physical and emotional burden on the woman. Clara Schumann (1819–1896) wrote in her diary that the numerous pregnancies took away her time for the beloved virtuosity on the piano, for concert tours and composing. Clara Schumann came from a better home, because the bitter poverty and need showed itself in the poor layers in the city and in the countryside. While children were a kind of later pension insurance in old age, they were unproductive eaters at the table in their childhood. Thus, the lack of nutrition, because simply nothing was available, led to insoluble problems and infants, predominantly illegitimate, were given up as foundlings at parishes and monasteries. Also, the abandonment of children is known, however, with the strong piety fewer and fewer children were abandoned in the last centuries. Nevertheless, it must have moved people, because in literature such as in Heinrich von Kleist's novella "The Foundling" or in George Eliot's "Silas Marner" these fates have been movingly told. Not a few remember the fairy tale "Hansel and Gretel", which shows the extreme hopelessness of parents.

Were parents who abandoned their children cold-hearted when they drove their children to certain death? Studies show that this was by no means the case. Just as the mother rejoices in the fairy tale to hold Hansel and Gretel again in her arms, so parents mourned very much for their lost children. The high birth rates left clear traces in women in all phases of life. The risk of dying as a complication of pregnancy or childbirth led to a 10% higher mortality compared to men. This societal trauma has also been taken up in literature in fairy tales like "Brother and Sister", "Snow White" and "Cinderella".

The loss of the wife in childbirth was equally traumatizing for the husbands. Adalbert Stifter (1805–1868), an Austrian writer of the Biedermeier period, lost his wife like Otto Modersohn (1865–1943), whose wife Paula Modersohn-Becker (1876–1907), who was a painter of early Expressionism, died of thromboembolism. The literary description can also be read here in the work "The Buddenbrooks" by Thomas Mann impressively, who describes the death of Thomas Buddenbrook's wife during childbirth.

Margaret Sanger—the Women's Rights Activist and Her Fight for the Contraceptive Pill

Margaret Sanger was born as Margaret Louise Higgins in the city of Corning in the state of New York and trained as a nurse. The start in her family suggests that Margaret had to develop into the woman she later was with her convictions. She was the sixth child of a strict Catholic, Irish mother and an atheistic and very political father, Michael Hennessey Higgins, who worked as a stonemason. *"I cannot look back on my childhood with joy"*, she wrote towards the end of her life. *"The difference between happiness and unhappiness was not money but rather being born into a large or small family"*[66], as the Arizona Daily Start wrote in an interview in 1961, five years before her death.[67] Her childhood must have shaped her later political path as a women's rights activist, as her mother was pregnant 18 times and had eleven live births. Her mother died in 1899 from tuberculosis and cervical cancer. This suffering may have been the motivation for Margaret to train as a nurse. She married William Sanger in 1902, who was six years older and had emigrated from Germany. He was an architect, had an artistic streak and was very politically interested. He would have preferred to become an artist, but he never really pursued this wish. His political commitment was left-leaning and he became a member of the Socialist Party of the USA at an early age member of the Socialist Party.

Only later at Manhattan Hospital for Eye and Ear Medicine was Margaret Sanger finally able to complete her training successfully. William Sanger suggested that she should work as a private nurse for a wealthy man on the Upper East Side, but she did not let this deter her from her desire to offer her medical and social services to the poor. Her marriage did not make her happy. After a fire in a suburb of New York, the Sanger family moved back to the city in 1921, a year before their divorce, and Margaret worked as a nurse in the slums of Manhattan. Today's visitors to New York, who walk in the Lower East Side, would not believe that here, where bars, restaurants and hip music clubs can be found, was the poor quarter of New York. Margaret Sanger was shocked by the social situation and poverty, and during her house visits she assisted in many births or helped to survive illegal abortions. She must have seen some women who tried to perform an abortion on themselves. Many women bled to death and were perhaps lucky in their misfortune, as women who survived often died painfully from infection later. We must remember that penicillin and other antibiotics were not discovered until thirty years later. Sanger felt that she had to make a change and wanted to educate women on how good contraception worked. She wrote columns for the socialist newspaper New York Call such as "What Every Girl Should Know", which were better guides to Safer Sex, as we would call it today.

That was not enough for her and she wrote the brochure "Family Limitation", which was sent out by US mail. This must have caught the attention of a postal worker and Margaret as well as William were charged, as they were violating the so-called Comstock Act of 1873. This federal law, written by Anthony Comstock (1844–1915) and introduced into Congress, prohibited the mailing of erotic writings and contraceptives. William, who distributed the leaflet in New York, was charged and jailed for thirty days. Margaret faced twenty years in prison and in 1914 initially evaded arrest by fleeing to London and the Netherlands. In London she discussed with sexual scientists and she visited the Netherlands to find out why the country had the lowest child mortality rate in the world. She used the time to prepare for the charge, as she planned to return to the USA. On her arrival in New York in 1915, she was greeted by the New York police and arrested immediately. Margaret was brought to court for publishing "indecent content". The pressure from the media, the advocacy of some New York bigwigs as well as the emotional appeal due to the early death of her daughter Peggy were too great and the charge was dismissed. But it was the first of many charges to come. Margaret Sanger was now the well-known feminist and advocate for birth control in the USA. She travelled the country for years, gave lectures and received more than a million letters from American women asking for information on birth control. In addition to words, actions were also needed and she founded the Brownsville Clinic for family planning and birth control in Manhattan with her younger sister Ethel Byrne (1883–1955) in 1916, for which they were promptly arrested. There was a fine of 50 dollars for distributing their brochures and both were sentenced to thirty days in jail for spreading information on birth control. Ethel Byrne felt this verdict was deeply unjust and went on hunger strike. She was successful and after ten days both were pardoned by the governor.

In 1923, Margaret Sanger, as patron, established the Clinical Research Bureau, which became the first legal birth clinic in the USA. In the same year, she also founded the "National Committee on Federal Legislation for Birth Control" and presided over it until its dissolution in 1937. She advanced her cause of birth control to the highest levels in American society. Margaret Sanger was friends with First Lady Eleanor Roosevelt, who was an advocate of similar ideas in the 1920s. Her husband and President Franklin D. Roosevelt did not want to financially support birth control policy, and the First Lady officially stayed out of politics. The political dynamite must have been too great in a conservatively thinking America. However, from the 1940s onwards, many meetings took place both privately and in the White House to explore possibilities for political and financial support behind the scenes. Over time, Margaret Sanger became very international and during a visit to India in 1935 met Mahatma Gandhi (1869–1948), whom she failed to convince of her idea of birth control through contraceptives. The great Indian moral teacher and pacifist could not reconcile himself with the idea that conception could be prevented by chemistry or other methods.

He recommended abstinence, which has never prevailed in the history of mankind.

What was Margaret Sanger's sex life like? She didn't think much of the then prevailing moral notion that a woman in marriage has to be and function for a lifetime. William and she grew apart very shortly after their marriage and she wanted a divorce as early as 1914, which he agreed to, but which was not finalized until 1921. She married a second time, but many sexual affairs became known during her first marriage and afterwards.

Margaret Sanger and Natural Product Chemistry

What makes Margaret Sanger so special for natural product chemistry? She was the woman who brought together important philanthropists, scientists, doctors, and women's rights activists in the 1950s to think about chemical contraception and put it into practice. One evening in December 1951 was to change a lot.

The old lady was waiting for her visitor on this December evening in 1951. The expected gentleman was twenty years younger than her, but his gray hair and mustache showed his advancing age. He was a highly gifted physiologist, whose reputation in the puritanical American society was as controversial as his scientific creativity. He held a professorship in Boston from 1951 until his death, and his scientific passion was hormones and contraception. Any of their common opponents who knew that the two were meeting so unnoticed despite their old age would have immediately concluded a sexual relationship of Margaret Sanger. Her many affairs and love nights with changing partners were not unknown, she was not the housewife and propagated free love, before it was completely normal in the Love-Buses from Wolfsburg 15 years later on the East and West Coast of the USA. But the purpose of the evening conversation was quite different. She wanted to discuss birth control and contraception using new chemical hormones with the gentleman. She had read that increased amounts of progesterone prevented ovulation and Mr. Pincus knew a way for hormonal contraception.

"Goody, come in. The weather outside is dreadful." In the door stood Margaret Sanger and Abraham Stone, the director and vice president of the "Planned Parenthood Federation of America" *(PPFA)*. Goody was Gregory Pincus (1903–1967), who was born in 1903 as the eldest of seven children in a Jewish community in New Jersey, the son of a farmer and teacher. Gregory Pincus was in a difficult situation. He needed money for his research and the company Searle, his previous financier, no longer wanted to support him due to lack of success. They saw him as an imaginative but also notorious researcher who caused a great stir with the artificial fertilization of rabbit eggs with sperm in a test tube and their re-implantation into the uterus of a surrogate mother. He was certainly a pioneer for artificial fertilization in humans, for which Robert Edwards (1925–2013) received the Nobel Prize in 2010. The New York Times published an interview with him under the title "Rabbit Babies from the Test Tube: Nightmare Vision Realized by Harvard Biologist" and caused much protest in a very conservative American society. A second interview in Collier's magazine with a photo of a sad-looking rabbit on his arm guaranteed that now also the animal rights activists or anti-vivisectionists, as they were then called, were up in arms against him. He did not have a good standing at Harvard University in the 1930s and his career wavered more than once. But the signs were good for him with the emerging steroid industry and experts like him were urgently needed in the industry.

Frustrated by the bureaucracy of universities, he and his friend Hoagland founded their own research institute, which was to focus on steroid research. The plan was to raise research funds that would cover the expenses in the lab and their own living costs. What is normal today, that young scientists raise their own salary from foundations or other donors and conduct research, was revolutionary in America in 1944. One can imagine that this concept had its teething problems. Pincus rented a small villa in Massachusetts with a $25,000 financial injection from a friendly Rabbi and business friends. He was also the caretaker in the animal stables, Hoagland mowed the lawn, and a hired Chinese scientist, who was caught by the Second World War in England and could not return home, lived in an adjacent barn because money was tight. Rumors circulated that a Chinese in chains was living in the house. Success soon followed. The Chinese scientist Min-Chueh Chang (1908–1991) received an award for a publication, the American Cancer Society and the company G. D. Searle as main sponsor again financed the work with $100,000.

Sanger's small foundation wanted to give Pincus research funds to explore hormonal contraception. "Mrs. Sanger, what a pleasure to see you and thank you very much for your invitation." His joy was not to last too long, as he must have made a frugal face when he heard the rather modest sum that was offered to him. He could understand that this foundation did not fund like the state NIH, but it was a start he wanted to take. Pincus reported to Sanger about the first works and he surely also told her that the financial means were not sufficient. Margaret Sanger discussed this matter a year later with her friend Katharine McCormick (1875–1967, Fig. 4), who increased

Fig. 4 Katharine Dexter McCormick, suffragette, April 22, 1913 (public domain, Wikipedia)2

the research funds fiftyfold from her private fortune. Katharine McCormick had received a fortune through her marriage to Stanley Robert McCormick, who in turn had received it from his father, who sold agricultural machinery. The marriage was overshadowed by a severe illness, as Stanley suffered from Dementia praecox, as hebephrenic schizophrenia was then called. In 1908, he was declared incompetent by court order, was considered incapable of doing business, and the care was transferred to Katharine and the family.

Katharine McCormick had a bachelor's degree in biology from the prestigious Massachusetts Institute of Technology (MIT) and gave up further study in favor of her marriage to Stanley. Katharine's commitment to women's rights was in her genes and during her studies she clashed with the administration of MIT. The so-called rules required that female students wear hats, preferably with feathers, which she refused on the grounds of flammability and safety in the lab. Margaret Sanger and Katharine McCormick were among the suffragettes, who strongly advocated for women's suffrage in the USA. The two crossed social boundaries in the 1920s. Katharine McCormick, like Margaret, was socially ostracized and legally persecuted when she tried to smuggle diaphragms from Europe into the USA through her counseling centers. The rebel Katharine McCormick therefore very gladly financed Pincus' research until the 1960s, before hormonal contraception became a billion-dollar business for the pharmaceutical industry.

The two ladies and Gregory Pincus did not know in 1952 that a young chemist was already much further along and had found the substance that would solve their problems. Carl Djerassi (1923–2015), whom we will get to know more closely, played an important role as a chemist at an unknown company in hot Mexico. Gregory Pincus and his Chinese colleague, who was to be forced into his luck, wanted to start again where a Doisy and Inhoffen supposedly ended. But everything was to come differently and much faster. But let's look at everything in order, how chemistry helped steroids to their bloom.

Female Sexual Hormones

The Path to the Contraceptive Pill

The contraceptive pill or short "the pill" development is based on important discoveries in the chemistry of steroids. This group of natural substances drives many a medical student, biochemist, or biologist to madness. So diverse and difficult to learn are the multiple rings, the complex stereochemistry, and the diverse biological activities. The name is derived from Cholesterol, which the French chemist Eugène Chevreul (1786–1889) discovered in gallstones and whale blubber and named cholestérine in 1824. Derived from the Greek word cholé for bile and stereós for solid, hard or hardened, later generations used the latter part of the word to describe these hard, fat-like substances that seemed chemically similar. The early fat researchers, who had no knowledge of the structure of steroids or sterols, were intuitively correct when they gave this class of natural substances the name steroids. No one suspected that cholesterol would be the starting substance for the biosynthesis of many steroids. The female and male sex hormones, but also other steroids such as bile acids, toad poisons and cardiac glycosidic steroids play an important role in physiology and as animal and plant toxins today.

The discovery of the chemical sterility of the woman through steroids and the elucidation of the menstrual cycle are inextricably linked. Before the discovery of sexual steroids, the female cycle and its physiology were discovered at the beginning of the 20th century. Already in 1905, the gynecologist Josef von Halban (1870–1937) had proven that the cycle begins in the female ovaries, also called ovaries. He worked with female baboons, from whom he removed the ovaries. From the immediate cessation of menstruation he concluded that not only the organ, but also chemical substances are released from the ovaries into the blood. He was not mistaken, because

Ernst Starling (1866–1927) and William Bayliss (1860–1924) introduced the term hormone for these substances, which is borrowed from Greek and means stimulate or excite. Franz Hitschmann (1870–1926) and Ludwig Adler (1876–1958) presented the cyclical change of the uterine lining as a normal and not pathological or inflammatory process. It is the merit of both that the changes in the uterus are no longer great mysteries, but normal periodic processes in women. Robert Meyer (1864–1947), endocrinologist at the Berlin Charité, reported on the corpus luteum, which arises from the burst follicle shell. He hypothesized that there must be two hormones (progesterone and estrogen) from the ovaries and the corpus luteum that control the changes in the uterine lining.

We want to briefly explain the menstrual cycle and start with the name. Menstruation means monthly, runs cyclically, and this cycle is particularly controlled by the hormones progesterone and estradiol. In this cycle, which lasts 21 to 35 days, an egg blister (follicle) matures in the ovary, in which there is a fertilizable egg cell. At the same time, the uterine lining is built up, into which the fertilized egg can implant. Between the 12th and 14th day, when the estrogen level is highest, ovulation occurs in the ovaries and the mature egg migrates from the ovary to the fallopian tube. If sexual intercourse takes place during this time and a sperm fertilizes the egg in the fallopian tube, it continues to the uterus and implants there. If no fertilization occurs, the cycle ends and the egg and uterine lining are shed with the menstrual bleeding. The hormone level changes again.

The contraceptive pill cleverly intervenes in the menstrual cycle. A combination preparation is administered that contains the two female sex hormones estrogen as follicle hormone and gestagen as corpus luteum hormone. In the brain, the two hormones simulate pregnancy in the woman and inhibit the release of further messenger substances, which are normally responsible for ovulation (Ovulation). Without ovulation, no egg can be fertilized, and the gestagen changes the mucus in the uterus so that the sperm can no longer rise and can no longer implant into any egg.

The Innsbruck professor of physiology Ludwig Haberlandt (1885–1932, Fig. 5) wondered whether the maturation of the egg in women could be chemically reversed by giving them corpus luteum hormones. *"There must be something better than a condom."* This is how Haberlandt justified to his wife why he wanted to turn from heart research to contraception.[68] *"I still remember so well how the basic idea came to me suddenly one February evening in 1919, as if by a higher inspiration, and immediately the far-reaching significance of the matter became fully conscious to me," he wrote* in the annual book, which can be found in the family's estate. It was a brilliant flash of

Fig. 5 Ludwig Haberlandt (1885–1932) (Permission, Museum for Contraception and Abortion, Vienna, AT)

inspiration and the construction principle of the pill.[69] Until the 1920s, it was known that the female cycle is controlled by hormones, but exactly which ones were unknown. Ludwig Haberlandt is today considered a pioneer in hormonal contraception and may be seen as one of the forefathers of the birth control pill, even though the hormones of the female cycle were not his field of research.[70] He studied medicine in Graz, received his doctorate and later conducted research in Berlin and Innsbruck, where he qualified as a professor in 1913. He went to Berlin with his father Gottlieb Haberlandt in 1910, who accepted a call for botany in Berlin-Dahlem. The son was immediately offered a call for physiology at the then Kaiser-Wilhelm-University, today's Humboldt-University. But he did not stay long in Berlin and returned to Innsbruck in 1911, where he qualified as a professor in 1913 and was appointed associate professor in 1919. He saw research in physiology *"as the highest task ... "that can befall the human mind,"* which is why he published diligently with great commitment.[71] Of his 134 publications, however, only 14 dealt with the hormonal cycle of women or contraception. He set his goals high, perhaps too high, because he wanted to finally give the initial hormone research before the First World War the importance in central life processes. He wanted to find and research three substances. The one that regulates the heartbeat, which we know as adrenaline, the second substance that determines temperament and does not exist, and last but not least, the hormones of reproduction.

His major field of research was the physiology of the heart and its internal secretion. He wanted to explore the nature of the heartbeat. Overarching this topic was the research on performance enhancement and the dawn of hormone research was tailor-made for him. In animal experiments with rats, he discovered in 1919 that pregnancy inhibits the further maturation of eggs in the ovary. This surprised him and he came up with the idea of temporarily making women infertile by administering pregnancy hormones. This idea surprised even him so much that he wrote with joy and astonishment that no one had thought of this before: *"I remember how the basic idea for this scientific venture suddenly came to my mind one evening in February 1919 and I immediately became aware of its significance. […] The thing is indeed so surprisingly new and yet so simple that it was referred to me from various sides as the 'Egg of Columbus'. Therefore, I am always surprised anew that it remained undiscovered so far."* The man had humor, comparing ovulation with the egg of Columbus compared.[72]

Haberlandt wanted to know exactly what inhibits further ovulation during an existing pregnancy. He transplanted ovaries from pregnant animals into non-pregnant animals and showed for rabbits that they did not become pregnant for 10 weeks. The corpus luteum in the transplant produced so many corpus luteum hormones that no pregnancy occurred despite mating with a male.

If that was already successful, then in the next step the administration of extracts from the ovaries of pregnant rabbits could have the same effect. He mixed the animals' drinking milk with the extract, but to his disappointment, the rabbits did not become sterile. This was a big surprise and Haberlandt repeated the experiments with ever larger amounts. *"In one case, the animal received 100 ampoules of the preparation within 13 days, in the other case even 500 ampoules perorally (with some bread) within 40 days, without temporary sterility occurring,"* as he reported in a publication in the Munich Medical Weekly in 1926.[72] He believed that the amount of hormones in the extract was too small, but he was wrong, as we will see later (Fig. 6).

After the First World War, there was a shortage economy in Austria and there were no materials and also no money for research. Haberlandt could

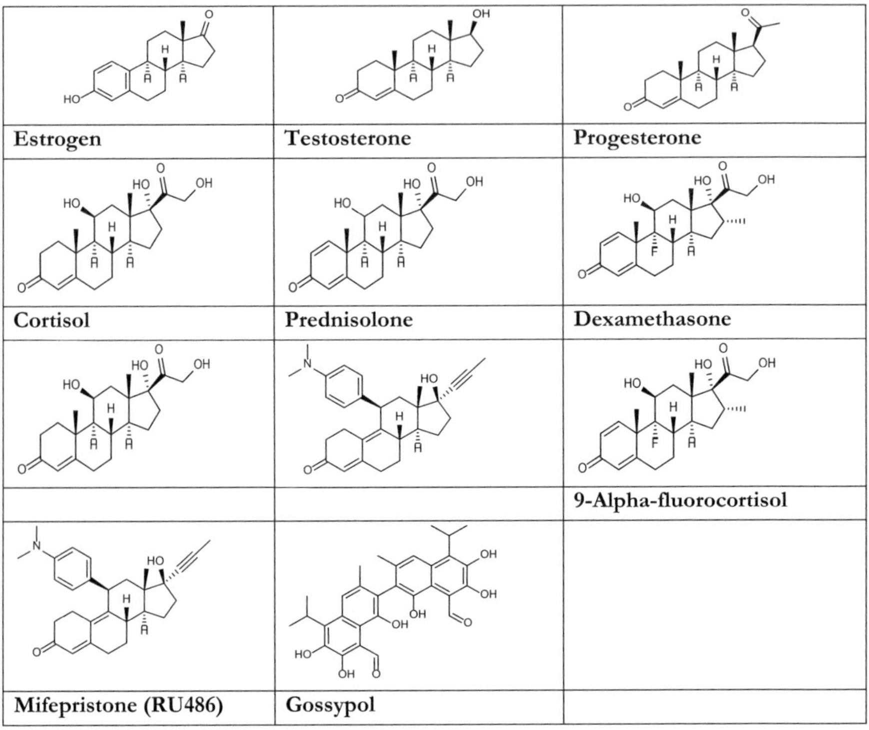

Estrogen	Testosterone	Progesterone
Cortisol	Prednisolone	Dexamethasone
		9-Alpha-fluorocortisol
Mifepristone (RU486)	Gossypol	

Fig. 6 Steroids and steroid-active natural substances *(Selection)* *(Work of the author)*

no longer afford the large amounts of rabbit ovaries, which is probably why he switched to small white mice: *"In the spring of 1919, I performed my first ovary transplantations, which have been carried out continuously since then. Due to the high prices for adequate animals and the immense amount of feed, we had considerable problems with animal procurement, especially since we needed ovaries from pregnant animals (cows and sheep), which were usually not slaughtered for economic reasons."*

However, in the 1920s there was interest from the pharmaceutical industry in the physiology of sexual hormones and they were very interested in testing sufficient quantities of the female sexual hormone, generally referred to as Progynon, in clinical trials. Haberlandt promoted his idea of temporary female infertility at various professional conferences in Innsbruck and Tübingen, but he drew a storm of hostility upon himself.

In addition to the Catholic Church, it was also colleagues like Alfred Greil from his own university, who wrote in the Central Journal for Gynecology in 1924: *"Do not spoil your father's fame with the corpus luteum!"*

After a lecture in Innsbruck on the topic "Oral hormonal contraception—consequences for medicine", Haberlandt was accused of his contraception preventing unborn life and this being in clear contrast to moral, religious and ethical values. At the local and national level, the reactions of contraception opponents were fierce and reminiscent of the hatred of today's radical abortion opponents. However, Haberlandt had every reason to believe that the time for a contraceptive had come in society. Mass unemployment, great mobility and social impoverishment after the First World War led to an ever-increasing acceptance of contraceptives. The condoms that the soldiers were allowed to get to know in the field had long since made the desire for family planning a reality. Why shouldn't chemical contraception also find a place?

Haberlandt came under pressure in the political discussion and this also affected his role as a scientist. He lost his scientific reputation completely unjustifiably and there was also scientific stagnation, as he no longer received enough ovarian extracts after the First World War to fulfill his contractual obligation with E. Merck as an industrial partner. I. G. Farben terminated the contract in 1930, but the idea was not dead. Haberlandt turned to the company Gedeon Richter in Budapest to get enough material for a clinical study. The company was founded in 1901 by a Jewish pharmacist. Haberlandt concluded a research contract in 1928 that allowed Gedeon Richter to commercially exploit the results. But what exactly happened is still unclear today. The company is said to have tested the drug Infecudin® in a women's clinic in Innsbruck in 1931. Results of this clinical study have never been published. Infecudin® is said to have been the first contraceptive to hit the market, but Gedeon Richter denied this in 1990 upon request. It is only certain that Infecudin® came onto the market in socialist Hungary in 1966. Gedeon Richter is still very active today as a public company on the Budapest Stock Exchange and was, alongside the Jenapharm in the GDR, the Hungarian company that produced and distributed the birth control pill in socialism.

In 1931, Haberlandt wrote a monograph titled The Hormonal Sterilization of the Female Organism (Fig. 7). He already saw the practical use early on when he *"describes the great practical application possibility in women [...]"*.[73] If you read his monograph and see the three main goals, you rather get an uneasy feeling. *"The direct prophylaxis for the sick woman, who could take serious damage from a possible pregnancy with all its consequences and should therefore be protected from conception for a certain time, i.e. until her recovery. But here the endeavor to avoid inferior offspring as much as possible and finally the directly related goal to give the temporarily sterilized woman*

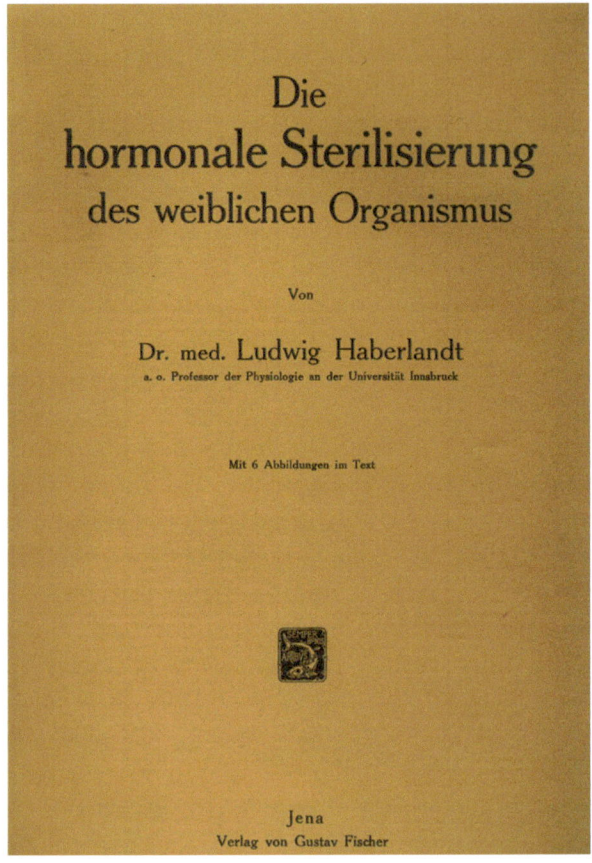

Fig. 7 L. Haberlandt, Publication: The hormonal sterilization of the female organism, 1931 (Permission, Museum for Contraception and Abortion, Vienna, AT)

the possibility to have healthy children after complete or at least relative recovery of her health. […] This opens up far-reaching eugenic and racial hygiene perspectives for the future; but also in the field of sexual hygiene, new points of view open up for both sexes, […]. The objection, […] against a future temporary, hormonal sterilization of the woman in the sense […], that probably too many women would make use of it, takes care of itself, since the sterilization preparation will only remain in the hands of the doctor and will not be generally available. […] That in the present time in all civilized states the correct form of reproduction regulation has become one of the most important problems for both the individual and the community is beyond question. […] it should be pointed out once again at the end […] that the temporary, hormonal sterilization of the woman proposed by me is by no means a random birth control, but a eugenic birth regulation, which must of course be reserved for medical judgment. [...] But it is obvious that its clinical introduction will avoid a lot of disease and misfortune due to the restriction of the numerous abortions."

Was Haberlandt a so-called "racial hygienist" of his time and did he play into the hands of the upcoming National Socialists? His biographers describe him as German-national, but not as a National Socialist. He was very conservative in values, but Haberlandt was also a child of his time, which openly discussed the issue of eugenics and racial hygiene like many others. Like Margaret Sanger, who fervently demanded contraception, he too saw in chemical sterility instruments to minimize the so-called *"inferior offspring"*. Such a publication seems very strange to us today. Also, the citation of Sigmund Freud in the same monograph, that it would be the *greatest triumphs of humanity if the responsible act of procreation could be elevated to an arbitrary and intended action,* is an inappropriate mixing of one's own ideas with the thoughts of the psychotherapist, who wished for more self-determination in sexuality.[73]

Haberlandt's personal hopes for a successful scientific career, in his view, did not come true.[70] The full professorship was denied to him in Innsbruck and applications to the universities of Jena, Graz, and Rostock were unsuccessful. He took his own life in 1932, leaving behind three children aged 17, 14, and 11 to his wife Therese. Many of his raw data from clinical studies were destroyed during World War II. Only late did Haberlandt receive the recognition he desired. Thus, Carl Djerassi, who produced the synthetic hormone Norethisterone, emphasized the great importance of Haberlandt's scientific work for him. Late in 1970, Haberlandt's work was rediscovered by Hans Simmer at the University of California in Los Angeles and received with great interest. An exhibition at the German Hygiene Museum in Dresden on the history of the pill honored Haberlandt's role as the grandfather of hormonal contraception and his statement: *"There must be something better than condoms!"*

Haberlandt's life ended tragically. He took his own life at the age of 47. But what would have happened if Haberlandt had not ended his life so early? Just two years after his death, pure progesterone was made available for pharmacological tests and Haberlandt would have found out that progesterone is active in mice, but not in rabbits or humans. There is a simple explanation for his failed attempts from 1926. Unlike the mouse, which has a very permeable small intestine, progesterone is not absorbed in humans and therefore needs to be replaced by a synthetic progestin that acts like progesterone but is better bioavailable, as the pharmacist calls the uptake into the body. This problem was solved by Carl Djerassi with the progestin Levonorgestrel 30 years after Haberlandt's death.

During Haberlandt's lifetime, his publications received considerable attention. The pharmaceutical company E. Merck from Darmstadt wanted to support him, concluded a research contract, and produced extracts according to his specifications. It can be assumed that Haberlandt and E. Merck entered into a partnership because he needed material for research and E. Merck was already thinking about clinical studies that Haberlandt also wanted. The company E. Merck wanted to catch up with the competitor from Berlin, who with the endocrinologist Eugen Steinach (1861–1944) brought an ovarian extract to the market in 1928 under the name Progynon®. Schering sold Progynon®-dragees for menopausal symptoms and for gender corrections, which among other things consisted of estrogen, which was obtained from the urine of pregnant women. It was not a pure substance, as besides the estrogen, other impurities such as proteins were present. The medication was prohibitively expensive because there were not enough women who voluntarily donated urine. Schering quickly switched to bovine placentas and was also open to all other animal sources. Between 1925 and 1935, particularly mare's urine was processed, pig ovaries were extracted and human urine was still used. With intelligent extraction and fractionation, estrogen, progesterone as well as testosterone were to be obtained. The demand grew and the need could not be met, a chemical solution had to be found and this was the hour of the great natural product researchers Adolf Windaus and Adolf Butenandt.

Adolf Windaus, the Lord of the Four Rings

One of the first steroid chemists was Adolf Windaus (1876–1959, Fig. 8), who remained faithful to the topic of sterols from his doctoral thesis on "New Contributions to the Knowledge of Digitalis Substances" and his habilitation on cholesterol until the end of his scientific career, and for this work received the Nobel Prize for Chemistry in 1928. During his school years, he was not convinced that he wanted to pursue a career in natural sciences. In his inaugural speech on the occasion of his admission to the Prussian Academy of Sciences in 1937, he told the audience: *"But as a young man, I had the idea that dealing with beautiful literature was a hobby, but not a profession."* Inspired by Robert Koch, Louis Pasteur and their discoveries, he wanted to become a doctor. Although his parents would have liked to see their son in the family cloth factory, they accepted his decision and Windaus was grateful to both for the freedom to decide for himself.[74] However, he soon realized that medicine would not fulfill him and he began studying

Fig. 8   Adolf Windaus (1876–1959) (Release, University of Freiburg, DE)2

chemistry in parallel. After passing his preliminary examination in 1897, he completed his chemistry studies and went to Freiburg, where he received his doctorate at the local university with "New Contributions to the Knowledge of Digitalis Substances". The start into the world of steroids was successful, as digitalis substances, which were taken as heart-active natural substances, belong to this chemical family. He wanted to stay in Freiburg and habilitated in 1903 with a work "On Cholesterol". That was probably it, because the research of a natural substance so widespread in nature promised little prospect of a steep scientific career. He wrote: *"The problem appears to be little appealing from an experimental point of view. No nitrogen, which gives stimulation and variety to the processing of alkaloids. Only carbon, hydrogen and a little oxygen, all in the traditional bond, which does not expect any surprising images. The task presents itself as a long, unspeakably tiring march through a barren structural desert."* Right, but that's often the life of a scientist. A lot of patience and perseverance coupled with a high tolerance for frustration are the pillars of success, as we have often been able to read. But Windaus was right when he recognized that cholesterol has a structural property. If we look at the structural formulas of the other beautiful natural substances in this book, we see a fireworks display of chemical atoms and arrangements of bonds. But here? The chemistry of cholesterol seemed to be as exciting as watching water evaporate in a chemical column. But Windaus was at the beginning of steroid research and could not guess what the iceberg looked like underwater if he only had the small tip in view. All the steroids in the plant kingdom, the bile acids, the heart-active steroids, the glucocorticoids and the sex hormones were just waiting to be discovered. He did not let go and dealt with the degradation of bile acids at the University of Göttingen. His friend Heinrich Otto Wieland (1877–1957) from Freiburg, who received the Nobel Prize a year before him for his work on bile acid, had introduced him to these. For almost thirty years, Windaus had researched the structure of cholic acid and cholesterol and in 1932 he finally presented the four-ring formula, which is now so self-evident in every textbook and is a horror for students of chemistry, biochemistry and medicine. Now everything made sense, and he saw that plant saponins, heart poisons and bile acids are based on the same four-ring system.

Unnoticed in the shadow of cholesterol research, an equally active steroid hormone research was taking place, which was less concerned with synthesis, but more with the question: How has nature constructed female and male steroid hormones? At this point, two more scientists must be mentioned who were active in the field of hormone research and made an immense contribution to answering this question: Adolf Butenandt (1903–1995, Fig. 9)[75, 76] and Heinrich Otto Wieland (1877–1957, Fig. 10), whom we have already met. Adolf Butenandt and Hans Herloff Inhoffen, who would later come onto the stage, received their doctorates from Adolf Windaus in Göttingen. Adolf Butenandt studied chemistry and biology in Marburg and went to Göttingen to earn his doctorate and habilitation. Unusual for the group of steroid researchers around Windaus was that Butenandt wrote his dissertation "On the chemical constitution of rotenone" in 1927, which is a fish poison from the chemical class of phenols. But he followed the path of his teacher and habilitated in 1931 with "Investigations on the female sex hormone estrogen", which he had already isolated in the summer of 1929. Two years later he discovered the male steroid hormone Androsterone and in 1931 the hormone progesterone. Both substances and his work were important for understanding cortisone synthesis and later chemical contraceptives. Some researchers therefore see Butenandt as the father of the birth control pill. His work was groundbreaking, which is why he was awarded the Nobel Prize in Chemistry in 1939 together with Leopold Ružička (1887–1976). Since the leader Adolf Hitler forbade all Germans to accept the prize after the award of the Peace Nobel Prize to Carl von Ossietzky (1889–1938), he

Fig. 9   Adolf Butenandt (1903–1995)

was not allowed to travel to Stockholm. It was not until 1949 that he was able to receive the certificate.

Adolf Butenandt's role in the Third Reich is controversial in historical reappraisal. One might think that the refusal of the Nobel Prize would have clouded Butenandt's relationship with the National Socialists. Thus, he was a member of some National Socialist organizations and the NSDAP. Later, he had all documents marked as "Secret Reich Matter" destroyed at the Kaiser Wilhelm Institute for Biochemistry. Anti-Semitic attitudes are not verifiable and he also helped individual Jews from persecution by the Nazis in the 1930s. However, his involvement in various so-called aviation research projects, his knowledge of ethically questionable projects on child epilepsy, and as a reviewer of the journal "Der Biologe", taken over by the SS-Ahnenerbe, is not undisputed. Nevertheless, historians have concluded that he was not involved in ethically questionable experiments.[77]

Adolf Butenandt was the father of German biochemistry and was interested in metabolic pathways and hormonal regulations in the human

Fig. 10   Heinrich Wieland (1877–1957)

body. During his doctorate as an assistant to Adolf Windaus, he isolated the female hormone estrone in pure and crystalline form from the urine of pregnant women in 1927. After his doctorate, he wanted to change the subject and found thyroid hormones exciting. His doctoral supervisor Adolf Windaus and Walter Schoeller waved him off and gave the young Butenandt the advice to work with steroids from ovaries or bull testes, as the actual topic of the structural elucidation of thyroxine from the thyroid gland was occupied by a publication of the English colleague Harrington. The decision certainly did not excite the young Butenandt. Even young doctoral students today know that a change of their doctoral subject in their stressful work means not only delay, but also a complete new start with again unknown methods. Butenandt may have thought that the new task of isolating the steroids would be difficult. The stuff was insoluble in water, the chromatography never properly developed and the countercurrent extraction was invented, but much too expensive. The young chemist spent his days with demixing attempts over seven stages and provided purified fractions for biological tests. But what was worst, the competitors Edward Adelbert Doisy

(1893–1986) and Edgar Allen (1892–1943) at the Washington University in St. Louis were also working in the field of steroids.

In 1929, Butenandt was able to report on the isolation of estrone, which he named Progynon. He had to hurry, and the publication was certainly intended by him for a later date, but he learned from visitors of the 13th Physiologists Congress in August 1929 that Edward A. Doisy made a brief announcement about the isolation of estrone from pregnant urine. If he did not act now, he would have been preempted for the second time in his dissertation. He had to prevent this. The publication of just two pages in the fall of 1929 only rudimentarily describes the isolation from crude oil after shaking out the pregnant urine, which was provided to him by Schering-Kahlbaum AG. One can tell how hastily the publication was written and how carefully another publication of eight pages followed in the same year in the German Medical Weekly. Someone had to stake out his territory. But he also wanted to be a fair winner and mentioned the work of his colleague Doisy from St. Louis very positively and paid him great respect. Both found estrone independently of each other, although Doisy only published his results a few months after Butenandt on April 12, 1930.

Let's take a look at the other side of the Atlantic. In St. Louis at Washington State University in Missouri, Edward A. Doisy (Fig. 11) and Edgar Allen (Fig. 12) were researching in the 1920s, who were pretty best friends. The two played on the same university baseball team, the "Bears", and were biochemists, if the job title could be used at that time. Unlike Doisy, Allen did not own a car, but he was invited by Doisy to ride with him in his "Tin Lizzy" to the university. Both often got into conversation during the ride and talked about their research on sex hormones and reproduction. They quickly found common ground and thought they could work well together in this field. They even complemented each other very well: Allen provided the ovaries and follicles, Doisy took care of the isolation of the substance that had the estrogenic effect.

Doisy and his neighbor, colleague and baseball teammate entered into this strategic partnership. Edgar Allen came to the university in St. Louis in 1923 and dealt with the anatomy and physiology of the female uterus in the mouse. He knew quite precisely about the physiological change in the cycle and knew the basics of the changes in the mucous membranes, ovulation, and the hormonally induced interactions. He suspected that there must be a chemical substance, but his mice were simply too small and did not provide enough biological material. Allen became a customer at the local butcher and got ovaries from slaughtered pigs. That's a damn lot of work, he thought to himself and dragged the ovaries home, put them on the kitchen table,

Fig. 11 Edward A. Doisy (1893–1986)

and asked his wife to help with the extraction. He injected the obtained extracts into his mice and hormonal differences showed up.

Doisy reached the end of his capabilities and urgently needed help to further process these extracts. He had already thought about the problem half the night. Tired, he stood lost in thought with his lunch box on the curb and waited for Edward. He had to talk to him on the ride, he could help him. He heard the horn, looked up, and saw the car coming. Was that Ed or who was coming towards him in this unknown car? In front of him was this brand new Ford T. He would have liked to have such a car too, but they had just arrived in St. Louis and had to invest the savings in furniture and the house. In addition, they were already saving the 364 dollars that the dealer wanted for a new car, but maybe a used one would do for now. "Hi Ed, everything okay, how are you? New car?" "Yes, great car and it's brand new. I got it unexpectedly from the dealer yesterday. Get in, we're a bit late, I have to go to the lab." Edgar got into the car and leaned back as the car started to move. "Ed, tell me, I have a problem with my experiments. I have obtained an extract from these pig ovaries and given it to the mice, but something is

Fig. 12 Edgar Allen (1892–1943), around 1933

not right." Ed listened, but looked intently at the road. "What kind of problem?" he asked briefly. "Can you tell me how I can better isolate the active substance from the extract? I need the substance pure and not so diluted." Ed thought. "I don't know, we do isolations, but we're looking for this insulin." "That's different because it's more water-soluble," he replied. Ed looked at the road: "Your stuff has the solubility of sand, doesn't it? I'll ask in our lab if the guys have a good idea, I can't promise you anything." Edgar nodded, both were silent until they arrived at the university parking lot and the gatekeeper greeted them with a military salute. The barrier lifted. "Will we see each other tonight?" "As always, sharp at 6:00 PM?" Both nodded.

Edgar was already standing in front of the tin can and saw Ed coming. "Sorry, I'm sorry I'm late, but this extraction just took too long. Come on, let's go." Both got in, the car bucked a little when starting and drove off the parking lot. The gatekeeper had opened the barrier and finished his shift hours ago. "Edgar, I did a little reading today. Over in Europe, there are two Krauts who isolate female sex hormones. Their names are Selmar Ascheim and Bernhard Zondek. Ever heard of them?" Edgar Allen shook his head. "So, the two of them isolate the substances from the urine of pregnant women." "Pregnant?", Allen was taken aback. "Where am I supposed to get those?" "Stay calm, I've looked at all this and find it interesting. Do you know the maternity clinic at the university? I know a nurse there. We met at a Christmas party once. Can we ask her to collect the urine of pregnant women for us?" "That smells. It's unappetizing if we have so much urine." "Edgar, don't be like that. That's just how it is. Think about it." The car stopped in front of the Allen house. "This will be great, just wait, the Germans and Austrians are doing something very exciting and we should be part of it."

Doisy took on the problem and understood that the amounts in urine are very small. He saw the problems that the two Germans reported in the publication. He also read that a new technique in chemical process engineering had been invented, called counter-current extraction, which would improve the yield. He had a brochure sent to him by the company and took it to his dean. "Hanau, take a look at this device. With this apparatus, you can isolate small amounts from large volumes much faster and easier." Hanau Loeb rocked in his chair and took the brochure. He saw his new calling, for Doisy he had really put himself out. He thought highly of him and was sure that Doisy would give the faculty the right momentum. "Ed, this looks good, but it's certainly expensive." "Not really, only 750 dollars." Hanau had to be careful not to slip out of the rocking chair in shock. "That much? That's our budget for the annual investment." "Boss, it's a lot of money, but it's worth it. We can use it for the insulin work, but also for the new steroid project." "I don't know." "Hanau, you wanted me," Doisy retorted, "that's just the cost. Give it a push." Hanau Loeb thought about it. The other colleagues would make a big fuss if the new guy just got this device. "Ed, okay, but not a word to the others. We'll call it a strategic investment to modernize the lab equipment. Understood?" Edward Doisy nodded. But he didn't care. The new counter-current system was delivered and set up. It took Edward Doisy a whole five years to improve the method and set up a multi-step process. He was successful and in 1929 estrone was isolated, which Doisy named Theelin, and ten years later estradiol followed.

With his charm, he convinced the nurse, who was surprised, but agreed to collect urine. She didn't want to have to carry the many bottles and Doisy organized a driver who drove to the clinic every morning with his flatbed truck. One morning the driver didn't arrive on time, which made Doisy curious. "Joe, what happened this morning, did you oversleep?" "No, Doc, the police stopped me because they thought I was a smuggler." Doisy made a questioning face. "You're a smuggler? Did they really believe you were transporting alcohol? How did you get out of that one?" The driver took off his hat and grinned. "I let them smell it." "Excuse me, you let the police smell the urine?" "Yeah, sure, they smelled two bottles and then held their noses. One of them said he'd never come across anything like it. Then I was allowed to drive." Doisy laughed. "You did a great job. I'd offer you a shot if we weren't in prohibition."

Doisy found estratriol and estradiol and described the physiological effects very accurately, which differ due to the absence of a hydroxyl group. Unfortunately, he was not chemically trained enough to perform the structural elucidation as quickly and well as the Europeans. At the end of the scientific rally around the physiology, chemistry, and medicine of sex hormones, there were two other winners of the Nobel Prize. One was Adolf Butenandt and the other was not him, but Leopold Ružička. The latter received the Nobel Prize for the synthesis and structural elucidation of testosterone and androsterone. It seems incomprehensible that Edward A. Doisy was not considered at all, but he had to wait four years until he was awarded the Nobel Prize in Chemistry for the discovery of Vitamin K in 1943.

From the collaboration between Doisy and Allen, an animal experiment was developed to determine estrogenic effects, which was later referred to in the professional world as the "Doisy-Allen-Test". After years of hard work, Doisy found crystals in Allen's extracts and presented the results at the aforementioned conference in Boston, which electrified Butenandt. Doisy also knew that what he had found was something very big, and jokingly told his wife that they would soon be very rich. However, he changed his mind because he was grateful to his dean Hanau Loeb for financing the counter-current system. He transferred all rights of his two submitted patents, which were granted on July 24, 1934, to the university, which still puts the proceeds into a foundation to support the biochemical faculty for research. How much money have the two patents brought in for the university to date? It is estimated to be 150 million dollars.[78] The university invested this money wisely and today its assets amount to 8 billion dollars, making it one of the richest universities in the country.

Estrone, which was worth more than gold, is one of four natural estrogens found in women, but also in men (Fig. 6). This may be surprising, but it is converted from androstenedione. This is now known in the bodybuilding scene as "Andro". In the canon of the three important sex hormones, the male testosterone and the progestogen progesterone, which Butenandt also found in 1934, must also be mentioned. It was a tremendous success of science that the three most important sex hormones were finally known. It was very hard work, as all steroid hormones had to be painstakingly isolated from thousands of liters of mare's urine or human urine. Once Doisy,

Butenandt or one of their doctoral students had a few crystals in their hands, the real work began to elucidate the chemical structure. Compared to today's methods and procedures, the structure elucidation must be imagined as very simple. A typical method was combustion, which was referred to as elemental analysis. The substance to be examined was burned and the elements carbon, oxygen, hydrogen and also nitrogen were converted to carbon dioxide, water and ammonia. In an apparatus developed from Justus Liebig's five-ball apparatus, the gases, liquids were bound to adsorbing substances and the weight gain was determined. As a result, the elemental quantities were put into relation from a defined starting quantity and given in a sum formula such as $C_{18}H_{24}O_3$ for estriol. However, the exact chemical arrangement of the atoms in the molecule remained a mystery. In addition to estrone and estratriol, the biologically more active estradiol was discovered in 1939 by Doisy and colleagues, but structurally elucidated by Butenandt. There was much confusion about how many rings the steroids have? Butenandt clarified the question and identified four rings, three 6-membered and one 5-membered ring. He was the lord of the rings.

The Birth of the Contraceptive Pill and its Midwives

The birth of a contraceptive drug was not worth considering in the pharmaceutical industry in the 1930s and 1940s, even though Ludwig Haberlandt frequently presented the basic idea at conferences until his death. Two problems arose that were serious. The first problem was the lack of bioavailability of steroids. Testosterone, like progesterone or estradiol, was not absorbed in the intestine when taken orally, unless it was given in exorbitantly large amounts. This was neither pharmacologically nor economically sensible. The second problem was the small amount available on the market. Isolation from the urine of pregnant women and pregnant mares yielded only small amounts, which were just enough for research. Thus, it was impossible to develop a drug that was supposed to be economically successful.

The history of steroid hormones is characterized by the isolation of the smallest amounts. It was clear to all involved that the sale of a hormone preparation would only work if sufficient quantities were available for an industrial process. In academic research and especially in the pharmaceutical industry, there was nervousness about how this huge business could work at all if the production of the substances was so difficult. On the academic side, German steroid researchers were striving for a basic synthesis of steroids, the Swiss around Reichstein for bile acids as raw material, which they isolated from animals. Despite all concerns, Schering in Berlin wanted to start developing a drug against female menstrual disorders in the 1930s. The term "contraceptive pill" and the goal of contraception were not envisaged at Schering AG in Berlin. No one had given a thought to preventing pregnancy or had any idea of the economic potential of such a drug. Treating women's ailments, as menstrual disorders were often called in the 1930s, was supposed to be the big business. But not only the Berliners, but all major companies in Europe had the same vision. Organon, Schering, Ciba formed a cartel, and in this steroid cartel, Schering with Butenandt was to provide for the estrogens and Ciba with Leopold Ružička for the testosterones. Let's first turn to Berlin.

The Austrian Walter Hohlweg (1902–1992) synthesized and tested different estrogens and progestogens in the main laboratory of Schering AG in Berlin. Although Walter Hohlweg was a chemist, his interests shifted towards the physiology of hormones. Walter Hohlweg's boss, Walter Julius Viktor Schoeller (1880–1965, Fig. 14), led the main laboratory for decades. Both knew each other well and had been working together trustingly for a long time. One day, Hohlweg stood in his office and suggested to the second Walter that they should try estradiol acid. Doisy and Allen in America were working with similar acids, and these substances showed good estrogenic effects. However, he could no longer synthesize and wanted to know if he, as the boss, could find a new chemist to take over this job. Schoeller was very taken with the idea and he immediately knew who he had to ask—Hans Herloff Inhoffen.

Hans Herloff Inhoffen (1906–1992, Fig. 13) went to school in Berlin-Schmargendorf and passed his maturity examination in 1926 at the Kleist Gymnasium.[79] He studied chemistry in Berlin, Bonn, and London, passed his examination in Berlin in 1930, and received his doctorate there a year later with the work "On Oxy Carboxylic Acids of the Cyclohexane Series". His doctoral supervisor was none other than Hermann Fischer, the son of Nobel laureate Emil Fischer. After his doctorate, Inhoffen was an assistant to Windaus in Göttingen until 1935 and worked with the Nobel laureate on the research of Ergosterol. Why exactly at this university? Perhaps it was due to the university town of Göttingen, the academic network of the big and small Fischers, and the fact that Windaus was a former student of Emil Fischer. But Göttingen itself promised greatness and fame. Perhaps it was also a bit more complicated, because in Berlin there was Walter Schoeller, who knew Inhoffen's parents well and recommended the large laboratory of Mr. Windaus to the young man. Schoeller also advised him years later to go to England, as a postdoc to Sir Edward Charles Dodds (1899–1973) at

Fig. 13 Hans Herloff Inhoffen (1906–1992)

the Courtauld Institute of Biochemistry, where he dealt with non-steroidal estrogens. One of these substances was the Diethylstilbestrol, which was synthesized in 1938 by Edward C. Dodds and Robert Robinson (1886–1975). Robert Robinson was also a unique figure in the world of chemistry. He was an enthusiastic chess player and participated in the Correspondence Chess Olympiad in 1940. However, his passion in chemistry was not for steroids, but for alkaloids. In addition to synthesis, it was the structural elucidation that gave him pleasure and for which he received the Nobel Prize in 1947. Thus, the Robinson-Schöpf synthesis for the extraction of Tropinone is well known to many pharmacists and chemists. The author was also allowed to reproduce this synthesis in his internship once with moderate success (Fig. 14).

In 1935, Inhoffen returned to Berlin, where Schoeller offered him a position in the main laboratory of Schering AG in Wedding (Fig. 15). Inhoffen remembers his first lab well: *"A chemical lab, a biological lab with an operating room, and a histological lab were at my disposal. There was a room with air conditioning and cages for three thousand rats, a room with boxes for*

Fig. 14   Walter J. V. Schoeller (1880–1965), 1935

rabbits, and a room for cages with baboons." Fourteen employees were available to Inhoffen, including lab technicians and animal caretakers. Schoeller may not have been entirely without ulterior motives with his generous equipment: He needed a specialist for the synthesis of steroids and therefore sent the young Inhoffen to the experts. The time investment was now to be rewarded, and Schoeller entrusted Inhoffen with the difficult task of synthesizing steroids. Schoeller was a strategist and had long prepared the field of steroids. As early as 1927, he worked with Windaus, who was the worldwide expert in the field of steroids. But Windaus was too busy with his vitamin D chemistry and handed over the work to his assistant Adolf Butenandt. Schoeller agreed, because he wanted to keep a foot in the door, as 55 employees were working on vitamin D, cholesterol, and steroids in Göttingen. Inhoffen was one of them and was to acquire the knowledge for Schering AG.[80]

Inhoffen's time as an academic apprentice came to an end and the scientific world believed that the end of steroid chemistry had come anyway, as the structure of cholesterol and ergosterol had been fully elucidated. The new challenge was to synthesize the first steroid hormone. A race

Fig. 15 Schering main laboratory, Berlin-Wedding, 1937

developed between him and Butenandt, who would be the first to synthesize estrone from a steroid precursor. Inhoffen achieved the impossible and defeated his mentor. Butenandt judged sportingly as a fair loser in 1942. Because Inhoffen had created estrone from cholesterol, he invited him to Göttingen to habilitate. He accepted the academic accolade, which surely helped him after the Second World War to get a professorship and a chair in Braunschweig.

The talented young chemist remained from 1936 to 1945 initially as department head in the main laboratory of Schering and fully met the expectations of his former mentor. At Schering AG, Inhoffen succeeded in synthesizing orally active progestins from testosterone by ethinylation. Ethinyl testosterone was only of importance for a few years, but it was the precursor of the large group of later oral progestins. These are steroids that lack a methyl group in position 19. The absence of this simple carbon atom was the cause for the discovery of norethisterone by Carl Djerassi, which later became the first active ingredient in the pill. Inhoffen also dealt with the aromatization of estrogens and developed a chemical method in 1940 with the help of a Chinese colleague, which you will learn more about shortly.

The two chemists Hohlweg and Inhoffen at Schering made a good team. Hohlweg, the experienced hormone specialist, and Inhoffen, the young chemist who took care of structure elucidation and syntheses. But what happened to Hohlweg, Inhoffen, and the promising estradiol acid that was to be synthesized? Inhoffen later casually mentioned in a lecture how Hohlweg had approached him with his problem. *"One morning my colleague Walter Hohlweg [...] comes to me and says: 'Make me a derivative of the follicle hormone, the 17-carboxylic acid; I believe it is orally active.' I look into the air for a few seconds and say: 'You can have the acid in two weeks. I attach acetylene to estrone and then ozonize.'"* Of course, nothing was easier than that. The first step should also succeed perfectly, as we have read above, but the conversion to acid was then postponed by another 50 years. The synthesis was much more difficult than thought and Inhoffen spoke of *"horrible efforts,"* that had to be made. Hohlweg did not get his acid, but 50 mg ethinylestradiol, which to the surprise of all involved had an enormous estrogenic effect on the animal after oral administration. Hohlweg, so Inhoffen tells, must have stormed into his office with the results of the animal experiment, so impressed was he. In their enthusiasm, they immediately discussed that testosterone must also be ethinylated and tested. Steroid chemistry was to be rewritten. In 1937, they synthesized a variety of very different estrogens and progestins. The goal was to synthesize known steroids and distinguish themselves from the globally competing academic labs like that of Ružička in Switzerland. The known chemistry of steroids was to be turned upside down and every carbon atom was the subject of considerations of what could be changed here.[81]

A scientist who is almost forgotten today also made a significant contribution to this chemical revolution at Schering. Ming-Long Huang (1898–1979, Fig. 17) a Chinese man who studied in Europe. In our modern world, it is completely normal and no longer noteworthy that foreign, and particularly Chinese, young people study or earn their doctorates here. In the 1920s, a Chinese man in Zurich, where he began his chemistry studies, as well as in Berlin, where he finished his studies, would certainly have stood out. In the yearbooks of the old Kaiser-Wilhelm-University, 200 registered Chinese men and women are officially listed. There was also a tiny Chinese quarter in Berlin, if you could even call it that, namely a 200-meter stretch of road at the Schlesischer Bahnhof. Not necessarily students lived there, but small traders, merchants who had been settled in Berlin for a longer time. Life was hard and most Chinese were impoverished. They lived off food stalls and worked in laundries and vegetable shops. A second Chinese street was Kantstraße in Charlottenburg, where, among many Russian shops, the only Chinese restaurant in Berlin could be found. In 1923, Mr. Lien, a former cook at the Chinese embassy, opened the China-Restaurant named Tientsin. Ming-Long Huang probably lived in the area around Kantstraße, as there was also an office for Chinese students at Kantstraße 118. The Prussian police always looked suspiciously at their small Chinese community, as many were suspected of being left-wing ringleaders, as the fight between the Communists under Mao Zedong and the Kuomintang under Chiang Kai-shek erupted in China in the 1920s.

While academic Chinese people, who were also enrolled as students, lived in Berlin, the small Chinese community in Hamburg's Schmuckstraße looked different. When ships from China arrived in Hamburg, the Chinese crew members were completely exhausted from working at the boilers. For weeks they had been working as coal haulers, stokers, or greasers, only the better crew members were lucky enough to work above as kitchen helpers or auxiliary waiters. When they were brought to the Landungsbrücken on Hafenstraße by ferry 7 after arriving in Hamburg, they did not take the path to the brothels on Herbertstraße. They went to Schmuckstraße, where almost exclusively Chinese people lived at that time. The Chinese from overseas enjoyed their stay in the Hamburg opium dens, where they also indulged in gambling (Fig. 16). Schmuckstraße was Hamburg's Chinatown, and a police report from 1923 summarizes what was there besides food stalls and laundries: *"The police are aware that there are a number of opium dens in Hamburg, in which not only the numerous coolies and other Chinese*

Fig. 16 Opium-smoking Chinese in a Hamburg opium den

who are staying in Hamburg, but also Japanese and Germans indulge in the enjoyment of this poison. The police managed to locate two of these dangerous places, namely Hafenstraße 126 and Pinnasberg 77.'[82] In Schmuckstraße, it was Mr. Ko Yen Kow who was betrayed by a tip. In 1923, larger quantities of narcotics were confiscated in his basement pub. The police listed 59 cans of opium in their report, but also army revolvers with ammunition. Mr. Kow had to go to prison for six months and pay a fine of 115,000 inflation marks. The peak of Chinese drug smuggling, which had already caused a stir in Hamburg's high society in 1921, was the smuggling of heroin in cans hidden in hollow gravestones from Shanghai. Commissioner Bley of the harbor police found that the gravestones were too light during a check and noticed a very thin seam at the bottom. The stones were promptly opened and 457 cans each containing 250 g of pure heroin were found. At a street sale price of two marks per gram, the Hamburg police pulled heroin worth 228,000 marks out of circulation. Drug smuggling remained a daily business for the Hamburg officers then as it is today, and the city was the gateway for all kinds of drugs into the German Reich and Eastern Europe. With the downfall of the Weimar Republic, Chinese life in Berlin and Hamburg ended. The Chinese communities were dissolved under the National Socialists, many died in concentration and labor camps. Nazi Germany waged war against the People's Republic of China and the Chinese living in Berlin and Hamburg were accused of espionage and fraternizing with the enemy.[83]

Among the Chinese studied in Germany was Ming-Long Huang (1898–1979), born in Yangzhou, who traveled the world as a ship steward with his father.[84] His older brother Shengbai inspired him to learn German, and he himself, who only went to school for six years, was a great autodidact who taught himself the knowledge of his time. His older brother and mentor sent him to Europe with German prisoners of war from the USA after his pharmacy studies in 1919. He accepted a doctorate in Berlin, which he successfully completed in 1924. As a doctorate chemist with a medical background, Ming-Long Huang was well trained to return to China after his studies and doctorate in 1925 and become the director of the Zhejiang Research Institute and a professor of chemistry. A meteoric rise in his homeland, considering the fragile and high career ladders in Europe. His 10-year older brother, who also became a professor of medicine, encouraged him in the 1930s to take a sabbatical to Germany with his second brother Mingju Huang. He liked the idea and embarked with dried plant material of the plant *Sen-Hu-So* in his suitcase, which we know better as *Corydalis ambigua* and which contains alkaloids. During a guest professorship from 1937 to 1940, he met Hans Herloff Inhoffen and worked with him and colleague

Zühlsdorff on the synthesis of estradiol. Inhoffen, who copied a partial reaction to estrogen synthesis from Windaus, asked Ming-Long Huang to continue it. It was later included in the chemistry textbooks as the Huang-Minlon reduction. The young Chinese enthusiastically accepted the offer and further developed the partial synthesis. It is also of interest that Carl Djerassi was pursuing the same synthesis route as a student under Alfred L. Wilds (1915–2002) at this time.

With the outbreak of World War II in 1940, Ming-Long Huang had to leave Europe and returned as a part-time professor to the Southwest Associated University and later as a researcher at the Institute of Chemistry of the Chinese Academy in Kunming, a city in southern China. Due to the turmoil of war and poor supply situation, he changed his research direction and isolated natural substances such as alpha-Santonin from plants he bought at the market, and developed chemical rearrangement reactions. After the war, he was drawn to the USA, where he again became interested in steroids like cortisone and its chemistry. He was a great chemist and significantly shaped Chinese steroid chemistry in the 1950s. He also made political appearances, as China wanted to slow down the rapid population development in its own country. Ming-Long Huang became deputy head of the family planning specialist group of the national science committee. The synthesis route to oral contraceptives was simplified at his institute in Shanghai and the industrial production of oral contraceptives was initiated. He can rightly be seen today as the father of steroidal contraception in the Middle Kingdom (Fig. 17).[84]

The extraction of estradiol or the still unknown progestogens was limited at Schering, as well as at colleagues Doisy and Allen, to isolation from the urine of pregnant mares. Schering was in a conflict, because they wanted to isolate the hormones themselves, but in Göttingen Butenandt was waiting, who was also promised an extract for isolation and further processing. Allen was clear that a large amount of urine had to be extracted, because the estrogens and progestogens were simply too low in concentration. The classic method so far was to evaporate the urine to an oily solution. This technique consumed vast amounts, as the evaporation of water is expensive. Alternatives were sought and the breakthrough was brought by organic solvents such as acetone, toluene or chloroform, which could be purchased in large quantities for the first time at a low cost. If solvents were still used for the production of explosives in the First World War, the market fell away after the end of the war and the recession in the young Weimar Republic demanded that the solvent manufacturers look for new markets. There were

Fig. 17 Hans Herloff Inhoffen (l) and Huang Minlon (r), Schering AG, Berlin, around 1936

now cheap solvents that offered a better alternative than the laborious and foul-smelling evaporation of urine in high vacuum. The amounts of isolated steroids could be significantly increased, they not only helped Schering to higher productivity, but they allowed Butenandt to catch up with his time lag to Doisy and Allen, who worked with counter-current extraction, and still get the Nobel Prize.

But what was actually the first solvent in history and what role do these silent helpers play in process engineering? This question also arose for the author, as the brown bottles with acetonitrile or toluene are always taken for granted on the laboratory bench. The first solvent in history was certainly water, especially hot water, which was appreciated for the extraction of plants for extracts and later tea or coffee. The first organic solvents suddenly appeared in chemistry and one has to think hard about which one was first used in larger quantities in its pure form. It was certainly alcohol, which could be distilled. On a technical level, it was benzene or benzol, as it is called today, which was described by Faraday in 1825 and technically used and probably came from coal tar production.[85] With the rectification of petroleum and the emerging chemical industry, our modern organic solvents were provided in large quantities. From helpers in the lab, they also became problems. Today, organic solvents account for 80% of the waste from chemical laboratories and inspire generations of chemists to think about green solvents.[86] The model is once again nature, which allows trillions of chemical reactions in an aqueous solvent, usually at room temperature, normal pressure and a neutral pH value.

From the mid-1930s, the structure of estrogen, progesterone and testosterone (Fig. 6) was known and sufficient quantities were made available for research purposes. The synthesis of estrogen was still too difficult, as the introduction of an aromatic ring did not work well methodologically. The synthesis of progesterone as a gestagen should be easier, thought Bernhard Zondek (1891–1966), who noticed early on in one of his publications that the chemical similarity of testosterone and progesterone is high. Both substances differ only by two carbon atoms and this puzzled him. Compare the structural formulas in the profile once. Even without deep chemical knowledge, you will surely notice in comparison to all the other already seen structural formulas that with lucky chemistry, testosterone could perhaps be made into progesterone. This was also recognized by Inhoffen and he read in the patents of his two colleagues Arthur Serini (1897–1945, Fig. 18) and Lothar Straßberger (1902–1968), that *"substances with the effectiveness of male sex hormones can be converted into those with the effectiveness of female sex hormones".*[87] Inhoffen could even see a picture in the German Reich patent 730.050 that the two colleagues made ethinylestradiol from estradiol using ethylene.[87] This was exciting and also the hint that with *"peroral administration a considerably greater effectiveness* [of ethinylestradiol, author's note] *than estradiol"* was seen. It was a pity that the yield was a meager 3% in the synthesis. A year later, another Scheringian should solve this problem. Dr. Josef Kathol filed as inventor for Schering on November 22, 1936 a patent in the field of ethinylation, which showed modifications of the synthesis. The discovery that liquid ammonia was excellent as a solvent and reagent led to an increase in yield of up to 90%. Kathol was very well-read and understood from the publication by Edward Curtis Franklin (1862–1937) "The ammonia system of acids, bases and salts" [The ammonia system of acids, bases and salts],[88] that the sodium acetylide of his two colleagues would dissolve much better in liquid ammonia. Despite the great interest in the history of ethinylation, who did what when, no significant facts can be gathered today

Fig. 18   Arthur Serini (1897–1945), around 1940

about the discovery of Josef Kathol. Many files were destroyed during the Second World War and the German Patent Office also routinely destroyed their documents and administrative procedures. Only the patent texts seem to have survived all these upheavals.

The ethinylation was not only a real breakthrough in chemistry (Fig. 19). In the journal Naturwissenschaften, Kathol, Logemann, and Serini published that a transition from a male to a gestagen hormone is possible with the help of ethinylation. In 1938, Inhoffen and Hohlberg added an article in the same journal with the title "New orally active female gonadal hormone derivatives: 17-Aethinyl-oestradiol and Pregnen-in-on-3-ol-17" about the outstanding effect of ethinylation in castrated rats, but they were very restrained in the methodological description.[81] The fear of competition was too great. Schering did not want to spread the chemical trick of ethinylation to the world, imitators had to be deterred. This is interesting because none of this information can be found in publications of the time. Only in the US patent of Inhoffen and Hohlweg was the synthesis and the phenomenal

Fig. 19 Estradiol pilot plant at Schering AG, Berlin, 1935. In the background is laboratory assistant Johannes Schultze, who carried out the ethinylations for Inhoffen

yield of 90% referred to, and this should not be overlooked by a Carl Djerassi.

The ethinylation of estradiol solved the major problem of bioavailability. The introduction of this acetylene group, certainly unusual in medicinal chemistry, allowed the administration of very small amounts that were excellently absorbed, had good physiological tolerance, and had better efficacy in the body. But where would the large amounts of steroids come from that could be ethinylated and sold? The need was certainly a few hundred kilograms per year and not in the upper ton range as today, but even this small amount was a challenge. The young Reichstein in Switzerland showed a technical process from bile acids that involved twenty steps in the chemical transformation. The biotechnical conversion and targeted manipulation of microorganisms, which shortened the synthesis to only eight steps, had not yet been invented and would not make production economical until the 1950s. Compared to industrial production today, in 1936 only 20 mg of progesterone was obtained from the uterine tissue of 50,000 goats or 200 mg of cortisone from 20,000 cow brains. The synthesis from cholesterol was too expensive, it needed a completely new approach.

About Black and Bright Heads

A man was to solve the problem and go down in the history of steroid chemistry as an eccentric and brilliant chemist. He was a scientist, of whom Carl Djerassi said that he deserved the Nobel Prize, but never received it. This highly praised chemist despised his profession, called his colleagues scoundrels, appeared on the front page of the New York Times in 1949, was described by his biographers as the father of the Mexican steroid industry, and out of frustration destroyed all laboratory records, letters, and prints of his 213 publications. This eccentric was Russel l E. Marker (1902–1995, Fig. 20) from Hagerstown, Maryland, in the USA.[89] As the son of a farmer, he saw the poverty and the hard work of his father as a tenant. He swore never to become a farmer. His father had a completely different opinion on this matter and did not want young Russell to leave the farm. His strong-willed mother showed her husband his limits and sent the boy to high school, which was four miles away by train. *"Just away from farm work"*, was his answer in an interview to the question of how he endured the efforts for so long. He persevered and finally wanted to go to university. *"Initially, I decided on a career as a chemical engineer. […] After the first year, I switched to regular chemistry. I didn't even know what a beaker and what a test tube was. I had to ask the man next to me at the table because he had learned something like that in school."* Marker struggled as a young student, bought a chemistry book, studied all summer, and solved all the tasks. *"Organic chemistry became a piece of cake"*, he remembered. Already in 1923, he completed his bachelor's degree and two years later his master's degree. He stubbornly ignored his father's letters urging him to finally return to the farm and *"make the best of it"*. This stubbornness and his unyielding spirit of resistance were traits

Fig. 20   Russell E. Marker (1902–1995), right image person on the right

that would make the young chemist more than well-known among colleagues. Incomprehensible to almost everyone, he dropped out of his doctorate because he believed that this annoying compulsory subject of physical chemistry, which was essential for the doctoral examination, was a waste of time. He told his doctoral father Morris S. Kharasch (1895–1957) that he had taught himself the subject matter and that every minute outside the laboratory was lost time for him. What a stubborn head, Kharasch must have thought. Now he's throwing away his scientific career because of an exam. He had no idea that this resistance had to do with the difficult relationship with Marker's father. He offered him another year of scholarship and warned him that without a doctorate he would only perform urine analyses as a laboratory assistant until the end of all days. Marker remained stubborn, he dropped out of the doctorate and left the university in June 1925.[89, 90]

Marker moved to New York, got married, and worked for the Ethyl Gasoline Corporation in a garage in the Yonkers district. The Ethyl Corporation wanted him to develop an anti-knock agent for gasoline. Because the automotive industry was booming and modern fuels had to be developed. He successfully solved this task, and we still know the term octane number and the system he invented for its evaluation. But he quickly became bored again and switched, surprisingly without a doctorate, to the prestigious Rockefeller Institute, today's Rockefeller University, where he sometimes stood in the lab for two days in a row and published thirty publications in the first year. It was a tradition for researchers at Rockefeller to go to the library for two hours after lunch to read the latest journals from Europe. Marker was particularly interested in contributions to steroid chemistry, which attracted great interest on both sides of the Atlantic. He read about the work of Nobel laureates Windaus and Butenandt, who were in competition with Allen and Doisy. He quickly realized that obtaining larger quantities of steroids would pose a huge problem. He was amazed, this progesterone was so expensive that not even a Rockefeller Institute could afford it. If anyone paid these astronomical prices, it was the racing stables to improve the fertility of their world-class racehorses. Doped like a racehorse, he felt called to research a synthesis. Germans and Swiss had good ideas, but no one was successful.

After the lunch break, he went to his boss, who was already on his way to the library. "Phoebus, do you have a moment?" Phoebus Levene (1869–1940) stopped abruptly. His superstar was standing in front of him. He couldn't just brush him off. "Hello Russell, what's up?" He wiped a crumb from lunch from the right corner of his mouth. "Are you also reading the reports about this new field of steroids and hormones? It seems to be interesting. Shouldn't we research these substances?" Levene flinched. He briefly considered what these steroids were again. In his world of nucleic acids, they only appeared on the periphery. "Russell, that's interesting, but how do you plan to extract the substances?" The answer came promptly. "From plants." As if that were self-evident. "Russell, that's not possible. Walter from the pharmacology department has already tried it, and he told me that it's impossible, without a doubt." Marker became angry. "That's nonsense. You mean Walter A. Jacobs? He doesn't know what he's doing. I'll try it again." Now Levene also became angry. Why couldn't this Marker accept authority? "Mr. Marker, we're not doing that, and besides, plant extraction is clearly assigned to the pharmacology department organizationally. You have no business there."

Marker couldn't stand this patronizing and turned to Simon Flexner (1863–1946), the director of the Rockefeller Institute. He had an idea of what was coming and invited the two for a discussion. "Mr. Marker, Dr. Levene, you both asked me to hear your dispute and share my opinion. I have dealt with the matter in advance and, like Dr. Levene, have come to the conclusion that extraction belongs in the field of pharmacology. I think…" Marker interrupted him. "Does that mean I'm not allowed to work on the steroids?" Marker became restless. "No, we just want to maintain order in our institute. Not everyone can do what they want. What would the colleagues from pharmacology think?" Marker couldn't accept that. "If it's the case that I can't work in the field of steroids here at Rockefeller, then I'll find a place elsewhere where it's possible." Now Flexner became angry and slammed his fist on the table. *"No one just leaves the Rockefeller Institute. What do you think you're doing? It's a great honor to belong to the Rockefeller. No one just leaves, unless their contract expires or they're fired. You will work with Dr. Levene, no matter what he says."* There was silence in the office. They heard the faint sounds of the first cars on the Upper Eastside penetrating the office. Marker had already resigned internally at that moment.

He wrote a letter to his old patron and supporter "Rocky", better known as Frank Whitmore (1887–1947), who was his colleague at Ethyl Corporation and now Dean of Physics and Chemistry at Pennsylvania State College. *"Whatever may come, he would support him and offer him a job."* He did, but financially and in terms of the lab's equipment, it was a step down. Marker had an excellent lab and a salary of 4400 dollars and an old lab with a kilogram of contaminated cholesterol in the basement. Marker was disappointed, but he switched anyway and fully immersed himself in steroid research. He was a man of action and in a short time, he managed to isolate 35 g of progesterone from the urine of pregnant women. A huge amount, which he brought to Parke-Davis in a glass bottle. The American company was already financing part of his work and was grateful for the offer. For 1000 dollars per gram, they bought the progesterone and were able to sell 15,000 doses of 2 mg each to women who regularly suffered miscarriages and could be cured with small doses of progesterone.

Marker saw how much work was involved in the isolation and could count on five fingers that isolation from urine didn't bring much. The Germans and the Swiss had also recognized this. And what about the plants? Walter at Rockefeller had looked at it, he thought it wouldn't work. This cholesterol, also called sapogenin in the plant, was quite simple to process. He just had to remove the side chain. Some work, but not impossible. He just needed enough material and he started with a few kilos of sarsaparilla plants that friendly colleagues had sent him.

Marker and his scientific assistant Ewald Rohrmann first isolated the Sarsasapogenin and worked their way through countless regulations like cooking instructions, which the Germans and their colleague Louis Frederick Fieser (1899–1977) had published. The samples were heated, mixed with acids and bases, and sent to New York to the Rockefeller Institute. Two weeks later, the analytical results from a primitive spectrometer came back, which both were thrilled about. The degradation of the side chain was successful. They were on the right track, and the successful trick was to heat sarsasapogenin in a sealed test tube in acetic anhydride to 130 °C. This Marker reduction is still used today and laid the foundation for the later multi-billion dollar industry.[91]

However, the chemical mechanism was not fully understood, and Marker's publications provoked resistance from major steroid chemists such as Louis Fieser (1899–1977, Fig. 21).[92] As a reviewer for submitted manuscripts at the scientific Journal of the American Chemical Society (JACS), Fieser had Marker's contributions particularly on his radar. Marker believed that all his forty manuscripts submitted to JACS were personally reviewed by Fieser. The successful author was invited by Arthur Lamb, the editor of the journal, to Harvard to get to know each other. On this occasion, he could also meet his colleague Fieser. Marker liked the idea and set off for Harvard. Fieser and his wife Mary invited Marker to dinner, which he gratefully accepted. In the end, however, Marker sat alone at the table with Mary, as he had contradicted Fieser's opinion on degradation in the preceding scientific discussion and remained stubborn. *"Fieser got so angry that he left me to go to dinner with his wife alone,"* Marker later recalled. However, he was to be proven right and find a second successful way of degrading the side chain. This path was all the more successful as testosterone could be obtained

Fig. 21 Louis Fieser (1899–1977), 1965

directly, which could be converted into estrone. This molecule, over which a competition between the Germans Inhoffen and Butenandt was raging, was now a simple synthesis in his laboratory and only one step away from estrogen.

"*We hereby confirm that the recipient of this letter, Professor R. E. Marker, is working on projects for hormone research that are closely related to national defense. [...] The most promising expectations are tied to certain plants in Texas. [...] Since these works are of utmost importance not only for national defense, but for science in general, we hope that you will find suitable means and ways to support Professor Marker in every conceivable way in procuring one ton of the following plants: [...].*" A list of the plants that Marker would have liked to have followed. With this letter of recommendation from his dean in hand, Marker traveled to Arizona, North Carolina, and Texas in search of sarsaparilla plants that promised to deliver him a sufficient amount of the plant steroid. Two unusual words in the letter of recommendation are interesting. Russell E. Marker became a professor and even received a doctorate. On the other hand, the word "national defense" is surprising, but the USA was not yet in World War II and the letter of recommendation was written one month before the Japanese attack on Pearl Harbor.

Marker was on the road with the letter, a doctoral student, and a retired botanist. The trio found many plant species and he later isolated many hitherto unknown sapogenins from the found plants in the laboratory. In the past, it was the privilege of the researcher who first discovered a new natural substance to give it a name. Today this is no longer common and one tries to strictly adhere to the internationally agreed IUPAC nomenclature. Marker was particularly imaginative and creative in naming. Thus, he named a sapogenin Pennogenin in reference to Penn State College, another Kammogenin after Dr. Oliver Kamm, who supported him as research director at Parke-Davis. This was followed by Markogenin after himself and Rockogenin after Dean Whitmore, whose nickname was Rocky, as well as Fiesogenin and Nologenin after his two colleagues Carl R. Noller and Louis Frederick Fieser.

He lay on his bed in the house of the retired botanist in Texas, where he stayed overnight, holding a book about the flora of Mexico in his hand. The night promised no cooling and he was browsing a book from his host's private library. He was interested in the yam genus *Dioscorea* and read in the book about the approximately 600 species in the family of yams roots. In a poor black-and-white photograph, Marker saw a huge root, which was called Cabeza de Negro in Mexico. He felt that this *Dioscorea mexicana* (Fig. 22) was the plant he was looking for. In the description, he read that the root could weigh up to 50 kg, and the piece shown was in the province of Veracruz in Mexico, where the road from Orizaba to Cordoba crossed the small Rio Jamapa. The next morning he called Whitmore. "Dean, I have now found the place in Mexico where I can find the plant to extract sapogenin." "Where, about in Mexico? You want to go to Mexico? That's not possible. We may be at war with the Mexicans soon. Stay here." "No, I will go and I need money. Send it to me." What times, that the dean of one's own faculty sent the professor money for research on call. In our current university landscape, this is unthinkable. Marker set off alone to Mexico City and the conductor on the train named a hotel where he could stay overnight. Upon arrival, Marker was surprised by the present guests, as he heard neither English nor Spanish, but another language unknown to him. It was the year 1940 and he would later learn that it was German and here the German spies had stayed during the war. He wanted to go to the capital to get a permit to search for and extract the *Dioscorea* plant, but it was delayed indefinitely. The authorities gave him a botanist who did not speak English, which is why an interpreter and the botanist's girlfriend set off with him. Still in the capital, Marker was very emphatically told that he should better go back to the USA because of the upcoming war, but he remained stubborn. The further it went to Veracruz, the larger the mostly verbal,

Steroids and the Pill   169

Fig. 22 *Dioscorea mexicana* 1837 (public domain, Wikipedia)

anti-American heckling against him became, which scared the botanist and he doubted whether it was wise to follow the Gringo. The journey on the country road by car or truck was abandoned and the small group returned to Mexico City. Without permission, but with a spirit of adventure, Marker alone got on a rickety regional bus and drove south overnight to Puebla, changed buses again and arrived in Veracruz. Marker remembered: *"There were pigs lying on the floor* [in the bus, author's note] *and the woman next to me had chickens with her. As it got light, I came to Orizaba and found another bus to Cordoba. When we crossed a stream, I asked the driver to let me out. In a small shop there, I was able to more or less make it clear to the owner, a man named Alberto Moreno, that I needed Cabeza de Negro. He recommended me to come back tomorrow. The next day he had found for me the first plants I wanted to collect."*[93]

Alberto Morino did not disappoint him. He brought the Cabezas, which weighed 2 to 5 kg. The Mexicans called the plant Cabeza de Negro or black head, because it looked as if a black man was buried up to his head in the ground. The plants seemed to grow everywhere and in infinite quantities, as there was no real use for them. Occasionally, a root was taken, chopped into small pieces, and scattered into the Rio Jamapa to fish. The root contained a fish poison that prevented the fish from breathing, and the locals could collect the dead fish a little further downstream with simple nets. Alberto told Marker that these Cabezas would be about 25 years old. Alberto stuffed about 30 kg into two of Marker's bags and shipped him and the plants on the next bus back from Veracruz to Orizaba. When the bus arrived in Orizaba, Marker could not find his two bags, his Carbezas were gone. With 10 dollars to the policeman, one of the two sacks miraculously reappeared. Marker, who did not speak Spanish, gestured that the other sack was missing. The policeman demanded another ten dollars, which Marker no longer had or wanted to give. In the end, he continued with one sack, which he did not let out of his sight until he reached his lab at Penn State. He extracted, repeated the synthesis exactly as he and Ewald Rohrmann had tried it, and he was successful. Progesterone could be obtained from this Mexican plant. Marker went to Parke-Davis Parke-Davis in Detroit, who paid for his research and demonstrated the synthesis once again to a large audience. The enthusiasm quickly turned into great rejection when Marker suggested, *that the huge masses of wild-growing plants could be converted into small hormone bottles, namely in Mexico.* Excuse me? In Mexico? *"No, no, no,"* exclaimed the doctor and President Alexander Lescohier (1885–1951) at Parke-Davis. *"Out of the question! You can't do scientific work or even set up production in such a backward country."* As proof of his opinion, he cited a very personal experience. When he was on vacation in Acapulco, he had an appendicitis and due to the incompetent treatment, he was very close to death. This example alone clearly shows that such a complicated procedure can never be established in Mexico. He concluded his monologue: *"Besides, you made progesterone from bull urine and that's enough for us. We will get a barn full of bulls here. A thousand bulls, if necessary. But scientific work in Mexico is out of the question."* Marker pleaded with him and wanted to set up the production himself. Just 10,000 dollars and a room in the new packaging plant in Mexico City, which Parke-Davis was just building. No, Lescohier also decisively rejected this time. Lescohier would regret this decision, because shortly afterwards Parke-Davis separated from its president, *"because it was too late for the people to bring the lucrative hormone industry to themselves, which soon developed in Mexico without their participation,"* Marker said later.

Marker spoke with Ciba, who was suspicious whether he and Parke-Davis already had a joint patent. Merck and Schering-Plough also did not like the idea, as they also considered Mexico to be completely unsuitable. He found no sponsor in 1942 and saw only one way, that his private account had to be plundered. He withdrew his savings and went back to Mexico on his own. In October 1942, he was again in front of Alberto's store and he agreed that he would deliver 10 tons to him. Alberto was diligent. He chopped the Cabezas into small pieces with a machete and dried them in the sun. Drying is a common procedure because in pharmacy it is preferred to extract from dry plant material. A fresh plant can contain up to 90% water, which Alberto evaporated by laying out the small pieces in the sun. Marker brought the small snippets to Mexico City, where he ground the material into a powder and extracted it with alcohol. He transported this extract as a concentrate back to the States, but not to Penn State, but to Norman Applezweig. Applezweig had his own lab in New York and made a deal with Marker that he would get a good third of the progesterone for the lab and his own work. Marker estimated that they would have about 2 kg in the end, but it turned out to be 3 kg with a sales value of 244,000 dollars. An unimaginable sum and at the same time the largest amount of progesterone in one place in the world.

Marker once again quit his job and even an offer to stay from Whitmore could not persuade him. He offered Parke-Davis to support possible patent applications and asked them to present these to him before his departure to Mexico. Only a year later, a senior employee came to Mexico to have him sign the patents. Marker was bitter and disappointed with how he was treated and refused. From now on, he wanted neither to have patents for a company nor for himself. Everything should be free and he wanted, as he said later, *"to keep the field open for anyone who wanted to enter the competition and produce, to push the price of the various hormones to a level where they would be available at reasonable costs for medical purposes."*

He went back to Mexico, a country that was considered backward in science at the time, and Russell's colleagues still didn't understand why he was seeking his fortune there. He had a vague idea, a few dollars in his pocket, and wanted to start a company. The man was a real adventurer and to start his company, he used the Mexico City phone book. He looked for a phone booth and opened the local telephone directory. With nimble fingers, he went through the book to the business pages and found a company in the chemistry section that dealt with hormones—Laboratorios Hormona. He dialed the number, but no one answered. Siesta, Marker wondered, he wanted to try a second time and after repeated dial tones, someone finally picked up. "Hello, is this Laboratorios Hormona? I would like to speak to the manager." "Perdoné, no hablo Ingles," came back. It was a stupid situation. Marker didn't speak good Spanish and the other party didn't speak English. "El jefe porfavor!" he shouted into the phone, but he was not heard. The click on the line ended the call. Sweat was pouring down Marker's forehead in the overheated phone booth somewhere in Mexico City. He tore the page with the phone number and address out of the directory and crumpled the sheet into his pocket. Then he would have to pay this company a personal visit. He took a taxi to the Laboratorios. "Jefe, esta un gringo a la puerta. Se muere por verte," the young secretary spoke through the crack of the opened office door. Frederic Lehmann sat in his office, smoking, chewing on his nails, and listening to Mozart. Annoyed by the uninvited guest, he rose from his creaking office chair, straightened his tie, forwent the jacket in this heat, and went to the door. "Do you speak English?" this man asked him. "Of course." "I found your company name in the phone book," and this sweaty guy waved the torn-out page in his right hand. "I have something that will interest you, cheap progesterone." Lehmann interrupted him. "Who are you?" "I am Marker, the steroid chemist." Lehmann thought he was crazy when Marker showed him a bottle of progesterone and explained how he made the substance from plants. He didn't take him seriously and left Marker waiting and disappeared back into his office. After a while, perhaps after a few phone calls, he came back and asked him if he was the Marker who had published many papers? He nodded. "Dear Dr. Marker, please come into my office, we can talk much better there." The tone became very friendly. "You must know that I am not the owner of this company. That is a Mr. Emerik Somlo in New York. I have already called him and asked him to come to us immediately. But it will take a few days."

Marker met with Somlo and learned that he had studied at the Sorbonne in Paris and emigrated to Mexico in 1928. There he was involved in the import/export business between the USA and Mexico with his company *Dr. E. Somlo S. A., Representantes Exclusives*. He was already trading in natural substances such as mescaline, psilocybin, hyoscyamine, and bufotenin, all of which were psychoactive substances. Somlo believed early on that there would be another war and that Mexico could no longer rely on Europe and the USA. Therefore, he wanted to have an independent pharmaceutical industry in Mexico. He placed an ad in the magazine "Die Chemische Industrie", to which Frederic Lehmann responded. Somlo went to Berlin with his wife in 1933, the two talked and became one, Frederic packed his belongings and moved to Mexico City. Frederic and his wife sailed on the *Orinoco* from Hamburg via Le Havre and Havana to Veracruz on September 16, 1933. Frederic or Frederico, as he now called himself, liked his new home. In the 1930s, the city had just 1 million inhabitants and no more than 15 million in Mexico as a whole. Born in 1889 in Marktbreit in Franconia, he set off for the New World with Hitler's rise to power in 1933, for he sensed what was coming up in Europe, as his son Pedro A. Lehmann wrote in a magazine article in 1992.[90] Lehmann studied medicine and chemistry in Würzburg and one of his visits in the 1920s even took him to the Rockefeller Institute, where Marker was working at the same time. Did they ever unknowingly pass each other in a corridor? It might have been possible, but no one could remember.

Somlo, Lehmann, and Marker decided to form a corporation, which they wanted to finance with the 2 kg of progesterone that Marker still had in the basement. The three quickly agreed, the new company belonged to Somlo by 52%, to Lehmann by 8%, and 40% belonged to Marker. To have a reference to Mexico, the first name Synthesis S. A. was discarded and Syntex was taken as a combination of *Synt* hesis and M*ex*ico. Marker traveled to New York, handed over 2 kg of the agreed progesterone to a doorman on one of his visits, which Somlo sold for 160,000 dollars and from the proceeds put 40,000 dollars as a deposit for Marker into the company. The financing was not well thought out and strange from the start. Marker received no salary and often had to ask Lehmann for 1000 dollars to pay for his and his wife's maintenance. Lehmann and Somlo were stingy and kept most of the money, which prompted Marker to first send his wife back to the USA and later to quit the job and his participation altogether. If the two had known a little more about Dr. Marker's psyche, they would never have done this and later regretted his departure. This came inevitably in May 1945.

What followed was worthy of a film and perfectly suited for a crime story. Marker, together with Norman Applezweig, founded a rival company called Botanica-Mex in Texcoco, a village about 30 km east of the capital. They commissioned root collectors in Orizaba, whose roots were always stolen overnight. Marker suspected that something was amiss. He drove to Orizaba and wanted to meet a collector for lunch. The collector fell ill and died the following day. The women employed in his lab were not doing any better. A lab assistant, who was the only one who could speak English, was brutally beaten on her way home at night. Marker took care of her and found out that it was a thug from Somlo whom the woman recognized. Marker felt uneasy and hired two bodyguards for himself. Both were armed and they stayed with him not only during the day but also at night. An attack on him was thwarted, but the attackers shot one of the bodyguards in the leg with his own pistol. Why was Somlo so upset and putting Marker under massive pressure?

Somlo was boiling with rage in the stuffy office of Syntex in Mexico City. Was this guy actually crazy? He couldn't believe it. Didn't Marker know that they had just struck a gold vein? Progesterone was worth its weight in gold and he, Emerik Somlo, owned the only company on this planet that could make progesterone from plant raw materials, but none of his people knew how. There were no patents, any halfway talented chemist could steal their idea and Syntex would be doomed. What was even worse was that the synthesis of the equally coveted cortisone from the same material was possible, and this Leopold Ružička in Switzerland was surely hot on their heels. He left the dark office and went out into the sun, looking into the courtyard. He pondered and searched for his cigarettes. He was deep in thought and didn't notice that the cigarette in his right hand was slowly turning to ash and he hadn't taken a puff yet. Damn, he was driven by sheer economic necessity. With Marker, all the knowledge disappeared. He had never taken notes or kept a lab book. All the knowledge was in his head. Even the solvent bottles and the chemicals were coded and no one in the lab could make heads or tails of this chaos. He had to talk to Marker again and invited Marker to a good French restaurant to persuade him to return to Syntex. They met at Paseo de la Reforma, talked until one in the morning, but nothing could persuade Marker. Somlo offered him 1% of the profits, but Marker turned down all offers. Then it had to be the hard way. Somlo hired a lawyer and drove with him to Botanica-Mex. He wanted all correspondence between the two of them. He didn't just demand, no, he also threatened. If Marker didn't give back his 40% of the shares to him, he would let Syntex go bankrupt. He would put him in jail and make sure he would never be able to travel to Mexico again in his life. The small office at Botanica-Mex filled with biting cigarette smoke, hot breath, and sharp threats. Marker gave in. After a conversation with his lawyer, he saw the risk as too high that Somlo, with his well-oiled contacts, would actually throw him out of Mexico. He gave in, lost his shares, 2 kg of testosterone that he still produced for Syntex, but he was free and rid of Somlo.

Peace prevailed, the employees were no longer attacked or raw material stolen from the yard. Botanica-Mex became a great success and the mastermind knew how to improve the extraction processes. He searched and found new *Dioscorea* species in the Tierra Blanca around Veracruz, which contained even more of the diosgenin and promised higher yields. Here there was the Barbasco root which he didn't tell Somlo about and he trimmed the company so much for efficiency that the price for 1 g of progesterone fell from 80 to 2 dollars in 1946. Botanica-Mex was a profitable company and there were interested parties who wanted to buy it. Gedeon Richter from Budapest bought his company in 1946 and Marker got a consulting contract for three years. In 1949, the company was named Hormosynth and shortly afterwards it was sold on to SmithKline & French, who changed the name to Diosynth. The last sale was in 2005 to Organon in the Netherlands and they became part of the AkzoNobel.

At the end of the 1940s, Marker decided to stop working chemically and to destroy his documents and publications. His last scientific contribution was his 213th published article in 1949. The behavior of this great chemist seems incomprehensible. Did his big ego stand in his way or was he not given the recognition he perhaps expected? The decision to retire from research in 1949 may sound all the stranger as the competition for cortisol was now starting and promised another billion-dollar business. Chemists worldwide set out to synthesize new cortisol derivatives. Groundbreaking work was coming up in pharmacology and physiology, but new syntheses were also being developed. Without the cortisol research at the end of the 1940s, there would certainly have been no hormones for contraception. Marker's time was not over, but he voluntarily made way for two Jewish exiles and chemists, who Hitler had driven out of Europe, and who were now to continue the steroid story.[94]

Somlo was up to his neck in water. He hastily made inquiries about where suitable chemists could be found on both American continents. He didn't have to search for long, as Hitler's racial madness had driven excellent chemists to the New World, and they were all looking for new jobs. He received names from his network and three chemists aroused his interest. Georg Rosenkranz (1916–2019, Fig. 23), S. Kauffman and J. Pataki, who all did their doctorates at Leopold Ružička's at ETH in Zurich and were proven steroid chemists. Rosenkranz had been stuck in Cuba for four years. He was fleeing National Socialism in Europe and friends had secured him a professorship in Quito, Ecuador. He was able to successfully complete the first part of the passage to Havana, but after the Japanese attack on Pearl Harbor he was stuck, as the USA did not let any ship through the Panama Canal. The Americans and Mexicans did not want to give him a visa as a potential enemy foreigner from Nazi Europe and his journey involuntarily stopped on the Caribbean island in 1941.

Rosenkranz was the ideal replacement for Marker, as he had been trained by the great Leopold Ružička and had already produced testosterone and progesterone from the sarsaparilla root in Zurich. Somlo had to act and tried to get in touch with Georg Rosenkranz, who was working in Havana for

Fig. 23 George Rosenkranz (r) and Luis E. Miramonte (l), 2001

a secretary's salary of 35 dollars. Although he was not a doctor, he worked in a laboratory for the treatment of sexually transmitted diseases, which did not particularly fulfill him. Somlo was excited and convinced that this Jewish refugee, who was also Hungarian like him, had to be the solution to his problems. He contacted Rosenberg and asked if he could imagine working in Mexico. Rosenkranz accepted Somlo's offer, even though he still had the warning words of his boss in Switzerland in his ears: *"One last piece of advice, Rosenkranz. Never touch a steroid, not even with a three-meter-long pole. Everything that could be done there has already been achieved."* Thus, even the greats of chemistry can be mistaken.

Rosenkranz arrived in Mexico by detours on August 6, 1945, the day the first atomic bomb was detonated over Hiroshima. The job interview was more than unusual for Rosenkranz. Lehmann asked him if he would recognize certain glassware in the lab and asked him to put on a lab coat and prove that he could handle the equipment. Rosenkranz did as requested and waited for the great Marker to walk through the lab door at some point. However, to his surprise, he was nowhere to be seen and Rosenkranz suspected that he was to be his replacement. He got the job and was faced with the coded bottles and chemical containers, which the employees tried to identify partly by smell. Such nonsense, that's just *"chemical archaeology"*, Rosenkranz said disdainfully. He was supposed to reconstruct the synthesis through these cryptic vessels? That was beneath the level of a chemist who had been trained by a Nobel laureate. He wanted to completely reset the synthesis.

Somlo and Lehmann expected him to synthesize testosterone and cortisol in addition to progesterone. That was simply too much work for one chemist and he spoke with Somlo about the urgent need for reinforcement. On site, Rosenkranz realized that no real start could be made with the local forces and he urgently needed more chemical adventurers. He wanted to hire Carl Djerassi, the chemistry student Luis E. Miramontes (1925–2004) and Alejandro Zaffaroni (1923–2014). Somlo just nodded and let Rosenkranz have free rein in assembling his team. It was a very good choice. The Holy Grail of cortisol synthesis was to be found by none other than Carl Djerassi (1923–2015), whom Rosenkranz had on his radar because of his dissertation. It had to be this promising young Austrian who had come to the USA as a Jewish refugee via Bulgaria and had completed his doctorate in Wisconsin with the help of a scholarship from Eleonore Rockefeller. The topic of his dissertation, which he wrote in 1945 at the age of 22, was "The partial aromatization of steroids and the dienone-phenol rearrangement". Rosenkranz was convinced that Djerassi was the right man, but how was he to bring the ambitious chemist, who was working at Ciba Pharmaceutical Products in New Jersey, into the country?

Carl Djerassi was surprised when he received a call from Mexico inviting him to visit the Syntex company. He was assured that all costs would be covered, and he was curious enough to take a look at the company before embarking on a predetermined career in the conservative pharmaceutical industry. With the task Diosgenin to convert cortisone, he found his way to Syntex in Mexico in 1949 at the age of 26.[95,96] The company was completely unknown to him and many of his colleagues from Harvard and Wisconsin, and his friends asked him if he was crazy to go to Mexico. Djerassi actually wanted to stay in science after his PhD at the University of Wisconsin, but at Syntex he was promised that he could publish his results, and that was ultimately the springboard back to a professorship in the USA. Djerassi (Fig. 24) was adventurous, wanted to learn a bit of Spanish, and

Fig. 24 Carl Djerassi (1923–2015), 2004

found the idea of working for an underdog exciting. The golden age of steroid chemistry was dominated by Schering, Ciba, Roche, and Organon, who did not notice that a small laboratory in Mexico was setting out to produce steroids and revolutionize their synthesis under the radar. The laboratory, according to Carl Djerassi, was very simple but well equipped, and most Mexican scientists did not have a PhD, but were very diligent. Carl Djerassi signed the employment contract and joined the small company.

Djerassi quickly synthesized cortisol, with which Syntex achieved a breakthrough in the pharmaceutical world. The market was huge and the estimated volume of cortisol was estimated to be several tons per year worldwide in the 1950s, which only Syntex could supply. After his first success, Djerassi wanted to produce estrogen and estrone from Diosgenin. Djerassi was really good and solved this task at a breathtaking speed. By the end of the decade, anyone who wanted hormones had to go through Syntex. The company was the global player in the pharmaceutical world like Apple or

Google today, which have a monopoly. But that was not yet the goal of Syntex, it was supposed to get even better.

With a designer hormone, as the management called it, they wanted to produce a progesterone that could be taken orally and no longer had to be injected. The painful injection was to be avoided and the patients would receive a tablet with the active ingredient. A bold dream? No, because in distant Berlin in 1938, Inhoffen and Hohlweg had shown by ethinylation that this was possible for testosterone. Djerassi knew the patents and now combined the ethinylation with his synthesis approach to synthesize novel progestins. He was amazed at how easy it was and converted the etisterone, which he had previously created himself. The effect was preserved and the molecule was chemically stable. This information awakened his inventive spirit to try something that Maximilian Ehrenstein (1899–1968) had discussed a few years ago in Helvetica Chimica Acta.

Ehrenstein grew up as the son of a pharmacist and food chemist in Göttingen and followed his father scientifically. Under the direction of Adolf Windaus, he dealt with steroids and remained loyal to the topic. He habilitated in 1931 and married his wife and pharmacist Elisa the same year, with whom he had been in a love relationship since 1925. Due to the politically very difficult circumstances in Germany, both emigrated to the USA, where they found positions in the scientific operation. He got a position at the University of Virginia and was later appointed professor at the University of Pennsylvania, where he stayed until his retirement in 1967. It should be briefly noted that it was not the same university where Russell E. Marker was once employed. Ehrenstein's interest was in the arrow poison Strophantin and syntheses to make a carbon disappear in position 19. He succeeded in this, but the isolation was difficult at the time, as there were two compounds that differed only in their stereochemistry.

What is meant by stereochemistry? In chemistry, two identical molecules can be chemically and pharmacologically different if they differ in the spatial position of their atoms or substitutes, as the chemist calls it, and cannot be superimposed. A vivid example is your hands. Put the book down and look at your left and right hand. Both seem to be the same. Now try to superimpose both hands and you will notice that you will not succeed. This phenomenon also exists in molecules that carry four different substitutes on one atom. The chemist can determine this peculiarity when he illuminates both substances, dissolved in a liquid, with light of a fixed wavelength. Both substances show their own and very specific light refraction. The observed angle is typical for the respective position of the spatial location of the substitutes in the molecule. Often, the substances refract the light to the left or right of an imaginary axis, which is why they are referred to as right-handed or left-handed. If both stereoisomers, as the molecules with the different spatial arrangement of the substitutes are called, are present in the same mixing ratio, then no light refraction can be seen. This stereochemistry plays a major role in medicine, because sterically different drugs can either not work at all or work very strongly, like the beta-blockers, or even be toxic, like the thalidomide in Contergan, which will be discussed later. The reason is easy to explain. The sterically different molecules do not fit equally into the binding pockets of the receptors. Here is a comparison from real life. If your feet were not chiral, then the right one would not fit well into a left shoe and vice versa as a binding pocket.

After his syntheses, Ehrenstein had two 19-norsteroids in front of him, which he could not separate. But in his 1944 publication, he was able to show that the absence of a carbon, which the chemist simply describes with the abbreviation *"Nor"*, increased the progesterone effect. Djerassi later said: *"It was Professor Ehrenstein's work from 1944 that aroused my interest in the 19-norsteroids and their possible medical significance."*[97] Carl Djerassi stood in the lab and looked over Luis E. Miramontes' (1925–2004, Fig. 23) shoulder. It was a warm day, this October 15, 1951, when Miramontes added some of the red chromium trioxide and formic acid to the approach of alpha-ethinyltestosterone, or better known as 19-norlestosterone. This conversion yielded an enol ether, which led to norethisterone with acetylene, potassium amylate. The young chemist, who switched from the Universidad Nacional Autónoma de Mexico to Syntex, now held a drug in his hand that was found in about half of the contraceptive pills sold in the 1950s and 1960s. The US Patent Office considered the invention of norethisterone so important that it ranked it among the 50 most important patents and placed Djerassi's name in the same line with Pasteur, Edison, and Bell.

Although the basic chemistry for the synthesis of improved progestins and other steroids was laid, research should continue without him. He wanted to develop further and Carl Djerassi left Syntex in January 1952 as a wealthy man to return to the USA. He accepted a call to Wayne University in Detroit, where he took up a professorship in organic chemistry. Carl Djerassi never completely left the industry. From 1957 he became Vice President for Research and later also President of Syntex. He no longer had to go to Mexico, because an escalating bureaucracy in research and accounting prompted Syntex to relocate to Palo Alto in the USA. The new tasks at Syntex gave him so much responsibility that he could only fulfill his second professorship at Stanford University as a part-time professor.

But that was only one life of Carl Djerassi, who saw himself as an academic polygamist according to his own words. He is also well known as an author of plays and books, which he aptly called Science in Fiction. All readers are highly recommended his book "Cantor's Dilemma". Anyone who wants to know how scientists tick, this book is a must. With the money that Syntex paid him for his discoveries, he bought a farm in California, which he called SMIP-Ranch—"Syntex Made It Possible". You can also exchange the word Syntex for steroids.

What do scientists do when they become rich with their patents and licenses? Leopold Ružička, who founded the Ružička Foundation with his money, bought old Flemish masters of the 17th century. Many of these priceless works by Adriaen Brouwer, Jan Brueghel, Joos van Cleve, Jan van de Cappele, Joannes Fijt, Jan van Goyen, Frans Hals, Meindert Hobbema, Willem Kalf, Aert van der Neer, Rembrandt or Rubens, to name just a few, can be admired today in the Kunsthaus Zürich. Not only the beautiful, the internet stars or the bankers become rich, but also some, if not many scientists can have a high fortune of up to several million to billions dollars. The richest among them is said to be the discoverer of DNA James Watson with 20 billion dollars. He is followed by the Biontech founder Uğur Şahin with 7 billion EUR, then two women and other, mostly Chinese scientists. They all have become wealthy with their spin-offs and companies in the biotech industry. As a society, we can only hope that there are more cultural polygamists like Carl Djerassi among them.

It is not entirely clear how Carl Djerassi's contacts with Margaret Sanger, Gregory Pincus and John Rock were established, but Gregory Pincus was one of the first biologists to get his hands on norethisterone and further develop the active ingredient for contraception. Norethisterone was patented by Syntex on October 15, 1951, but there was also a second patent application by a Frank B. Colton (1923–2003), who worked as a chemist at G. D. Searle in Chicago. G. D. Searle also filed a patent for the substance Norethynodrel as an orally bioavailable progesterone in his name in 1953. Carl Djerassi was never able to shake off the suspicion that they had stolen his idea, because Norethynodrel was a prodrug that was not stable in acids and quickly converted to norethisterone in the stomach at low pH. This was the Syntex product and would, in Djerassi's view, violate the patent. He urged Parke-Davis to take G. D. Searle to court, but the management declined. There was too much at stake. G. D. Searle owned the antihistamine Benadryl®, which Parke-Davis was allowed to distribute under license and was much more important than the small fish Norethynodrel.

Frank Colton also felt no guilt, because in a promotion at the University of Wisconsin, N. A. Nelson had found a new method in his work to remove the C19 in the steroid and he had already told Carl Djerassi about this in 1950. Why he would now be upset, he could not understand at all. There was no good relationship between the two and each only assumed the worst of the other. Four decades after the disputes, the two irreconcilable characters met at a symposium of the American Chemical Society in 1991 in New York and the following exchange is recorded, which they both gave at the symposium:

F. B. COLTON: *"After hearing Dr. Djerassi's story this morning, I wonder, what am I doing here? Djerassi has done everything alone in the field of oral contraception."*

C. Djerassi: *"As you present the chronology of events, your patent was granted in 1955, but the Syntex patent in 1956. Of course, the fact is that the dates of the granted patents are meaningless. What matters is the date of application. Why didn't you say: 'Syntex 1951' and 'Searle 1953'? Why did you apply for your patent only one and a half years after we first reported? Why have you never published your results, if you say it was the most important thing you ever did? While we published our results immediately! If you claim today that the chemical properties were coincidentally the same, that's nonsense."*

F. B. COLTON: *"One point for you, Dr. Djerassi. We are indeed negligent in publishing, and I'll tell you why. Syntex was a company with the atmosphere of a university institute. At the Mayo Clinic, where I worked, it was like a university. Our work was published immediately. But Searle is not a company with a university atmosphere. The more important a preparation was, or if it was evaluated as such by management, the more difficult it sometimes was to report on it publicly—for competitive reasons. As for me, there is another reason: I conducted most of the experiments myself, and frankly, I don't like to write. I simply don't like to neglect everything else for it. Looking back, however, I wish we had published more diligently."*

C. Djerassi: *"If that's true, why did you file the patent in 1953, one and a half years after we had published?"*

F.B. COLTON: *"The filing of patents was in the hands of patent attorneys. And often it happened that they, instead of immediately filing the application, sought to extend the term of the patent by first describing the compound and later filing the application when the substance was on the verge of use. It was not a sinister conspiracy. It was the way our lawyers handled it."*

Why am I telling this story? Because once again it was an incredible coincidence that Gregory Pincus and his research institute were funded by G. D. Searle from 1944 onwards. He did not know that there was a department at

G. D. Searle that was dealing with new and better progesterone derivatives. It is likely that Searle deliberately did not want to inform his client, because Albert L. Raymond, the head of Searle, was very dissatisfied with the research results that Pincus provided him. If Raymond was right, Pincus was burning money in his lab, and since 1944 nothing important had arrived at G. D. Searle. He wanted to turn off the money tap and Pincus' request for additional money for a new biotechnological process for steroid production was reason enough. Pincus wanted to use animal gland material again and let it flow through with a steroid precursor to produce progesterone.

Raymond sat in his comfortable leather chair, a cigarette in his right hand, Pincus' research application in the other. Had this Pincus completely lost touch with reality? This method had already failed with cortisone, he had already shown that before. And then the justification! Searle should bring a pill to the market that can prevent pregnancy. With the company name on the package!! He might as well declare his company bankrupt. He shook his head in disbelief. Thank God he had a real expert in the house with Frank B. Colton. He would sort it out. Before Pincus came up with any other stupid ideas, he should not tell him about Colton and the progesterone derivatives. He crushed the cigarette in the ashtray as if it were Pincus he was extinguishing, and pressed the button on his intercom. "Sweetie, can you come for dictation?"

Dear Dr. Pincus:

So far, you have brought us nothing to justify the half million in investments we have put into your work. Well, to be fair, there is still a chance that … the perfusion process will prove useful. But to date, everything you have to show in the interest of the economic goals of the Searle Company is a miserable failure full of false leads, regrettable misjudgments, and assurances on your part that have proven to be mistaken. And yet you have the nerve to ask for more research funds? You will only get more if a stroke of luck gives us something from your team that brings us profit. If I had unlimited resources, I would set up a large program in the field of steroids, but I do not have these resources, and the results so far do not justify a large program.

Yours sincerely

Albert Raymond

The money tap was turned off, Pincus had to find a new golden goose. Therefore, Pincus met with Margaret Sanger. But let's go back to that cold

December night in 1951 when Margaret Sanger, her friend Katharine McCormick, and Goody sat together. The agreement on research funding was sealed with a handshake, and Goody knew that only a few problems were solved. He had no idea that there was a man in Mexico who held the solution to his problem in the form of orally bioavailable norethisterone. He returned to Massachusetts and persuaded his colleague Chang to now deal with progesterone. Chang was initially not enthusiastic. He was the sperm expert, and constantly injecting progesterone into rabbits to see if they became pregnant was boring. He had great doubts, because many rabbits had to be killed to autopsy them and check that the eggs were not fertilized. A pregnancy test did not exist yet. Organon International introduced it in 1971 as an antigen test in Europe and Canada. The USA hesitated until 1977 because they feared the introduction would lead to a decline in sexual morality.

The antigen test also had a significant impact on environmental ecology, as previous tests were conducted in mice and later in the 1950s with frogs. With knowledge of the effects in rabbits, pregnancy in women was tested by injecting urine into a mouse. After a few days, ovulation occurred, which was checked by dissecting the animal. To avoid unnecessary killing of animals, the pregnancy test was later performed in the African clawed frog *(Xenopus laevis)*. This frog was also referred to as the pharmacist's frog, as in designated pharmacies, some of the frogs, into which urine had been injected into the lymph sacs, croaked. Female frogs began to spawn after a few hours, and the males produced sperm. The frogs survived the experiments and could be used again in cycles of two weeks. Unfortunately, the worldwide shipping and use of the clawed frog (Fig. 25) in the 1960s caused a catastrophic destruction of biodiversity, as they led to the spread of the

Fig. 25 *Xenopus laevis,* African clawed frog

fungus *Batrachochytrium dendrobatidis*, which resulted in a decrease in the population of 501 frog species and the extinction of 90 other species.[98, 99]

Chang later found interest in research on contraception and overcame his reluctance. He did not delegate the many animal experiments to a technician and carried them out himself, arguing, *"if you let the lab assistants do all the work, you lose the fun of it. Would you let someone else play tennis or chess for you?"* The animal experiments with Russell Marker's synthetic progesterone numbered in the hundreds. Anne Merrill was hired to assist Chang, and from 1953 McCormick increased the research budget fiftyfold to accelerate the research. Pincus, Chang, and Merrill caught up, but also encountered the insoluble problem that progesterone had to be administered orally in far too large quantities. But then came the lucky break that burst the knot. In 1952, Pincus met an old acquaintance and researcher from Harvard at a gynecological congress, whom he had last seen twenty years ago—Dr. John Rock from Brooklyn, Massachusetts. John Charles Rock (1890–1984) worked just a few kilometers from Pincus' lab near Boston. Unlike Pincus, he pursued a completely different goal. By administering progesterone and estrogen, he wanted to help women become pregnant. John Rock was the luminary of gynecology, his tall, slender appearance, his enormous knowledge, and his extremely friendly manner made him the godfather of gynecology, as his colleagues called him. His word was law, even if it was not correct, and his judgment had to be accepted. He brushed aside any supposed moral or ethical concerns about the use of hormones, pointing out that miscarriages, the absence of pregnancy, and the lack of a recognizable cause for infertility were a great suffering for women. His theory to help these women was the short-term prevention of pregnancy and the induction of a pseudopregnancy by chemical means. He believed the cause to be an immaturity or maldevelopment of the fallopian tubes. If pregnancy was prevented for several months, Rock believed, the body would be *"tricked into believing"* that it was pregnant, could thus get used to this situation and recover. He tested his working hypothesis with hormones on eighty *"frustrated but heroically risk-taking women"*. However, he had to be very careful in his choice of words and could not speak of contraception, as there was still a law in Massachusetts that punished the dissemination of information about birth control with imprisonment. He obtained progesterone and estrogen from Syntex in Mexico and gave it to the women, who indeed had a hormonal pseudopregnancy. Even more: At the end of the study, after four months, 13 of the 80 women actually became pregnant. This "Rock's rebound effect", as it is called in medicine, could not be clearly explained, however.

Rock exchanged this knowledge and his experiences with Pincus at the gynecological conference, and a pale Pincus sank deeper and deeper into his chair. Sweat appeared on his forehead. All the work and all the toil with the rabbits had been in vain. Rock had long since shown that pregnancy can be chemically prevented. Although there were only a few women and the objective of these small studies was not contraception due to the strict laws in Massachusetts, he immediately knew that chemical contraception was possible and that this was enough reason to expand the studies. He had to show the safety of contraception in many women, but there were huge obstacles. The moral acceptance in North American society was not given, the religious reservations were immense, the legal situation could put him directly in jail, and there was still not the perfect progesterone that could be taken in small quantities as a tablet.

When researchers often don't know what to do next, they ask for help in their network. Pincus asked his colleagues for any kind of progesterone derivatives that could be made available to him for pharmacological testing. Researchers are not vain and like to share, as they see science and the gain of knowledge as the highest good. His call found a great echo and the packages piled up with new compounds that he and Chang wanted to test. To his great surprise, his old financier G. D. Searle also sent about 200 derivatives, among them was Norethynodrel. Another package came from Mexico from a certain Carl Djerassi. He had never heard this name before. Chang was to take care of it. No sample was labeled with the hint that it might be used for contraception, but after the tests, Chang found that Searle's Norethynodrel and the Mexican Norethisterone were the best candidates. *"The chemistry of the pill he was researching had long been invented,"* Pincus later surprisingly found out. The studies in rats and rabbits were very successful and Pincus now wanted to conduct a study in women and approached Rock in Brooklyn, Massachusetts.

Margaret Sanger was beside herself: *"Rock, this Catholic doctor, was to conduct the studies on contraception? He would not dare to advocate for research aimed at contraception while remaining a Catholic."* McCormick reassured her friend, for she saw in Rock a *reform Catholic,* who *"believes that religion has nothing to do with medicine or medical practice, and that as long as the church does not interfere in his affairs, he will not interfere in hers."* She was proven right and Sanger later revised her opinion about Rock. Rock was conservative and Catholic, but his faith had been shaken at his wedding. Before he married his wife Anna Thorndike, he wanted to confess one more time and told the priest who took his confession that as a doctor he had saved children's lives through cesarean section in the past. The priest was horrified.

Only vaginal birth was just and he denied Rock absolution. The young Rock was upset. He believed that the wedding had to be called off and wanted to tell his future father-in-law that now nothing was possible. He looked puzzled and furrowed his brow. "Dear John, entrust yourself to my priest. I know him from my student days and today he is the Archbishop of Boston." John Rock accepted the offer, the Archbishop granted him absolution and left an even more insecure Dr. Rock behind. How could two priests be so different, wasn't the word of God always the same and were the rules so different from priest to priest?

John Rock signed a petition against the anti-contraception law of the state of Massachusetts in 1931. He explained to his medical students how they could prescribe the birth control pill, even though he was against its use in women. After the fall of the anti-contraception law, he inserted pessaries in women and was threatened by Catholic colleagues that he might be excommunicated. Before the invention of Norethisterone, he wrote a marriage guide in 1949 in which he educated couples about contraception. He was certainly not a radical, but a person with common sense. This John Rock now worked with Pincus to research contraception with Norethynodrel. They chose the Searle product on the grounds that Norethisterone from Syntex might have a masculinizing effect. That would be the death of any birth control pill. The truth is certainly that Pincus felt obligated to his old sponsor and patron. The project ran under the code name PPP—Pincus Progesterone Project and was later disrespectfully changed to the Pee-Pee Project, as a lot of urine from the 50 test subjects had to be collected and analyzed during the studies.

The results amazed everyone and showed the outstanding effect of the ethinylated progesterone derivatives. When the studies were on Pincus' desk, he couldn't stand it and reported directly at a conference in Tokyo about their success. Pincus did not publish the data in a scientific journal, but in the Ladies' Home Journal. This was a tabloid aimed at housewives. Certainly, it is not the scientific target group, but scientific journals refused to publish contributions in the 1950s if the fertility cycle was interfered with. Rock was the thoughtful doctor, patient and able to question himself. A few months later, Rock considered the data sufficient to present them to hormone researchers at a conference in Canada. At the Laurentian Conference on Endocrinology, he almost overwhelmed the attendees with data, facts, and pictures and came to a sober conclusion that was typical for him: *"We are entitled to assume that ovulation is interrupted, at least in a very high percentage of cases."*

Why wasn't the pill brought to market earlier? Ethinylestradiol was synthesized from estrone as early as the 1930s and tested in rats. Its effect in women was well known and clinically well documented since 1939. After the publication by Hohlweg and Inhoffen, it would take until 1949 for ethinylestradiol to be introduced to the market as Progynon C and for Carl Djerassi to discover the effective norethisterone in 1953. The reasons for the delayed introduction are manifold. On the one hand, World War II, which made resources necessary for other drugs, the political attitude of the National Socialists, who were interested in children, future soldiers, and not in contraception. In many countries, the sexual morality, which was strongly influenced by the church, also played a role. Other reasons for the delay may have been uncertainty about whether ethinylestradiol might have too many side effects and competition with the popular stilbenes. Schering had supported Charles Dodds in London in stilbene research and did not want to give up this market prematurely. It was known that stilbenes also have an estrogenic effect.

Another important aspect was the so-called brain drain, as the National Socialists drove many high-ranking scientists abroad or killed them in concentration camps because of their descent from 1933 onwards. The madness of the National Socialists was so bad in Vienna that in 1938 alone, 100 scientists at the medical faculty of the university were forced to emigrate. The conduct of clinical trials of estrogens and progestins was hardly possible under these political conditions. The Second World War also had an impact on Schering. Many employees were drafted to the front. In the first two years of the war, operations in the main laboratory continued largely undisturbed, as Hohlweg recalls. The Wehrmacht also drafted him at the end of 1943, but he was released on May 1, 1944, due to his scientific qualifications. Inhoffen survived the Second World War because he was commissioned in 1945 to go to Braunschweig to explore possibilities for a new production site after the capitulation. This exploratory trip saved his life. Unlike Arthur Serini, who was killed by Russian soldiers in 1945, he was lucky. Neighbors later told Inhoffen that Russian soldiers also wanted to pick him up.

On the package inserts of Progynon C and later Anovlar®, these were listed as medications for menstrual disorders and the absence of pregnancy under risks and side effects. Anovlar® was the first European pill to hit the market in 1961, when cosmonaut Yuri Gagarin flew around the Earth. "A historic day," the STERN headlined, and the pill was sold for 8.50 marks in the pharmacy. Although nothing about contraception was noted on the package insert, 95% of women took Anovlar® for exactly that purpose.[100]

Obtaining the pill was problematic and was only prescribed to married women if they already had two children. It was clear that the *pill* was being offered at significantly inflated prices on the black market. It goes without saying that the then company Schering in Berlin carefully considered the market launch in Germany. The pill touched on the sensitive area of sexuality and morality, and no one wanted to provoke a storm of indignation from politics and the church.[101] The assessment was cautiously successful with the first market launch in Australia, where only two Catholic doctors publicly complained, but the women were very positive. This success prompted Schering to conduct surveys among the German medical profession, and it turned out that of the 1370 doctors surveyed, only 15% showed interest and the majority had little or no interest. Only gynecologists and selected physicians were approached with uncertainty. The STERN got wind of it and published a large article about the introduction of the pill in Germany. Only now were a broad range of physicians informed, and Pro Familia also helped with acceptance, prescribing the pill.

The first pill approved in the GDR was Ovosiston®. The later approval can be explained by the procurement of the raw materials, as the import of yam roots from the capitalist abroad was expensive. Alfred Schubert (1915–2000) started looking for alternatives in 1953 and began the synthesis with bile acids from pig bile. Even in socialism, sexual morality was an important argument against the introduction of the pill, because the *"lust of women"* did not necessarily correspond to the morality of the new socialist human. The lack also shaped the production, because in 1965 there were no dragees in the GDR, as the dragee technique and the auxiliary materials were not available. A glass tube with 21 green tablets and 7 blue placebo tablets were sold, which had to be taken exactly in order. The starting material for mestranol, as the drug was called in the GDR, was estrone, which was imported. Paying foreign exchange for imported raw materials was not in the interest of the GDR. Other ways had to be found to avoid the diosgenin used in the USA and Western Europe.[101] Necessity is the mother of invention and at the Institute for Crop Research in Gatersleben, the potato *Solanum auriculatum* was grown on 3.5 ha, to Solasodin to win. Cuba was asked if wax from the sugarcane could be provided, which contained 8 to 9% phytosterols. The extraction from natural sources did not convince, which is why a total synthesis was aimed for, which the Russian J. V. Torgov developed at the Soviet Academy of Science in Moscow. In just two years, a team of 30 scientists from the University of Jena and the Academy of Sciences managed to set up a total synthesis.[102]

Does the Country Need New Men?

What are the new challenges for modern contraception? Many women do not have sex daily or regularly. However, the pill should be taken daily throughout the year. For many women, it would be interesting to take a pill for a month or better shortly before planned intercourse. Would a biochemical test also make sense, in which a drop indicates the fertile days? Thinking differently, why is there no pill for men? Can sperm function be reduced by a drug or by vaccination? When it comes to men, vasectomy, a surgical procedure for male sterility, has been discussed. However, this is more about birth control and not about temporary contraception, as the cut through the vas deferens is practically irreversible. Vasectomy is popular among Chinese and about half of the married Chinese with completed family planning are sterile. However, the figures look different in other nations. While vasectomy enjoys a rational popularity among a third of male Americans, it is rather 6% in Germany and Austria and only 1% in Italy. Do Europeans believe that it leads to a loss of desire as with castration? A loss of libido is hardly noticeable in men, as the concentration of hormones remains the same and sperm make up only 4% of his ejaculate.

Contraception plays an increasingly important role in a world with a growing population. Not so much because the absolute number of people is steadily increasing, but with it also the number of abortions, which are regrettable, painful and often lethal for women. A small calculation example may illustrate the dimension. How often the world sleeps with each other can be estimated from the annual birth rates of the WHO, and it is realistic if a maximum of 3.5 billion people all non-sexually mature children, seniors and sex muffles are subtracted. With an assumed world population of about 7 billion people, about 130 to 150 million people have sex every day. This may seem very low at first, but if you consider that Germans statistically only sleep with each other once a week, this number fits. Not every sexual intercourse leads to fertilization, which is seen in 1.2 million women every day. Unfortunately, half of these pregnancies are unwanted and of these about 600,000 unwanted pregnancies about 120,000 abortions are performed per day, of which a third are illegal. The dimension and importance of contraception can be very clearly grasped with these last numbers per day.

Male Sexual Hormones

> **Chronology of testosterone**
>
> **1889** – First description of the androgenic effect by Arnold A. Berthold
> **1921** – Explanation of the physiology of testosterone by Ernst Laqueur
> **1935** – Isolation by Organon employees
> **1939** – Structural elucidation by Butenandt and Hanisch
> **1939** – Synthesis of testosterone by...
> **1939** – Nobel Prize for Ružička

The fact that there are biologically two genders is as God-given as the Amen in the church, but nobody had an idea until our more recent times that it is hormones that shape our male and female being and consciousness. The first considerations arose in ancient Greece when men and boys were castrated. Dear men, beware, castration and sterilization are not the same. In castration, which to this day has been one of the most frequently performed operations in history, the germ glands and the testicles are removed. In sterilization, the vas deferens are cut.

The high number of castrations in history can be explained because it was a simple operation in which the testicles were cut off, chopped off, or crushed. In China, castration was offered to prisoners of war as an alternative to the death penalty, which often still led to death. The testicles of the prisoners of war were smeared with feces and bitten off by dogs. Often the men bled to death or died of gangrene, which is a type of blood poisoning. If the castration was professionally performed by a so-called knife man in China, he took this penis and scrotum in his left hand, looked the man in the eyes and asked if he really agreed with the operation. If he answered yes to the question, the knife man cut off both parts with a curved knife and treated the wound.[103] In a well-performed castration, the penis and scrotum were cleanly cut from the body, a kind of silver cork was inserted into the urethra for hemostasis and wound healing, and the man's best parts were preserved in vinegar and given to the castrate. In the case of simple men like slaves, wound care was less professionally carried out. Hot sand from the midday heat, for example, was applied to the wound to stop the bleeding. If this did not help, the castrate bled to death within a day. If he survived the first day, the chances of recovery were high, unless a life-threatening infection still carried him off. The operations were crude and technically not at the high level we know today. A surgeon needed a lot of skill and a slave trader who rather craftily castrated his male slaves in the desert of Africa could injure the sphincter or the prostate, causing the castrates to become incontinent and lose urine drop by drop. In China, as in the Ottoman Empire, the eunuchs had a kind of oversized, blunt needle with a knob that they pushed into the urethra. This could stop the flow of urine and widen the urethra, which always threatened to close due to the scar.[103]

The physiological consequence of castration is an irreversibility of fertility and hormone production. In contrast to castration, sterilization is only the severing of both vas deferens (vasectomy) in men or fallopian tubes in women. Sperm die off and are broken down by scavenger cells in the tissue. However, sterilization does not affect the man's hormone balance. Because the surgical techniques were not sufficient until modern times to specifically sever the two vas deferens, the testicles were also removed in a painful operation with sometimes high blood loss and a high death rate. The consequences of the absence of hormones were very early to be seen in the eunuchs, but only in the 17th century did one see the physiological, hormonal effect more clearly, such as a smaller larynx and a high singing and speaking voice.

In singing, the eunuchs took over the soprano voices of the women, who were not allowed to participate in church choirs. Also in musical works composed by Georg Friedrich Handel (1685–1779) or Nicola Porpora (1686–1768), eunuchs took over the high female voices in the 17th and 18th centuries. Some castrates like Carlo Farinelli (1705–1782), Domenico Annibaldi (1705–1779) or Alessandro Moreschi (1858–1922), of whom only one singing recording is known, thrilled the audience. Farinelli had a beautiful boy's voice, which the parents wanted to preserve and agreed to a castration in his youth. He toured Europe and warmed the hearts of women in particular. The depressive, Spanish King Philip V. was thrilled and had himself sung to every evening.

What happens biochemically during a castration? The production of the male sexual steroid testosterone ceases. The change from youth to man does not take place, men have a higher body size and the typical hairiness of the skin does not occur. Of course, the libido is reduced and some eunuchs became impotent, depressed, developed osteoporosis at an early age, and tended to be obese. Eunuchs could always be quickly recognized by their bulky stature, the smell of urine, hunchback, hump, and voice.

Eunuchs played an important role in history, particularly in Chinese politics. At the imperial court in Beijing, they could hold high political offices over 23 generations. They were like Admiral Zhèng Hé (1371–1435) involved in expeditions, in diplomatic service or in high administrative positions. The last surviving eunuch from the Forbidden City was Sun Yaoting (1902–1996), who was castrated at the age of eight by his own father. Eunuchs were valued because they were seen as loyal, cultured, and well-organized, and we're not talking about a few, but hundreds of thousands who were active in the Chinese civil service.

Are there still castrations today? The last castrations, which are now legally prohibited, were carried out until the 1970s on serious sex offenders and homosexuals. The most famous case was that of the Briton Alan Mathison Turing (1912–1954, Fig. 26), who cracked the Wehrmacht's Enigma code during World War II. He developed the Turing machine, which was a precursor to modern computers. Turing was one of the most talented scientists of the 20th century and was homosexual. Homosexuality was punishable in England until 1967 and Alan Turing was charged and found guilty in 1952. He was given the choice of going to prison or being chemically castrated

Fig. 26 Alan M. Turing (1912–1954)

with estrogen. He chose chemical castration in 1952. It was not until 2013 that the British government apologized and rehabilitated Turing.

The Discovery of Testosterone

The English doctor and anatomist Thomas Willis (1621–1675) recognized the physiological connections, but he had no idea what hormones were and spoke of ferments that were missing and were given from the testes into the blood. It was not until the Göttingen physiologist Arnold Adolph Berthold (1803–1861) described in 1849 in the golden age of physiology in Germany the secretion of substances with an effect on the gonads and demonstrated by the reimplantation of the testes in capons, which are castrated roosters, a regression of hormone production after. Gonads are the germ cells in which the hormones are formed and released. He was the first physiologist who not only demonstrated a hormone-like effect, he also showed that the site of action and the site of supposed biosynthesis do not have to be identical. However, the scientific community greeted his results with great skepticism. Even his boss in Göttingen, Rudolf Wagner (1805–1864) thought little to nothing of his work. Only 50 years later Moritz Nussbaum repeated his experiments in frogs and Albert Pézard in Paris again in roosters and confirmed their correctness.

Shortly after, Franz Leydig (1821–1908) at the University of Würzburg, showed that certain cells in the testes, which were later named after him, are responsible for the production of testosterone. The idea that there must be chemically different hormones was developed in close cooperation between Ludwig Haberlandt and Eugen Steinach (1861–1944), the leading endocrinologist, in the 1920s. Steinach worked in the field of human rejuvenation as a professor of physiology and transplanted testes into female and castrated male guinea pigs. He found that the females became more masculine and the castrates normalized hormonally. It became even more bizarre when Steinach described testicle transplantation as a therapy for homosexuality. In the 1920s, some homosexual men, in the false hope of no longer being homosexual, went under his knife. This intended sexual transformation did not succeed and left many mutilated castrates behind. Serge Voronoff (1866–1951) was a student of Steinach's who wanted to build up testicle transplantation as a big business. He charged 100,000 gold francs for each procedure. However, he took testicles from monkeys and transplanted them into the scrotum of men. The success was less than moderate and many scandals in Paris, where he had his practice, made him quickly leave the continent. He founded a transplantation practice in Algiers, where his patients from all over the world came to meet. When testosterone was found and made available, he believed in a rehabilitation of his ideas about testicle transplantation, but it was never proven in his sense.

One last idea of Eugen Steinach was the rejuvenation of man by clamping off one vas deferens. His theory was as simple as it was wrong, that by severing only one vas deferens, fewer sperm would be formed in favor of an increased testosterone production. His most famous patients of this therapy, which was referred to as vasoligation, were Sigmund Freud and Adolf Lorenz (1854–1946), an Austrian orthopedist. Despite some therapy forms that sound strange today, he was nominated eleven times for the Nobel Prize in Medicine, which he never received.

In the discovery of the "testicle hormone" testosterone as one of the important androgens, the scientists played an important role who also made a significant contribution to the female steroid hormones. Adolf Butenandt not only addressed the question of female, but also male hormones. After he accepted the call to the Technical University Danzig in 1931, he had developed scientifically and technically to the point where he isolated androsterone and progesterone. At a conference of Northwest German chemistry lecturers in Hamburg, Butenandt reported on a new male sex hormone on October 23, 1931, which he isolated with Kurt Tscherning 15 mg from 15,000 l of urine supplied by young policemen in Berlin. In reference to earlier experiments with extracted substances from men's urine, a masculine effect was determined in the rooster comb test. *"Since then, we have been continuously involved with the closer characterization […] ,"* as Butenandt announced in his first publication of a forthcoming trilogy in 1934/1935.[104] Success came and with three fundamental publications in the same year, the physiology and effect in animal testing (rooster comb test), the chemistry, but not the chemical structure, were clarified. Here too, Butenandt showed that he was the steroid expert and not Doisy in the USA or Ružička in Switzerland. His work and a few contributions from others from his point of view. He also replied to the *"Swiss authors"* on their publications that he does not share their opinion that the difficult accessibility to a higher availability does not allow constitution clarification. That was no problem for Butenandt, then the substance is simply synthesized. This great self-confidence of a genius reminds of Albrecht Dürer, who was asked with which brush he would paint such fine portraits. His answer was laconic: *"Give me any, I can do it with any."*

Fig. 27   Bernhard Zondek (1891–1966)

However, there were other scientists who were interested in testosterone and found the question of chemistry and physiology very exciting. At the beginning of the 1930s, Bernhard Zondek (1891–1966, Fig. 27) noticed the chemical similarity between estrogen and testosterone. He was a gynecologist in Berlin before the Nazis drove him to Stockholm because of his Jewish faith, and from there to Palestine into what would later become Israel. Zondek's merit was that he discovered the pituitary gland as the highest command center of the gonads. Although not particularly well trained in chemistry, he noticed that estradiol could simply be formed from testosterone. In a letter to the editor of the prestigious journal Nature he wrote in 1934: *"I am also of the opinion that the metabolism of sex hormones in both sexes is essentially the same. First, the male sex hormone is synthesized from as yet unknown substances and the male hormone is then converted into the female one."*[105] This letter was read very attentively by Steinach and Kun, who wanted to get to the bottom of the matter. They injected rats and later five men with testosterone and looked in the urine for how much estradiol could be found. To their surprise, no testosterone was detectable and the estradiol level rose from 36 to over 1200 rat units. How exactly this conversion worked and that an enzyme called aromatase was involved, was not published until 1962 by Bernard B. Brodie (1907–1989), who fundamentally clarified the biochemical mechanism.

As we learned for estrogen and progesterone, the synthesis of testosterone from biochemical precursors in mammals was not successful. Butenandt was able to obtain the hormone from dehydroandrosterone and Ružička from *trans*-dehydroandrosterone, but an economical process was also unthinkable here. The need for a new synthesis became increasingly urgent, as Butenandt and Kurt Tscherning had to process and extract 25,000 L of male urine to obtain just 50 mg of crystalline androsten.[106,107] For this outstanding achievement, Butenandt and Ružička were to receive the Nobel Prize together, with the work on testosterone explicitly mentioned as the reason for the Swiss.[108]

After testosterone was found and christened with this name, which Butenandt did not like at all, the race for the first synthesis began, which he wanted to win. His competitors were again Ružička and Wettstein in Basel and Zurich, whom he left behind in 1935. But it was not that easy for the master either, because the structural elucidation as a scientific finish line was not achieved by him, but by the Breslau-born Ernst Laqueur (1880–1947) in the same year, who was a professor at the University of Amsterdam and founder of Organon.[109, 110] He achieved this small miracle with 10 mg from 1000 kg of bull testes.

Laqueur was German, but he lived in the Netherlands. Before the First World War, he conducted research at the Rijksuniversiteit in Groningen. At the beginning of the war, he volunteered and survived the war without permanent damage to mind or body. He was taken with the Netherlands and sought a professorship in Amsterdam. Interestingly, he received his call and the equipment of the chair from the hand of the mayor of Amsterdam. This is quite unusual and nicely shows how the appointment as a professor differs in different countries. Thus, German professors usually receive their appointment certificates from the hand of their president or rector, in Poland the certificate is handed over by the President of the Republic, in Denmark formerly by the King, and in the Netherlands, at least in Amsterdam, it used to be the mayor, but today it is also the rectors of the universities.

But back to Mr. Laqueur. He was a German university professor in Amsterdam and was supposed to get an appropriate laboratory. The mayor was generous and provided 240,000 guilders for the conversion of a new laboratory in the decommissioned Oostergasfabriek on Polderweg in Amsterdam. The man who was very committed to the conversion was the city inspector Jacques van Oss, who later at Organon would be a partner. Two years later there was a big celebration and the new laboratory was opened in 1923. But what was he supposed to research now? Laqueur had read a year ago that two Canadians had found insulin. He was convinced that great successes could be achieved with hormones and their isolation from animal organs. But for this, the new laboratory was already too small and where were the animal organs supposed to come from. Here van Oss offered himself, because he knew a slaughterhouse operator who could certainly help. He introduced Laqueur to Mr. Arnold van Zwanenberg (1856–1941) and the two became friends. "Meneer Laqueur, can you imagine that these hormones or whatever the substances are called, could also have economic significance?" He affirmed: "Of course, they will have, but it is a lot of work and we will need many organs." That should not be a problem, thought Zwanenberg. A contract was signed with the German and with van Oss, and Organon was founded on July 9, 1923.[111]

Zwanenberg held 80% and the rest was shared by Laqueur and van Oss. Laqueur became an advisor, a member of the board of partners, and was responsible for pharmacological testing. Good years came and the company was to develop splendidly until the time when German megalomania invaded the small Netherlands. Laqueur was Jewish and suspected that he would have many problems as a German. Married to a Dutch woman, he decided to give up his German citizenship and adopt Dutch citizenship. It didn't help much, because in 1941 the Nazis forced him to resign his professorship and give up his shares in Organon, which were now managed by Schering AG. They promised him and his family not to refuse emigration from the Netherlands, but he stayed in his new adopted homeland throughout the war and survived this second war of his life. It is tragic that he died in 1947 of a heart attack while trying to help an injured person in an accident on a trip to Switzerland near Gletsch. This was too much for his heart, which, as the doctors had repeatedly told him, needed rest.

But how could it be that three groups in Gdansk, Basel, and Amsterdam almost simultaneously elucidated testosterone? It happens frequently and regularly in current research and history of science that several groups publish the same results in a very close time frame, with all the laurels always being given to the first—the winner takes it all. An important reason was the European steroid cartel between Ciba, Organon, and Schering, which had learned to share resources and profits after the expensive competition for progesterone and estrogen. As early as 1935, Ciba and Schering decided to cooperate and inform each other about progress. Both companies supplied their research group with extracts and provided their infrastructure. Only two years later, this testosterone cartel was expanded and the companies Boehringer, Chimio Roussel from France, and Organon joined.

Gossypol—the Pill for Men?

In addition to testosterone, another natural substance must not go unmentioned, which was seriously researched as a contraceptive. However, this natural substance was intended for men. Gossypol was isolated from the oil of the seeds of the cotton plant *Gossypium hirsutum* (Fig. 28).[112] It is a yellowish-red dye that was tested in China in the 1970s as a supplement to the pill for men. About 10,000 Chinese men received Gossypol in a field study to

Fig. 28 *Gossypium hirsutum* (public domain, Wikipedia)

better assess its usefulness as a substance for temporary male sterility, but almost all men remained impotent forever. But how did the idea come about that Gossypol could be the pill for men?

As so often, the story of Gossypol also began with a poisoning. In the 1960s, cotton was grown in the Chinese provinces of Hubei and Hebei and some of the field workers complained of allergic reactions and pain in various parts of the body. A Chinese doctor reported that many of the farmers sat in the shade on stones and did not want to move anymore. This poisoning took on a large scale and the political officials changed their opinion that it must not be the laziness of the farmers, but that it was a real disease. The cause was quickly found. It lay in the private extraction of oil from the seeds of cotton, which were of no interest to the collectives and were given to the farmers. The political leadership acted quickly and banned private oil extraction, and allergies and poisonings quickly declined. However, no one suspected a chronic, i.e., long-term effect of the cottonseed oil. In both provinces, the number of births also dropped sharply after a few years and many men were impotent. Some of the older doctors remembered that in the 1930s and 1940s in the village of Wang in the province of Jinagsu no children were born over a period of more than 10 years and the reason was the change in diet from soy oil to cotton oil. As in Jinagsu, the investigations showed the same picture. In many of the impotent men, the number and mobility of sperm decreased significantly. The farmers also did not recover when they avoided contact with cotton oil. The doctors were surprised because apart from infertility, the men were physically healthy and the mortality rate was not above average. The first ingenious doctors came up with the idea of using Gossypol, which was later identified as the responsible substance, as an active ingredient for the male pill. Research projects, workshops, animal experiments, and field studies followed, but the idea slowly lost its appeal and was completely abandoned by the end of the 1980s.[113] Most men remained permanently sterile and a later desire for children could no longer be fulfilled.[114]

Glucocorticoids

Brief profile – Glucocorticoids
Structural formula:

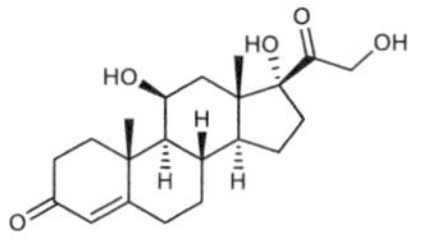

Cortisol

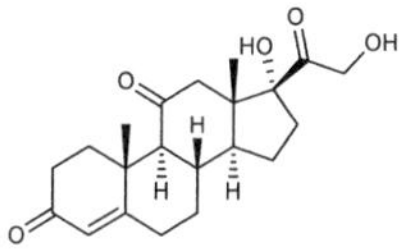

Cortisone

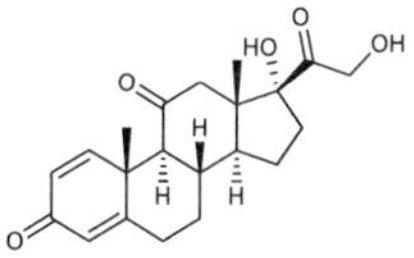

Prednisol

Modern drugs (selection):

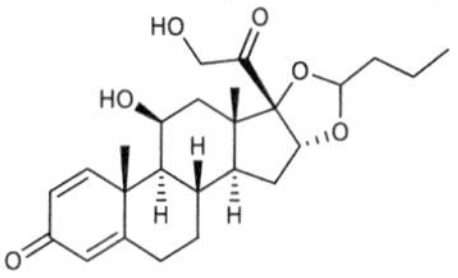

Budesonide

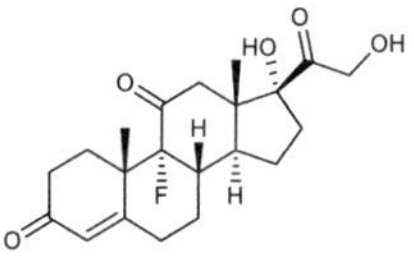

Fludrocortisone

Natural Product Group:
- Glucocorticosteroids
- Mineral corticoids

Effect:
- Anti-inflammatory drugs
- Anti-rheumatic drugs

The train station hall on this 12th of April 1934 in Zurich was drafty. Spring was forcefully breaking free from the grip of winter, and the sun was melting the remaining ice on Zurich's streets, which here and there had caused one or the other to fall during the winter. However, there was no real anticipation of spring, as the great neighbor of Switzerland was feeling world-powerful again and could become dangerous for the small, neutral Switzerland. The newspapers had been reporting excitedly for weeks about the new political Germany and the leader. The train from Basel had arrived on time, he strolled through the entrance hall and looked to see if he could already see his colleagues with whom he would discuss joint research. He saw no one on the platform. No, they must already be in the café. They had a lot planned. New research areas such as insulin, the new sexual hormones, pituitary hormones, Vitamin B1 and D, and blood pressure-lowering drugs were on the agenda. Everything was exciting for the young 37-year-old Tadeus Reichstein (1897–1996), who wanted to meet with his former fellow student, friend and promoter Gottlieb Lüscher (1897–1984), also a chemist and managing director of the Swiss food manufacturer Haco, and the chemist and pharmacologist Ernst Laqueur, professor in Amsterdam and co-founder of the pharmaceutical company Organon. He went into the café, which had been agreed upon as the meeting point. The cold April air drove out some of the smoke from the café as he opened the door and an older lady with a large suitcase was struggling to get through the door. She absolutely had to catch the train to Geneva and was already much too late, as she called out to him, apologizing. Tadeus Reichstein (Fig. 29) let her go first and stood in the door. He blinked through his glasses, which were fogging up because of the warmth, and saw a table with two older gentlemen who seemed to be waiting for someone. One of them looked directly at him and raised his arm. "Ah, Mr. Reichstein, welcome," Ernst Laqueur called out to him. "Come, sit down." He took a seat and saw that a lot of paper had already been spread out on the small table. "See, we have already started and discussed things, but surely we still need your advice." Cups of hot coffee were brought, they bent over all the paper, thought and got excited about which interesting natural substance would have the best future. The three gentlemen met to devise strategies for new modern active ingredients. Insulin was discarded, as it was a protein and the Canadians around Professor Banting (1891–1941) were already very far along and they did not know how many pancreases they would need to obtain sufficient insulin. Probably tons and could the butchers even distinguish between those from the cow and the pig, if purity was important? No, it should not be insulin. Sexual hormones? Certainly a good choice, but then only the male ones,

Fig. 29 Tadeus Reichstein (1897–1996), circa 1950 (l) and 1970 (r)

Fig. 30 Leopold Ružička (1887–1976)

as the female hormones could already be provided chemically in sufficient quantities by Butenandt and Schering in Berlin. Wouldn't that be something for the young Reichstein? He politely declined, as he did not want to quarrel with his doctoral supervisor and superior at the Swiss Federal Institute of Technology, as Professor Ružička had been researching the testicular hormone since his appointment in 1928.

Leopold Ružička (1887–1976, Fig. 30) was a professor of organic chemistry and one of the great powers against whom one should not compete.[115] He systematically built up the laboratory until 1945 with funds from the Rockefeller Foundation and had cooperation agreements with Ciba, based in Basel, which offered him an annual turnover participation of 3 to 5%

for all his research results. These conditions were ideal for hormone and vitamin research in Zurich. Both research directions led to a molecularization of biochemistry, as extracts and extracts with enriched hormones such as adrenaline, cortisol or insulin were used so far. The departure into the pure representation of hormones and the successful cooperation between a university and industry, as Eli Lilly and the University of Toronto demonstrated in the field of insulin research, moved other companies like Ciba to engage more or to found themselves, like the Dutch Organon. The company was founded in 1923 by Saal van Zwanenberg, a director of the Amsterdam slaughterhouses, and Ernst Laqueur in Oss, Netherlands. Van Zwanenberg, as a diligent Dutch businessman, had sought a way to further utilize his slaughter waste and not process it as dog food. Laqueur was the chemist who contributed the knowledge and implemented van Zwanenberg's ideas. After just a few years, Organon was experienced enough to isolate high-quality hormones and became a leader in the world. Laqueur was not a chemist who liked to deal with syntheses. He was more of a process engineer who tinkered with isolation methods, such as how a substance could be isolated in tiny amounts from a large batch of pancreas, kidneys or thyroid glands. Both realized, however, that their isolations would eventually be overtaken by efficient chemistry. The discovery of cortin, which today is a synonym for corticosteroids, was the starting shot into a new field of drug research. With a lot of money, new start-ups, as we would call them today, investors and specialists, for whom we have invented the name consultants today, the big companies wanted to have their piece of the cake. Exactly at this time, Reichstein, Lüscher and Laqueur met at Zurich station to make their cut in this steroid boom. The golden age of medicinal chemistry had come and they did not want to miss the trend of producing synthetic substances and asked Reichstein for a conversation to benefit from his knowledge. Why exactly the young Reichstein? He was well networked at that time in the leading working group of medicinal chemistry and Reichstein had an affinity for the Netherlands, as he was married to a Dutch woman and he, like Laqueur, was a Jew who emigrated with his parents after pogroms in Russia in 1906.

The discussion continued, of course, amid thick cigar smoke and coffee, and they heatedly debated. Vitamins were mentioned. *"Have you read about the ravages of rickets, that would be an important vitamin to help the world,"* Laqueur found. *"The world's oceans cannot provide so much cod liver oil, we need a different approach."* But he could not prevail because Reichstein interjected that his colleague Windaus in Göttingen had recently synthetically represented vitamins D_2 and D_3 and even got crystals. Should it be

Fig. 31 Edward C. Kendall (1886–1972)

the pituitary hormones that led to a breakthrough? It was a back and forth like pros and cons. They did not come to a good end. The last cigarette was extinguished and it was agreed that no decision could be made today and it would be better to adjourn. On May 28, the gentlemen would meet again in Utrecht and make a decision.

Just a few days after this meeting, Reichstein read a publication on April 25 by Edward Calvin Kendall (1886–1972, Fig. 31) from the USA, who was researching the isolation and crystallization of a substance from the suprarenal cortex at the Mayo Clinic in Rochester, Minnesota. He described the cortine of the adrenal gland, but he could not specify an exact structure as it was a mixture. To keep it simple, he alphabetized the substances he found. Substance E stood out and Kendall found it exciting and named it Cortisone.[116] In animal experiments, he showed that mice without adrenal glands survived a little longer after injection. Reichstein wondered, perhaps this was a new exciting field of research? He had to discuss it with Lüscher and Laqueur. As agreed, the three met in May at Organon, and they were joined by Marius Tausk, who was to advise them as the plant manager. Marius Tausk was born in 1902 in Sarajevo, Bosnia-Herzegovina, and studied medicine in Graz, Austria. His father was a lawyer in his first life, studied medicine in Vienna, was very interested in psychiatry and thus came into contact with Sigmund Freud. His mother was a women's rights activist, she was a member of the Styrian parliament and a member of the Socialist Workers International. Tausk came to Amsterdam as a young man and student because he took the opportunity of a trip by the socialist youth to their congress to introduce himself to Ernst Laqueur with a letter of recommendation from the pharmacologist Otto Loewi from Graz. Laqueur liked it and offered a guest position for a few months, which led to a doctorate.[117] After his dissertation, Tausk switched to the newly founded Organon, where his actual life's work was to begin.

Reichstein presented the cortine and the field was found to be exciting, even though the Americans had a four-year lead, but they wanted to try it. A handshake and a short note were to record everything: *"First of all, an attempt should be made as quickly as possible to obtain a crystalline hormone product. For the isolation, one should work hand in hand. No claims arise from this. If crystallization is possible and the product proves to be effective, Reichstein will be provided by Organon with the necessary quantities of the pure product or concentrations that are to serve for experiments for constitution clarification and possibly synthesis. If the work results in a usable success, Haco is involved in the result in an appropriate manner."* It can be that simple.

Reichstein was satisfied, but could not foresee that he would receive the Nobel Prize for Medicine for this deal and this new research together with Kendall. It was the right time for this project, as he had been able to complete the work on the Vitamin-C- synthesis a year earlier, and he was looking forward to a new challenge. He returned to Zurich knowing that he would not get in the way of his superior and that he now had a lot of work ahead of him to catch up with Kendall. Like Kendall, he was convinced that this cortisol or substance E, about which both knew little, would be really exciting. Organon assured their support and developed an animal experiment for testing cortin effect in rats. With this Everse-de-Fremery-test, Reichstein could quickly check all new substances. This should help him catch up with Kendall. Another improvement was a new process for processing slaughter waste.

For the first separation attempts, Reichstein only had 150 cm^3 of an alcoholic extract from 50 kg of gland material available. Far too little, considering the traces in which the cortin was present. But there was no help for it and he had to get the maximum out of it by clever processing. Reichstein evaporated the extract until a viscous, greasy syrup remained. This was rinsed with hot water and then mixed with ether to separate the ineffective and contaminating components. This was not pleasant work, as it was associated with unpleasant smells and poor handling, and the steps just mentioned were only the first of many that had to follow. It was a bit like alchemy, which Reichstein, but also Kendall applied. Reichstein wrote to Laqueur and Tausk that he would need much more. They would have to process more of the adrenal glands, otherwise they would have no chance. Reichstein now had to fight not only against the upcoming bottlenecks, but his race was also against the weather in Zurich. Tausk and Laqueur discussed how to solve the problem. The driving force was the man from Sarajevo, who as a doctor developed the Everse-de-Fremery-test in addition to the isolation. Without Tausk, Reichstein would never have been able to work so quickly towards his Nobel Prize.

In response to Reichstein's letter, Laqueur offered to send him the adrenal glands directly or to collect them from local slaughterhouses. Reichstein was therefore expecting a delivery of 1000 kg of glands from Oss, which was likely to be delayed. Concerned about his plan, which was in danger of failing, he wrote to Laqueur again: *"You were so kind as to promise me 1,000 kg of glands some time ago. [...] It would be extremely important for me to know as soon as possible when we can expect the material, in order to be able to schedule the work accordingly. We would very much appreciate it if you could send it as soon as possible, as every single working day is now important."* Reichstein rightly assumed that with rising spring temperatures and the coming summer in Zurich, extraction in the courtyard would no longer be possible and the material would rot. A cleaning stage could only be carried out at a maximum of 0 °C and there was no thought of cooled rooms in the institute.

Tausk and his chemists at Organon processed enormous amounts of slaughterhouse waste and concentrated the extracts from the adrenal gland for Reichstein.[118] This undertaking was massive and could not have been accomplished in a small university lab. A calculation example may illustrate this, which Reichstein presented on the occasion of a lecture at a medical conference in Oss on June 13, 1939: *"For example, we have processed the extracts from about 3 tons of bovine adrenal glands in total. Since the two adrenal glands of a cow weigh about 50 g, 1 kg of gland material corresponds to about 20 cows and the 3 tons to a herd of 60,000 head. Kendall in America had larger quantities available and had already processed 70 tons of gland material about two years ago, corresponding to 1.4 million cows [...]."* But how could Kendall process so much gland material? It was due to the strategic location of Rochester, where he conducted his research. The city was two hours by car from Chicago, where the country's largest slaughterhouses were located. Thousands of cattle were slaughtered every day and the adrenal glands were collected.

Nevertheless, it was a tremendous effort for both researchers, as there is only a small amount of the sought-after hormones in the adrenal gland. This has physiological reasons, as the adrenal glands do not store any hormone, they are very heavily perfused and the hormone is immediately released into the blood. Reichstein found a very apt comparison for the function of the adrenal gland and justified the high amounts to be processed with a witty and catchy image: *"If one filters an extract out of a gland during ongoing production, this procedure is somewhat like, for example, if we were to raid, destroy and rob a rapidly working car factory and finally examine what has fallen into our hands. Apart from finished cars, we would probably find a whole range of semi-finished products and other products."* To isolate just a few milligrams of an unknown substance from several tons of slaughterhouse waste was a real art. Kendall used bisulfite for separation, which was particularly useful for separating aldehydes, but drastically reduced the yield of the desired cortin. Reichstein pursued a different strategy and used the new reagent from Girard. These Girard reagents were trimethylammonium acetic acid hydrazide chlorides, which allowed much better purification and isolation.[119] André Girard (1901–1968) and his colleague Georges Sandulesco developed novel reagents specifically for the isolation of hormones. Perhaps this reagent was one of the reasons why Reichstein was able to catch up with Kendall.

In 1936, Reichstein was able to isolate a substance using Kendall's guidelines that had similarities to the competitor's substance E. Reichstein also recognized that it must be a steroid, thus motivating many research groups to be the first to give this substance E a structure. It is not surprising that similar substances were published several times in the following years. Thus, cortin was isolated three times and referred to as Compound F by Wintersteiner, and Reichstein called what he found, by analogy, Substance Fa. Reichstein was to be the winner, and in 1937/1938, cortin became cortisol and hydrocortisol, which differed only in one oxygen atom. It was a difficult path, which found its pitfalls not only in the then almost impossible isolation, but also in the fluctuating accuracy and reliability of the animal experiment, which Organon established to determine an effect. In the extract mixtures of Organon, there were too many structures that resembled the sought-after hormone and had similar effects in animals. *"They are extraordinarily closely related. They all have the same carbon skeleton and the substituents are all in principle in the same places […]"*, Reichstein complained. Organon tried to get as much adrenal gland as possible and isolated like crazy, but by the mid-1930s, even slaughterhouse waste was scarce. What was previously virtually given away for free by the slaughterhouses now cost money, as the slaughterhouses saw a lucrative business and demand increased significantly. By the end of the 1930s, the adrenal glands cost *"as much as a beefsteak of the same weight,"* as endocrinologist George Thorn (1906–2004) laconically noted in his review of the history of adrenal hormones in 1968.[120] The pressure on Reichstein increased and he saw himself surrounded by problems. The constantly rising prices of the slaughterhouses, the complex purification and extraction, which constantly produced new substances and crystals, as well as the time pressure to be ahead of Kendall in order to be able to file patents, made him nervous. Reichstein was visibly annoyed by the animal experiments at Organon, as ineffective fractions were later active again and from one extract initially only one then several substances crystallized again, to which he could not assign any effect. He abandoned the hypothesis of one gland equals one substance and discarded his previous image of the hormonal organism and thus some of his research goals.

In the 1930s, animal testing was the only way to find a biologically active substance. Modern structural elucidation was unknown, and in-vitro experiments with cells were not introduced as a standard in drug research until 30 years later. Since the animals and also the extracts behaved differently to the chagrin of Reichstein and Kendall, a strong standardization had to be pushed for. The animal testing model of Everse-de-Fremery was simple, but also built with great biological variance and was supposed to work as follows: Rats had their adrenal glands surgically removed and suffered from hormone deficiency, from which they died after a few days. The animals were kept at 25 °C and received the hormone or a test substance over 20 days. In the best case, 80% of the animals survived this period. In this animal experiment, also called a survival test, the only criterion was that the animals survived. If one now had an extract or a substance that was given to these animals and compensated for the deficiency, it was clear that this replaced the missing hormone(s). Reichstein criticized the animal experiment because he produced substances with high chemical precision that did not provide reliable data in an imprecise experiment on animals. *"It is very regrettable that the exact investigations of the chemist ultimately have to be checked by such an imprecise instrument,"* as Organon had to admit in a letter to Reichstein in 1935. The Everse-de-Fremery-Test was refined: Anesthetized rats were strapped to a board and the left hind leg was attached to a pulley. An electrode was inserted into the hollow of the knee and the animal had to pull a weight with its left hind leg after stimulation. The pulling force and the fatigue of the contraction differed significantly between the control animal and the animal lacking the adrenal gland. In this fatigue test, the decrease in contraction at the second, third, and fourth stimulation was measured. Four to five animals were grouped together per trial run and the experiment was repeated over four days. But this procedure was also fraught with errors, as the depth of anesthesia, the experimenter's experience, the size of the rat, and indeed the purity of the substances could distort the result. A fierce dispute erupted between Kendall and Reichstein, in which both claimed that the famous substance E (Kendall) or the substance Fa (Reichstein) was not active at all, as they had not been tested in their own animal experiments. Both only trusted their own experimental models.

Worldwide, physiological laboratories were looking for alternative animal testing models. Thus, the effect of cortin was tested in dogs, and in rats the cold test, the swimming test, or the glycogen test according to Verzár was applied, who recognized that rats without adrenal cortex lose a lot of water and salt when given glucose. Organon wanted to conduct all animal tests in their own laboratory, as they feared losing knowledge if their own substances fell into foreign hands. But Reichstein prevailed, knowing that setting up new tests would take forever and the lack of standardization would again lead to poor results. Organon gave in and sent Fritz Verzár (1886–1979, Fig. 32) in Basel some substances, which to everyone's delight produced clear results with only a tenth of the amount required in the

Fig. 32 Fritz Verzár (1886–1979)

Everse-de-Fremery test. Reichstein now urged Organon to make even greater use of control substances, more about whose purity and identity was known. When the standardized tests now worked well, Reichstein largely adhered to the Organon test.

What did the internal struggle at the Institute for Organic Chemistry of the ETH look like besides the competition between Reichstein and Kendall? Leopold Ružička let his assistant have his way because he led him to believe that this cortin was not a steroid. Reichstein justified this with the significantly different solubility of his cortin, which deteriorated during chemical degradation in extracts. But he knew better: *"From the sum formulas, taking into account the function of the oxygen atoms and the degree of saturation, the presence of 4 + 1 rings became apparent. Under these circumstances, it is obvious to think of possible membership in the sterin series,"* Reichstein wrote in his fourth communication in 1936.[121] He had to backtrack because it was a sterin, and he saw a problem with his boss coming up. Not only would he clash with Ružička, but Reichstein also faced the possibility of being fired. The situation could not escalate, and he could not allow what the three discussed that April morning at Zurich's main train station, that there would be a dispute in Leopold Ružička's enclosure. How was the problem to be solved? Not only Ružička, but also the Ciba standing behind him was large and powerful compared to the Haco and Organon companies. Ciba had no intention of talking to Reichstein and his industry friends. They saw him in the environment of Ružička and believed they should also have access to his research results. After all, the whole shop was financed by them. Reichstein knew that cortin and cortisone are definitely steroids, and these were his superior's specialty. He had to talk to Ružička.

"Reichstein, what can I do for you so early in the morning? You asked for a meeting. Do you have interesting news for me?" "Yes, you could say that. But that doesn't make things any easier. You know that I've been on the trail of cortin for some time now, and it turns out that it's a steroid." Ružička looked up, puzzled. "Let's sit down briefly and talk about it. You know that this is my field. I didn't think we could get in each other's way here." He gestured invitingly to the chair in front of his desk and disappeared into the comfortable leather armchair behind it. "Professor Ružička, please don't misunderstand me. I didn't want to poach. But that's how it turned out. Neither Kendall nor I believed in steroids. But it is so." "Do you know the exact structure?" "No, that's currently being worked on. I know that the basic framework is correct, but the substituents are different from testosterone or estradiol." Ružička nodded again, pondered, and shifted a little in his chair. "Dear Mr. Reichstein, this is all very nice, but also complicated at the same time. You know who we owe the lavish equipment, the biomedical research laboratory to, and who will not be thrilled if he cannot participate in the results of the research." "Yes, Professor, that makes it so difficult for me too. My research is funded by Haco and Organon. They also want to market it urgently." Ružička rose from his armchair and crossed his arms. He became serious. "I don't want a withdrawal of Ciba. That would be the death of our research and the end of my chair. I can only give you one piece of advice. Talk to Ciba, I will stay in the background, but you have to convince Ciba to find a way to apply cortisone with Haco and Organon. I also won't tolerate you passing on your steroid knowledge to Organon and foreign competitors. I wish you a nice day."

Reichstein was aware that Ciba was in a very comfortable position and had a strong negotiating position. The important patents for the synthesis of steroids and their chemistry were in their possession. On the other hand, his work on cortisone was so new that Ciba absolutely wanted to have it. They were the ideal extension of the previous product development. In early autumn 1936, Ciba approached Haco. They wanted to know if Reichstein and they were interested in a collaboration. Gottlieb Lüscher sent a lukewarm yes, they were interested, and yes, they also represented the interests of Reichstein. But they were not the right contact and referred to Organon in the Netherlands. Haco remained tactically silent, they wanted a collaboration with the big Ciba, but they did not want to scare off Organon. It was also a question of money and their own capabilities, because Haco was not a pharmaceutical company with a large financial cushion, and a large-scale production of cortisone was not in sight at all. They probably also had doubts about the performance of the Dutch partner. That the future lay not in organ extraction, but in synthesis, was clear to Gottlieb Lüscher.

But Organon hesitated. Was it due to the proverbial cunning of the Dutch merchants, or were van Zwanenberg and Laqueur actually considering setting up a chemical laboratory for the synthesis of cortisone? Reichstein became nervous. His own career was at stake with Ružička and he wanted to finally overtake Kendall with patents and work. To exert a little gentle pressure, he wrote to his patron at Haco: *"Dear Lüscher, I have thought about the Ciba matter again and have come to the conclusion that it would probably be advantageous for Organon at least to seriously consider the offer."* Reichstein argued that the synthesis at Ciba was very well established, that the precursors were already available, and that he feared that Ciba could simply prevent the synthesis of cortisone with their existing patents. Organon agreed to the arguments and declared itself ready to enter into negotiations with Goliath.

Another disagreement also existed between the parties, as the agreement between Organon, Haco, and Reichstein from the Zurich train station meeting in the spring of 1934 contained no clear terms. Should a crystalline hormone ever emerge from a synthesis, Haco should be appropriately involved. What was a utopia two years ago had now become reality. Gottlieb Lüscher's hands were sweaty. The word "appropriate" could not bring Haco any financial success. He urged Laqueur to make a concrete contract, and as quickly as possible. He wrote on October 27, 1936: *"[…] According to the current state of research, the isolation of the active ingredients has been successful, and the elucidation of the constitution is making good progress, so that we can now turn to synthesis, and we have no doubt that our attempts will be successful. If this case occurs, it is absolutely necessary that our business relations be regulated by contract, and we therefore suggest that you make a corresponding agreement today."* Organon could agree with these lines. But they had to think much more about the following lines, because what did Mr. Lüscher want to tell them? Should they sell to Ciba at a discount? More lines followed: *"[…] If one considers the financial expenditures that have been invested in the work up to today, you will agree with us that these are high and that the share of Haco-Reichenstein is rather higher than yours."* Was it so that Lüscher wanted to push Organon to hand over all rights and work to Ciba? The two Dutchmen looked puzzled as Lüscher asked them to take over the patent costs and the costs for the defense of the patent. Did he already know that Ciba would move heaven and hell to prevent the patents? In the event that Organon should succeed in setting up a production, Haco wanted to be involved in the turnover for a period of 15 years with 25%. That was quite a lot and practically covered the upcoming patent term. The small Organon, which had recently been founded, did not want to be put under pressure. After all, there was another partner in distant Berlin. They could also work in partnership with him. In the Reich capital, which had just awakened from the frenzy of the Olympic Games and its narcissistic self-love, there was Schering AG. They too had been dealing with steroids for several years. They contacted the Berliners and found that the Germans were much more willing to negotiate than the Swiss. In mutual interest, friendly letters were exchanged, phone calls were made, and they wanted to meet for contract negotiations. This development was viewed with great concern in Switzerland, and Reichstein asked in a letter to Laqueur dated November 7, 1936, not to make the wrong decision. *"By the way, I am convinced that the so far not very pleasing negotiations with Ciba are more due to the clumsiness of subordinate gentlemen and an agreement would be fruitful for you. In any case, I would be grateful if you do not conclude anything definitive with*

Schering before you have spoken to Dr. Lüscher about the matter and before it is certain that you will not receive the same advantages from Ciba. [...] In the current political situation, it is absolutely impossible for me to work directly for a foreign factory."

Whether it was wise to negotiate with Schering is doubtful. The situation was coming to a head. Reichstein, being Jewish, did not want and could not work for Nazi Germany. Ciba was snubbed by the negotiations and wrote clearly to Laqueur and van Zwanenberg on November 9, 1936: *"[...] We would like to emphasize once again that for obvious reasons we would have to refuse to make agreements with a foreign company about cooperation with a professor of our Technical University in Zurich."* This sentence speaks volumes. Ciba saw ETH at that time as an extended workbench of their company. Van Zwanenberg and Laqueur, who until this point had never personally met any Ciba employees, decided to travel to Basel and boarded the train on November 24, 1936. It was the beginning of tough negotiations that would drag on for months.

Reichstein was Organon's trump card. Van Zwanenberg and Laqueur argued that the work of their academic partner was much further advanced. They could find another industrial partner at any time. Ružička did not like these statements at all. He believed that Reichstein was not playing with open cards, was withholding information from him, and was deliberately endangering the relationship between his working group and Ciba. *"The Ružička-Ciba-Organon matter is very unpleasant and testifies to great annoyance on the part of Ciba,"* Lüscher wrote in a letter to Reichstein in January 1937. The annoyance was probably not only with Ciba, but also deeply with Ružička. For him, it was clear that Reichstein had to go. He could not allow his steroid empire to collapse. But there was also a estrangement between Reichstein and Ružička, who ironically turned the Causa Reichstein around: *"The affair with Ružička is now provisionally patched up, but still not 100% in order. Mr. Laqueur has made great efforts through his repeated telephone interventions, and Dr. Goldberg has done everything in his power."* But one must expect surprises. Laqueur and van Zwanenberg, who suspected that the negotiations would soon be broken off, rowed back and tried to mediate. There was a ceasefire, which was converted into a contract, but with which no one was really satisfied. The ideas were simply too far apart. Organon was extremely skeptical about direct cooperation with Ciba. Although Laqueur emphasized in a letter to Reichstein at the end of March 1937 that any Ciba money would be welcome to advance research, but only that from Ciba, as double financing by Haco would be problematic, as one would have to *"serve two masters"* and that would not be a good idea. Poor Reichstein was now caught between all chairs. Ciba put pressure on Ružička. He passed the pressure on to Reichstein. Ciba wanted Reichstein's change into their services, which was not possible for Reichstein because of his friendship with Lüscher.

Ciba prevailed, and Lüscher was disappointed that his friend and his knowledge turned away from him. Reichstein tried to reassure him that he would soon be able to *"make decisions more freely, without having to take too direct account of Ružička."* On May 31, 1937, a contract was concluded. Ciba was granted the right to use all of Reichstein's inventions in the field of adrenal cortex. In return, Ciba committed to giving Organon a 6% share of the turnover of all products. What was agreed for Haco? It received 3 to 5% royalty fees on all marketed Ciba preparations. Organon was granted exclusive rights to all inventions related to cortin, and Haco, as she demanded in one of her letters, was amicably involved with 25% of Organon's turnover over 15 years. This made it clear that Haco and Organon could no longer count on Reichstein: He had de facto entered the service of Ciba. Reichstein apologized in a letter to Organon in October 1937: *"You know my fundamental opinion about the whole matter and know very well that I did not agree to negotiations about it entirely voluntarily. That especially in the spring it was a matter of making the best possible compromise out of a rather difficult situation."*

The weakness of the contract would show when Reichstein took up a new professorship in Basel and new interesting steroids were discovered. The bracket of the contract threatened to burst, and the questions related to cortin and the new corticosteroids led to negotiations and discussions that dragged on for years. The contracts were bad because they only provided a good basis for the current state of knowledge in 1937, but did not regulate who should own which knowledge in the event of later application. For Reichstein, it was time to leave. He had to separate from Zurich and Ružička. At the end of 1937, Reichstein received a call to a professorship at the University of Basel, which he accepted. For him personally, there were no reasons to remain in the shadow of his former boss and Ciba. The ETH considered half-heartedly whether they should keep Reichstein and make a counter-offer, as Ružička had promoted his student to titular professor in 1934 and to extraordinary professor in 1937. Was Reichstein, when he was elected associate professor, *"undoubtedly the most significant of the young Swiss researchers in the field of chemistry"* in the minutes of the appointment on November 20, 1936, the opinion changed that they could do without him: *"We do not need Prof. Reichstein under all circumstances,"* as can be read two years later in the minutes of January 7, 1938 in the school board protocols. But despite all the praise, at ETH it would only have been about a new title and not an improvement of his material situation like higher salary or even his own laboratory.

In contrast to ETH, the University of Basel made him an offer he could not refuse. In Basel, he continued to research the synthesis of cortins and the emerging corticosteroids, as we scientifically name them today. Reichstein thought through the syntheses and tried to find new starting materials and developed the so-called etio-cholan acid series, which should be intermediates for cortin synthesis. These works were watched with suspicion by Ciba, as they feared that further structures could be found beyond the contract area, which Reichstein would offer to the competition. The fears were justified, as the chemical group of progestins was emerging and the competition in Berlin had taken the lead in this field in the meantime. Reichstein also had a small lead over the working group of Ružička, who was also dealing with the progestins, but had not yet found a good precursor for the synthesis. Reichstein wrote to Organon: *"The situation, which has already had unpleasant consequences once, is therefore still not clear."* How right he was, and he saw the shadow of his former boss looming over him: *"To explain it again, I just want to emphasize that any attempt I would make to get to oxy-etio-cholan acid in a simple way, which is absolutely necessary as a starting product for cortin-like active substances, would at the same time compete with Ružička's work to a certain extent, as progesterone could also easily be produced from this acid."*

The web of interests proved to be so intertwined that Reichstein found no real solution for all the partial interests of his partners. In December 1937, he even wrote to Kendall that he could not continue the work in the coming months. Reichstein and Haco approached Ciba. The situation was escalated to the board. The proposal to continue handing over all cortin-like results of Organon for patenting found no echo, as Reichstein and Ciba would have jointly developed the results. The conditions that had been agreed upon in May 1937 seemed to hinder each other time and again. An end to the complications was not in sight. Because how should what be regulated if the cortin-like substances became important for the production of progesterone and testosterone? In the end, a solution was found that was as simple as it was pragmatic. Ciba was to take over the production process for the etio-cholan acid series. In return, it paid Organon a settlement of 7000 francs, which today corresponds to around 100,000 EUR when adjusted for inflation. For the old Ciba, this was a good deal. The new deal came at the right time. Because more and more structures were being synthesized in Reichstein's lab. A woman was responsible for this. Marguerite Steiger (1909–1990, Fig. 33) made the breakthrough in 1937: She discovered a

Fig. 33   Hermine Raths (1906–1984) (r) and Marguerite Steiger (1909–1990) (m), 1972

substance that was five to seven times more potent than the cortisol used up to that point.

Marguerite Steiger was an exceptional woman. The native of Zurich became a chemist and was the first woman to earn a doctorate at ETH in the then Department IV (Chemistry). She did her doctorate on sugar, more precisely on ribose conformations. In organic chemistry, conformation refers to the way certain chemical groups—in the case of the sugar ribose, these are the hydroxyl groups—are spatially arranged on the carbon chain. She did this work under the direction of Ružička, so it was God's work and her contribution that flowed in here. Her work on ribose was of great importance, as Reichstein was later able to clarify the synthesis of artificial vitamin C in this way, and this would not have been possible without Marguerite Steiger. She stayed with Reichstein until 1938 and worked with him in the field of steroids, then moved to Alexander R. Todd (1907–1997) in London and Manchester, who studied the biosynthesis of vitamin B_{12} and received the Nobel Prize for it in 1957. What luck that she was able to work with future Nobel laureates twice. With the outbreak of World War II, she left England and thus also academic research. After her return to Switzerland, she founded the company Opopharma AG, which distributed the products of Organon. She and Hermine Rath had good entrepreneurial genes. Because Opopharma grew quickly and employed 190 people at its peak as a pharma distributor and generated a turnover of 80 million Swiss francs. After her death, Opopharma was closed and all subsidiaries were sold. What remained was the Opo Foundation based in Zurich for the promotion of scientific projects related to Switzerland.

Marguerite Steiger synthesized a substance that was later Desoxycorticosterone, shortly Doca, because she wanted a comparison to understand the hypothetical structure of corticosterone. The only difference was the absence of a hydroxyl group at a certain position, namely at the carbon atom with the number 11. That was the reason why Reichstein gave this new structure the prefix Desoxy. The "a" in Doca is explained by Reichstein's chemical modification of the substance by adding acetic acid to the molecule. This acetate was reflected in the extended name Doca. In experiments on rats, Doca had a strong effect to everyone's surprise. A year later, he was able to prove with difficulty: Doca was actually in the cortin mixture and had to be a natural substance. This electrified him and he wrote to Laqueur in the Netherlands that he saw a great significance for Doca, *"if it can be proven that this substance is actually a naturally occurring, thus a real hormone."* Since this laboratory substance was not seen as a real hormone with a striking effect, the euphoria did not last long. On the arduous path to the corticosteroids, it was a precursor and a step in the right direction. Nevertheless, Ciba tried to further develop Doca into a drug and launched it in 1937 to replace Addison's disease organ therapy. In 1940, Doca was approved as a drug in Switzerland.

Thomas Addison (1793–1860) was a British doctor who first reported on a disease characterized by symptoms such as skin browning at the onset of the disease, followed by fluid loss, fatigue, exhaustion, weight loss, and much more. As a doctor, he initially worked as a surgeon in London, but then turned to practical medicine. His superiors are said to have directed his interest towards skin diseases. In fact, he devoted himself intensively to dermatology throughout his entire professional career. Perhaps this was also the reason why he noticed the change in skin color in Addison's disease, the later term for adrenal insufficiency. At first, he believed it must be anemia, but he himself recognized that the adrenal gland must be involved in the disease process. He did not manage to provide a clear description of the disease he discovered during his lifetime. However, as a brilliant diagnostician, he described the symptoms and the course of the disease excellently.

The former US President John F. Kennedy (1917–1963, Fig. 34) was one of the most prominent patients with Addison's disease. The young John was

Fig. 34 John F. Kennedy (1917–1963)

rather frail and physically no giant, quite unlike his very athletic father. He spent many days of his youth and boarding school time in the sickroom, and his parents were shocked when the frail boy came home for the holidays. One of the best-kept secrets of his life and presidency was that he suffered from Addison's disease. The adrenal gland did not function properly and produced too little cortisol. Although the drug Doca was still new and expensive, the wealthy Kennedy family could afford it and John was injected with Doca under the skin. The chronic administration certainly resulted in early osteoporosis and forced John to wear a support corset due to his severe back pain. This circumstance may also have been the reason for his death in Dallas in November 1963. The first shot that hit him was not fatal according to the Warren Report, but the corset kept Kennedy in position in the open car and he could not flee into the car before the second shot killed him. Was Doca partly responsible for the young president's death?

Ciba had a clinical trial for Doca in the planning stages and there was internal disagreement about what should happen with the results. It was a battle between the scientific-technical department and the propaganda, as marketing was called back then. As is the case today in the pharmaceutical industry, scientists are not trusted as dry fact freaks to make the results interesting for the medical profession. As it was said in a meeting of the merchants, "*rational implementation requires a certain commercial instinct*". "*What is important is not only that testing is done, but that a usable test report comes out. If the testing were left entirely in the hands of the technical-scientific department, there would be a risk that the commercial side of the problem would be neglected in favor of the scientific one,*" was the clear opinion of the propaganda department. The tricky part was that they didn't know any pharmacologists or doctors. They had to rely on Reichstein's recommendations. As a chemist, he couldn't help. By chance, however, he learned that there were two cases of Addison's disease at the cantonal hospital in Zurich. The treating doctor, Dr. Koller, was willing to test the new synthetic substances. Treating two patients, as already mentioned, cannot be considered a clinical trial by today's standards. Neither Reichstein nor Koller and the doctors at Ciba were clear about the dosage of Doca. Quite pragmatically, Ciba suggested dissolving 50 mg of Doca in 0.5 cm^3 of pure alcohol. This ampoule was accompanied by a second one. It contained 4.5 cm^3 isotonic saline solution. The doctor was instructed to combine the two ampoules immediately before administration to the patient, shake them, and quickly inject the resulting suspension under the skin.

Here, a fundamental problem of pharmacy is revealed, which has not been well solved to this day, namely the poor water solubility of medicinal substances. Why is this a problem? A basic law of pharmacy is that only dissolved substances can be absorbed by the body. In today's pharmaceutical development, many natural substances as well as synthetic substances are discovered to be highly potent, but their development into a medicinal substance is not pursued further because they are not soluble in water and also not in organic solvents. It is estimated that 40% of newly discovered natural substances are therefore not further developed, which is very depressing for the therapy of difficult diseases without good alternatives. Unfortunately, we cannot do without many medicinal substances and must look for ways to make them biologically available to our bodies despite their poor solubility. The field of Pharmaceutical Technology uses many tricks and techniques here. Medicinal substances are packaged in emulsions or liposomes, the substances are ultra-finely ground into micro- or nanoparticles or dissolved in polymers. Well-known substances include steroids as well as the cancer drug Taxol® with the natural substance Paclitaxel, the immunosuppressant Ciclosporin and the microbial natural substance Sirolimus, which were given a second life thanks to Pharmaceutical Technology.

Ciba began initial human studies in 1938. The timing is remarkable, as a sufficient amount of Doca had to be produced first. Reichstein pushed, as he always felt he had to drive Ciba, which moved powerfully but slowly through the seas like a supertanker, to do everything. His fear was of Schering AG in distant Berlin, which announced in March 1938 that it would bring to market a synthetic corticosterone derivative with good clinical effect via I. G. Farben. Reichstein wrote bitterly to the Ciba management: *"The best way to prevent patent circumventions is still to bring a good product to market as early as possible."* Referring to the procurement problem, the response was brief: *"[...] we ourselves have the greatest interest in not falling behind in any way, but on the contrary in being at the forefront."* Little impressed by what the Germans were doing, they wanted to focus on their own strengths and considered how to expand the circle of doctors to include interesting patients. Unthinkable for our current ideas, an advertisement was to be placed in the newspapers indicating that they had enough medication available for interested doctors. The propaganda department found this uncharming. Because this way they would not have full control over the potential patients. Many considerations were made and discarded until Ciba sent its Doca to London to have it tested by the Therapeutic Trials Committee of the British Medical Research Council and by George Thorn in Baltimore, who was researching at Johns Hopkins Hospital and had a lot of experience with Cortin. He was satisfied with Doca, which eight of his Addison patients received. However, in a 1939 publication, Thorn critically noted that Doca could not yet replace the good old Cortin, but was an excellent substitute. It was clear to all involved that the replacement of Cortin by synthetic steroids was only a matter of time, and they had an imminent market launch in mind. But Ciba was very hesitant to bet on the new horse. Because in the Ciba sales committee, they did not know how to make money with a rare disease like Addison's disease. There was doubt and a Dr. Brodbeck asked in the minutes of the meeting whether it would be economically worthwhile to invest more money in the production of Doca. They didn't even know what this Cortin was all about. It was clear that there were more active substances than Doca in the amorphous fraction, which had to be tested in the dog test. All of this was very premature. If Doca came on the market and a better preparation was found, everything would be obsolete. Dr. Brodbeck referred to a statement by Reichstein, which he had made at a doctors' congress in Oss, where he had said: *"First, the entire side chain of cholesterol is oxidized by a brutal oxidation. [...] This is an unfortunate reaction, in which about ten grams of the required 3-oxyethiocholenic acid are produced from one kilogram of cholesterol."* The sales committee believed in the future of glucocorticoids, even though the indication areas were still quite manageable. Against his advice, the expansion of production capacities was decided, because in addition to the glucocorticoids, they also had their eyes on progesterone, with which the Berlin-based Schering wanted to make money.

At the end of 1938, shortly after Reichstein was appointed to the University of Basel, Ciba Percoten® was launched on the market, which, in addition to Doca, contained the first synthetic adrenal cortex hormone desoxycorticosterone. It was intended to become a prestige and research preparation that was far ahead of the competition. Ciba did not make real money. According to Reichstein's records and estimates, one to two kilograms of Doca per year would suffice, enough for 40,000 to 80,000 packs per year. As Reichstein wrote in a letter to Organon, there were not so many Addison patients, and most of the drugs were needed for research purposes.

The activities already led to tensions between Ciba and the small partner in the Netherlands. Organon felt more and more marginalized. *"Of course, we are in a strange situation here, as we are competitors despite all our desire for cooperation,"* Tausk wrote to Reichstein. *"We do not want to put you in a moral conflict and of course we welcome it if more material is available through the efforts of Ciba. However, we will naturally have to try to maintain the great lead we had over Ciba."* Was it just concern or a great disappointment that Reichstein was working for Ciba, or was it also a disappointment that Reichstein was now working for Ciba and it was seen that the 1937 contract was poorly negotiated?

It may have been somewhat reassuring that Kendall in the USA in 1938 had relativized the effect of the amorphous fraction of cortin: *"According to this, the most powerful amorphous fraction from adrenal ducts in dogs is only half as strong as doca"* [Thus, the strongest amorphous fraction from the adrenal ducts of dogs is only half as strong as doca]. This did not banish the evil spirits that Organon had always seen. But it showed that money could be made with cortin. The question of the effect and the final structural elucidation of cortisol should no longer be decisive, but the question of how to quickly establish a technical process and which substrate allows cheap production as a precursor. A synthesis from cholesterol with less than one percent yield was not affordable in the long run. It remained exciting. The race for this exciting group of natural substances was conducted both in process technology and in chemistry. The two greats, Reichstein and Kendall, wanted to bring the structural elucidation and synthesis to their advantage to an end. Let's take a look at the laboratories on both sides of the Atlantic.

Reichstein researched Doca, which had meanwhile been elevated to the status of a drug, but he was convinced that he would find better chemical substances, *"which were richer in oxygen,"* as he said. For some inexplicable reason, glucocorticoid research did not really get off the ground. He believed that additional oxygen atoms must be present in his molecules. These were missing and he attributed the effect to them. In Basel, they agreed that position 11 in the basic structure is decisive for the effect as a glucocorticoid and differs from the steroid hormone. Since the structural elucidation was still difficult, all further ideas about the structure of the molecule remained vague. In November 1941, Reichstein reported on the new dehydrocorticosterone, but doubted whether he was on the right path to the discovery of the much more effective glucocorticoids.

At the same time, things did not look better for Kendall on the other side of the Atlantic. As a result of the Second World War, the Americans were looking for new drugs to treat soldiers with severe burns, surgical shock, and similar severe injuries. The society urged Kendall to propose a structure, a synthesis, and thus to circumvent the Ciba patents. Kendall's substance A, which turned out to be dehydrocorticosterone, was to be produced industrially as it seemed easiest to synthesize. Reichstein learned in 1942 that Kendall was working on substance A and seemed to have succeeded in synthesizing it. However, he did not believe it was something big. He saw it as a small project of his colleague. But one could not be sure, which is why he urged Ciba to apply for patents for his 11-oxy and 11-keto substances. The synthesis of dehydrocorticosterone was also attempted in Basel at Ciba, but it was a complete disappointment and left the chemists frustrated. André Lardon, who was tasked with the synthesis, later wrote that the synthesis was successful, but *"the yield by this method is very low."* He considered industrial implementation to be completely hopeless. He was wrong, because later we read that small microorganisms took over the work of the chemist. Nevertheless, Ciba decided to follow Reichstein's suggestion and to start the synthesis and testing of other steroids.

Did Ciba, as suggested by Reichstein, apply for further patents on the 11-Oxo and 11-Keto structures? Yes, they did. Patents play a huge role in the industrial and academic landscape. Just like yesterday, companies today pursue the same strategy with patents to ward off annoying competition. It is patents that can decide the rise and fall of new groups of active substances, but also entire companies. In the chapter on biopiracy, I go into detail on the question of what patents are, how they influence the pharmaceutical market and what they can do. At this point, I just want to mention briefly that the application of patents by Ciba was a matter of course. But what are patents or protective rights, as they are called in official language?

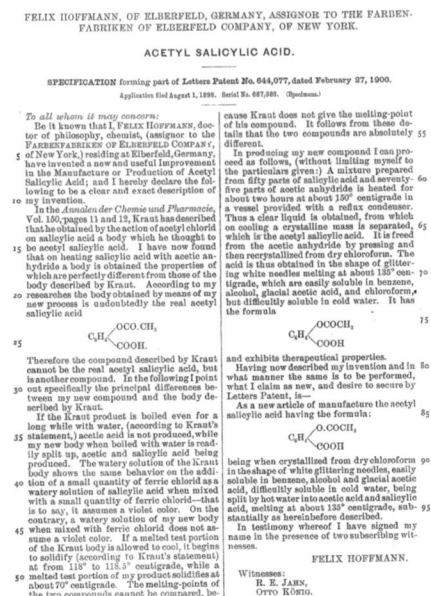

Fig. 35 First page of the US patent for acetylsalicylic acid by Bayer AG, 1900

Patents, which are granted by a government agency, give the patent holder the right to a monopoly of knowledge for a period of twenty years. No one else may use the same pharmaceutical substances that have the same effect as described in the patent during this time. Often the synthesis is also the subject of a patent (Fig. 35). The patent attorney then speaks of a process patent. However, this is not as strong as an application patent, if the pharmacological effect was discovered for the first time. Of course, the competition is not stupid and tries to find new substances with the same effect to circumvent the patent and secure a part of the lucrative market. To counteract this, the first company applies for further patents that describe further structures and new effects. These similar patents are grouped together into patent families. Thus, the field of intellectual property becomes ever broader and more solid. The goal is to make it very difficult for the competition to get a piece of the pie. In addition, potential competitors should be deterred from finding the pie interesting at all.

Microorganisms in white coats

It was difficult to obtain cortisone by chemical means. Although Reichstein showed a synthesis from bile substances, it gave the technical chemists headaches. There were too many synthesis steps and too high purification losses. In particular, one spot in cholesterol was a challenge. In position 11, an oxygen must be introduced, which is important for the effect as a glucocorticoid. Synthetically it is possible, as the company Ciba showed, but it was also expensive and associated with poor yield. An alternative synthesis had to be found. At Ciba, the whole thing initially remained lying and, much worse, opportunities were missed.

A windy American named Russell Marker from Mexico had contacted them and offered Diosgenin for the synthesis, but one could not seriously believe that he wanted to produce cortisol from a plant steroid. The American pharmaceutical company Upjohn in Michigan, which merged with Pharmacia in 1995, saw it quite differently. Upjohn did not take Marker seriously, but the company Syntex, from which he had left, did. The two biochemists Dury Peterson and Herb Murray came up with the brilliant idea at Upjohn in 1952 to put progesterone into a bioreactor and ferment it together with the mold *Rhizopus nigricans*. The special thing was not that Upjohn produced cortisone like the competition, but how they did it. Where others struggled with organic synthesis, they used the biosynthesis performance of a microorganism. After the penicillins, industrial microbiology was perhaps reinvented here, replacing difficult synthesis chemistry.

What triggered this development? The great potential of steroids was recognized by Frederick W. Heyl Frederick W. Heyl (Fig. 36) at Upjohn as early as 1933. His motto was not to dabble, but to go all out, and as head of the research department, he hired ten scientists. All ten were specialized in hormones and his foresight would pay off in the 1940s. Whether it was steroids, glucocorticoids or mineralocorticoids, Upjohn had a solid foundation and was able to stand up to the steroid cartel in Europe. The Second World War acted like a catalyst for American cortisol research. President Donald S. Gilmore of Upjohn also showed foresight when he quadrupled the budget of his research department in 1945. *"We want cortisone, don't spare the horses,"* Gilmore told his senior staff. By 1950, Upjohn had rebuilt and renovated the chemical and microbiological laboratories. Now, sparkling clean, silver

Fig. 36 Frederick W. Heyl (public domain, www.upjohn.net)2

shining fermenters stood in the labs, ready to be used. Cortisone apparently did not have the highest priority, America and the world needed antibiotics, but in the end, it would turn out differently.

At Upjohn, there were no research departments that knew nothing of each other. All four teams working on steroids were in regular exchange from 1950 onwards. In one of the four groups, which dealt with the synthesis of progesterone from stigmasterol, oxidation proved to be a challenge. This alternative plant cholesterol from the soybean was available in large quantities and was to become the raw material for glucocorticoids. For years, Upjohn had been isolating stigmasterol from soybeans. Upjohn wanted its own process, which had nothing to do with that of Syntex, Parke-Davis or the Europeans. To solve the problem, Gilmore offered the young chemist R. H. Levin from Wisconsin a position in the company. Levin arrived in 1941 and turned the entire research upside down. Until the enzymatic conversion with microorganisms was discovered, all Upjohn employees worked on classical chemistry, but in 1949 the flow of information between the departments paid off. The fact that microorganisms not only produce antibiotics but also other natural substances was still new at the time, but only a few scientists were concerned with the potential of bacteria and fungi.

With an article from a scientific journal about the hydroxylation of cholesterol in position 7 by the fungus *Proactinomyces roseus* in hand, David I. Weisblat came to Murray in 1949. *"Watch out, they're doing something special here,"* he called out to him. Murray may have wondered what his department head wanted now, as Weisblat stood in the door. But he knew that Weisblat was one of the experts in the field of microbiological conversion and listened to him. Weisblat was to be proven right, as the same year the working group of O. Hechter at the Worcester Foundation described the hydroxylation of cholesterol in position 11 during passage through adrenal tissue.

Once again, sport brought them together. Peterson and Murray played handball together. During a game break, Peterson asked his colleague if he had read the publications on metabolism in the kidney? Murray nodded, and the two came up with the idea that mold could be the solution to oxidize cortisol in position 11, as in the kidney. They informed their boss Weisblat. He was thrilled and put together an interdisciplinary team in November 1949. In addition to Murray and Peterson, Lester Reineke, an expert in paper chromatography, was appointed to the team, which was to be led by Levin. Levin also invited S. H. Eppstein and A. Weintraub, who were already working on a similar question. One percent inspiration and 99% perspiration, as so often, make such a project successful. Even Levin as the boss was later under no illusion: *"The strategy was developed over time through the collaboration of the group. There was never a flash of inspiration. The developed strategy essentially consisted of accelerating the screening of a large number of microorganisms by working in the micro area and quickly determining by means of paper chromatography whether a change in the selected steroid substrates had taken place. Scale-up and isolation studies were limited to screening results that indicated significant amounts of only a few transformation products. It was inevitably a cooperative project."* The bottleneck was the screening, because it was time-consuming and there were no references that would have been useful as markers. They were lucky, because Alejandro Zaffaroni from the University of Rochester published an important article on analytics in January 1950, which dealt exactly with the problem that was burning under the nails of the Upjohn chemists. Alejandro Zaffaroni (1923–2014, Fig. 37) was just doing his PhD on the topic "Synthesis, Isolation and Measurement of Corticoids". Later he would work on the development of the birth control pill at Syntex and then become the CEO of the company.[122]

Levin sent Peterson and Eppstein to Rochester to learn the technique and after some adjustments they had an analytical technique that significantly accelerated the screening back in Wisconsin. Murray was not picky when

Fig. 37 Alejandro Zaffaroni (1923–2014)

it came to selecting microorganisms for screening. In the late 1940s, there were no strain collections as we know them today, where you can buy thousands of microorganisms. He was on his own looking for many so-called wild strains. The microbiologist goes about this in a very simple way. The samples are taken as they are found. Samples from the earth, under the shoe, on plants, from ecological niches such as the stomach of horses, the intestinal tract of fish, from thermal springs or from oil puddles at gas stations. There are no limits to creativity.

Progesterone from their own production was immediately used for screening. Why conduct experiments with cholesterol when progesterone was available? After just one year, Murray had success. A sample showed two spots in the paper chromatography. One was the progesterone that had not been converted by the mold, and there was another spot that Murray and Reineke found very interesting. They decided to give Peterson a little work. They cultivated the progesterone again and gave the culture approach to the chemist for further purification and structure elucidation. To everyone's surprise, he found not only the sought-after 11-oxyprogesterone, but also a second steroid that carried an additional oxygen at position 6. What should be cause for joy in the lab is usually a nuisance, because this impurity must be removed. This can not only be costly in an industrial process, but also significantly reduce the maximum yield of the desired product if the by-product is numerically superior.

The apparent passion of molds for the indiscriminate oxidation of everything that comes into their hands, however, had another unexpected advantage. The mold *Rhizopus nigricans,* found on the bark of rotting melons, carried out the first oxidation at position 11 of the steroid. In further screenings, it was noticed that *Streptomyces fradiae* and *Cunninghamella blakesleeana* incorporated the second missing oxygen at position 17. This was ideal. With the help of two molds and a small chemical reaction, it was possible in 1952 to produce cortisol and cortisone from progesterone on an industrial scale. Upjohn filed a patent: "Oxygenation of Steroids by Mucorales Fungi", which was granted in 1952 under number 2.602.769. The company tried to obtain a sufficient amount of progesterone from stigmasterol through its own syntheses, but this conversion did not work well. The purchase of progesterone from Syntex was an alternative. John Hogg, a chemist at Upjohn, studied at Penn State and wanted to do his doctoral thesis with Marker. But Winterthorn, dean of the faculty, thwarted his plans. Professor Marker would soon leave the university to start a company in Mexico. John Hogg remembered this company and placed an order for 10 tons of progesterone. The letter with the order fell out of Somlo and Rosenkranz's hands upon arrival in Mexico. It was an incredible amount, but since Upjohn did not want to pay more than 68 cents per gram, it was not going to be a big deal. This was a massive price drop from 1000 USD per gram to a few cents in just five years. Somlo sold, but he knew that Syntex was getting into a ruinous price war.

In addition to Ciba, the German Schering AG also developed glucocorticoids after the Second World War, which certainly helped the Berlin company to economic size and reputation. However, the Berliners did not want to depend on the Swiss and looked for alternative sources. They found them in Mexico, where precursors of the birth control pill were also isolated from the yam root. The company Syntex was contacted, and contracts were concluded to have these important key substances delivered. In order not to become dependent, the Berliners also looked at what Upjohn was doing and developed semi-syntheses from sitosterol. Microorganisms also played an important role here. Strain collections were established and almost every carbon in the sitosterol or cholesterol molecule could be specifically altered. A broad chemistry also developed, which made it possible to produce semi-natural cortisol derivatives that were many times more effective and even water-soluble for infusions. A stable basis was created, and in addition to the mentioned drugs, Schering was able to bring further profitable drugs to the market in the 1950s: Scheroson® with cortisone (1953), Scherisolon® with prednisolone (1956), Scheroson® F with hydrocortisone (1957) and Ultrala® with fluocortolone (1957).

The unique achievement of the Upjohn company and its microbiological helpers was outstanding, which is why the American Chemical Society (ACS) on May 1, 2019, designated the production site in Kalamazoo, Michigan for its innovations in the field of steroid medications with a National Historic Chemical Landmark (NHCL): *"Until the mid-20th century, steroid medications were only available as expensive, impure, natural extracts. This changed in the early 1950s when scientists at the Upjohn Company made several advances and provided the world with affordable, high-quality steroid medications. The innovations included the formation of multidisciplinary research and development teams for rapid problem-solving, the use of sustainable materials such as soybeans as raw materials, and the use of microorganisms to carry out critical conversions in combination with new chemical reactions for efficient production."*

The interdisciplinary teams led by David I. Weisblat and Robert H. Levin enabled the rapid development and marketing of cortisone (from 1952) and hydrocortisone (1953). Upjohn had a monopoly and dominated the market. Many pharmaceutical companies decided not to invest anymore in research and development of cortisone synthesis and to buy directly the precursors or the cortisones from Upjohn. Until 1990, Upjohn was leading with another 30 new steroids and their analogs. Economically, the company was strong and was an interesting takeover candidate. Pharmacia bought Upjohn in 1992, the new Pharmacia & Upjohn merged with Monsanto to Pharmacia in 1995, which was taken over by Pfizer in 2002.

Cortisol as a Miracle Drug

What happened to the miracle drug, as Kendall at the Mayo Clinic liked to call cortisol? Research on cortisone or 11-dehydro-17-hydroxycorticosterone was eagerly pursued worldwide. Not only for the US Army, but also for the National Socialists, anti-inflammatory effects and possibilities of body regulation seemed to be of importance. The possibility of reducing the stress of pilots, submarine drivers and tank commanders appeared very attractive. This led to massive state interventions in research in England, the USA, and the German Reich. Kendall at the Mayo Clinic had a research project with Parke-Davis since the severe recession of the late 1920s, which was promoted with the utmost urgency by the Department of Defense during

World War II. The reason was that the German Luftwaffe was able to let their pilots fly at an altitude of 10,000 m with the help of corticosteroids. For this reason, the US government gave the highest priority to the research of cortisone for a period of more than two years. Chemical cortisol and cortisone were hard to come by for Kendall. But he also got it from the slaughterhouses in the area. In one of his later publications, he stated that he had needed the adrenal glands of 40,000 cattle to obtain 600 mg of cortisone.

The anti-rheumatic effect of cortisone was not published until late, in 1949, which is surprising given the nearly twenty years of research. Hadn't anyone tested the substances yet? Yes, of course, they were tested as cortin. But there was no pure substance in sufficient quantity to make comparisons. Philip S. Hench had the idea to test cortisol in patients with severe rheumatism. A woman with particularly severe rheumatoid arthritis was Mrs. G., as she was encrypted in the patient records. Hench administered her twice the dose of 50 mg intramuscularly in September 1948. The very next day, Mrs. G. was so lively and mobile that she could get up and go shopping for three hours. A cortisone miracle had happened, which astonished everyone at the Mayo Clinic. The experiment was soon extended to a larger number of patients, all of whom were very satisfied. On April 13, 1949, the excellent effect was presented by researchers from the Mayo Clinic in a lecture to doctors. It was a triumph and the demand for broad application became loud. For their groundbreaking work, Kendall and Reichstein were awarded the Nobel Prize in Medicine as early as 1950. The fact that cortisol and prednisone received such a great response on April 13, 1949, was also due to the fact that William L. Laurence (1888–1977) was among the listeners at the lecture. This man was not a doctor, but a journalist who, a few days later, on April 21, published an article on the front page of the New York Times with the headline "Aid in Rheumatoid Arthritis is promised by new hormone". William L. Laurence or "Atomic Bill", as he was called, was the only journalist who was allowed to document the ignition of the atomic bomb in New Mexico and over Nagasaki in a B-29 bomber, was enthusiastic and reported euphorically. He compared the discovery of cortisol and cortisone in their significance to the discovery of insulin or penicillin. Laurence was so popular that every American wanted to read his stories. This gave the popularity of cortisol a huge boost.

Cortisone was supposed to help many rheumatism patients: For the medical profession, it became a miracle weapon against inflammations, even psychoses and depressions were supposed to be treated with cortisone. Today, there is no longer any medication that contains cortisol, which has been replaced by new cortisones. New chemical variants with new names have replaced this, but the term "cortisone" remained firmly anchored synonymously in the minds of patients. Today, cortisones are standard medications not only in dermatology and in the treatment of asthma and allergies. Many diseases, where an anti-inflammatory and also analgesic effect is required, are classic areas of application for glucocorticoids. Even though much is complained about side effects such as weight gain, weakening of the immune system up to osteoporosis or glaucoma, they are indispensable in therapy under medical supervision.

Vitamin D? Then rather Cod Liver Oil!

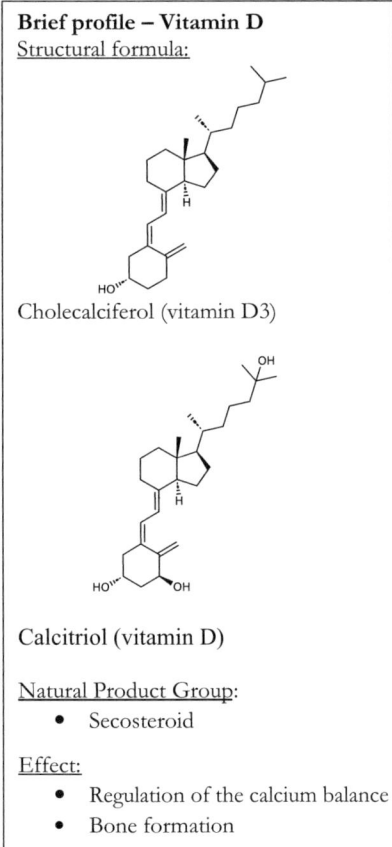

Brief profile – Vitamin D
Structural formula:

Cholecalciferol (vitamin D3)

Calcitriol (vitamin D)

Natural Product Group:
- Secosteroid

Effect:
- Regulation of the calcium balance
- Bone formation

We have already met Adolf Windaus as the discoverer of the structure of sterines and cholesterol. The Göttingen natural product researcher was also of great importance for the cure of rickets. This disease is a deficiency of vitamin D, which has a fundamental effect on bone structure, the immune system, and the general hormone balance. In history, this disease usually occurred in cases of malnutrition as a result of wars. It was described in 1650 by the English anatomist F. Glisson (1596–1677), but even Egyptian mummies showed the typical bow legs or saber legs.

Rickets was also a theme in art. Albrecht Dürer painted "Maria with the Pear" in 1512. The little Jesus child shows the known characteristics of a deficiency, such as a protruding receding forehead as well as thickened hand and foot joints. The painting can be admired in the . It is not known what prompted Dürer to draw a rickety Jesus child. Perhaps he took the lying putto or Andrea del Verrocchio's (1435–1488) "Madonna with Child" as a model, which show a similar symptomatology of vitamin D deficiency.

Chronology of vitamin D

1919 – Discovery that rickets can be treated with UV light

1921 – Mellanby shows that rickets can be cured by cod liver oil (...)

1922 – McCollum gives the name vitamin D to an antirachitic substance with an unknown structure

1928 – Nobel Prize for the discovery of vitamin D to Windaus (...)

1935 – Synthesis from 7-dehydrocholesterol

1936 – Correct structural elucidation of vitamin D

1971 – Discovery of the hormonal system in the human body for vitamin D

1973 – Discovery of the vitamin D receptor by

"Then rather cod liver oil" is a German short film from 1931, and Emmerich Pressburger wrote the screenplay. Watch the film on and you will find that , in addition to its medical benefits, also had an educational effect, through which parents in the 1920s and later wanted to demonstrate their care. Kästner and Pressburger caricatured the administration of the thick, bitter-tasting oil as an educational task. The dear Pippi Longstocking later took it to the extreme when she told adults that no one could force her to take cod liver oil if she preferred to eat candies. What began as a necessary therapy against the rickets of the bourgeoisie and the state turned into a sign of the good thriving of children. But do we succeed as parents? We want to give our children the best with a balanced organic diet with little sugar. But the offspring's rebellion against the healthy has remained. It is probably timeless, showed us early and Astrid Lindgren later in their books.

But when did rickets in our modern society become not only conscious but also political? Rickets was mostly known due to poor nutrition, especially in the poorer population groups, but with industrialization, this was to change and the rich social classes also discovered the problem with their children. Let's first stay with the socially poor classes of the developing industrial society of the 19th century. In the drawings of , a connoisseur of the Berlin milieu, there is *"not a single child that is not sick"*, noted Kurt Tucholsky about the horrors of rickets among . Vitamin D deficiency became politically the occasion and meaning of the trade union struggle for the social improvement . For the reform movement around the turn of the century, which we know today under the terms nudism, healthy nutrition, movement in light and healthy air, public health became a state task, but also an individual responsibility to take care of one's own body.[123]

Rickets was not described as an English disease (Morbus Anglorum) because of the Englishman Glisson, when the first cases occurred around 1620 in the counties of Somerset and Dorset (Fig. 38). In later Victorian

Fig. 38 Children suffering from rickets. In: De Rachitide sive morbo puerili ..., Leiden, 1672

England at the beginning of industrialization, rickets occurred much more frequently. The high number was due to the strong absorption of UV radiation by smog and smoke in the air. Smog and smoke were released by the emerging young industry in England. But there were also rickety children in the upper class, as overglazed winter gardens came into fashion due to air pollution, where the children played.

The benefits of cod liver oil, a fish oil rich in vitamin D, were recognized early on (1824). How is it that this oil, which is mainly obtained from the liver of cod and haddock, is so rich in vitamin D? Fish usually do not sunbathe on the water surface. The high content of vitamin D is due to the animals' diet. They feed on plankton, which can be a mixture of various animals, plants, but also bacteria, such as fungi. Many of these organisms swim on the water surface. Light rays reach them and they form vitamin D. In fish that prefer to feed on plankton, the vitamin D accumulates, so that it reaches us through the food chain.

Recently, mushrooms with an increased vitamin D content have also been offered to us consumers. How is this possible? After all, mushrooms are grown without light. Do the vitamin D mushrooms keep the promise that one can eat a sufficient amount of vitamin D? In principle, yes, says a well-known consumer protection foundation in Germany. Mushrooms, but also other fungi such as shiitake, can form 30 times more vitamin D from the plant sterol ergosterol under UV radiation than ordinary cultivated mushrooms in the dark. If the vegan vitamin D in the mushrooms is too expensive for someone, they can expose their purchased ordinary mushrooms to daylight. This way, ergosterol is also converted into vitamin D, although the amounts are likely not to be as high.

Berlin Air and Light Against Rickets

The chemistry of vitamin D as a second field of research came rather accidentally to Adolph Windaus and it should not go unmentioned that it is related to his steroid chemistry, which Windaus did not suspect.[124] Alfred Fabian Hess (1875–1933), who worked as a pediatrician at Rockefeller University in New York, asked Windaus in a letter for support in the question of whether vitamin D could also be chemically represented. Hess was a researcher and worked at the same time in the slums of New York, where he saw many slum children with typical symptoms of rickets such as bow legs and too thin bones. After the First World War, up to 75% of children in the cities of the northern hemisphere suffered from vitamin D deficiency.

He knew that it was a deficiency of vitamin D, which was already found in cod liver oil in 1914. The disease could have been cured quickly, but cod liver oil was too expensive for the poor of the Lower East Side and Hess thought about a cheaper alternative. He believed that the anti-rachitic factor must have a great similarity to steroids and suspected a chemical relationship with cholesterol. He was right, because if you rearrange the rings in the vitamin D molecule and introduce another bond, you get a four-ring system, as is also found in cholesterol. Try it out! Windaus was skeptical. Yes, the analogy was obvious, but should the vitamin be contained in cod liver oil of all things? Then there was the publication of the Berlin pediatrician Kurt Huldschinsky (1883–1940), who reported with the proof of an X-ray (Fig. 40) that rickets in children was cured by irradiation with ultraviolet light (Fig. 39).[125]

The effect of ultraviolet radiation in light was demonstrated by the Berlin pediatrician with a simple and astonishing experiment for us. He first put a welder's protective goggles on three children in the winter of 1918/1919 at the Oskar-Helene Hospital in Berlin-Zehlendorf and irradiated the children for two minutes each from the front and back with a mercury vapor

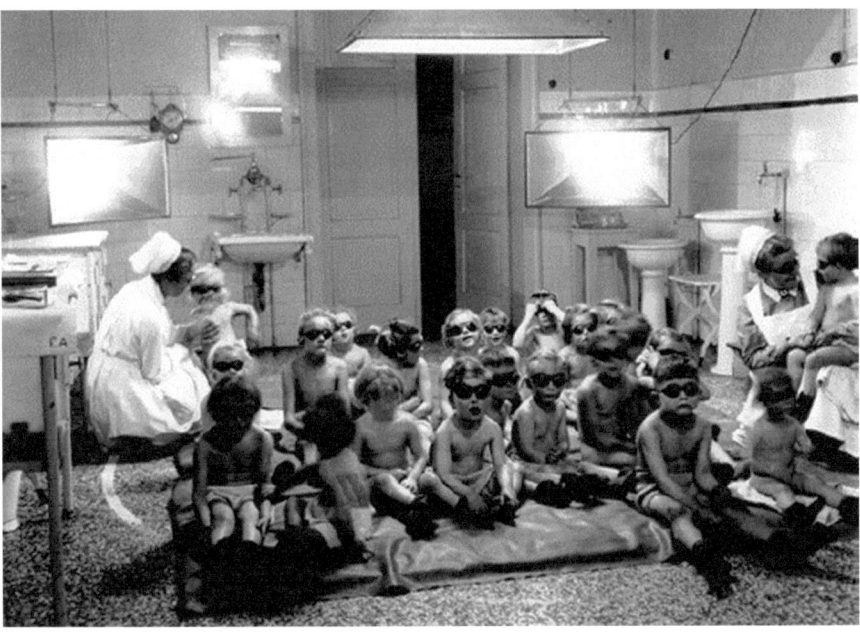

Fig. 39 UV irradiation of children of factory workers in the BASF "nurse ambulance" for rickets prophylaxis, 1937

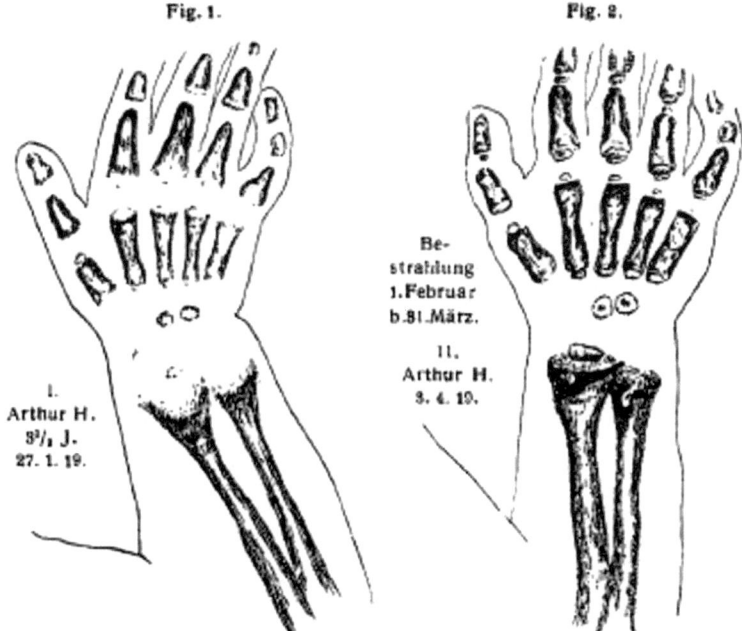

Fig. 40 X-ray of the right hand of the boy Arthur H. before irradiation (27.01.1919) and after three months of irradiation on 09.04.1919

lamp. Hudschinsky himself was surprised by the success. "*Child lies gray in the crib*", he wrote at the beginning in the patient's file. But the child, a boy, was so far healed after three months that he could sit upright in bed. Originally, he assumed that the light would only have a local effect. In a second attempt, he only irradiated one arm, but rickets noticeably decreased in both arms. If both arms were getting healthy, he concluded, the light must have activated a chemical substance that spread throughout the body. The Berlin AOK found this study so good that it set up light bath institutions throughout the country.

In the 1920s, there were two major theories to explain rickets. One was a lack of light or a deficiency in the diet due to the absence of an anti-rachitic factor. The researchers of the time were not sure, maybe it was not a vitamin at all. The body produced it itself. Windaus also remembered a publication from London. The author, McCollum, thought that these light theories were wrong and that rickets was simply an infectious disease. That was the second theory. In the end, the first theory of light should prevail and we will clarify the confusions and complications over the years.

Fig. 41 Elmer McCollum (1879–1967)

Elmer McCollum (1879–1967, Fig. 41) was a young scientist who was determined to find the anti-rachitic substance and analyzed many fats and butter. In 1919, he found a molecule in butter that he named Vitamin A. Today, we give this name to a different vitamin. But it was the pioneering era of vitamin research and a nomenclature was not yet thought of. He was unable to isolate the substance and in 1920 it was found that everything decomposed in the pan when heated and the effect disappeared. The fed rats died after forty to fifty days from xerophthalmia, a vitamin A deficiency disease of the eye. McCollum persisted and re-examined the butterfat in 1922. The American biochemist realized that there must be another substance besides the suspected Vitamin A that he wanted to isolate. But he did not make any progress. At least he gave the unknown substance the name Vitamin D.[126] Vitamins B and C were already known.

Edward Mellanby (1884–1955) from King's College for Women wanted to know exactly and developed a diet for his puppies in 1922 to induce rickets in them. The young puppies he raised were to be the children of the poor quarters, and he fed them a similar diet—bread and skim milk. The growing dogs in his lab showed the same symptoms: soft bones, crooked legs, curved

Fig. 42   Edward Mellanby (1884–1955)

back, and deformed chest bones. He fed his dogs cod liver oil and they got healthy. Edward Mellanby (Fig. 42) believed that Vitamin A was the reason for the cure, which his colleague McCollum also believed to see in similar experiments with rats and the administration of butter. He heated the oil to destroy the Vitamin A, which he succeeded in doing, as the dogs became night-blind. However, a cure for rickets did not occur. Mellanby fed the puppies with mushrooms and gave them orange juice to see if B vitamins or Vitamin C could cure rickets. Both researchers were mistaken and good advice was expensive.

Edward Mellanby remembered the old folk wisdom of the 1920s that fresh air and sunlight were good against rickets. He also remembered that there was this pediatrician Huldschinsky in Berlin who put the Berlin brats in front of the sunlamp, and after three months the children were much better. This should be repeatable in animal experiments. Edward Mellanby made his rats rachitic and put them in the sunlamp—rats in the solarium. With Alfred F. Hess (1875–1933) and Mildred Weinstock, new rat experiments were planned. The rachitic rats were taken out of the cage and irradiated and compared with the control group. The cages were also irradiated with ultraviolet light, and after each stay in the sunlamp, all animals were returned to their cages. To their surprise, the irradiated and the non-irradiated rats recovered equally well. This surprised them and the two researchers attributed this to the irradiated food that remained in the cage. They were right! They irradiated not only the food but also other foods like milk and butterfat and fed the rats. Without knowledge of the structure, they simply named the supposedly anti-rachitic substance "Accessory Factor". It was a success and Alfred Fabian Hess from Columbia University now set about isolating this ominous additional factor himself and reported on his irradiation experiments with ergosterol and sitosterol, which was later called phytosterin. Sitosterol, as "plant cholesterol", only occurs in plants and only proved to be anti-rachitic after irradiation. However, Hess did not trust his own results, and as early as 1926 he had an almost prophetic gift when he put forward the hypothesis in his publication: "*It seems quite possible that cholesterol* [it is 7-dehydrocholesterol, as we know today, author's note] *in the skin is normally activated by UV irradiation and made anti-rachitic—that the sun's rays and artificial rays can cause this transformation. This view considers the superficial skin as an organ that reacts to certain light waves, and not as a mere protective cover.*"[127]

Harry Steenbock (1886–1967, Fig. 43) from Wisconsin worked with Mellanby. He was a biochemist and researched vitamin D and later the "anti-rachitic factor". Together with Mellanby, he conducted animal experiments and immediately recognized the economic benefits of irradiating food with ultraviolet light. In 1924, he filed a patent and paid the fees and his lawyer with 300 dollars himself. His foresight proved to be ingenious, as milk and bread were now irradiated in the USA. By the end of World War II, when the patent protection expired, rickets was virtually unknown. Steenbock could have become a rich man: The food manufacturer Quaker Oats offered him 1 million dollars for his patent. He declined and founded the first university knowledge transfer office in the USA for the marketing of patents with some former students. His conviction was that the money should benefit the University of Wisconsin, where he was a professor and researcher.[128]

Hess could not make further progress with the chemistry of vitamin D and asked Adolf Windaus for help. He found the question of vitamin D synthesis exciting and wondered how he could solve the problem.[129] Since the publication by Hess in the USA, he knew that it had to do with ultraviolet radiation, and here too a physical method and a previous experiment he conducted with Robert W. Pohl (1884–1976) in Göttingen helped the

Fig. 43   Harry Steenbock (1886–1967)

natural product chemist. Pohl helped him in the analytical distinction between cholesterol and ergosterol and both had a joint publication titled "On the optical detection of a vitamin".[130] Was irradiation the key to the sought-after vitamin D? Windaus became alert and ran to the library to look at his old publication again. He turned to page 435 in the journal and read: *"Since then, the identity of cholesterol as an anti-rachitic provitamin was considered proven. It only appeared as a problem to clarify the chemical structure of the photochemical reaction product, i.e., the actual D-vitamin."* That was clear, thought Windaus, and he looked again at the illustration of the irradiation experiments on the same page. He had to examine the unirradiated and irradiated cholesterol again and asked his students Reuss and Hilsch to measure the return samples again. A few days later he received the results. Both had done good work. Was the cholesterol now the provitamin? No, the results showed a different picture and were misinterpreted at the time. Hess had been mistaken in his 1924 publication. Hess had assumed that it was pure cholesterol, which had previously been isolated from the brains of rats. However, it must be assumed that impurities were present. He did not know that it was 7-dehydrocholesterol, which he mistakenly took for cholesterol. Windaus was also not aware of this and suspected that there must be another cholesterol derivative that was the actual provitamin. The new experiments put the old work in a proper light and forced a reevaluation.

Windaus' interest was now aroused for the new vitamin. A good cooperation developed between him, Hess in the USA, and Sigmund Otto Rosenheim (1871–1955) in London.[131, 132] He sent his colleagues Rosenheim and Webster new material to make a pure cholesterol sample. This was recrystallized and chemically treated so that no impurity remained. Only when the boss was sure did the release come from Göttingen. With this sample, irradiations could be carried out. It turned out that the irradiated cholesterol was not an antirachitic. So cholesterol was not the precursor, but what substance was it? As is often the case in science, a rocky road began and Windaus decided to synthesize thirty cholesterol derivatives. Of these synthesis products, ergosterol proved to be very effective after being irradiated with UV light. This substance is also found in the fungus ergot, which gave the sterol its name. Windaus and his doctoral students believed they had finally found the precursor, but they only got one step closer to their goal. Rosenheim and Webster in London or Reerink in van Wijk at the N. V. Philips light bulb factory in the Netherlands did not fare any better in 1931. Windaus named the new substance, which was obtained from ergosterol, vitamin D2 or calciferol, and he was impressed that only one hundredth of a microgram cured rickets after two months. Windaus remained skeptical, as he could not explain how a substance from a fungus, which does not occur naturally in humans or other animals, could be converted into vitamin D in humans with the help of light. This question could only be answered in 1937, long after Windaus received the Nobel Prize for Chemistry in 1928 for researching the sterol structure.[133] By the way, he shared the prize money with Alfred F. Hess in New York, as the Nobel Prize had been awarded a year earlier, among other things, because of a joint publication.

Windaus, together with Bock, isolated 7-Dehydrocholesterol from pig skin, which was later also found in human skin. For Windaus, it was not an unknown substance, as he had already synthesized and irradiated this compound in 1935. 7-Dehydrocholesterol was broken down by light, but did not yield Vitamin D2. Windaus was puzzled, as it was a different Vitamin D derivative, which he named Cholecalciferol or Vitamin D3. The 7-Dehydrocholesterol, which was isolated from human skin in 1936 by his later successor Hans Brockmann along with Vitamin D3, was the effective provitamin and exactly matched the chemical structure of Windaus' synthesis products. The chemical work was a fundamental breakthrough that led to the synthesis of inexpensive Vitamin D. Windaus, who always saw high science at the forefront of his work and did not ask about practical use, now had something very practical in hand. Together with I. G. Farben and the chemical factory E. Merck, he signed a contract that brought him 50,000 Reichsmark, a 25% share in the net profit, and the consortium the production of Vitamin D from ergosterol by irradiation. It remained a difficult path to market Vitamin D as Vigantol® worldwide. Numerous patent disputes and product overlaps by third parties had to be contested by E. Merck. The author Jochen Haas lists in his book "Vigantol" about the history of Vitamin D that in Germany over 16 and internationally at least 34 patent infringements were complained about.[134]

Was that the end of the chemistry of Vitamin D? No, but it would take another 20 years until Velluz had understood all light-dependent chemical processes and had chemically clarified them outside of humans.[135, 136] What exactly happens biochemically in the skin interested Holick. He deciphered the last biochemical secrets in the 1980s.[137] The product Vigantol®, which initially consisted of irradiated ergosterol, came onto the market in the 1930s. To increase acceptance—today we speak of compliance—among the target group of children, Merck and I. G. Farben launched Vigantol® lozenges with a sugar coating, Vigantol® dragees or Vigantol® with a chocolate coating. Unfortunately, the stability of Vigantol® was not particularly good and further chemical improvements led to a replacement from 1931 and only Vitamin D2 with excellent stability and better effect was offered in Vigantol®. After Hans Brockmann, head of the Biochemical Department of the Chemical University Laboratory in Göttingen since 1935, had discovered Vitamin D3 in the oil of tuna, the composition was changed again in 1952. The ingenious chemist Windaus had succeeded in producing this from the 7-Dehydrocholesterol known to him. A technical synthesis met the demand only after the Second World War and solved a medical problem for all time.

Caffeine and Viagra

Caffeine, Theobromine, and Theophylline

Profile - Caffeine

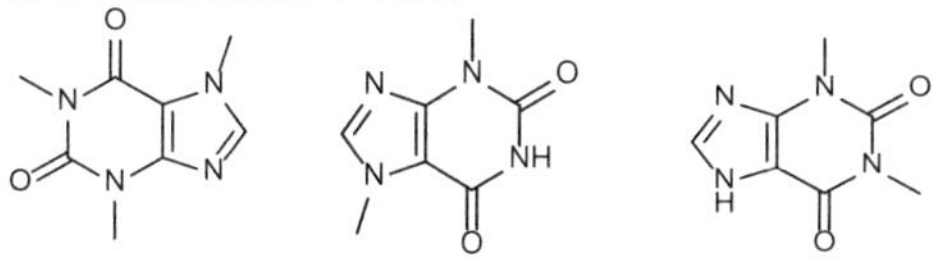

Caffeine   theobromine   theophylline

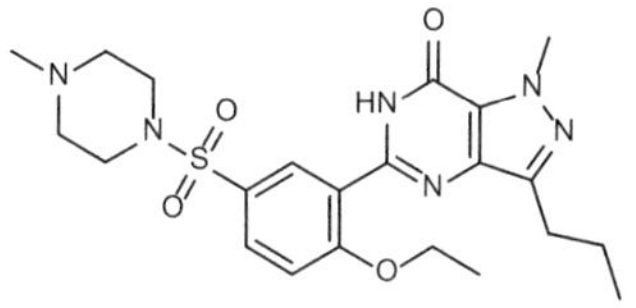

Sildenafil
(Viagra®)

Effect / Indication

- Stimulant (caffeine)
- Asthma medication (theophylline)

Natural Product Group

- Purine alkaloid

Who doesn't know this? To get going in the morning, you need a little booster. Completely legal is the psychostimulant that is consumed most frequently worldwide and is highly valued by many not only in the morning. An estimated 100,000 tons are consumed worldwide each year, uniting humanity to take a break together, treat themselves to something, or simply have a moment of peace. We're talking about caffeine, which we like to consume in coffee, tea, or energy drinks. Chemically, caffeine is a purine

O. Kayser, *From the Maluku to Molecules*, https://doi.org/10.1007/978-3-662-69923-2_5

alkaloid, and two other purine alkaloids should also be mentioned. One is the Theobromine, which is mainly found in chocolate, the second alkaloid is theophylline, which we consume with tea. To bring some clarity to the terminology right from the start: The frequently mentioned tein or theine in the literature is not an independent natural substance, but theophylline, which is chemically very similar to caffeine.

Coffee (*Coffea arabica*, Fig. 1) was a drink of the Arab upper class and was known even before the siege of Vienna by the Turks, which is often mistakenly given as the date of its discovery by Europeans. In the Middle Ages, traders brought coffee beans to Europe, but it remained a luxury drink for which a high price was paid. It is interesting to know that the caffeine kick was also enjoyed with tea all over the world. In Asia, especially in China *(Camellia sinensis)* (Fig. 2) and Japan *(Camellia japonica),* it was green tea and in South America the mate tea *(Ilex paraguariensis).* Even more interesting is that both tea plants, before they were consumed as a pleasure drink, were also plant-based remedies. Asian tea came to Europe almost simultaneously with coffee. On the European mainland, with the exception of Russia,

Fig. 1 *Coffea arabica*

Fig. 2   *Camellia sinensis*

the hot drink was never as popular as on the British Isles. The reason for this can be seen in the trade relations with China and later with India, which were dominated and monopolized by the British East India Company. The domestic market was supplied and the demand was large enough. The historical consequences, which were described in this book in the chapter on opium, are now well known to us. The Opium Wars, the cultivation of cinchona trees, and the colonization of India are just a few examples.

About Chocolate Men in the Snow

Much has already been written about the trade and enjoyment of tea and coffee. Therefore, let's take a look at cocoa and mate, which came from the New World to the West. In South America, cocoa was highly valued by the indigenous people as early as 1500 years BC. The Olmecs loved the bitter taste and passed on this knowledge to the Maya on the Yucatan Peninsula, who also planted and cultivated the cocoa tree there. The Toltecs and Aztecs adopted the custom, drank the cocoa water and also used the cocoa beans as a means of payment. The Aztecs paid 200 cocoa beans for a turkey or 100 beans for a slave. The Spaniards, who landed in Yucatan and Central Mexico, were interested in the cocoa, which was made without milk. Hernán Cortéz (1485–1547) found it strange that cocoa was almost excessively drunk at the court of Montezuma. Cocoa beans were crushed, mixed cold with water and whipped with a wooden whisk. The brew was considered intoxicating, sent by the gods and therefore unsuitable for women and children. It was preferably drunk by warriors, nobles, and priests. Christopher Columbus probably brought cocoa beans to the Spanish court on his fourth South American journey in 1502, but it was not until 1544 that it was reported that the court also drank cocoa. It did not taste good to the royal palate and was made palatable with sugar or honey. The Aztecs remained true to their habits and added vanilla and sometimes tobacco to the cocoa. Other Europeans still liked the cocoa and from the 17th century onwards it was traded and cultivated as a product of the colonies. The French tried it on their Caribbean islands, the Spaniards in Venezuela and the Dutch in Brazil.

The production of chocolate is a demanding process, which the Swiss have perfected with the technique of conching. Chocolate also makes the hearts of microbiologists beat faster, as it is a complex fermentation process. The cocoa mass is not only extracted, but made accessible in terms of taste by microorganisms in the first place. However, biting into the cocoa fruit, which looks like a small football, does not provide a taste experience. The shell is firm and leathery, the pulp sweet and fruity. In South America, it is also often used to make an alcoholic cocoa schnapps. The popular cocoa butter is found in the up to 20 seeds, which are dried and fermented in the sun after peeling. In addition to baker's yeast, it is *Hanseniaspora opuntiae* in a colorful and diverse society of many microorganisms that decide on taste and quality after fermentation.

Drinking cocoa (*Theobroma cacao*, Fig. 3) with milk or eating it as chocolate is an invention of the Europeans. The first milk chocolate was made in Dresden in 1839 from donkey's milk, as cow's milk was too expensive. The chocolate as we know it today was a further development of the Swiss Daniel Peter, who in 1867 first mixed cow's and later condensed milk with the chocolate and heated it. The consistency did not want to set and the tinkerer needed a few more years until 1875, until the melt and taste were improved and the customers were thrilled. But what triggers the enthusiasm, why do we like to eat chocolate? On the one hand, the sweetness and

Fig. 3 *Theobroma cacao*

its ingredients such as the tryptamine, which brings us the cheerful mood. But also the cocoa butter, which has a special physical property in the cocoa beans, as the seeds are called. This cocoa butter consists of a special fat that melts at 37 °C, approximately body temperature, on the tongue. This melting property was used in pharmacy to make suppositories. But cocoa butter is not stable, it becomes rancid and crystallizes quickly. Cocoa butter can occur in six crystal forms, all of which have their own melting point between 17 and 36 °C. Only the beta crystal form is perfect and melts at body temperature, but during storage it changes to the less stable crystal forms. You may know this if you want to eat the Santa Claus only at Easter. Suddenly you see a white layer, as if the chocolate man was standing in the snow after all. Often this is considered a microbiological flaw or it is suspected that the chocolate has gone bad, but it is the same chocolate just in a different crystal form.

Maté Tea—the Drink of the Gods

Maté tea has been growing in popularity in our latitudes in recent years. However, it has been drunk in South America for centuries, especially by the Guaraní, who saw something divine in the plant and the tea. Originally, drinking maté was a ritual of the Avá people in the area that today comprises part of Paraguay, northeastern Argentina, and southern Brazil. When a member died, their mortal remains were buried and the maté plant was planted on the same spot on the grave. When the plant was fully grown, it was harvested, an infusion was prepared, and it was drunk with the family. These rituals were performed because people believed that in this way the spirit of their relatives buried there would grow in the maté plant and pass into their bodies through the maté tea to stay with them. The Avá were practically inclined and also planted various types of vegetables next to the maté bush because they believed that this would promote the growth of the plants. They were right, as there was no better fertilizer than that of their deceased ancestors.

The Guaraní are an indigenous people from central South America. They live today in Brazil, Argentina, Paraguay, and Chile. They spread the infusion in gourds, which they called Mati. They came to Paraguay and cultivated the bush, as it had both invigorating and healing properties. The priest Pedro del Montenegro wrote: *"God has helped this poor land with this medicine, for it is more beneficial to him than chocolate, and it came to his natural inhabitants as well as cocoa in the east. For these very hot and humid lands cause serious slackening of the limbs, and we see that they usually sweat excessively, and wine and warm things are no remedy to suppress it. But Yerba Mate is, taken in hot weather with cold water, as the Indians use it, and in cold or mild weather with hot or warm water."* Parallels can also be drawn to the coca plant, which was used by the indigenous peoples of Peru to increase performance. The Spaniards forced many indigenous people to hard labor in the mines or to other physical work, which they could only endure with maté. Many of them carried small leather bags, the Guayacas, in which they stored maté leaves, which they either chewed or put in a gourd filled with water and took the broth with a straw, with the closed teeth serving as a filter.

In South America, per capita consumption is just under 7 kg, outside South America the tea has not yet found so many friends. Could maté tea have replaced Assam tea from India? The English actually thought about maté tea as a replacement in the 18th century. Because it was stronger and above all cheaper than the tea from the British colony. But they decided against it. They feared the monopoly of the Jesuits and did not want to make their arch-rival Spain even richer. The existing tea trade with India and Ceylon—today's Sri Lanka—brought in enough money, so the idea was dropped again.

Coffee and Caffeine—a Popular Drug?

Who doesn't know them, the colleagues in the office who drink liters of coffee. Are these people coffee addicted or caffeine dependent? According to current research, coffee addiction is not clear, and the smallest common denominator is that we do not become addicted at everyday amounts. The potential for dependency is low. At high doses, our reward system is not stimulated and we feel rather uncomfortable because the tannins make themselves felt. But it is undisputed that there is withdrawal: those who stop abruptly complain of headaches, shaky hands, and fluctuations in blood pressure. The good news, however, is that these withdrawal symptoms disappear after a few days and coffee no longer rules our lives. But do we want that?

With a cup of coffee we take in between 50 and 100 mg of the psychostimulant, so about half compared to a cup of espresso at our favorite Italian place. If we eat chocolate, we find on average 90 mg in 100 g of semi-sweet chocolate and still 15 mg in the same amount of milk chocolate. In chocolate, we also find another purine alkaloid, theobromine. Particularly beautiful is the derivation of the name, because "Theo Broma" is the translation from Greek and means "food of the gods". Who would want to contradict here?

Friedlieb Ferdinand Runge (1794–1867, Fig. 4) discovered and isolated caffeine for the first time. Encouraged or requested by the great Johann Wolfgang Goethe, Runge investigated the great scholar's question of whether the antidote to atropine could be found in coffee beans.[138] As we know, he did not succeed in this, but he found caffeine in large quantities and isolated it. Runge was from Hamburg and initially studied medicine, then chemistry in Jena and Berlin. It was not a given that the young Friedlieb would become a chemist. As a poor pastor's son, he could only attend public school because there was not enough money. This was followed by a pharmacy apprenticeship and after graduation, he went to university. With a thesis on the dye indigo, he graduated in 1826 and was very much a child of the then so popular dye chemistry. The thing with caffeine was not his field of research. Rather, he tried to find technical applications for coal tar, which was then a waste product in the production of coal and

Fig. 4 Friedlieb Ferdinand Runge (1794–1867)

coke. Goethe's influence on Runge is not well documented. Jena was not far from Weimar and the two were friends. Perhaps the idea, which Runge enthusiastically adopted, arose in a conversation between the two over a cup of coffee. But coffee was more of a medicine in Goethe's time and was often taken as an antidote. But here Goethe was mistaken, because he considered it a good antidote for treating the side effects of atropine from the deadly nightshade *(Atropa belladona)*.

Caffeine was not discovered only once. The great French natural product researchers Pierre Joseph Pelletier, Joseph Bienaimé Caventou, and Pierre-Jean Robiquet repeated the isolation and also had the white caffeine in hand in 1821. The dissemination of scientific knowledge took a little longer, as we could read in the chapter about morphine. The three Frenchmen did not know their German colleague. The multiple discovery of caffeine and the purine alkaloids is not surprising from today's perspective.

It is interesting that caffeine in its chemical structure resembles the building blocks of deoxyribonucleic acid (DNA) in our genetic material and is probably one of the oldest alkaloids in evolutionary history. Today, purine alkaloids are known in over 100 plant species and are much more widespread in the plant kingdom than other alkaloids. The caffeine content in the leaves of the coffee plant, at up to 1.7%, is even below that of the tea plant *Camellia sinensis* with 2 to 3%. You have now learned that coffee, chocolate, and tea contain the three purine alkaloids, and you may have wondered whether our body incorporates the three purine alkaloids into the DNA or whether they are dangerous. In long-term experiments, it could not be determined that coffee or purine alkaloids have a toxic effect. Although 16 mutagenic substances are known in coffee, there is no clearly defined case of cancer. The purines have effects on our psyche, but neither they themselves nor their metabolites are harmful. Some coffee consumers ask whether coffee can trigger gout attacks because of the purines. No! The good news for heavy drinkers is that other metabolites are formed and excessive coffee consumption does not lead to a gout attack, as uric acid is not formed from caffeine and Co. The purines cannot be metabolically converted into each other in the body. Theobromine does not become caffeine and that is not broken down into theophylline.

But coffee is not just a pleasure product. It has amazing effects on our body, as we will see later. But let's stay with the history of coffee for a moment when it comes to its medical significance. We all know that coffee and coffee beans came from North Africa via the Arabian Peninsula to Europe and from there to the whole world. After its cultural discovery, it swept over us in three waves. With industrialization, the young industry in Europe needed workers who were fit and not drunk. Because the drinking water was too bad and made people sick, everyone, no matter what class they belonged to, drank beer or wine. This was also accepted and what could happen? The worst thing was when someone fell drunk from a horse or fell asleep during field work. The new world of industrialization turned everything upside down. Drunken workers were a danger to themselves, to others, and to the expensive machines whose operation was their task. Mistakes could be devastating, yes, deadly. But the new working class also had to remain efficient, because the 16-hour working day, which was later reduced to 10 hours, drained their strength, and the new black hot drink made the tired workers awake again. In this first wave, which also captured the intellectuals who sometimes met conspiratorially in coffee houses, which the spying authorities did not like at all, coffee became a mass product.

The second wave is associated with the Dutchman Alfred Peet, who was supposed to enter the family coffee trading business. But Peet was not interested and found the idea of traveling the world more exciting. In this gap year, as many high school graduates do before starting their studies, he traveled to the USA. There he was shocked by the poor quality of the coffee and decided to do something about it. Alfred founded Peet's Coffeebar and brewed his own coffee. Taste and quality prevailed and he sold his roast blend to other companies. Among them was a company from Seattle, which later built the largest coffee chain in the world. With Howard Schultz, who first as an employee and then as CEO of Starbucks brought the corporation to the top of the world, a huge consumer market was opened up, offering all coffee specialties. Coffee became a cult and a consumer good that is indispensable from the modern to-go world. This strong market penetration was the beginning of the third wave that swept over us.[139]

Real connoisseurs and coffee purists got their money's worth. Away from the standard bean, towards coffee-to-stay and specialties, which also opened up a market for luxury. There were now local roasteries and small brands for individual taste. This market experienced a huge upswing, especially through online trading. Today, it is the organic coffee from fair trade, prepared with rituals reminiscent of Japanese tea ceremonies. French Press, Cold Brew and the new barista feeling are in vogue. The disposable paper cup is out.

The author of this book is also a passionate coffee drinker, even though he unfortunately has to drink decaffeinated coffee for health reasons. One of his favorite coffees is stomach-friendly and very digestible. It is made from a coffee bean, the production of which sounds rather unappetizing—Kopi Luwak. These beans have a special journey behind them. They are semi-digested and are obtained from the excrement of the free living civet in Indonesia. Genuine Kopi Luwak is not easy to get in Indonesia. The cat-like and nocturnal civets eat the fruits of the coffee bushes and excrete the core of the coffee fruit through the intestine. At this point, we take the time to correctly classify the coffee bean botanically, because the botanist refers to the so-called bean as a stone fruit, because it is a completely hardened core surrounded by pulp. But I stick to the colloquial term coffee bean, because it should not sound too scientific here. This coffee bean now passes through the entire intestine of our civet and undergoes a biological fermentation. First, acids act on the beans, later digestive secretions also have an effect. Especially lactic acid bacteria nibble at the bean and break down tannins such as protein and mucus, which usually lie on the coffee bean. It seems to be the lactic acid bacteria that provide the special taste. The sugar in the bean is converted into lactic acid, which creates the natural flavors that we perceive as taste. At least for most of us, because the British comedian John Cleese described the taste after a cup of coffee as *"earthy, musty, mild, syrupy, full-bodied, with jungle and chocolate notes"*. Interesting!

If you have acquired a taste for it, pay attention to which coffee you buy. The demand is high and the supply is small. To get the fermented beans, thousands of civets are kept in captivity and fed one-sidedly with the coffee fruit. The animal welfare organization PETA rightly describes this keeping as animal cruelty. Alternatively, animal-free fermentation is practiced in the Erlenmeyer flask. The microbes are obtained from the animals' intestines and cultivated. 1.5 kg of beans are added to this culture suspension of 2.5 l and kept at 35 °C for 12 hours. At predetermined time intervals, the pH value is adjusted, almost exactly as one would imagine the change during intestinal passage. [7, 140, 141] Does it taste good? You have to try it yourself.

Back to the medical significance of caffeine and coffee. How does caffeine affect us humans? The effect is a sophisticated biochemical system in which caffeine occupies a receptor and displaces the actual substance that wants to dock at this receptor. It is the adenosine that can no longer dock at its receptor and no longer ensures the release of glutamate and dopamine. This effect makes you awake. But caffeine also works behind the cell wall, which no longer allows the messenger substance in the cell to be broken down. Its concentration increases and prolongs the effect of adrenaline, we stay awake and active longer.

One more thing: In a country that is rather hesitant about coffee with its more than 1000-year-old tea tradition, the Japanese have found another medical application and enjoy bathing in roasted coffee beans. A bath is recommended for the treatment of skin diseases, as not only the caffeine but also the tannins certainly have a positive effect on skin diseases. So don't throw away your cold coffee, but pour it into your next bath, you don't usually treat yourself to anything else.

Caffeine and Theophylline—a Strong Coffee Against Asthma

When one wonders in which countries tea is predominantly consumed, England, India, and Japan come to mind. All these countries have distinct tea cultures. When we think of Japan, we think of green tea, served with mindfulness and reverence in Buddhist monasteries. If coffee was the quick pick-me-up, tea in our cultures is the mindful "chiller" or "downer". In addition to these unique cultural habits, in the second half of the 18th century,

Swedish King Gustav III, who despised coffee, wondered if there was also a medical difference? He was convinced that coffee was a devil's brew, and he wanted to prove it. For this purpose, he chose a pair of twins from among his subjects who were sentenced to death, and ordered that one should drink only tea and the other coffee under medical supervision for the rest of their lives. They remained in custody and each had to drink three pots of coffee or tea every day. This was probably the longest long-term study ever. It did not end when the doctors or Gustav III died, but with the death of the tea-drinking twin at the age of 83. Unfortunately, it is not recorded when the coffee-drinking twin had his last cup of coffee.

It was not until the 19th century that coffee was taken more seriously as a possible herbal medicine, and the birth of the rational medical research of coffee alkaloids can be seen with the recommendation of the English doctor Henry Hyde Salter in 1860 in an essay in the Edinburgh Medical Journal, who recommended strong coffee for asthma: *"One of the commonest and best-reputed remedies of asthma, one that is almost sure to have been tried in any case that may come under our observation, and one that in many cases is more efficacious than any other, is strong coffee."* [One of the most common and well-known remedies for asthma, which is almost certain to have been tried in any case that comes to our attention, and which in many cases is more effective than any other, is strong coffee.][142] Salter did not understand why, as he freely admitted in the same publication. He can be forgiven, for in the middle of the 19th century, neither adenosine receptors were known nor was it understood biochemically, what asthma actually was.

This strong coffee that Dr. Salter recommended contained, in addition to caffeine, the purine alkaloid theophylline, which was not discovered until late, in 1888, by Nobel laureate Albrecht Kossel (1853–1927). Whether Kossel was a tea or coffee drinker, we do not know exactly, but that was not the reason for the discovery, as he was studying nucleic acids. Deoxyribonucleic acid consists, among other things, of the two purine building blocks adenine and guanine, which are very similar to theophylline. He is considered a DNA pioneer who provided the basic insights into the chemical structure of DNA and later paved the way for Watson and Crick. The chemist Emil Fischer and his assistant Lorenz Ach confirmed the chemical structure of theophylline, the simplest of the three purine alkaloids, by their own synthesis, and Wilhelm Traube (1866–1942) described in 1900 the industrial synthesis of purine alkaloids, which is still used today. Traube was a gifted chemist who brought many ideas, was appreciated by many colleagues, and was appointed head of department at the Chemical Institute of the University of Berlin and later as a full professor on the advice of Emil Fischer in 1929. His life ended tragically in Gestapo custody. As a German of Jewish faith, he was arrested on the morning of September 11, 1942, beaten and mistreated for defending himself against his tormentors. Colleagues like Otto Hahn and Walter J. V. Schoeller intervened on his behalf, but their plea for his release came a few hours too late, as Traube succumbed to his injuries on September 28.

Friedrich Blume felt older than he was when he climbed the stairs to his beautiful third-floor apartment in Sandhof out of breath, as his wife said, shaking her head. She was already standing in the door, her apron tied, watching the World War veteran and Iron Cross recipient gasping up the stairs. He didn't want to admit it to himself at 52, but it wasn't just the high blood pressure that was bothering him, but also the shortness of breath that tormented him even with the slightest exertion. Once he reached the top, she helped him, the cavalry captain who had served bravely, out of his coat. He had often wondered if it was really the high blood pressure that the young Dr. Hirsch had diagnosed at the city hospital, or if it was the French gas attack and a leaky gas mask that were causing his shortness of breath. He sank into the leather armchair, thought, and heard his wife working noisily in the kitchen. Had this young doctor even served? He was fresh from university. He would have to ask him next time. His thoughts were interrupted by a call from the kitchen. "Do you have the medicine?" she called. He answered with a loud yes. Just as he had learned in the Reichswehr, clear and distinct, like the response to a command. "I have it in my coat pocket. Can you bring it to me?" She didn't answer, it became quiet, and he knew she was searching in his coat. "It's a miracle that there are still medicines in the pharmacy in these troubled times." She handed him the pack of Theocin®, which he looked at closely. He recognized the Bayer cross. "Bayer!" he said briefly and looked at her from his leather armchair. "It must be good if it's from Bayer!" He waved with the box. "Take one," she replied just as briefly. "I'm making dinner. Please be at the table in 20 minutes." He nodded and asked for a glass of water.

"Doctor, with the new medication you've given me, I'm feeling much better." Mr. and Mrs. Blume were sitting with him in the consultation room. The old cavalry officer had come to him because he was suffering from severe heart complaints. Although he was just over 50 years old, his gray face made him look like a man of 60. He could never sleep, he complained. But now he could close his eyes for a few hours at night. Samson Raphael Hirsch (1890–1960) nodded, leaned forward, the chair creaked with every movement. He looked at the medical record: *Has allegedly been suffering from bronchial catarrh and heart complaints for 2 years. Digitalis and a Nauheim spa cure were unsuccessful. Complains of severe shortness of breath, which makes walking impossible, has not slept for weeks. Nightly asthma attacks."* He nodded again and looked at him. "Can you breathe better now?" he asked back. "Yes, much better," Mrs. Blum quickly answered for him. "My husband has struggled so much." The husband nodded. "Good, then I'll continue to prescribe it. Mr. Blum, shall we see each other again in four weeks? I want to know if the theophylline continues to do you good."

He had dismissed the two from his consultation, they were the last before lunch break. Hirsch sat in his chair and thought. He had not heard that theophylline works against asthma. He would go to the library tonight and see if there were any publications on it. He followed scientific research, but did not want to rule out that he was not completely up to date due to work and long shifts. Should he call his colleague from the pulmonary department? Yes, he was interested, he picked up the receiver and dialed 231. "Hello, this is Samson. Do you have time and feel like going to the cafeteria with me? I have a question about theophylline. I would like to discuss it with you." A short pause ensued. "Good, then see you soon. There's sauerkraut and sausages."

Encouraged by the success with the old cavalry officer, he also administered theophylline to Adolf Rademacher, who worked as a house servant for a family on Gartenstraße in Frankfurt. The young man was not yet of age at 18 and suffered from severe shortness of breath. The young man appeared weak at the first visit, had rickets and had been suffering from cough and cold for a year. The colleague from the pulmonary department gave him theophylline on his suggestion, and to the surprise of both, the clinical picture improved significantly. His greatest success was Lilly Bahnemann, who had been suffering from severe rheumatism for three years and had been bedridden for a year. She could no longer work as a secretary, she barely made it to the hospital and could hardly stand on her feet. She strictly rejected his suggestion to take Spasmopurin® in suppository form, as this would be an unrefined therapy. He made it available to her, and she must have overcome herself, because later she practically urged him to give her more suppositories, as she was doing exceptionally well.

Hirsch published his results in 1922 in the Clinical Weekly and received great resonance and recognition for his case studies. He showed that theophylline helped old and young people and that its antispasmodic effect was far superior to that of caffeine. While it was known from individual book chapters and reports that coffee helps, theophylline was used in medicine as a blood pressure lowering or diuretic. Almost simultaneously, two physicians from Johns Hopkins University published studies in 1921 and 1922 on the effect of theophylline in pigs, confirming Hirsch's statements. David I. Macht and Gin-Chin Ting conducted the studies on isolated bronchial muscles of pigs because they were surprised by the unexpected activity of various drugs that were actually prescribed to lower blood pressure. This included theophylline, which immediately caught their attention. Theophylline was the first drug to act on the kidney and vasodilation, and became the prototype of modern blood pressure reducers, even though these have a completely different mechanism of action and a completely different chemical structure.[143] The medical profession of the 1920s remained skeptical. In the 1930s, the prevailing opinion was that asthma could be explained by irritation of the phrenic nerve or a vagus neurosis. Others suspected mucosal changes as the cause. Still others believed in a simple catarrh.

In asthma therapy, the purine alkaloids could not really convince, and yet it is astonishing that they were prescribed for almost four decades until they were replaced by salbutamol in the 1960s. The background can be speculated about excellently. The publication by Macht and Ting did not receive sufficient distribution, Hirsch ended his scientific activity in 1927 and stopped research in this area, a more effective drug, ephedrine, came onto the market, and the erroneous theory of vagotonia by Eppinger and Heß, who assumed that the autonomic nervous system functioned pathologically in an unknown way, was not accepted by most doctors.

Theophylline and the other purine alkaloids have a major disadvantage, which we briefly mentioned above with caffeine, the substance is very poorly soluble in water. What is not soluble cannot be well absorbed by the body, which is why there was a search for better soluble derivatives. Even Kossel, who managed to isolate theophylline, was aware of the poor solubility. It had to dissolve 300 mg in 60 ml, which was acceptable for oral intake, but decidedly too much for injections of a maximum of 10 ml.[144] The pharmacists of the time tried to solve the problem by forming salts with sodium acetate and sodium salicylate, but the pH became too alkaline and the tissue irritation during injection was too great.

In 1908, Reiner Grüter from the Chemical Works in Berlin managed to solve the problem through a chemical modification. He combined theophylline with ethylenediamine and, according to his own statements, found a *"surprisingly high solubility in water."* Since the chemical substances of amines can be toxic in medicine, he had his new compounds tested by Carl Neuberg (1877–1956). Neuberg was one of the most important biochemists of the 20th century and headed the Department of Animal Physiology at the Kaiser Wilhelm Institute for Biochemistry in Berlin. His animals felt well, he could not detect any toxic effects. The new derivative was marketed under the name Euphillin®. Because of his Jewish faith, the Nazis forced Carl Neuberg into retirement in 1934. His successor was Adolf Butenandt, whom we got to know in the last chapter. Further purine derivatives followed, such as Aminophyllin in 1937, Pentoxiphyllin, Pentifyllin in 1951 and Dipyramidol in 1959, all of which pursued the same goal, namely to achieve better solubility in tablet form. With the exception of Dipyramidol, all purine derivatives have left the big stage of pharmacology. Only Dipyramidol is usually given in combination with acetylsalicylic acid for coronary heart disease and as a platelet aggregation inhibitor, especially in patients with heart valve replacement.

Where does the path of caffeine, theobromine, and theophylline lead? Is there a path or is it a dead end? The idea of theophylline is being revisited today, as this purine has anti-inflammatory properties and synergistically improves the effect of corticosteroids. This is good for people suffering from chronic obstructive pulmonary disease like COPD. Caffeine can embark on a new career as a drug against dementia. Rodents performed better with caffeine and showed a positive effect on their long-term memory. Three cups of coffee a day are supposed to have these positive effects on us humans as well, but current studies also show that accompanying symptoms such as restlessness, aggression, depression, or anxiety are intensified with increasing coffee consumption. The problem with all these studies is that they often do not differentiate between coffee as a multi-component mixture and caffeine. For example, drinking four cups of coffee a day, which is supposed to be normal, can be good for us because we also take in other accompanying substances such as tannic acid, polyphenols, and many more. With the isolated caffeine in energy drinks, we lack their positive effects and suddenly everything is different and also opposite. What is recommended? Do what is good for you, eat a balanced diet, and enjoy everything in moderation.

Sildenafil—from Waterfalls and Hard Facts

Did it have to do with the sexual liberation of the 1960s that interest in erection and its dysfunction increased? While erectile dysfunction was only dedicated a single page up to the third edition of the Campbell Textbook of Urology in 1972, it was already 180 pages in 1998, in the eleventh edition. In 2000, the Ig Nobel Prize was even awarded for the recording of the function and movement of male and female genitals during coitus in an MRI machine.[145] Between us, this is the anti-Nobel Prize, which really nobody wants and which honors questionable or funny science. Check out the website of the Ig Nobel Committee, you will be amazed.

How erection works was almost unknown until the 1980s. It was simply a taboo, and who would have given money for the research of the cardiovascular conditions on the penis? Nobody! The ideas that came up throughout history to explain what physiologically happens in the body during an erection were rather amusing. In ancient Greece, it was assumed that air is pumped into the penis, which is produced in the liver. It was not until Leonardo da Vinci doubted this theory and suspected that blood flowed into the penis. The Dutch doctor Reinier de Graaf (1641–1673) confirmed this theory by letting water flow into the penis of corpses he dissected. However, he was more concerned with the anatomy of the female sex organ than with erection and became the first describer of the sensitive zone, which the German gynecologist Ernst Gräfenberg (1881–1957) anatomically located as the G-spot and vaginal pleasure center. It took another 200 years until the physiology of erection was researched on a molecular biological level. Now it was clear that an erotic idea in our head, a manual stimulation of the penis leads to a signal that forces the blood flow into the corpus cavernosum and then seals it for an appropriate time.

At first glance, caffeine and sildenafil (Fig. 5), the active ingredient of the drug Viagra®, don't have much in common. Or do I have to worry about getting an erection when I drink my morning coffee? Certainly not, but if you look closely at the structural formula of sildenafil, you will see that it is caffeine. Coincidence? Most likely not, because all three substances have a similar physiological mechanism of action. This has long been known for caffeine, and the researchers at Pfizer were surprised that sildenafil is so similar to caffeine and prevents the messenger substance cyclic adenosine monophosphate from being broken down. Caffeine is a mild stimulant that

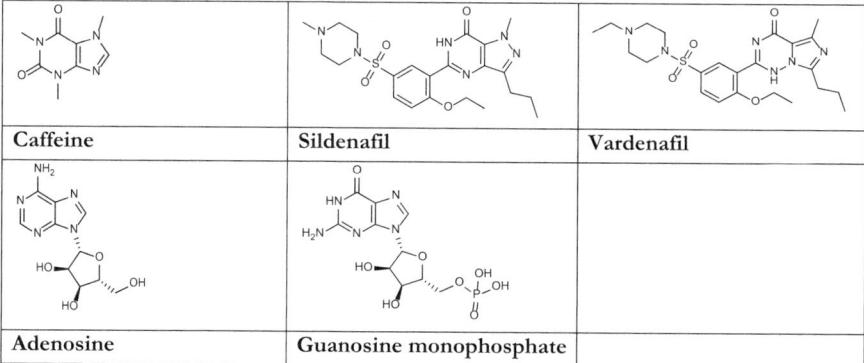

Fig. 5 Caffeine, physiological Messenger substances and derived Potency drugs

is indispensable for many people to start the day. It affects the brain, but also other organs and muscles of our body. It stimulates the heartbeat, increases blood pressure, and dilates the bronchi. Caffeine can also change our mood, or is our mood sometimes so bad because there is no coffee, our favorite drug, nearby? In any case, caffeine is not among the top twenty common drugs with potential for physical and psychological dependence. This list is led by alcohol followed by heroin.

The reason caffeine works so well in our body is due to an ancient biochemical regulatory principle, because our body has the messenger substance adenosine. This adenosine is somewhat larger and more complex than caffeine, but it has a similar substructure to caffeine. When we drink coffee, we take in caffeine, which is an antagonist of adenosine. In our brain, signals are constantly exchanged via the small gray cells, regardless of whether we are sleeping or working hard. These messenger substances are important for function and coordination in our head and this miracle of neurobiology is explained in one of the following chapters on drugs. In these biochemical actions, our nerves consume energy. In all our nerves and also in all our other cells, this energy is adenosine triphosphate (ATP). When energy, i.e. ATP, is consumed, adenosine is produced as a breakdown product, which protects our brain from overexertion and has its own binding sites that allow our brain to relax. The more work, the more adenosine and the easier the brain goes into its rest mode. Adenosine also slows down the release of other messenger substances in the brain that are important for signal transmission. Now caffeine comes into play. It binds as false adenosine to the receptors and blocks the signal transmission. The adenosine does not reach the nerve cells and no signals are transmitted. The nerve cell believes it should continue working because no adenosine is calming it down. If more coffee is drunk, the caffeine also penetrates the cells and inhibits an important enzyme there that is responsible for throttling the cellular messenger substance. This enzyme is phosphodiesterase, which inhibits the breakdown of the cellular messenger substance cyclic adenosine monophosphate (cAMP).

Sildenafil acts on the related phosphodiesterase 5 (PDE5), which prevents the breakdown of a messenger substance also related to cAMP and cGMP. The two cyclic relatives do not differ noticeably, but adenosine is replaced by guanosine (see Fig. 5). This small chemical change makes the difference that sildenafil is not our chemical coffee, but works specifically in the best part of the man. Exactly here, in the corpus cavernosum, the erectile tissue of the penis, a high concentration of cGMP releases nitric oxide (NO) from the amino acid arginine, which leads to a dilation of the vessels. Blood flows in and the penis erects. Sildenafil is a very selective inhibitor of the

NO/cGMP-specific phosphodiesterase. Men with potency problems, scientifically described as erectile dysfunction, were grateful and Pfizer had found another gold mine.

It may be surprising why Pfizer was researching in the field of erectile dysfunction. After all, it is not one of the classic diseases for whose cure humanity has long and hopefully waited. That's true. With Viagra®, a lifestyle drug has been found that elevated erectile dysfunction from a fashionable disease to a serious indication. What fashionable diseases are is discussed in the chapter on modern drug research, but let's take a moment to find out how chance helped make sildenafil a blockbuster.

Was it worth it?, the young student wondered. He was uncomfortable with what was happening below his beltline. But he needed the money, the phone bill had to be paid, and he wanted to fly to Thailand with Sandra this winter. For the blood donation, he got just 50 pounds, which was far too little, and he also had to deduct the costs for getting to and from this study. Participating as a volunteer in the drug trials was more attractive. The agency offered him 1200 pounds. He had to take a blood pressure reducer several times, which a pharmaceutical company wanted to bring to the market. He felt healthy, regularly exercised at college, it wouldn't knock him down. It was just too bad that he was here in this clinic. He was allowed, no, he had to stay this weekend and he was bored. He could watch TV with the others, whom he didn't know, play table tennis, or just stay in his room. According to a plan that the head nurse explained to the young subjects in a lecture upon arrival, blood was taken from them, urine samples were collected, and there was only standardized food, which was the same for everyone. The nurse had inserted a cannula so that she didn't have to constantly stick him in the arm. She was careful when inserting the venous cannula and he hardly noticed it at all. This medication was strange. The nurse explained that it was a blood pressure reducer, but he felt his blood pressure increase and only in the part of his body below the beltline. Was that normal? It was uncomfortable for him when she always came into the room and he was lying on the side with an erection in his pants. He could have slept with her too, because the erection in his pants made him wild. Maybe they had a few tablets left and he could try it with Sandra?

"Hello Ian, do you have a moment? Some of the subjects are asking if they can take a few tablets home?" Ian Osterloh looked at her questioningly. "Take home?", he asked confused. "What do they want with UK-92480? They don't have a problem with high blood pressure. Besides, the stuff hardly works. Pfizer wants to stop the clinical tests. They no longer believe in it." The nurse shrugged her shoulders. "I don't know, the question came from several. But I noticed that some of the guys were always lying on their side when I entered the room. I think, they have a boner." Ian Osterloh leaned back in his chair and looked at her seriously. "We know from the phase-I studies that some of the men got an erection, but we didn't take it seriously." "Maybe you should." She left him in the office. Osterloh shifted in his chair and scratched his chin. Not a bad idea, he thought. But at Pfizer, they had no experience with impotence. That was a taboo in our prudish society anyway. Should he really go to the board with this absurd idea? He did and convinced Pfizer to finance another study. This time with men who couldn't get it up anymore. Was that a disease and if so, what could it be called? He turned on the computer, the modem beeped and he had access to the internet, as they called it. With Netscape, he logged into PubMed and did a little search with various keywords like disorder, erection, penis, impotence, to see if there was anything useful. There it was: erectile dysfunction. Yes, that was the right disease for UK-92480.

Ian Osterloh took a liking to the research topic, because impotence seemed to be a male problem that pharmaceutical research was hardly interested in.[146] Some natural substances were found that were known through the ethnomedicine of some indigenous peoples around the world, such as ginger or *Muira puama* from northern South America. In Indian ethnomedicine, the alkaloid yohimbine is still extracted today from the Indian snake root (*Rauvolfia* species). Another alkaloid was papaverine. The French doctor Ronald Virag (1938) accidentally injected it into an artery that supplied the penis with blood during an operation. The anaesthetized patient got an erection that lasted two hours. Later, he injected papaverine into his own best piece and was convinced that erectile dysfunction could be treated with papaverine. Enthusiastically, he published his results in the medical journal Lancet in 1982.[149]

It is certainly questionable whether this method improves the mood of a couple who want to sleep together, and whether the injection of men is accepted will. One man—Giles Brindley (*1926)—did not let this stop him from demonstrating the pharmacological effect of papaverine in a self-experiment very offensively. The British physiologist and doctor, who also turned to the fine arts and invented the electronic bassoon in the 1960s, was invited to the annual meeting of the American Society for Urology in Las Vegas. He was to give a lecture on impotence and its treatment. Anyone who has ever been to a scientific conference knows that the topics are usually presented very dryly and that some scientists lack the gift of inspiring the audience. Not so Giles Brindley, who wanted to explain the effect of papaverine very vividly and injected the substance into his penis 15 minutes before his lecture in his hotel room. The course of the lecture was legendary. Let's let Dr. Laurence Klotz, Professor of Urology at the University of Toronto, report from his point of view: *"About 15 minutes before the lecture, I took the elevator to go to the lecture hall. On the next floor, a slim, older-looking and bespectacled man entered the elevator, wearing a blue tracksuit and carrying a small cigar box. He seemed quite nervous and shuffled back and forth. In the elevator, which was slowly filling up, he opened the box and began to examine and rummage through the 35mm slides inside. I stood next to him and could vaguely make out the content of the slides, which appeared to be a series of pictures of an erect penis. I concluded that it was indeed Professor Brindley on his way to the lecture, although his clothing seemed inappropriately casual. The lecture took place in a large lecture hall, with a raised lectern separated from the seats by a few steps. It was an evening program, sandwiched between the day sessions and an evening reception. The lecture was relatively poorly attended, perhaps 80 people in total. Most of the participants came with their partners, obviously on their way to the reception. I sat in the third row and in front of me were about seven middle-aged male urologists with their partners in full evening dress. Professor Brindley, still in his blue tracksuit, was introduced as a psychiatrist with broad research interests. He began his lecture directly. He had, as he said, hypothesized that the injection of vasoactive substances into the body of the penis could induce an erection. Since he did not have easy access to a suitable animal model and in view of the long medical tradition, he took himself as a research object, and he began a series of self-injection experiments, including papaverine, phentolamine, and several others. His slide lecture consisted of a large series of photos of his penis in various states of tumescence after injection of various doses of phentolamine and papaverine. After seeing about 30 of these slides, there was no doubt that the therapy, at least in the case of Professor Brindley, was effective. Of course, one could not rule out that erotic stimulation played a role in achieving these erections, and Professor Brindley admitted this. The professor wanted to present his arguments as convincingly as possible. He pointed out that in his opinion, no normal person finds the experience of giving a lecture to a large audience erotic, stimulating, or erection-inducing. Therefore, he had injected papaverine into himself in his hotel room before coming to the lecture and deliberately wore loose clothing to be able to demonstrate the results. He walked around the podium and pulled his loose pants tight around his genitals to demonstrate his erection. At this point, I, and I believe everyone else in the room, was stunned. I couldn't believe what was happening on stage. But Prof. Brindley was not satisfied. He looked skeptically down at his pants and shook his head in dismay. Unfortunately, the result is not clear enough. He promptly dropped his pants and shorts and revealed a long, thin, clearly erect penis. There was not a single sound in the room. Everyone had stopped breathing. But the public display of his erection from the podium was not enough. He paused and seemed to think about his next step. The sense of drama in the room was palpable. He said with a serious face: "I would like to give some in the audience the opportunity to confirm the degree of swelling." With his pants around his knees, he waddled down the stairs and approached the urologists and their partners in the front row to their horror. As he approached them, with his wobbling erection in front of him, four or five of the women in the front rows threw their arms in the air and screamed loudly. The scientific merits of the presentation were outstanding due to the nature of the lecture and the presentation* [of the genital, author's note]. *The screams seemed to shock Professor Brindley, who quickly pulled up his pants, returned to the podium, and ended the lecture. The crowd dispersed in a state of stunned confusion. I imagine that the urologists who were present with their partners had a lot to explain. The rest is history."*[147] Giles Brindley remained dedicated to the research of erectile dysfunction and published the results of 17 more erection-inducing substances. It is not hard to guess which penis was the subject of study.

The data from the two clinical trials at Pfizer confirmed that treating angina pectoris or hypertension with UK-92480, as the Pfizer code for Sildenafil was, would be a pure waste of money. Although Peter Ellis, Nicholas Terrett, and Ian Osterloh did not fully understand the biochemical mechanism of erection and the literally outstanding effect of UK-92480, they sold the idea of erectile dysfunction to the board, which approved the money for another study on men with ED, as the disease was abbreviated. There was a control group without Sildenafil and a verum group with Sildenafil. But how should the erection be standardized? In the words of Ian Osterloh, it went like this: *"In the first study, the men watched erotic videos while a device monitored the circumference and hardness of their penises. The initial results were encouraging and showed that the drug was much more effective than a placebo."* That UK-92480 had a demonstrable effect was clearly shown by the results. The study provided hard facts in the truest sense of the word. Pfizer wanted to seize the opportunity to establish itself as a pioneer in the field of impotence. Action had to be taken quickly and money was no object to bring UK-92480 to market. UK-92480 also got a proper name. It was a portmanteau of Vigor and Niagara. The name was meant to symbolize the vitality and power of the endless waterfalls on the border between the USA and Canada.

In 1985, there were only 15 scientists working on Sildenafil, but by 1997, when Pfizer applied for market approval, there were already 500 employees. Just one month after FDA approval on March 27, 1998, Viagra® was on the shelves of US pharmacies and was prescribed 300,000 times in the first quarter. In these three months, Viagra® already generated sales of 400 million USD and by the end of the year, the blockbuster broke the billion-dollar barrier. From 2003 to 2016, the drug annually brought in 1.7 billion dollars for the pharmaceutical company. Viagra® was so successful and profitable that even drug cartels counterfeited the drug. Especially China, which does not adhere so much to existing patents, became the world's chemical laboratory for Sildenafil, and it is estimated that at the peak of the legal sale of real Viagra, about 75% of the Viagra® tablets offered on the internet were counterfeits.

Pfizer was fortunate to bring a socially controversial drug to market at the right time. Two more fortunate circumstances were also to boost the business. In 1998, the FDA lifted the ban on drug advertising, which had prevented pharmaceutical companies from directly addressing patients. Pfizer could now directly address men with erectile dysfunction. The success was overwhelming, and numerous advertising campaigns followed. The second fortunate circumstance was the Clinton-Lewinsky affair in the Oral Office of the White House. Former President Bill Clinton and his intern Monica Lewinsky found each other more than sympathetic. In a very broad public, not only theoretical but also practical and very vivid discussions about possible sexual practices of the two were held. Prudish America had to deal with sex and the Monica-Gate increased the acceptance that men bought the little blue pills in the pharmacy.

In addition to its medical applications, Sildenafil has also brought some unexpected benefits for species conservation. The sex life of pandas, which hardly reproduce in zoos, was invigorated by the intake of Viagra. Viagra was also beneficial in protecting endangered species such as seals and tigers. In the Chongqing Zoo, a pair of tigers from southern China were rekindled to mate. The two had lived in the artificial world of the zoo since their birth and had lost interest out of boredom. It is remarkable that the exhausted sex life of these animals is supposed to symbolize strong, Chinese manhood. Unfortunately, the genitals of these animals are still illegally traded as potency enhancers in China today. The result is a senseless slaughter of these animals, pushing them to the brink of extinction. The price for these senseless and ineffective animal genitals dropped when Chinese men believed more in Viagra® than in Traditional Chinese Medicine to cure their erectile dysfunction.

Quinine

Rheumatoid arthritis

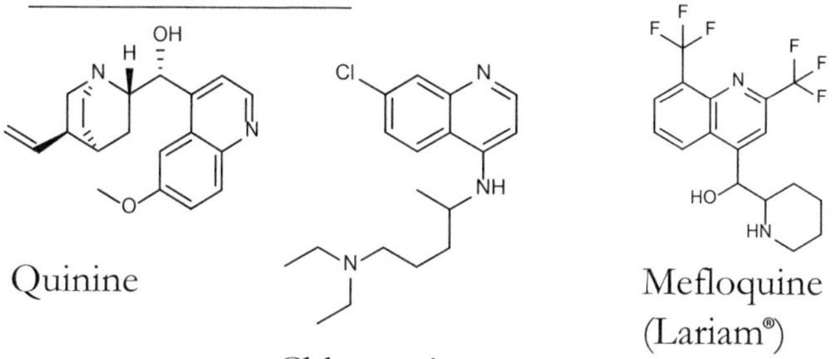

Quinine

Chloroquine (Resochin®)

Mefloquine (Lariam®)

Natural Product Group

- Quinoline alkaloid

Effect / Indication

- Antimalarials
- Rheumatoid athritis

Malaria has all but disappeared from the memory of the West. We still flinch when our travel destination is on the list of dangerous tropical diseases, but we believe we are safe with the drug Artemisinin for malaria prophylaxis. Hardly anyone knows that every year 219 million infections occur and we have to mourn more than 400,000 deaths. A number that is higher than that of the flu combined. Fortunately, the number of cases is decreasing and

O. Kayser, *From the Maluku to Molecules*, https://doi.org/10.1007/978-3-662-69923-2_6

every decline since the beginning of the 20th century is associated with the development of new active ingredients such as Chloroquine (from 1940), Artemisinin (from 1990), but also repellents and insecticides like DDT from 1945. In 1910, 40% of people still lived in a risk area, but 100 years later it was only 24%, and not only in Africa and Asia, but also in the USA and Latin America.

The treatment of malaria has always been associated with natural substances and goes far back in the history of Europeans and their colonies. The most famous and successful natural substance for treatment is quinine, which was isolated from the bark of the cinchona tree in Peru. It is astonishing that quinine served as a model structure for many antimalarial drugs and antipyretic drugs for more than 200 years. Only in the 1960s did the young Chinese scientist Tu Youyou (*1930, Fig. 1) find a substance against malaria in Project 523 of the Chinese government that, unlike quinine, contained no nitrogen and was not an alkaloid. For her discovery 40 years ago, she received the Nobel Prize in 2015 together with two colleagues.[148] Now we are already far into the present, but let's take another look at the past before we approach Artemisinin.[149]

Fig. 1 Tu Youyou (*1930) (r) with her tutor Lou Zhicen (1920–1995) (l) at the Chinese Academy of Medical Sciences, 1951 (public domain, Wikipedia)

Kina Kina—the Miraculous Fever Tree

Chinchón is a small town southeast of the Spanish capital Madrid. Tourists rarely stray into the town of just over 5,000 inhabitants, but alongside local handicrafts, tourism is one of the sources of income for the residents. In this sleepy nest, it gets hot in the summer and hustle and bustle is not welcome, but once a year the residents really let loose and hold a bullfight on the Plaza Mayor. When everything is over, the tranquil calm settles back over the village like stirred-up dust, you can discover a monument on the northern edge of the square in front of the Plaza de Palacio with the inscription "El pueblo de Cinchon a da Francisca Enriques de Rivera, Condesa de Cinchon Virreina de Peru, descubridora de la quina, en 1629" [The village of Cinchon was given to Francisca Enriques de Rivera, Countess of Cinchon, Viceroy of Peru and discoverer of the cinchona bark, in 1629]. The citizens of Chinchón commemorate Francisca E. de Rivera, the Countess of Chinchón as the discoverer of the cinchona bark in 1629. Francisca E. de Rivera was the second wife of the Viceroy of Peru, Luis Jerónimo Fernández de Cabrera Bobadilla Cerda y Mendoza. She was with the Viceroy for eight years in Peru and spent most of her time planning banquets, balls, and fiestas.

The story of the young countess is supposed to be the following, but unfortunately, there is not much truth to it. In 1638, the lady fell slightly ill with malaria, without suffering the typical fever attacks. At that time, malaria was not only widespread in South America but also in Europe, and it was only a matter of time before a mosquito bit her in Peru and injected the Plasmodium pathogen under her skin. The governor of the province of Loxa in southern Peru is said to have learned of the illness of his boss's wife and sent a package labeled "kina kina" to the court. Kina is the old Peruvian word for a particularly valued tree bark. This reddish-brown bark was crushed and boiled by the countess at the court in Lima. The fever disappeared, she regained strength and became healthy. The red powder from the Andes seemed to be worth its weight in gold. The countess had more bark brought for the inhabitants of Lima and wanted to take it with her to Spain in 1639 when her husband's service in Peru ended. However, she died on the return journey in the port of Cartagena in today's Colombia.

This story pleased the Swedish botanist Linnaeus so much that he gave the genus of cinchona trees the name *Cinchona* (Fig. 2). For easier pronunciation for non-Spanish speakers, he left out the first "h". However, the likely story is much more sobering, and it can be assumed that the Jesuits, who

Fig. 2 *Cinchona officinalis L. (public domain, Wikipedia)*

came as missionaries to South America, were the actual discoverers. In a very systematic investigation, they searched among the Indians for unknown medicinal plants. The Augustinian monk Antonio de la Calancha (1584–1654) already reported in 1633 in his work "Cronica Moralizada del Orden de San Augustin en el Peru" [Chronicle of the Morality of the Order of Saint Augustine in Peru] about a tree, which they called fever tree. He also gave the dosage with the weight of two silver coins in his book. However, it must have taken a lot of overcoming for the patients to swallow the brew, as it must have tasted disgustingly bitter. Anyone who has ever drunk our modern tonic water and found it bitter should consider that the dilution is 1:200,000. The brew cooked at that time was extremely concentrated.

It is also interesting to ask whether malaria and the malaria pathogen existed in South America before the arrival of the Europeans. Probably not, because in the old records of the Aztecs and Mayas, which are still preserved today, no disease is described that resembles our malaria. Science assumes that the Plasmodium pathogen came to South America with the European settlers and slaves from Africa. In Africa, the situation was quite different. Here, malaria was and is endemic, but there were no native medicinal plants that helped as well as the South American cinchona bark. Whether the Africans could have been helped with quinine was irrelevant to the

European colonial powers. This is already shown by the numbers of their low appreciation of the people in their colonies: in 1905 there were only 21 doctors in French Africa, by 1946 there were 310 European and about the same number of African doctors in 23 hospitals. It should be noted that they were very unevenly distributed among the hospitals where either Europeans or Africans were treated. In the British colony of Nigeria, there was only one doctor for every 200,000 inhabitants and one hospital bed for every 12,000 Nigerians.

Without knowing the biology of the malaria pathogen, the Jesuits knew that the tree could heal, and they shipped the cinchona bark to the Old World, where it was accepted by the people. As a precursor of salicin and acetylsalicylic acid, the bark became known as a fever-reducing agent throughout Europe. Cinchona bark was always taken when fever set in, and not just for malaria. With Napoleon's continental blockade, the economic decline of the cinchona bark trade began, as the British attacked the Spanish ships from South America and the goods could no longer be delivered to Europe in sufficient quantities. Now the search for native plants for fever reduction began and the rise of the spirea bush and aspirin® began. For the cure of malaria patients, the plant and quinine continued to be of great importance. Malaria was a major problem in Western and Southern Europe until the 19th century. Even in Germany, the disease was endemic and feared. A severe malaria epidemic was documented in parts of Northern and Central Germany in 1557, and at the beginning of the 19th century, every second child in the far north of Friesland is said to have been infected with plasmodia. But even at the beginning of the 20th century, individual cases still flared up (Fig. 3). A last flare-up of malaria was reported after the Second World War in 1945/1946, when bomb craters stood full of water in the warm summer months, became breeding grounds for mosquitoes, and infected returning soldiers and prisoners returned to their hometowns.

In addition to Spain, England was the second colonial power interested in quinine and the cinchona tree. The Empire expanded spatially and economically, as industrialization sucked the raw materials from the colonies and sought markets for its products. British soldiers and administrative officials were sent to areas of the Empire, where the sun never set, that often had a massive malaria problem. The British government therefore had a great interest in breaking the monopoly of cinchona bark. With the help of the Royal Botanical Gardens in Kew, near London, seeds were to be brought into the country and trees grown. The huge glass dome of the Palm House, which can still be admired today on a visit, was built for this purpose. If the cultivations were successful, seeds and seedlings were sent to the colonies,

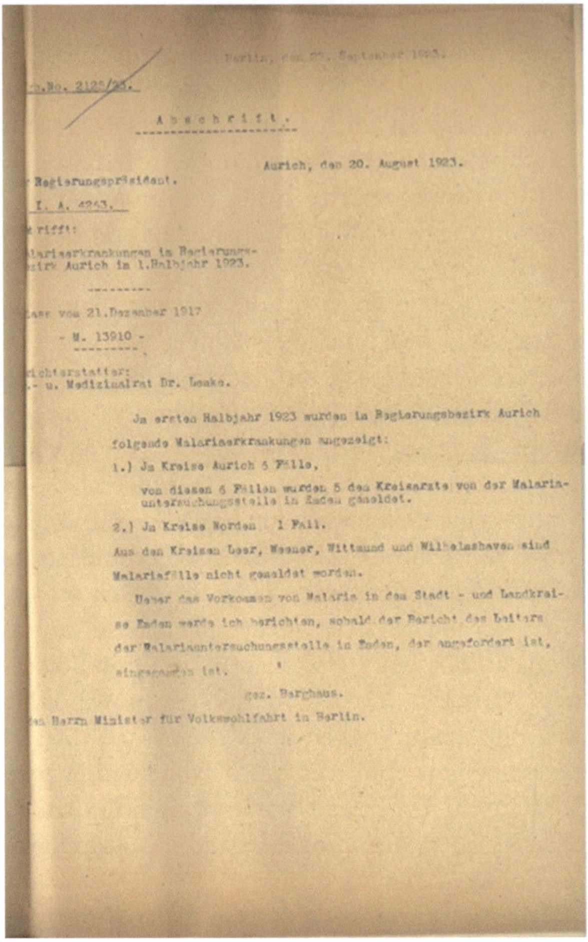
Berlin, den 22. September 1923.

..h.No. 2125/23.

Abschrift.

Aurich, den 20. August 1923.

..r Regierungspräsident.

I. A. 4263.

..triff:

..lariaerkrankungen im Regierungs-
..zirk Aurich im 1.Halbjahr 1923.

..lass vom 21.Dezember 1917

- M. 13910 -

..richterstatter:
..- u. Medizinalrat Dr. Leake.

Jm ersten Halbjahr 1923 wurden im Regierungsbezirk Aurich folgende Malariaerkrankungen angezeigt:

1.) Jm Kreise Aurich 6 Fälle,

von diesen 6 Fällen wurden 5 dem Kreisarzte von der Malariauntersuchungsstelle in Emden gemeldet.

2.) Jm Kreise Norden 1 Fall.

Aus den Kreisen Leer, Weener, Wittmund und Wilhelmshaven sind Malariafälle nicht gemeldet worden.

Ueber das Vorkommen von Malaria in dem Stadt - und Landkreise Emden werde ich berichten, sobald der Bericht des Leiters der Malariauntersuchungsstelle in Emden, der angefordert ist, eingegangen ist.

gez. Berghaus.

..n Herrn Minister für Volkswohlfahrt in Berlin.

Fig. 3 The Aurich government reports on seven cases of malaria in Northern Germany, August 20, 1923 (Federal Archives, Koblenz, DE)

especially to India and then Ceylon, now Sri Lanka. Kew Garden officially became the imperial site of biopiracy, and three secret expeditions were dispatched.

How Tonic Water and Gin Became Pretty Best Friends

Sir William Jackson Hooker (1785–1865, Fig. 4) was the director of Kew Garden and was commissioned by his government to find ways to cultivate cinchona trees in their own colonies. The industrious director was looking

Fig. 4 William Jackson Hooker (1785–1865) (public domain, Wikipedia)1

for a suitable candidate, whom he found in Clements Markham (1830–1916, Fig. 5). He persuaded the researcher to travel to Peru and South America a second time to bring seeds and seedlings of the cinchona tree to Kew. Hooker found Markham more than suitable, as he had conducted anthropological studies among the Quechuas in South America on his first trip in 1852 and 1853 and already knew where the trees were. But the man in whom Hooker placed such great hopes was not a botanist, and his second journey from 1859 to 1861 took place during a time of great political instability, as the South American countries had just broken away from the Spanish crown. To solve the problem, Markham took his future brother-in-law Richard Spruce (1817–1893) from India, who was an experienced moss researcher and botanist and had explored the flora along the Amazon from the mouth to deep into the jungle for more than 15 years. Spruce (Fig. 6) was also well suited, as an official of the India Office in Nilgiri, India, he knew exactly about the climatic conditions of the future plantation site.[150]

Fig. 5 Clements Markham (1830–1916)(public domain, Wikipedia)1

The men set off on a secret mission, as heavy penalties were imposed for smuggling seeds. The two were a perfect complement, as Markham knew where the best places in Ecuador were, and Spruce was the botanist. In the first year after their arrival in Ecuador at Chimborazo, they collected 637 plants and more than 100,000 seeds. That year, some seedlings were grown next to their tents in the jungle, which were to make their way to Kew Garden. The two researchers, their seeds, and seedlings survived this journey by luck. Peru issued a decree stating that any cinchona smuggler should have their feet chopped off. Markham heard of this and was very concerned that this ban could also be enforced in Ecuador. He urged his porters on, and they carried the 637 plants and several thousand seeds on their backs to the Rio Babahoyo, where they were expected by Robert Cross, who was responsible for the transport to England as a gardener in Kew Garden. It was a wild river ride, and the valuable cargo threatened to be lost, but Cross knew his craft and secured the plants. In the coastal city of Guayaquil, the mysterious cargo was transferred and brought to England by steamship. The effort for this theft and violation of Bolivian law was enormous, the success modest. The plants only had a low quinine content of about 0.2%, which promised little economic viability in the Indian cultivation area Nilgiri. Few

Fig. 6 Richard Spruce (1817–1893) (public domain, Wikipedia)1

plants were then brought to the botanical garden in Calcutta, where they perished in the unsuitable climate. This could have been the end of the theft and cultivation of the cinchona tree outside South America.

The alpaca breeder Charles Ledger (1818–1905, Fig. 7) from Chulluncayani in southern Peru had his own problems. He was tormented by the worry of how he could circumvent an export ban on alpacas to Australia. The Peruvian government banned the export of these animals under threat of a ten-year penalty. But he had been living here in the Andes for 20 years and the Englishman was used to problems, because in addition to the wool of his animals, he traded in ores, furs, leather and sometimes also with cinchona bark. The man was well-versed in all matters, as he brought his alpacas over the Andes to Chile in six years to circumvent the ban and ship them from there to Australia. Once there, the business had collapsed, his partners were either bankrupt or had returned to England. A small sum of money remained, too much to die and too little to survive. What happened to his alpacas is unknown to this day.

Fig. 7 Charles Ledger (1818–1905) (Permission of the Wellcome Foundation, London, UK)1

But he was a resilient man and returned to Peru in 1864, where he now wanted to make his fortune by smuggling cinchona seeds. Ledger read in the newspaper how expensive cinchona bark was and that it had been possible in Europe to isolate the even more expensive quinine. The prices of quinine shot up on the world market and he wanted a piece of the pie. Greed ruled among the cascarilleros. These bark collectors felled the trees to peel off the bark. It is reported that at the height of the boom, 25,000 trees were felled, debarked and about 900,000 kg of bark were exported in 1860.[151] No one adhered to the old Jesuit rule of planting five new trees for every one felled. Ledger foresaw that soon no tree would stand on the Amazon. The British feared that no more supplies would come from South America to meet the demand in their own colonies in Asia. The British government sent a second secret expedition to the Andes, and the three men Markham, Spruce and Core met the alpaca breeder. An employee of Ledger was an Inca Indian, without whom the mission would not have been successful. Manuel Incra Mamani (†1871, Fig. 8) was a kind of partner for Charles Ledger and worked with him. The Englishman sent him into the Andean forests of Bolivia. Mamani knew all species and could distinguish them like no other

Fig. 8 Manuel Incra Mamani (?–1871) (public domain, Wikipedia)1

based on the shape of the leaf and the shape of the seed. When Mamani returned from the jungle to civilization, he had two sacks with about 25 kg of seeds with him. That was a huge amount of about 6 million pieces. These are only a few millimeters in size and a seed weighs less than 3 mg. Ledger wanted to circumvent the export ban with a trick and scattered the seeds like dirt on chinchilla furs and sent them to his brother in London. Hoping to make a good deal, Charles' brother brought the seeds to Kew Garden, where no one was interested. He went to other botanical gardens and finally to traders of cinchona bark, but no one could get excited. He couldn't believe what he heard and panicked that the seeds would dry out and not germinate. In his distress, he sold a pound for 100 guilders to the Dutch consul in London. The rest rotted and was eventually thrown away. The second catastrophic business of Mr. Charles Ledger in his life.

Why the British did not want to buy the seed is unclear to this day. Was the price too high or did they not trust Charles Ledger's brother? From today's perspective, it was definitely a big mistake, because the Dutch bought the seeds at a favorable price and thus established the cultivation of cinchona bark in their colony Java (today Indonesia). The special thing about this species was that it contained up to 10% of the coveted substance quinine. This species is also cultivated in Africa today and represents the industrial source. It is pointless to point out again at this point that this was also a special case of biopiracy, which was only made possible by the help of the local Inca Indian Manuel Incra Mamani. The story of Mamani ended unhappily. He was arrested by the police for aiding smuggling, beaten and died in prison from the consequences of the abuse. Against this background, it is hard to believe that in honor of Charles Ledger (1818–1905) the stolen plant species with the high concentration of quinine was named after him *Cinchona ledgeriana*.

The theft of the cinchona bark was a great loss for Peru. It was so significant for the economy and self-understanding of Peru that the cinchona tree was even included in the national coat of arms. Export was forbidden, as the trade brought a lot of money into the country. However, the monopoly disturbed the colonial powers and biopiracy was supported from the top. The stolen seeds made their way from Europe to South India and to Java, where cultivation was attempted due to supposedly similar climatic conditions. The British planted the cinchona trees in India in the Nilgiri Blue Mountains near the city of Udhagamandalam. Why this particular place? It seemed perfect for the botanists. The duration of sunshine throughout the year and the altitude of 2210 m was identical to the location in the Andes, and Nilgiri was almost at the same latitude with two degrees closer to the equator. Great effort was made. The English forced many residents of the Nilgiri region to clear and cultivate. Inmates in local prisons were released and put to work in the fields, as the prison cells were needed as rooms for drying the bark. In the end, the yield remained below expectations.

The Dutch were to be more successful in planting cinchona trees. With better seeds and more suitable climatic conditions, they planted cinchona in Java near Bandung. The city was an old seat of the sultans, located higher in the mountains, and due to its more pleasant temperatures, the city was also preferred as an administrative seat to the fever-ridden Batavia on the Java Sea. Batavia, the old name of Holland, was then a small nest with a few hundred inhabitants. The sailors avoided it because of the dirt and mosquitoes. The city was therefore also called the "graveyard of Europe", as the death rate for Europeans was very high. Anyone visiting the city today, which is now called Jakarta and has a good 20 million inhabitants, should go to the harbor and the old town, which still reminds a little of the long past colonial era of the Dutch and their trading company VOC.[152]

When cinchona trees are planted in plantations, the bark can be harvested for the first time after six years. The trees have the highest quinine concentration from the age of ten to twelve years, and they are usually peeled at the age of 18. For a state, these are too large time frames, which are rarely politically maintained. The clever Dutch came up with the idea to privatize the business. They gave the seeds to two farmers, who bred seedlings and plants between 1850 and 1860. From 1880, the Dutch had a monopoly, supplied 90% of the world market with the East India Company, and dictated the world price.[153] The Dutch are known as clever businessmen and it is not surprising that they had a global cartel until 1913 and dictated the world market price with 100 guilders per kilogram.[154] It was the first global pharmaceutical cartel, which only broke up with the occupation of Indonesia by the Japanese in 1942.

The two Frenchmen, Pierre-Joseph Pelletier and Joseph Caventou, also tried their luck and successfully isolated quinine from the bark they brought with them in 1820. The discovery came at the right time, as the European colonial powers were dividing the world among themselves. However, they could only be successful if they could treat or, even better, prevent infectious diseases such as sleeping sickness, leishmaniasis, and malaria in their soldiers in their tropical colonies. Both Frenchmen worked as pharmacists and chemists at the Ecole de Pharmacie in Paris. Pierre-Joseph Pelletier (1788–1842, Fig. 9) spent his entire career at the Ecole, where he

Fig. 9 Pierre-Joseph Pelletier (1788–1842) (public domain, Wikipedia)1

completed his studies at the age of 22 and became a professor at the age of 37. Interestingly, in addition to his academic career, he also worked as a pharmacist at 48 Rue Le-Pelletier, where he made most of his discoveries in his laboratory. It was the heyday of French natural product chemistry. Almost every year, Pelletier published a groundbreaking paper, as interesting plants were abundant from the colonies. He described morphine, emetine, caffeine, quinine, and the constituents of curare, even if he was not always the first to discover them. In his short life of 54 years, he is said to have handled hundreds of natural substances before he died of colon cancer in 1842. His academic twin, Joseph Bienaimé Caventou (1795–1877, Fig. 10), also studied at the Ecole, but was a military pharmacist with Napoleon and only returned to Paris after the lost Battle of Waterloo. The two complemented each other well in natural product chemistry and together discovered many natural substances in the short time of their collaboration from 1817 to 1821 in Rue Jacob, which have already been presented in various chapters. However, he was more interested in biological applications and was

Fig. 10 Joseph Bienaimé Caventou (1795–1877) (public domain, Wikipedia)1

particularly fascinated by tuberculosis. Like Pelletier, Caventou also became a professor, but for toxicology. He was the first to isolate the green chlorophyll and the highly toxic alkaloids strychnine and brucine.

The process for obtaining quinine is simple and has remained similar in its basic features to this day. Pelletier and Caventou suggested determining the quinine content of the bark after crushing it in a first step, allowing the bark to swell in water, and then alkalinizing it with milk of lime and caustic soda. This makes quinine and other alkaloids poorly water-soluble and lipophilic. The two Frenchmen then added an oil that extracted these alkaloids. The oils turned red. Modern methods do not use oil, but toluene, which can be worked with more reliably on a technical level. To extract the alkaloids from the oil again, the two added sulfuric acid, which converted the quinine into a salt. Salts are highly water-soluble and return to the acidic, aqueous phase. In principle, this change between lipophilic and hydrophilic phase can be repeated, but from the usually third or fourth repetition, by-products are increasingly enriched, which greatly complicate later processing. The white quinine is obtained by multiple crystallization from alcohol and water and can then be dried and sold.

Both researchers are to be credited with dealing with the chemistry of at least six cinchona species. In 1820, they isolated quinine in their pharmacy laboratories, which was not easy. The reason for the difficult purification lies in the close chemical relationship to quinidine, which is a stereoisomer, and cinchonine. Depending on the plant species, one finds more of one substance or the other. Thus, yellow cinchona bark contains more quinine, gray more cinchonine, and red bark contains both alkaloids in equal parts. The two Frenchmen renounced patents and a large fortune because they were inspired by the idea of serving free science. However, they were honored by the French Academy of Sciences and received a prize money of 10,000 francs.

Quinine, which has been freely available to medicine since its isolation in 1823 by Pelletier and Caventou, had a tremendous impact on the beginnings of the pharmaceutical industry. It was the natural substance that led to the founding of the chemical and pharmaceutical industry. Between 1840 and 1890, factories were established that isolated quinine, such as Merck AG, which we still know well today. The South American natural remedy for fever was in high demand, and anyone who had access to the plant or a small plantation could suddenly become rich. The Dutch East India Company definitely wanted this and started the first cultivation attempts on Java. As with coca, it did not go well and was a failure. Only the cooperation of government, extractors, and scientists at home with the farmers on Java made the business successful. The Dutch farmers learned to deal with the plants from the New World. From 1870, the cinchona tree could be successfully cultivated on the plantations, but it did not displace other economically lucrative plants such as tea, rubber, and coca. The Dutch took on an additional business. They knew that the cultivation of coca and the cinchona tree was crisis-proof, no matter what the state of the economy, because people always get sick.

He was known as the Humboldt of Java, so renowned and respected was the botanist Franz Junghuhn (1809–1864, Fig. 11), who, as a German, was commissioned by the Dutch government to manage the plantations and the Botanical Garden in Bogor. Despite his knowledge and his will to plant millions of Cinchona plants in Java, he did not have a green thumb. Either the plants did not grow or they did not produce quinine. Back home, people were impatiently looking at the colony and the Dutch Parliament in The Hague also dealt with the Cinchona question several times in the 1860s. During his lifetime, which ended in 1864, the venture remained

Fig. 11 Franz Junghuhn (1809–1864) (public domain, Wikipedia)[1]

unsuccessful and his successor Karel Wessel van Gorkom came up with the idea in 1866 to provide willing plantation owners with seedlings free of charge. These seedlings were made from so-called cuttings from mature cinchona trees by cutting off the side branches. The seedlings sprouted from the seeds of Charles Ledger, which promised a high quinine content. With this variety, not only the propagation and cultivation succeeded, but finally also the increase of the content to 10 and sometimes up to 20 %. The cultivation was so successful that from 1883 the free allocation of plants was stopped and the state support from the motherland was reduced to a minimum.[155] There were no more subsidies and by 1900 the Kinabureau, as the Dutch central administration in Amsterdam was called, already managed 120 plantations.

Sales boomed and with the trees, profits also grew, until the downturn came at the end of the 1880s. The farmers were insecure, demand kept falling and the price per kilo was only 20 guilders for the kilogram of bark. If the competition of the many growers remained high, it would be bad for everyone, which is why the cartel "Kinaver" was founded. This "Vereeniging ter Bevordering van de Belangen van de Kina-Cultuur" [Association for the Promotion of the Interests of Cinchona Cultivation] was an interest group of 100 farmers and the first pharmaceutical cartel in history.

Kinaver demanded that the abundant cinchona bark should now also be sold to the local population, financial support must come from The Hague and the extraction as a value-adding process step to Bandung be brought. Minister Lovink came to negotiate with the producers on site in Java. He saw the problem and promised to work out a plan for the factory. The scientist Pieter van Leersum (1854–1920, Fig. 12) was commissioned to develop this plan and in 1910 he worked out a business plan for a state quinine factory with an annual production of 60 t, which was mainly intended for the local market. The factory, named Bandoengsche Kinine Fabriek BV, was built west of the city and you can still see the building today.

The Kinavera Foundation wanted to build a second factory and excited its members and, the Minister for Overseas Territories however not. The second quinine factory was never built, as Colonial Minister J. H. de Waal Malefijt stopped all further planning and spending. The already existing first factory in Bandung was kept in operation, but was shut down a few years later. The cinchona bark was to be pressed out in the Netherlands. He distrusted the calculations and expectations of van Leersum and as a frugal official he was also not willing to spend so much tax money. The government made the smallest compromise with the farmers. In the Cinchona Treaty of 1913, which was extended every three years until the Japanese invasion, the

Fig. 12 Pieter van Leersum (1854–1920) (public domain, Wikipedia)1

plantation owners were guaranteed a fixed state purchase price and de facto 525 t of cinchona bark was taken off the free market.

With the First World War, quinine production fell to less than 20 % of pre-war levels in 1917, as the Allies blocked trade routes. After the war, there was hope for increasing production volumes, as public health care took on the problem of malaria. None other than Robert Koch (1843–1910) recommended the prophylactic administration of quinine to infected people in malaria-infested areas to break the life cycle of the parasite and eradicate the disease. This method of Koch's was taken up by the Italian doctor Angelo Celli (1857–1914), who convinced his government to eradicate the plasmodia in the lower boot of the country. This initiative was successful and prompted the League of Nations, as the precursor of today's World Health Organization, and the Rockefeller Foundation to support eradication worldwide.

To seriously implement these state measures, quinine was demanded from Java, and the price rose within a few years from 8 guilders (1913) to

25.96 guilders (1923). The monopoly seemed to work, but it was accused of exploiting an emergency situation after the World War. This may be true, but there would have been no more quinine if the Dutch government had not artificially kept the market alive. Despite the global economic crisis in 1929, production and demand continued to rise until 1933. The Dutch government wanted to continue to control trade and cultivation. A production at a lower level, stable sales markets, and a better foreign political reputation were the guidelines from The Hague.

Learning from Quinine does not Mean Learning to Win—the Lost Battle Against Malaria

In this wild history of the Dutch cartel, we can also see the origin of worldwide programs for the discovery of new synthetic antimalarial drugs, which were later widely used in the Vietnam War.[156] Until the 1940s, people had relied too long on quinine from Indonesia. With the occupation of Indonesia by the Japanese in World War II, all deliveries to Europe and the USA collapsed. The Japanese occupiers immediately seized the quinine factory in Bandung and shipped the tablets to Japan for their own troops, who also had a problem with malaria in the Pacific War. This dangerous dependence on only one supplier showed the Americans that they urgently needed an alternative for supplying their troops in World War II. In the USA, there has been a major malaria problem in the southern states since the importation of infected Africans as slaves. It is little known that more European settlers died from plasmodia than from tomahawks or arrows of the indigenous Indians. The US Army has been struggling with the weakening of its military power by this parasitic enemy since its foundation. The first purchase of cinchona bark for George Washington's troops for 300 dollars is as well documented as the efforts to supply the up to 80% infected Unionists with cinchona bark during the American Civil War.

The first attempts at synthetic antimalarial chemistry began with dye chemistry and England's realization that it needed to promote chemistry in order not to lose touch with continental Europe and especially Germany. In the 1840s, England saw itself being left behind by continental competition, especially Germany. A lot of money was spent on a program to raise chemical research to an international level. Personalities like Gladstone, Disraeli, and even the German-born Prince Albert felt called upon to promote applied chemistry in the country. Prince Albert (1819–1861), Prince Consort of Queen Victoria, persuaded the German August Wilhelm

Hofmann (1818–1892) to found the Royal College of Chemistry and to start his teaching activities in London. Hofmann was a gifted teacher and although he was German, he spoke almost perfect English. His lectures and experiments were even attended by the royal family. Hofmann, who was particularly appreciated for his didactics in experimental lectures, stayed in London for 20 years. After the death of his patron and benefactor Prince Consort Albert, he was called to the University of Bonn, which built a new institute for him according to his plans. The Friedrich-Wilhelms-University in Berlin was even more generous and offered more than the province. Hofmann moved from London to Berlin, where he stayed until his death.

Even before Hofmann traveled to London by ship and accepted the call, a young man was hanging around the capital as an autodidact. William Henry Perkin (1838–1907, Fig. 13) learned painting and was enthusiastic about mechanics. He built his first small machine and was fascinated by how the small parts moved and the gears meshed. But what he never let go of for the rest of his life was chemistry. A friend showed him a few chemical tricks, like how crystals became solutions. As if by magic, beautiful, ordered crystal structures were created, which fascinated the young Perkin, and he enrolled at the City of London School for Chemistry. The young man stood out less

Fig. 13 William Henry Perkin (1838–1907) (public domain, Wikipedia)1

for his student pranks than for his seriousness, and Thomas Hall, his teacher, immediately recognized the potential and saw that he would soon be bored here. He recommended Perkin to the young Hofmann, who was only 20 years older than his 28-year-old student. The two got along, and Hofmann, not yet knighted by the queen, took him on as an academic foster son and even handed over the management of the laboratory to him a few years later. But what should he do with the prodigy, what task should he assign him? The talent should try to synthesize quinine. That was a real challenge.

Perkin tried a lot and yet failed. He did not succeed in fully synthesizing quinine and it was not until 1944 that Robert B. Woodward and William von Eggers Doering succeeded, but even here historians still have questions and are not convinced. On the way to quinine, however, Perkin synthesized the dye mauveine or aniline purple in 1856. During his work, he did not yet wear gloves, his fingers turned purple-violet and Perkin liked the color. We do not know if it is true, but the first piece of clothing that was dyed with his dye was his sister's white blouse, which shone in a beautiful purple. Whether she screamed with joy is not recorded. Queen Victoria of England (1819–1901) wore a mauveine-dyed dress to the opening of the World Exhibition in London in 1862, which immediately attracted the attention of the press and the public. It became the cult color of that time and everyone wanted a piece of clothing in this color. Perkin became a wealthy man, as described in the chapter on Aspirin®.

The first attempt to produce synthetic quinine or new synthetic antimalarial drugs was not successful (Fig. 14). It was not until 50 years later that a small team from the Bayer company in Elberfeld near Wuppertal made the breakthrough. Fritz Schönhöfer (1892–1965) was a young chemist who the company recruited after his doctorate at the University of Heidelberg and hired at Bayer in 1921. In the 1920s, Schönhöfer developed various syntheses of aminoquinolines and in 1927 the company launched Plasmochin® on the market, which Schönhöfer had discovered in 1924. It is interesting to learn why the German company launched an antimalarial drug on the market. Because with the end of the First World War, there were no more German colonies in Africa where German colonists or soldiers were exposed to a risk of malaria. Perhaps it was the still ongoing research on methylene blue and azo dyes with antimicrobial effects that Paul Ehrlich discovered at the beginning of the new century. The young Schönhöfer was part of a team led by Werner Schulemann (1888–1975), who joined Bayer in 1919 and dedicated himself to malaria research. It is interesting that as a hobby ethnologist he acquired astonishing knowledge about the art and culture of Tibet and as a culturally interested person joined the NSDAP in 1937. As a

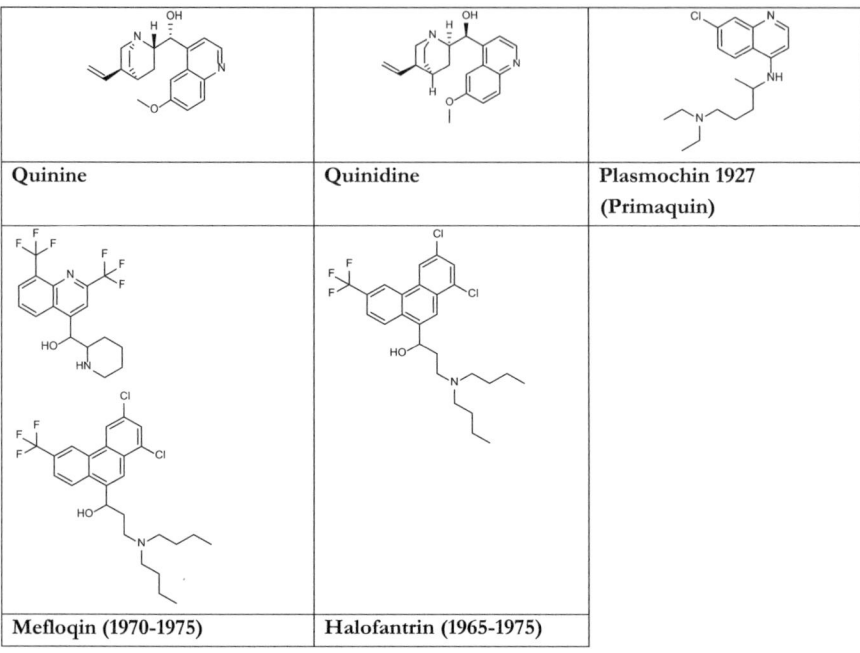

Fig. 14 Quinine and derived antimalarial drugs (selection) (work of the author)

follower with a Persil certificate, he was able to return to university service at the University of Bonn after the Second World War, which he left with his emeritus status in 1955.

Between 1921 and 1923, Schulemann led his research group with great enthusiasm, consisting of Schönhöfer, August Wingler (1998–1960), Fritz Mietzsch (1896–1958), and Wilhelm Roehl (1881–1929). The latter was the pharmacologist among the chemists who conducted the animal experiments. The basis of all considerations was that quinine has an alkaloid component that seemed important for the effect. They wanted to preserve this in the syntheses and where this damn complicated double ring system was, the Bayer chemists wanted to introduce other chemical elements. There was a division of labor between Schönhöfer and Wingler. Schönhöfer produced the alkaloid and developed ideas on how it could be chemically altered. Wingler implemented these chemically. During the work with the precursors, Wingler and his assistants fell ill because the occupational safety and ventilation were still inadequate. Nevertheless, they succeeded in producing various aminoquinolines, which the pharmacologist Roehl tested.

It is remarkable that the Bayer company and its employees were able to conduct such important research during the time of hyperinflation, the civil war-like conditions in the German Reich, and the eternal shortage everywhere. For example, Roehl lacked an animal model and could not quickly organize animal experiments in Elberfeld. Roehl was passionate about his work. *"I see it as one of the highest tasks of my life,"* he was later quoted as saying and was very creative in how the shortage could be overcome. He remembered the Greek doctor Phokion Kopanaris, who experimented with canaries at the Bernhard-Nocht-Institute in Hamburg before the First World War. He succeeded in infecting the small birds with the malaria pathogen and reproducing the disease as it could also occur in humans. Roehl picked up the idea and ordered female canaries, which were cheap and available in large numbers. The male birds were expensive because they were very popular in the domestic living room because of their singing. Roehl set up a bird station, infected the animals, and treated them with the synthetic antimalarial agents he developed in his chemical laboratory. Many of the substances worked excellently and were many times more effective than quinine.

Now it was time for the test on humans. Franz Friedrich Emil Sioli (1882–1949) from the Psychiatric Clinic in Düsseldorf volunteered: He first conducted simple toxicity tests and in a second step administered the substances to his patients, whom he deliberately infected with the malaria pathogen. Sioli's behavior is highly unethical from today's perspective and would have been so in the 1920s as well, had psychiatrist Julius Wagner-Jauregg (1857–1940) not made fever therapy known in 1917. The ambitious psychiatrist's idea was to kill the temperature-sensitive spirochetes responsible for syphilis by inducing fever. That syphilis was not only seen as an infectious disease will be shown in the next chapter. Because of the syphilitic effects on the brain and spinal cord, many infected people were also admitted to psychiatric clinics. As a psychiatrist, Sioli suggested injecting the patients with a fever-inducing substance, and the malaria pathogen was known for its fever bouts. This therapy was considered so important and groundbreaking that Julius Wagner-Jauregg received the Nobel Prize in 1927. Only with the discovery of penicillin was this type of therapy for syphilis patients discontinued. This "malaria therapy", which did not aim at killing the malaria pathogen, now offered a possibility to test the new active substances from Elberfeld and to eliminate the plasmodia. The clinical trials in Düsseldorf were successful, and the then colonial power England also showed interest. The drug Plasmochin®, called Primaquin® in England, brought hefty profits to Bayer AG, which was now operating under the umbrella of I. G. Farben.

Sioli's results encouraged Bayer AG to focus on improving the synthesis. Schönhöfer and Wingler had developed a synthesis with a lean yield of 3%. Bayer chemist Karl Schranz helped out and increased the synthesis to 30%. Now the marketing began and in a special supplement of the journal Archiv für Schiffs- und Tropenmedizin, the importance was pointed out. It was included in an important reference work for clinical drugs, the Martindale—The Extra Pharmacopoeia, and the then still living editor William Martindale could not resist putting the footnote *"England should have done it"* under the entry in 1932 out of sheer disappointment.

The first serious study on patients was conducted by Peter Mühlens (1874–1943) at the Bernhard-Nocht Institute for Tropical Medicine in Hamburg. He successfully tested Plasmochin® in 1925 and with this good data basis, Bayer expanded its cooperation with the United Fruit Company in the USA and the Royal Army Medical Corps. The United Fruit Company was just expanding its location in Costa Rica and wanted to produce fruits in Cuba and other South American countries and import them into the USA. The local workers suffered from malaria and the banana trader saw in Plamochin® a drug to stabilize the harvest and trade. Unfortunately, the drug fell short of expectations and showed too strong side effects in broad application. The effectiveness seemed to be limited. Today we know that Plamochin® only worked against the developmental stages of the malaria pathogen in the liver, but did not kill the stages in the red blood cells. These results were later confirmed in India and from 1930 Bayer provided a combination preparation with quinine, which they called Chinoplasmin®.

The team around Schulemann changed. Roehl did not live very long, as he died shortly before his 48th birthday from a staphylococcal infection. Wingler went to the USA to New Jersey to organize the business with Aspirin® for Bayer together with Winthrop Chemical, which Bayer had to give up after the First World War. He was replaced by Hans Mauss (1901–1953), who took over the syntheses together with the remaining Fritz Mietzsch. The two wanted to deal with a new class of chemical structures that had already been positively noticed in antibiotics. Julius Morgenroth, assistant to the great Paul Ehrlich, had discovered the acridines in 1911. They had already been successfully tested against amoebae and showed chemical similarities with quinine. However, the acridines showed high toxicity and attacked the optic nerve, which is why Morgenroth kept all substances under lock and key and they were forgotten for another 20 years. But for the Bayer chemists, there might still be something to gain. They synthesized 12,000 new acridine derivatives and handed them over to Walter Kikuth (1896–1968), who had been with Bayer since 1927 and replaced the deceased Roehl. The two chemists had a good nose when they transferred the plasmochin chemistry to the acridines. Because the chemical Erion was very active and Bayer brought a much better successor for Plasmochin to the market in 1932 with Atebrin®. The new active ingredient could kill all stages of the plasmodia and had fewer side effects. But one peculiarity worried Kikuth: All patients had yellow skin and yellow eyes. Should Atebrin® damage the liver and the patients suffer from jaundice? Fortunately not, because it was the typical yellow color of the acridines. Perhaps you have been prescribed Rivanol® for external skin disinfection in your life, then you will certainly remember the bright yellow color of the solution. The yellow Atebrin® was improved again in the mid-1930s and with Chloroquine brought to market under the trade name Resochin®. It was colorless, equally safe and remained standard until the 1970s.

By the way, it is not so rare that drugs can color the urine. Not seeing the usual yellow color of his urine causes a shock in patients, especially when it is a red coloration. Known for the red coloration is the tuberculosis drug Rifampicin, but also more well-known substances such as Phenytoin, Doxorubicin or Sulfamethoxazole can cause this. In the color spectrum, it continues cheerfully with the colors orange (Isoniazid), brown (Paracetamol, Nitrofurantoin), blue (Cimetidine, Propofol, Methylene blue and others) and at the end the black urine caused by herbal laxatives, iron supplements and Metronidazole. Some substances like Rifampicin also have the unpleasant effect that tears, sperm or vaginal fluid can be red.

Atebrin® did not play a major role for the Germans, the market was in England, India and later in the USA. However, during the Second World War, the substance became important for the USA, as no quinine was available due to the Japanese occupation of Java and the Germans stopped the delivery of Atebrin® to the enemy. For the Americans, it was the most important malaria drug in Southeast Asia and vital for their own soldiers. The Allies suspected as early as 1937 that Nazi Germany wanted to go to war and they could not count on the delivery of Atebrin®. The US company Winthrop and the English companies Imperial Chemical Industries (ICI), Boots and Wellcome were commissioned to produce Atebrin bypassing the German patents. This was not easy, because the German patents were vague and only a few pieces of information could be taken from them. A new synthesis had to be developed, which was achieved within two years. When the Second World War broke out, the chemists from ICI had not only completed this, but also a new factory in Blackley in North Manchester. Production began tentatively in 1939 with 10 kg. In 1943, it rose to 45,500 kg, from which 32 million tablets were produced. The success was to help the Allies in Southeast Asia a lot, because in 1943 almost every soldier was infected, with an incredible 750 infected per 1000 soldiers in action. With the use of Atebrin®, which was now called Quinacrine® in the USA and Mepacrine® in England, this number dropped to only 25 infection cases by 1945.[157]

What happened to Resochin®? The US Army did not benefit from the new antimalarial drug, which was only available to the troops after the Second World War. The reason lay in the clinical tests that had begun at the Bernhard-Nocht-Institute in Hamburg in 1938 and were completed with the start of the Second World War, but were withheld from the Allies. The Allies recognized the fundamental importance of researching drugs against bacterial and parasitic infectious diseases. They founded a consortium with the uninformative name Therapeutic Research Corporation. It was supposed to produce the new penicillin, test sulfonamides from German enemy territory and deal with drugs against tuberculosis in addition to malaria drugs. One infectious disease alone would have been a mammoth task. The large pharmaceutical companies synthesized and tested over 14,000 new substances. A huge task that was spread over three continents.

The British initially held back in the new consortium, they were a little afraid of collaboration and information exchange, because who should get the patents and thus the marketing rights? The Americans were more pragmatic, like Edward Mellanby, whom we have already met as the discoverer of the "Accessory Factor" [additional factor], which was later called Vitamin D. He wrote on April 17, 1943: *"The Americans are always a bit more relaxed about such things than we are. I suspect they believe that the primary goal is to get better malaria drugs, and it doesn't matter if someone makes a fortune out of it with public expenditure. I think such an attitude would not be in stark contrast to American standards of public life in general."*

Despite the great successes, not everything went smoothly among the Allies. On May 7, 1943, the Allies captured Tunis and the German Africa Corps surrendered along with their Italian allies. As war booty, the Americans got their hands on 5000 Sontochin® tablets. The leading military doctors immediately recognized the treasure, but did not know which chemical substance the tablets contained. In a hurry, the tablets were sent to the Rockefeller Institute in the USA to clarify the identity. Almost simultaneously with the realization that it must be 3-methylchloroquine, an astonishing effect was discovered for the candidate SN-183, which the US company Winthorp had already synthesized in 1941, but which had more or less been lost among the thousands of new substances. By March 1944, it was clear that SN-183 and Sontochin® were chemically identical. Hamilton Fairley confirmed the good effect on volunteers in the summer of 1944. Only with the occupation of Germany after 1945 did the Americans get all the documents about Sontochin®, which Walter Kikuth handed over to them in the bombed-out Wuppertal.

It seems to remain a sad truth that important drugs against malaria were discovered because of wars. The story of Mefloquine and Halofantrin cannot be told otherwise. The USA was in the Vietnam War and Vietnam is a tropical to subtropical country where malaria was and is a constant disease in the population. 128 US soldiers died of malaria and more than a million sick days were recorded in the troops, even though chloroquine was available. One reason was that the local mosquitoes carried the much more dangerous pathogen *Plasmodium falciparum* which was multi-resistant to chloroquine. The US government commissioned the Walter Reed Army Hospital in Bethesda near Washington D.C. to develop a new malaria drug with Mefloquine. But it only came on the market in 1973, shortly before the end of the war. It was too late for use in the war. For this reason, Mefloquine was tested on prisoners in the Stateville state prison in 1975. The WHO was thrilled and convinced the US government to release Mefloquine, which was hesitant to take over the costs for further developments and production. Michel Fernex and his team at the then Hoffmann-La Roche in Basel brought Mefloquine to market maturity together with the WHO. In 1985, it was released under the name Lariam® for the Third World.

The First Will Remain the Last

What remains of the discovery of quinine from the bark of the cinchona tree? Without quinine and the bark, the Panama Canal would not have existed or would have come much later. The Americans distributed 40,000 doses of quinine daily to the canal workers. England would not have become a nation of tea lovers because there would have been no tea plantations in India. The Empire would never have existed and without quinine, India would not have remained a colony for so long. There would be no oil production in Venezuela, no gin and tonic[158, 159], and no colonization of Africa, which was considered the grave of the white man. The official death rate of British soldiers on the Gold Coast, now Ghana, was 638 out of 1000 soldiers in 1836.[151] Cinchona bark became one of the Big Five along with rubber, sugar, tea, and tobacco. There is much more to quinine. The global rise of the pharmaceutical industry also has to do with cinchona bark and its extraction. Cheap and reliable, quinine was the remedy for malaria, but also for fever, before it was displaced by acetylsalicylic acid and paracetamol. A lot of money was made with cinchona bark and economic history was written.

The most beautiful gift of quinine is tonic water. It was invented, or rather conceived, by the British who wanted to combine the pleasant with the useful. In the hot colonies, refreshing drinks were always welcome. The bitter taste of quinine gave the bland water a better taste and the quinine was supposed to be a malaria prophylaxis at the same time. A good idea, but unfortunately, no one reached the therapeutic dose with the small amounts of quinine in the drink. In a classic drink, 80 mg/l are dissolved, but the therapeutic dose is between 500 and 1000 mg. The bitter drink was made palatable with sugar or gin and later also became part of the standard equipment of the British soldiers of the Indian Army. It was intended as a tonic for strengthening, from which the term "tonic water" was derived. The mixture of gin and tonic water was the birth of the well-known cocktail today.

What remains is a break in the field of antibiotic research. What began promisingly at the beginning of the 20th century with Salvarsan® and quinine is no longer of interest to today's pharmaceutical companies.[160] No company is willing to spend large sums on a remedy for malaria, amoebas or sleeping sickness if no money flows back into the coffers of the corporations in the end. Obliged to the shareholders, only projects are tackled that enable the famous return on investment in our western affluent society. We live in the luxury of having more than 80 beta blockers in the pharmacy, remedies for erectile dysfunction and dementia in the pharmacy, but we still do not have many effective and safe medicines against parasites. Even in 2024, people suffering from African sleeping sickness will be treated with melarsoprol, an arsenic-containing drug from the time of Paul Ehrlich. The first will remain the last for a long time, unless global climate change brings many of the so-called tropical diseases to us in Europe and forces us to rethink.

Coumarins—Rat Poison for the President

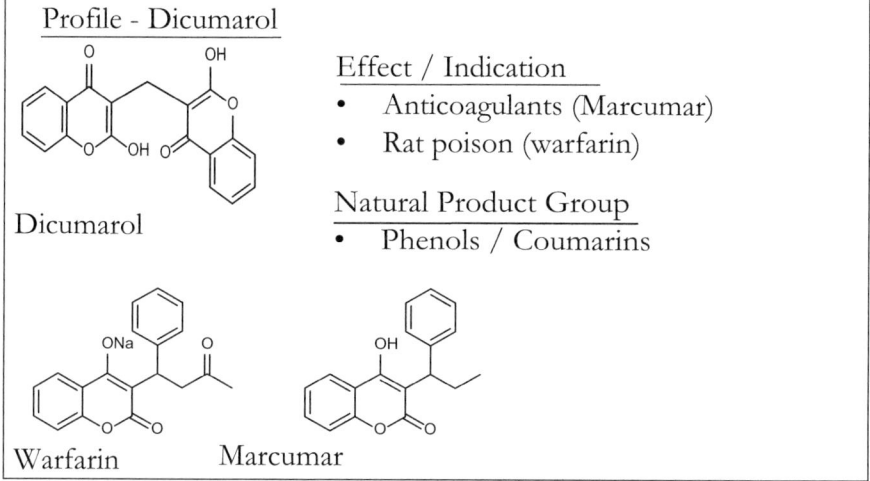

The veterinarian was quite surprised. He stood in the barn of the small farm to which the farmer had called him. It was drizzling and too cold for the winter of 1921. He put his hands on his hips and heard the cows mooing softly, a sign that they were suffering. "Ed, what have you done differently? Have you changed the feed?" he asked Ed Carlson in front of him, who put his hands on the belly of the dead cow. "No, nothing at all, the cows have been standing here for days eating what the Lord gives them." The fact that people here in North Dakota were still deeply religious was evident from the number of churches, which could not be found in any other state in the USA. Rudolph Lincoln had been living here as a veterinarian for a long time. He pondered: "What exactly have you given that God has put in front of your door?" The old farmer did not look up, for the dying cow was a fortune in this miserable farm crisis sweeping the country. His left arm lifted, and the paw with which he could have easily loosened the wheel bolts of his tractor pointed to the haystack in the gable. He looked up, a drop of water fell into his eye, and for a moment he was blind. Money was lacking everywhere and the roof was leaking. The farming family had been hit twice. They had sacrificed their son Theodore, God's gift, in this miserable war in Europe. He had died just before the end of the war in 1917 in a trench in Belgium from poison gas. Now came the agricultural crisis, which caused prices for everything to plummet. The Carlsons had borrowed money from the North Dakota First, bought new farm machinery, and electrified the farm. Hoping that it would take a long time for agriculture to recover in Europe after the war. Everyone had been mistaken, but he did not know exactly what had gone wrong. But something had gone wrong. "Hey, Doc, she's dead." The farmer's hoarse voice brought him back. "I'm sorry, but we need to know what's going on. I'll have the cow picked up and autopsied. I'll also take some of the hay." The farmer stood up and a gruff man stood in front of him. "If it doesn't cost me anything, do it." Rudolph Lincoln shook his head.

He had the dead animal taken away for examination. The animal was too large for the operating room, he put it in his garage and began to cut it open. The cow showed typical signs of internal bleeding. How was that possible? He looked at the cow's organs, everything was normal. All organs were in good condition, no pathological changes or injuries. The cow also died peacefully and seemed to have had no pain. He could not explain any of this. Was it a lack of feed, as for some years the farmers had been sowing European clover to make the barren land in North Dakota more productive? He picked up the sack of hay. The hay did not smell as typical as he knew it. Not the pleasant smell of coumarin, no, it smelled, or rather, it stank. Were the animals being fed rotten hay and had an infection? Many theories shot through his head, but here was the end of the knowledge of a practicing veterinarian.

Three years later, in 1924, Rudolph Lincoln read in the newspaper and later in his professional journal, the Journal of the American Veterinary Association, that more cows were dying in North Dakota and across the border in Canada. They showed the same symptoms as the one he had autopsied, and the doctors called the disease Sweet Clover Disease—sweet clover poisoning. This disease spread to Minnesota, Wisconsin, and even California. More and more cases were reported and now sheep and goats were also affected by the disease. But there was also hope, because the Canadian veterinarian Frank W. Schofield (1889–1970) now claimed to know what the problem was. In his opinion, it was not, as suspected, an infection, as no pathogenic bacteria or fungi could be found in the cultivation of tissue samples. He could also rule out malnutrition, as all parameters in the lab spoke against malnutrition. He was sure that the cause lay in the diet. Schofield wanted to continue his research at the University of Ontario and applied for research funds, which were however rejected by the management. So his life took an unexpected turn and the research on coumarin took a longer break. Annoyed, he quit and emigrated to Korea. There he taught bacteriology at the University of Seoul, lived as a missionary, never returned, and was buried in his new home with the highest honors.[161]

Lee M. Roderick, a practicing veterinarian in Canada, independently examined the causes of sweet clover poisoning in his lab in 1928. He collected samples from farms and fed them to rabbits, which in the lab showed the same symptoms as cows and died from coagulation disorders. He transferred the blood of healthy cows to the blood of sick cows, but no success was achieved. He believed it was an acquired coagulation disorder, which he referred to as a plasma prothrombin defect. He advised farmers to avoid sweet clover (*Melilotus officinalis,* Fig. 1) in the feed, which actually stopped the dying. Not all farmers followed the recommendation and when the Great Depression hit the USA in 1929, sweet clover was fed out of necessity and the dying began again. One of the farmers, however, had had enough and took matters into his own hands.

What the Hell—Blood, Blood, Damned Blood

"Hey boss, there's a farmer in the yard with a dead cow on his truck." Karl Link (1901–1978) was sitting at his office desk on this Saturday afternoon, more preoccupied with securing his lab than dealing with a crazy farmer, as a blizzard was due to hit Madison in a few hours. "Send him away," he hissed at Eugen W. Schoeffel, "you see, I'm busy." Eugen Schoeffel was his senior student from somewhere in southern Germany. He had once told him, from Swabia he believed, but he had forgotten again. He liked Eugen, he was quirky, but very diligent and loyal. Eugen came to the USA in 1926, worked in the warehouses of Chicago and began to study veterinary medicine. You couldn't converse with him, his English was more Denglish. But he was intelligent and freely quoted from Goethe's Faust, Shakespeare, the Bible, and other European literature he had never heard of.

Fig. 1 *Melilotus officinalis* (commonly known as, wikipedia)

Eugen Schoeffel ran down the stairs into the yard and wanted to send the farmer away, but Link knew that these farmers were stubborn and Eugen's Denglish would not impress them. He should have bet that Eugen would not be able to get rid of the farmer. Five minutes later, both were in his office. The farmer, who introduced himself as Ed Carlson, held a milk can in his hand. "Doc, I've had enough of this nonsense and want to know why my cows are dying. I've read that you work with blood and research blood clotting. I've brought you something." He put the can on the table and opened the lid. Red blood came to light. "I've driven 200 miles, and I had to fight my way through the storm all the way. It's important to me that you find out the reason for the expensive cows dying in this country. And one more thing, I've also left you some rotten hay at the door."

"Vat da Hell, a farmer shtruggles nearly 200 miles in dis Sauwetter, driven by a shpectre and den has to go home vit promises dat might come true in five, ten, fifteen years, maybe never. Who knows? 'Get some good hay-transfuse.' Ach!! Gott, *how can you do dat ven you haf no money?",* Schoeffel couldn't calm down. The farmer was on his way back to his farm. Schoeffel stuck his finger in the red milk can and rubbed his index finger against his thumb. *"Dere's no clot in dat blook! BLOOD, BLOOD, DAMNED BLOOD. 'The people touch me in their days of lament.',"* he quoted Faust. Link rolled his eyes and wanted him to bring the can to the lab, but Schoeffel was upset. *"Vat vill he find ven he gets home? Sicker cows. And ven he and his good voman go to church tomorrow and pray and pray and pray, vat vill dey haf on Monday? More dead cows!! He has no udder hay to feed—he can't buy any. And if he loses de bull he loses his seed. Mein Gott!! Mein Gott!! Vy didn't ve anti-shi-pate dis? Ya, ve should haf anti-shi-pated dis."*[162] Link and Schoeffel forgot the upcoming snowstorm, brought the blood to the lab, and experimented into Saturday evening. Link wanted to leave, but Schoeffel held him back. *"Before you go let me tell you something. Der is a deshtiny dat shapes our ends, it shapes our ends I tell you! I vill clean up and gif you a document on Monday morning."*

Eugen Schoeffel's courage and enthusiasm were not enough to quickly find out what in the rotten hay was causing the clover poisoning. It was not until six years after the farmer's visit, in January 1933, that dicumarol was found. Schoeffel returned to the control lab of the medical association in Chicago, new colleagues came. There was a lot to do in these years. Improving animal experiments on rabbits was the order of the day. The QUICK test for determining blood clotting in the lab was adapted and a strategy for isolating and processing hay was developed. *"Boss, this hay is a grab bag full of unknown structures,"* said Harold Campbell, Link's new student. He was the one who fractionated the extracts and discarded almost all isolated substances as inactive. He developed theories that the active substance might be a porphyrin, similar to the red blood pigment that forms during the breakdown of chlorophyll. But that was not true either. But everyone knew that it was only a matter of time and diligence and that they would find the substance one day. That day was June 28, 1939.

Karl Link came to the lab early that morning. The summer was already restless, war was threatening in Europe, and two weeks earlier the refusal to let 900 Jews off the SS St. Louis in Florida had shocked the nation. Roosevelt was personally asked to help, but he declined. It was, as always, a political matter, and everyone knew that the people back in the Reich, as Schoeffel always said, would have a very hard time. Everyone sensed that troubled times were ahead. On his way to his office, he passed Campbell's lab. The lab was deserted, no, he saw feet on the sofa, and recognized Campbell by his shoes. His assistant had pulled another all-nighter. Boyles, a former guard who had helped Campbell with the animal experiments, sat right behind the door. "Isn't it a bit early for a drink?" Boyles poured himself a sip from the lab bottle into the beaker in front of him. "I'm celebrating, Doc. Campy hit the jackpot." "Tell me!" Link urged him. "Campy was sitting in front of the microscope this morning and was all excited. He said he had crystals, a whole six milligrams of the substance you've been looking for all this time. He immediately checked the blood clotting and the stuff works. Let him sleep, you know him." Link knew he wasn't supposed to address Campbell. Campbell was too much of a scientist and wouldn't say a word until he had all the evidence in his hand. If he wanted to check everything, then after so many years a few more days didn't matter. Link tipped his hat and said goodbye to Boyles. He just nodded.[162]

Two days later, Campbell stood in front of Link's office door. "Boss, do you have a moment for me?" "Sure, come in, what's so important?" He already knew, but he didn't want to spoil his joy and tell him that Boyles had already told him. He handed him a small bottle and Link could see white crystals that sparkled in the light. "So what?", he asked boredly. "This is H. A.", Campbell called out to him. H. A. was the magic abbreviation for Hemorrhagic Gene. That's how they abbreviated the unknown magic substance in the lab. "Seriously, if that's the case, I'd like to respond festively with a few lines of German poetry that are appropriate: *"So finally I hold it in my hands and call it pure in a certain sense."*" Campbell looked puzzled. "Goethe! Schoeffel taught me. We should also write to him about the success." That good old Schoeffel could still experience this. "Do you have an idea of the structure?", Campbell shrugged. "The molecular formula is $C_{19}H_{12}O_6$ and the melting point is 288–289 °C, no, I don't know the exact structure, but I'll give it to Huebner, he can do that." Charles Ferdinand Huebner was a gifted chemist who loved to solve tricky problems with skill and creativity. It took some time and on April 1, 1940, everyone thought it was an April Fool's joke when he came into the lab and presented the substance, a 3,3β-Methylenebis(4-hydroxycoumarin). He had also synthesized the substance, it was chemically and pharmacologically identical and was later successfully tested on rabbits, guinea pigs, and dogs.[162]

It turned out that dicoumarol was not a plant-based natural substance, but was formed in rotting hay by mold from two molecules of an oxidized coumarin. This could only occur if the hay was stored incorrectly on the farms. Moisture and warm temperatures start a fermentation process, and plant coumarins are converted into toxins by fungi like *Penicillium nigricans*. The work of Link and his team showed that mice and rats, unlike cows and chickens, are very sensitive to dicoumarol. The blood clotting effect can be stopped with vitamin K as an antidote. Link wanted to introduce dicoumarol as a new active ingredient for the prevention of thrombosis and as an anticoagulant in the clinic, but met with great resistance.

The fact that dicoumarol was considered dangerous for humans because of its history annoyed him greatly and he had to do a lot of convincing. Even the argument that with vitamin K a very effective antidote was in the hands of the doctor, could not really convince. Link did not give up, saw the great potential of the dicoumarins and synthesized over a hundred derivatives in the 1940s, all of which he tested. At a lecture of the Harvey Society in January 1944 titled "The Anticoagulant from Spoiled Sweet Clover Hay", the successes of Karl Link were acknowledged with approval. However, the listeners agreed that it was still a long way to go before dicoumarins could be used in the clinic.

It was infuriating, couldn't there be any success? Link, who was becoming increasingly exhausted after years of intense work, took a longer vacation with his family in September 1945. They wanted to go canoeing and everyone was having fun until they got caught in a heavy thunderstorm. Everyone got soaked, but they shrugged off the cold. Not Karl Link, though, who neglected the impending illness that hit him full force two weeks later. The doctor thought he was diagnosing pneumonia, but Link told him that he had similar problems as an exchange student in Switzerland and that it was tuberculosis. Reactive pulmonary tuberculosis was now in the doctor's letter, which referred him to Wisconsin General Hospital for two months and then to Lakeview Sanatorium for another six months. He accepted that nothing was happening in the lab. Because of the war, it was quiet, his doctoral students were at war in Europe. He had a lot of time and used it to structure all the results of his lab work. He grabbed the stack and sat down on the terrace where he could work undisturbed and breathe fresh air at the same time, as the doctor recommended. He sat on the edge of the terrace where the sun couldn't blind him. On the table were some magazines that a guest had forgotten or finished reading. There wasn't much variety, and he picked up the magazines to flip through them briefly. He was familiar with Reader's Digest and it was also available at the institute. He picked up the magazine and discovered the article "Certain Death for Rats" by Paul De Kruif. He read the article, and it occurred to him, could his dicoumarins also be good rat control agents? He flipped through his documents. Yes, rats are much more sensitive than other animals, but the amount of dicoumarol seemed too high to him. Each rat would tolerate 2 mg well, and the dose might not be sufficient because natural vitamin K from the feed could balance the effect. But what about the other dicoumarins they had synthesized? Link went through his tables. There were still two interesting candidates with the numbers 42 and 63. He flipped back and forth. Made notes and thought about all the other structures. "Mr. Link, your dinner is ready. Come on, it's getting late." Nurse Annie stood next to him. He had lost track of time. "What's for dinner?", he asked, even though he already knew. "As always, cod liver oil and a bottle of beer." He made a face.

In addition to Karl Link, who was released from the sanatorium, L. D. Scheel was also discharged from the US Army in Germany and returned to Link's laboratory in Wisconsin in 1946. He asked him to review the list of synthesized coumarins from number 40 to 65 again. Link believed he had an interesting candidate, but he wanted it checked again. Scheel did as he was instructed to find the chemical "mouse trap," as his boss called the dicoumarins. Indeed, he was able to confirm the substances of both numbers as highly effective anticoagulants. The chemical compound with the number 42 was not the answer to everything, but it was ten times stronger than dicoumarol. Number 42 became the well-known warfarin, and the name warfarin does not derive from warfare, the war against rats, but from the initials of the Wisconsin Alumni Research Foundation, the scientific transfer office of the university, with the ending "arin" for coumarins. That's where the money for the research came from, and it expressed gratitude for the great support. The transfer office was to get more than its investments back. In the 1950s alone, 70 million baits were sold in the USA.

After initial skepticism from the medical profession in the 1940s, the picture changed when in 1951 a former soldier of the US Army tried to take his life with warfarin. He swallowed 567 mg of the rat poison d-Con, which contained warfarin. When the veteran was admitted to the hospital, he complained of back pain and nosebleeds, and the doctors gave him vitamin K as an antidote, which saved his life. After six days he could be discharged, and the doctors later stated that one could not intentionally poison and kill oneself with warfarin. Was this the accolade for the rat poison? Perhaps, because four years later warfarin was considered safe enough to prescribe to the president. The most prominent patient and probably the only president of the USA who was treated with rat poison in 1955 with the best of intentions was Dwight Eisenhower after a heart attack. *"What was good for a war hero and the president of the United States must be good for everyone, even if it is rat poison,"* one of his doctors is quoted as saying. The administration of warfarin was unusual because the standard was heparin. However, unlike warfarin, which was available in tablet form, heparin had to be injected. Despite all the euphoria, treatment with warfarin was not easy, as the therapeutic range was narrow and overdoses could quickly occur. Alfred Winterstein (1899–1960, Fig. 2) was responsible for developing a dicoumarin that was

Fig. 2 Alfred Winterstein (1899–1960) (gemeinfei, wikipedia)1

better tolerated by humans. He introduced phenprocoumon, which was brought to market in 1955 by Hoffmann-La Roche under the trade name Marcumar®. Marcumar is still used today in about 1% of all patients with blood clotting disorders and in about 10% of those over eighty. Who was its discoverer? Alfred Winterstein (1899–1960) was a trained chemist who followed in his father's footsteps and studied chemistry at the ETH in Zurich from 1917 to 1921. His scientific career took many interesting turns with his stay in natural product chemistry and the breakthrough of chromatography for the isolation of natural substances. His work to separate chlorophyll with powdered sugar in a chromatography column to isolate the green leaf pigment is legendary.

The entire research and development of dicoumarol from poison to active ingredient is very astonishing, as at no time was industrial research involved and it was heavily financed by the university transfer office. This model was forward-looking but understood late, as many universities only set up patent transfer offices in the 1980s or later. But the success also lay in the enthusiasm of Karl Link and his students: *"I think the success has three reasons. They never expected a miracle, they never gave up, and they knew they would do something good for humanity and not destroy it."*

Natural Substances as Starting Materials for Drug Synthesis

Of Sharks and Christmas Biscuits

What do sharks have to do with coronavirus vaccines? At first glance, it seems far-fetched, because what does a shark contribute to the biotechnological production of a modern vaccine? On second glance, the shark, or more precisely its liver, is unfortunately an important raw material for the extraction of squalene.[163] This natural substance is used as an adjuvant and auxiliary substance for vaccines. Although the pharmaceutical industry only requires about 1% of the world's squalene compared to the cosmetics industry (Fig. 1), even this is unnecessary, as squalene can also be produced from olive oil or biotechnologically in fungi. However, the easiest way to obtain the coveted auxiliary substance is to catch sharks, cut their liver out of them, and then throw them overboard in agony. The shark's liver is a large organ that contains a lot of oil and up to 80% squalene. The name comes from the genus name *Squalus* for shark, in whose liver it was first discovered in 1906 by the Japanese chemical engineer Mitsumaru Tsujimoto (1877–1940). He isolated the oil Abura-Zamé (Japanese: Zamé means shark) from two sharks of the species *Squalus mitsukurrii*, which, at just under 1.2 m in length, belong to the smaller sharks. His specimens from the Suruga Sea near Japan were male and quite small at 92 and 72 cm.[164] The liver can weigh up to 20 kg in large sharks and about 3000 animals need to be caught to produce one ton of squalene. Apart from the senseless killing of sharks, the liver oil is often contaminated with environmental toxins such as dioxin and pesticides and is not infrequently infected, which must be tested and ruled out

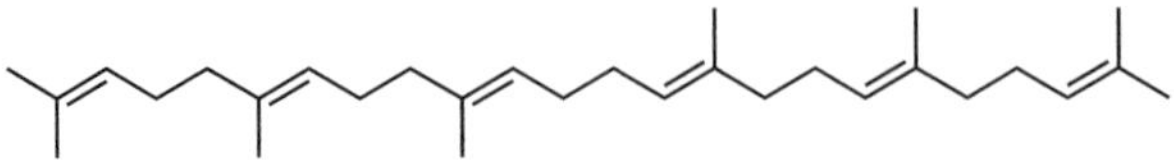

Fig. 1 Squalene (work of the author)

for further use in a vaccine. The desired alternative cultivation of plants from which squalene can be obtained is not easy, as one hectare yields just 50 kg of the coveted auxiliary substance, which must be laboriously cleaned. A biotechnological alternative would be ideal and could solve the sustainability problem. This is also urgently needed, as almost all modern vaccines against influenza and hepatitis contain oil-in-water emulsions with squalene to enhance the vaccine effect.

Modern vaccines today forego the old aluminum phosphate salts as adjuvants, which were added since the 1930s to enhance the vaccine effect. Every vaccine always consists of two components: the antigen, which has the medical and immunological effect, and the adjuvant, which helps to enhance the immune effect and save on the amount of antigen. If vaccines were administered without adjuvants, the antigen dose would have to be dramatically increased, which is hardly affordable. Adjuvants have no therapeutic effect of their own, but help to present the antigen to the immune system. Modern vaccines contain lipid nanoparticles or virosomes. Lipid nanoparticles are tiny fat globules on or in which the antigen is located or bound. Virosome is a portmanteau of virus and liposome. Virosomes are liposomes that no longer contain complete viruses, but only parts that make the virus's reproduction in the body impossible. Both taxis, which now transport an antigen, are physically particles that are particularly liked to be taken up by our immunological phagocytes. Our immune system has learned over the course of evolution that particles like viruses, bacteria, or fungi can generally be dangerous, which is why all foreign particles like the modern vaccines are indiscriminately fished out of blood and lymph.

But I don't want to digress into molecular biology, but rather show that natural substances also play an important role here. Let's ask ourselves again what role natural substances can play not as active ingredients, but as raw materials in pharmaceutical production. And is there a sustainable green chemistry or pharmacy? Green chemistry is not a term that only arose in our modern times. It existed earlier and the current search for raw materials from plants or microorganisms to produce in an environmentally friendly and energy-saving way is not an idea of the young bioeconomy. We have already encountered an example in the chapter on steroids, when diosgenin was isolated from plants and reused. Unfortunately, the pharmaceutical industry is struggling with the issue of sustainable production of medicinal substances, as 85 to 90% of all chemicals used to manufacture active ingredients are of fossil origin. The production of a medicinal substance in a chemical apparatus or fermenter is cheap compared to processing, isolation, and purification. It is precisely these purification steps that cause 80 to 90% of the costs and up to 90% of the solvents are consumed here. To put it simply and harshly: To produce 1 kg of a drug purely in the pharmaceutical industry, an average of 100 kg of raw materials and 300 l of water are consumed and the efficiency is 1%.[43, 165, 166]

The challenges are enormous and in addition to saving energy, water, toxic chemical catalysts and solvents, we are looking at what nature has to offer. It is a dream for many chemists that chemical reactions in the cell work reliably in water, at room temperature and normal pressure. Renewable raw materials and climate-neutral productions could revolutionize the pharmaceutical world, without leading the fight tank against plate. What does this mean? In the first generations of bioenergy plants, crops such as corn and wheat were used for ethanol production as fuel, which were no longer available as food in the producing countries. Today, it is ethically no longer justifiable that foodstuffs such as corn are grown for the production of biofuel when people are starving. In 2007, there were riots in Mexico because the corn for the traditional tortillas was no longer sufficient. The price per kilo doubled, the poor population could no longer afford the staple food, because in the northern neighboring country the Fords, Chevrolets and RAMs were driving with bioethanol in the tank. But our view does not have to go across the big pond, because here in Europe too, the butter ran out at the end of the decade and from rapeseed oil biodiesel was made.

The pharmaceutical industry is less driven by the prices for bioethanol or biodiesel, but it also occasionally competes with food and plant products of our daily life. The biotechnological industry needs sugars like glucose and residues from food production for fermentation, so as not to have to buy expensive amino acids as a nitrogen source. For standard production, pure glucose as sugar is much too expensive, which is why mash and molasses are purchased. For some syntheses, plant natural substances like scopolamine, podophyllotoxin or shikimic acid are needed, but they do not compete with food. Their synthesis would be too complex and isolation as precursors for further synthesis is cheaper. The required quantities are not as large as in other industries, so it is hardly noticeable when the pharmaceutical industry purchases them. However, shikimic acid was an example of a massive purchase by the then Roche AG at the beginning of the millennium to produce the antiviral drug Tamiflu®.

Many people who experienced Covid-19 infections in recent years still remember the outbreak of bird flu in 1997, which came from Hong Kong with the frightening abbreviation H5N1. Science was not advanced enough at the time to develop the now well-known RNA vaccines. The pharmaceutical industry offered the antiviral drug oseltamivir as a drug in Tamiflu®, which blocks viral neuraminidase. The starting material was shikimic acid, which can be found in almost all plants on this planet. Since it is a plant precursor for the construction of aromatic amino acids, lignin and other aromatic plant substances, it is quickly metabolized and is rarely detectable in high quantities. *Illicium verum* (Fig. 2) is a mega producer. The Chinese star anise grows as a small tree or shrub and stores a lot of shikimic acid. From 30 kg of the plant, 1 kg can be quickly isolated.

Fig. 2 *Illicium verum Hook* (public domain, wikipedia)

Natural Substances as Starting Materials for Drug Synthesis

The bird flu H5N1 was getting closer and the US government was getting nervous. They instructed their agencies to stockpile 300 million tablets for emergencies. With the amount of 75 mg of oseltamivir in one tablet, 23 tons of the active ingredient had to be produced and 8 tons of shikimic acid had to be provided as a starting material, which is why 840 tons of star anise were bought on the market.[167] The market was swept clean and we haven't even considered the demand from the rest of the world. Anise became very expensive and at Christmas, the popular star anise was only available at inflated prices or not at all. Shrewd businessmen came up with the bad idea of replacing and selling the Chinese "real" *(Illicium verum)* with the Japanese star anise *(Illicium anisatum, Fig.* 3). Both look similar, they also contain anethole, which produces the beautiful smell. With the Japanese star anise came the danger, because it also contains the natural substance anisatin (Fig. 4), which caused epilepsy. The authorities warned against the "false" star anise, but in the Netherlands in 2001 and 2002, 60 people were poisoned with cramps and epilepsy.[168, 169]

Fig. 3 *Illicium anisatum L.* (public domain, wikipedia)

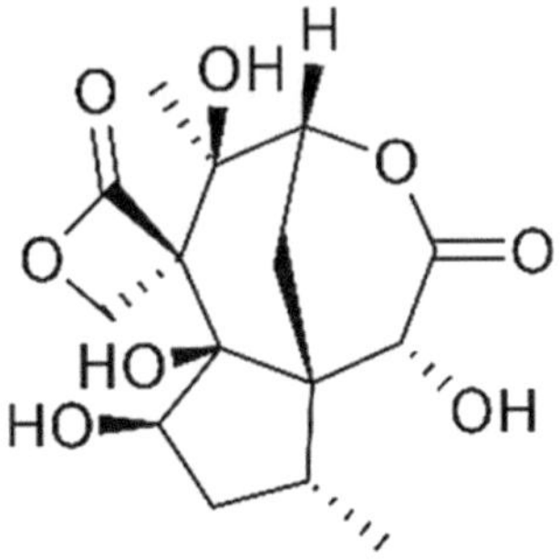

Fig. 4 Anisatin (work of the author)

Scopolamine—About Truths and Veracities

The KGB agent leaned over the slightly creaking wooden table. He was in a dark room somewhere in Moscow. There were no windows, it was damp and smelled of stale Herzegovina Flor cigarette smoke. He knew that the KGB agent must hold one of the higher ranks, because Herzegovina Flor stank, Stalin smoked them, and they were not so easy to get. Sweat was on his forehead and he had been stuck in the same dirty pants and dirty shirt for days. What did they want from him, he had said everything. Stalin was conducting one of his purges again, and this time he must have fallen into the KGB's net. Did a younger colleague want his professorship at Lomonosov University or was his research area, the extraction of rubber from the Russian dandelion, already political? He did not know. He didn't know anything anymore, because the guards denied him sleep. Sleep deprivation was torture, he couldn't take it anymore. He was thirsty and just wanted to go back to his wife and the two children who were taking the Matura. Had they been arrested? Maybe they were also in this prison or in one of these Gulags? There was so much speculation, but if only half of it was true, then… He was abruptly torn from his thoughts. "Comrade Jurentschik, for days you have been refusing to tell the truth. With your stubbornness, you are harming socialism. Which antisocial elements of the capitalist world are you in contact with? Who was the man in the gray suit in front of your office that Comrade Gemalda saw?" A man? Gray suit? He couldn't remember. Comrade Gemalda was the secretary of his dean in the office across the way. Who had she seen? Damn, he couldn't think of anything or anyone. "I don't know," came softly over his lips. He looked at the table, where the smoke from the KGB agent's cigarette spread like ground fog. "Well, you are persistent, Comrade Jurentschik. We've been hearing that for three days."

Angrily, he put out the cigarette in the ashtray. "We now have to use other methods to enable the victory of the working people. Comrade Murias," and he pointed to another officer who stood behind him in the semi-darkness of the room, in a corner where black mold was eating away at the ceiling, "has brought a light drink that you can drink and you will feel very good." "No, certainly not," he called back. Jurentschik wanted to get up, but his bound arms and legs wouldn't let him. "You want to kill me, I won't take anything," he screamed. "Calm down, comrade," the KGB agent replied calmly, "no one wants to kill you. We have just found a truth serum that lifts your mood and makes it easier for you to talk a little more openly. It's a plant extract from Hyos…, Hycos… or Humus…", he stammered the word fragments out of uncertainty about the name, looked at the second one behind him and broke off. But he knew the name of the plant, *Hyoscyamus niger*, a nightshade plant, which contains scopolamine and causes amnesia. He would fall into a kind of delirium and later be able to remember nothing. Later they would know all his secrets, and he knew they would ask not only about the man in the gray suit. The man from the dark corner was not wearing a uniform, he seemed to be a scientist. Did he know him? No. As he walked, he unscrewed the brown bottle and stood next to him. He shook his head vigorously, they shouldn't be able to pour that disgusting stuff into him so easily. The man next to him looked at the KGB officer, who nodded with the still cold cigarette in the corner of his mouth and thoughtfully got up from the chair. He went to the other side of Jurentschik, held his nose with his right hand and fixed his chin with his left. He could no longer breathe through his nose and when his mouth was opened, the plant extract was poured in. He coughed and spluttered, but most of it ran down the esophagus into the stomach. He had lost. The two let go of him and he became calm. In a few minutes, he would be in another world and the scopolamine in the extract would take effect.

Truth serums, which chemically open people's consciousness and let the truth bubble out of them uninhibitedly? Is this possible, and how does it work? Unfortunately, I cannot describe exactly what happens biochemically in our brain, but the intake of drugs or benzodiazepines (Midazolam or Diazepam) can lead to retrograde amnesia in us. This is a backward memory disorder in which the person can no longer remember contents or points in time. After taking it, the affected people feel remote-controlled and unable to follow their own will or control themselves. During the Cold War, the CIA and the KGB had a great interest in drugs that were administered to victims to find out the truth. While LSD was mainly used in the West, the KGB was more taken with Scopolamine.[170] To this day, little is known about these works in the East and West from the 1950s, but the experimental use is well documented and processed in literature. Who doesn't know the magnificent film One Flew Over the Cuckoo's Nest with Jack Nicholson. Ken Kesey wrote the book for the film in 1962, incorporating his experiences as a temporary nurse at the Veterans Hospital in Menlo Park, California. There, on behalf of the CIA, psychogenic effects are said to have been researched in the MKULTRA program. In recent years, all possible substances have probably been tried out as truth serums in the USA, Israel, the former USSR, North Korea, and other countries. The success is doubtful, because even under the influence of drugs or medication, liars continue to lie and anxious people can fall into psychoses. Interrogation specialists used these truth drugs more to uncover contradictions in what was said and to start there. Unfortunately, these consciousness-altering drugs and medications are also used in the civilian sector. It is known that inheritance hunters have administered benzodiazepines such as Midazolam to their old master in order to change or rewrite wills. Rapes were also covered up in the club scene, where women were drugged into submission. Recently in South America, Scopolamine was mixed into tourists' drinks to get their credit card PINs. The victims were asked to withdraw money from the ATM against their will with their credit card. These knockout drops, known as Burundanga, can also be offered in food or cigarettes. They make the victim just as submissive.[171] Be careful if you are offered invitations or drinks too generously.

Scopolamine—from the Witch's Kitchen to the Pharmaceutical Lab

The natural substance Scopolamine has far more good pharmacological properties, even though I have described it as dangerous as a truth serum. Its history is closely linked to that of its sibling adrenaline. Both natural substances are alkaloids from nightshade plants, which have been known for centuries. The German name for nightshade plants already hints at their effect on the brain. Nightshade plants do not grow at night in the shadow of the moon. This old German word was often used to explain that one had a nightmare. This nightshade must have been observed often when *Atropa* or *Hyoscyamus* plants were ingested. The intake of extracts could go wrong and overdoses were not uncommon. The dose had to be adjusted by trying it out, and this was mainly done by healing women, who were often referred to as witches. Every Harry Potter fan knows Professor Sprout, who holds

Fig. 5 *Hyoscyamus niger L.* (public domain, wikipedia)

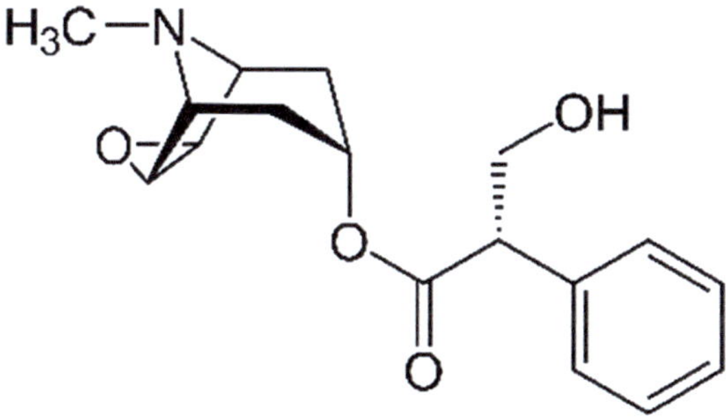

Fig. 6 Scopolamine (work of the author)

up a mandrake root *(Mandragora officinarum)* in the film "The Chamber of Secrets". Not without reason, because the roots were considered a ritual magic tool because of their active ingredient Scopolamine. *Hyoscyamus niger* (Fig. 5), also known as witch's herb, was similarly popular in the Middle Ages. Because of the highly effective alkaloids like Scopolamine (Fig. 6) in their extracts, it was valued by the healing women. The women made ointments from the extracts, which were applied to knees, crooks of arms, neck, and palms. The intake already seemed dangerous at that time, as the unknown dose varied and could be lethal. Scopolamine is a fat-loving substance that is well absorbed in the fatty ointments. Applied to the skin, the Scopolamine penetrated through the skin and got into the blood. Much more exciting, however, was to smear the Scopolamine ointment on the genitals. In the witches, Scopolamine and Atropine evoked an aphrodisiac feeling and they masturbated with a broomstick. This arousal probably led to the belief that they could fly. These recipes were soon called "flying ointment" in common parlance, and our current image of witches flying on brooms has persisted to the present day (Fig. 7).[172] Scopolamine is indispensable in modern medicine. It is used directly in the prevention of travel sickness and is contained in chemically modified form in the drugs Buscopan® and Spiriva®.

Fig. 7 Witches' Sabbath, Luis R. Falero, 1880 (public domain, wikipedia)

The desire for intoxication was widespread among the medieval population, and having to stay away from witches did not mean wanting to give up on these miracle plants. Large parts of the population wanted the intoxication and resorted to a trick. They hid the herb in a legal intoxicant. There was an intoxicant that was approved by the church and the authorities, and its effect could be wonderfully enhanced. The low-alcohol beer of the Middle Ages was fortified with henbane and now really hit the spot. This medieval "alcopop" became very popular in Europe, and we still know cities today where henbane was grown and traded: Bilsengarten, Bilsdorf in Saarland, Bilsen in the Netherlands and the Czech Pilsen, which became one of the most famous beer brewing cities.

The princes and territorial lords did not like the goings-on with the fortified beer. For the beer produced rebellious subjects, whose tavern brawls were the least of the problems. But other substances such as cat brain, caraway and other hallucinogenic plants were also allowed to ferment in the beer by the brewers. The brewing regulations issued in various Bavarian and Thuringian cities, such as the Bavarian Purity Law of 1516, were intended to put a stop to this and ensure the quality of the beer, so that *"no roots, neither Zermetat nor anything else that is harmful to man or may bring disease and pain, may be put into it,"* as the Landshut Brewing Regulation aptly states.[173] Not everyone seemed to adhere to the Purity Law, which Dukes Wilhelm IV and Ludwig X had issued in Ingolstadt and initially only applied to Bavaria. Johann Theodor Tabernaemontanus railed against all adulterators and angrily summed up the accusations in his herbal book 1644: *"Those who strengthen the beer with henbane seeds and other such harmful things should be rejected and damned, ... those who adulterate the beer with such harmful arts, as declared enemies of the human race, as thieves and murderers of body and life should be punished. ... No one should drink beer with henbane seeds, for those who have forfeited life, for they bring brain rage, insanity and sometimes sudden death."*[174]

Unfortunately, extracts from nightshade plants are still taken today, especially by teenagers and young adults. Particularly popular is the drinking and even smoking of extracts from the South American angel's trumpet *(Brugmansia vulcanicola),* which is widely found as an ornamental plant in Germany and Europe. The effects are always the same and lead a third of consumers directly to the intensive care unit, according to reports from the poison control centers. Some bizarre cases are known, such as that of a 14-year-old girl who had delusions after taking it. She believed she was being pursued by killer tomatoes. No less interesting is the case of the young man who, after an emergency admission to the intensive care unit, began to collect imaginary mushrooms.

Henbane and later scopolamine were used from the 12th century in combination with opium in early anesthesia. Scopolamine was not used for pain relief, but to dampen the perception of pain. A possible explanation for the use of scopolamine is anterograde amnesia, where memory loss and a reduction in perception cause the patient to forget that they are in pain. The administration of a sedative such as a benzodiazepine to patients before surgery to save on anesthetics is in some ways the modern application of scopolamine in today's anesthesia practice, as some strong benzodiazepines also lead to anterograde amnesia. Recipes from the 9th century describe the production of a sleep sponge *(Spongia somnifera)* which was moistened with the juice of mandrake, henbane and also water hemlock and then dried. If surgery was necessary, this soaked sponge was moistened again and placed on the mouth and nose of the patient to be operated on. The anesthetic dripped through the mouth and nose into the stomach and small intestine, where it was absorbed. The use of water hemlock (*Cicuta virosa,* Fig. 8) with

Fig. 8 *Cicuta virosa* L. (public domain, wikipedia)

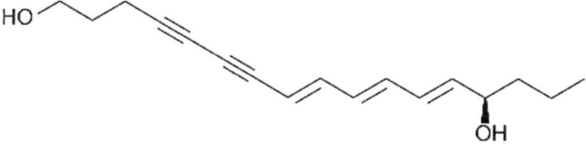

Fig. 9 Cicutoxin (author's work)

the polyin cicutoxin (Fig. 9) to calm the muscles during operations is also interesting.

Scopolamine is indispensable in modern medicine. It is used directly in the prevention of motion sickness and is present in chemically altered form in the drugs Buscopan® and Spiriva® contained.

Scopolamine—from Drug to Psychotherapy

The effect of scopolamine and nightshade plants on the human psyche was recognized early on. When witches or their patients took these strange extracts, they went into a trance and talked nonsense. This also interested the first healers or doctors who were fascinated by the human psyche. Could people suffering from mental problems like depression or schizophrenia be helped with scopolamine or nightshade extracts? Before this question can be answered, it is interesting to take a quick look at the history of psychiatry: What medications were available to doctors in mental health institutions?

There is much literature about the accommodation, treatment and care of mentally ill or so-called "mad" people in history. From the Romans we know that mentally ill people were treated well and enjoyed care and humane treatment. While many in antiquity believed that these people were in good or bad contact with the gods, the image changed drastically in the Middle Ages. Now people believed that the devil had taken possession of them, that evil lived in them. The only remedy was penance or self-flagellation, as recommended by the church. If this did not help, lifelong imprisonment was also considered justified.

At the end of the 19th century, psychiatry experienced a first decline, as the classic methods no longer worked. The medical progress seen in surgery, microbiology and drug development was not reflected in the treatment of mentally ill people, and the "age of sedation" set in from the middle of the 19th century until the beginning of the First World War, as Peter Ansari describes in his dissertation from 2013.[175] The first pharmaceutical companies grew, the heyday of physiology paved the way for pharmacology and drugs were tested for their suitability in psychiatry. However, the understanding of the biochemistry of mental illnesses was far too small to find suitable drugs. Microbiology and surgery galloped ahead and psychiatry strolled leisurely into the 1850s, until pharmacology came with the whip and got the horse moving.

With opium and scopolamin, the era of alkaloids was supposed to end and be replaced by the first synthetic drugs such as chloral hydrate and barbiturates. What they all had in common was that they sedated the mentally ill. This seemed so effective that the company Merck offered morphine, narcotine, and strychnine as psychiatric drugs. As a result, the mental health institutions, as they were called alongside the less friendly term "madhouse", filled up drastically. Sedated patients could be cared for with little staff in large numbers, so the number of residents housed in institutions in the German Empire rose from only one in 5300 people in 1911 to one in 500 people.

Already at the beginning of the 19th century, alkaloids were integrated into treatment plans, as reports from the Siegburg asylum in the Rhineland show. In good sanatoriums, the administration of these extracts was also personalized and adapted to the patient's gender, age, and temperament. The aim was not yet sedation, but the enhancement of self-esteem and social responsibility. Scopolamin was used against both schizophrenia and depression, even if the diagnosis and the disease patterns certainly did not correspond to our current definitions.[176] After the Viennese doctor and pharmacologist Karl Damian von Schroff (1802–1887, Fig. 10) had described the calming effect of scopolamin, doctors reached for it to administer extracts or the pure substance to agitated patients. From 1880, scopolamin was the standard drug for the treatment of depression and schizophrenia. It remained so until the 1930s: *"The sovereign remedy for the rapid combat of severe maniacal excitement, which is also indispensable in the hands of the practical doctor when admitting the mentally ill to the madhouse, is scopolamin. It almost never fails. If it does, the error usually lies in the fact that too small doses are injected."*[175]

With modern psychotropic drugs, the chemical straitjacket of scopolamin finally disappeared from psychiatry. A new view and doctrine prevailed. Antipsychotics are no longer pharmacological stirring spoons in the brain, they act specifically in neurobiochemistry at certain receptors. These drugs help without sedating consciousness and they should not affect his or her intellectual abilities. They help patients recognize their illness and distance themselves to perceive their subjective state as less burdensome.

Fig. 10 Karl Damian von Schroff (1802–1887) (public domain, wikipedia)

Today, natural substances no longer play a role. Only St. John's wort extracts *(Hypericum perforatum)* are still used for mild depression. The active ingredient seems to be hyperforin, which inhibits the reuptake of certain neurotransmitters. Even though extracts are available without a prescription in pharmacies, caution is advised. The extract contains light-active substances that can lead to a strong reddening of the skin similar to a sunburn when exposed to the sun. Also, women have become pregnant despite taking the pill. This is because St. John's wort extracts stimulate liver metabolism, causing the contraceptive to be broken down faster and become ineffective.

Scopolamine in Therapeutic Systems

Who doesn't know the uncomfortable feeling when the bus vibrates for hours on the way to Rome, or the ship turns the stomach from right to left in heavy seas. Where is the place in our body that triggers nausea? This travel sickness ends in the stomach, but it starts in the balance organ behind the ear, and that's exactly where a strange patch sticks, which only looks or is so by name. The small patch, no larger than a one-euro coin, releases the active ingredient scopolamine. What is barely visible behind the ear is a Transdermal Therapeutic System (TTS), whose inventor, Alejandro Zaffaroni we have already met. The very fat-loving scopolamine penetrates through a thin film from the reservoir of the patch into the skin and acts directly at the right place. The concentration is so low that no severe side effects occur. Confused speech or dry mouth, sedation are the great exception, but one notices the absorption by the large pupils that the travelers have.

How was scopolamine discovered as a remedy for seasickness? In the search for remedies for sea and air sickness, the US Navy found scopolamine shortly after the war. Without knowing the exact dose, scopolamine was successfully tested on young recruits of the Navy and the Air Force. But the administration was not safe, because even slightly increased doses could cause critical side effects. The idea remained exciting, the US Navy did not further research scopolamine, but the pharmaceutical company Alza took up the matter. The company's employees found out that the skin behind the ear is thin enough for a substance to easily penetrate and affect the sense of balance. Alza developed a transdermal patch that contained only 1.5 mg of scopolamine and released the active ingredient evenly over a period of three days. The first test subjects were company employees who were invited to a boat trip in the Bay of San Francisco in stormy weather in 1979. Half got a placebo patch and hung over the railing, the other half with the real patch seemed to have had a reasonably good boat trip. The results were so encouraging, that NASA went into business with Alza to test the small patches for future Spacelab missions.[177] Of course, further double-blind studies were conducted and the success proved the company right. Since the 1980s, the small scopolamine patches have been on the market.

The idea of patches coated with active ingredients is not so new. The well-known ABC patches against rheumatic complaints have been in practice since 1928 and contained extracts from arnica, belladonna and capsicum. The inventor, Professor Raubenheimer, combined the initial letters of the three medicinal plants and the name ABC patch® was born. Today, however, the patch would have to be called C-patch. Of the three letters, the "B" was first removed as an extract, as belladonna extracts, which also contained scopolamine, were considered questionable. Then followed the "A", because arnica was also considered problematic due to possible allergenic effects. This left only "C" for capsicum, which was probably always effective on its own.

Tiotropium and How Nuclear Power Became Something Good

Without knowledge of the pharmacological effect of scopolamine, tiotropium probably would not have existed. The question that occupied medicinal chemists was again: How can side effects be avoided and a drug synthesized that helps patients breathe better? A look at butylscopolamine bromide (Fig. 11), an older drug that has been used for a long time for spasms of smooth muscle, helped here. The trick why butylscopolamine does not work in the brain is easily explained. The chemists have provided the nitrogen with so many substituents that it physically became a salt. Because of its high water solubility, it could no longer pass the blood-brain barrier. This was the first hurdle for the upcoming tiotropium bromide (Fig. 12). The second step was to limit the effect to the lungs. The goal was to use it in people with COPD. The English abbreviation, which is

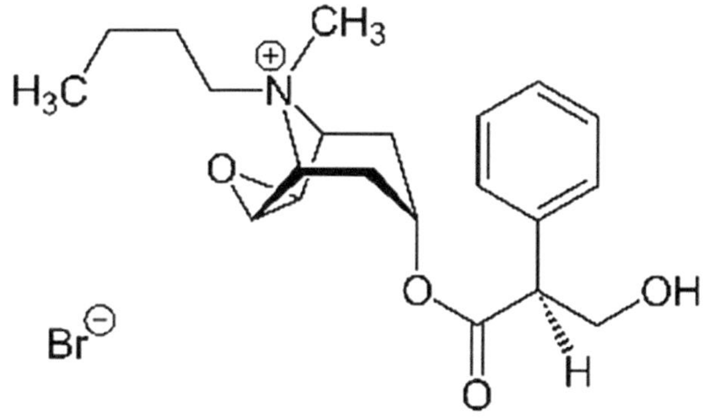

Fig. 11 Butylscopolamine (author's work)

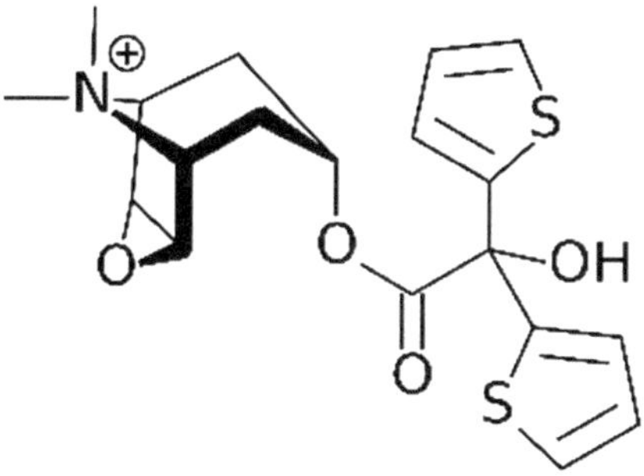

Fig. 12 Tiotropium (author's work)

also commonly used in German, stands for chronic obstructive pulmonary disease. This is an inflammation of the airways that narrows them so much that the patient can only inhale very little air and suffers from shortness of breath. When many people still smoked in Western countries, this disease was also referred to as smoker's lung, today the air pollution from car and industrial exhaust is the number 1 risk factor. Many COPD patients have never smoked in their lives, yet the number of those affected is increasing, especially in developing countries.

The synthetic precursor of tiotropium is scopolamine, as it is more practical to build the synthesis from a natural substance than to start from very simple chemical precursor chemicals. This project is being implemented by the company Boehringer Ingelheim, which cultivates several hectares of the plant *Duboisia leichhardtii* (Fig. 13) native to the state of Queensland on plantations in Australia and Brazil. The author had the opportunity to visit a *Duboisia* farm on one of his research trips. Local farmers grow the bushy shrub on behalf of the German company. They harvest the leaves and ship them to Germany, where they are extracted (Fig. 14). The cultivated plants

Fig. 13 *Duboisia leichhardtii* (public domain, wikipedia)

Fig. 14 Chemical worker, plant manager and master in the alkaloid operation at Boehringer Ingelheim (Permission Boehringer Ingelheim, Ingelheim, DE)

are high-performance plants from their own breeding, which contain up to 5% scopolamine. Compared to the nightshade plants common in Europe with a content of 0.2 to 0.5%, this is a lot. Unfortunately, global warming does not stop at Australia either. On the contrary, the summers are getting even hotter and it rains even less. This is hard on the plants, and despite breeding efforts, yields have been declining for years. Alternatives could be the transfer of scopolamine biosynthesis to related species such as tobacco or the biotechnological production in yeasts, as scientists have already impressively demonstrated.

After the arrival of the dried plant material at the Ingelheim warehouse, extraction is necessary. This process is very demanding and begins with the tearing open of the bags. The dust of the plant material already contains so much scopolamine that inhaling it can lead to pupil dilation and discomfort. Therefore, the imported plant material is extracted in as closed systems as possible to provide the obtained white scopolamine for production. For the upcoming synthesis, the natural substance is chemically split into the alkaloid part scopin and the phenyllactic acid. The latter is no longer needed for the further process. Scopin is now the part of the old scopolamine that carries the alkaloid nitrogen. Scopin is reacted with dienthienylglycolic acid and converted into the water-soluble quaternary tiotropium in one of the last steps.[178] The finished drug could be pressed into a tablet, but it is obvious that there is a better way for drugs that are supposed to act in the lungs. The uptake by inhalation of very fine particles or aerosols, as described in the chapter on asthma medications, is very efficient. Inhalers, or briefly "inhalers", are used, which transport the strong active ingredient in very low concentrations directly and with few side effects into the lungs. This is very advantageous for the patients, as only a few micrograms are released with each draw.

The inhaler used for the treatment of COPD is the Respimat Respimat®. It differs from classic inhalers. In the past, propellants such as the unwanted CFCs were used, but since the Montreal Agreement, which bans these propellants because of the destruction of the ozone layer, alternative inhalation devices are in demand. Since the 1950s, the technology of inhalation devices has fundamentally changed because it had to understand something typically human—inhaling! It sounds strange, but it is actually not easy to operate an inhaler correctly and inhale an active ingredient. Patients must inhale with the flow of the dispensed medication so that it reaches deep into the lungs. Children and older people sometimes do not understand this and do not inhale at all. The medication then remains on the mucous membranes in the oral cavity, the active ingredient is swallowed with the saliva and acts after absorption in the small intestine. Asthmatics may have too low lung volume and cannot inhale so deeply. The medication remains in the upper lung and does not reach the fine branches. The inhalers had to learn to adapt to normal breathing, not to be perceived as disturbing and to be propellant-free.

The development of Respimat® has an interesting history that finds its origin not in medicine, but in nuclear technology.[179] In the 1980s, uranium had to be enriched for the operation of nuclear power plants. This process is very complicated and demanding. From tons of uranium ore, also called pitchblende, the uranium isotope is isolated. Marie Curie was the first to process pitchblende from Austria with her husband Pierre in Paris and systematically isolate uranium, radium (1898), later polonium (1898), and discover the latter two. The laboratory was a mix of stable and potato cellar, as the German chemist Wilhelm Ostwald (1853–1932) noted during a visit. They managed to obtain just 100 mg of radium from several tons of pitchblende in 1898, which glowed wonderfully at night in the lab. Both stood in awe of the miraculous light, from which we would run away today because of the radiation. They were delighted with their successful work and the magic of the light. The radiation in their small lab was so strong that Marie's notebook, which was auctioned off in 1984, is considered unreadable.

In the beginning of the atomic age, uranium was of great interest for both energy production and military purposes. The worldwide demand was and is so great that the mining of uranium ores in Australia, Russia, Namibia, and the USA still seems to be indispensable. In the 1980s, new techniques were invented to improve uranium enrichment. One of the companies was STEAG microParts in Dortmund, based in Essen. This is the so-called nozzle separation process, which was discovered by Erwin Willy Becker at the then Institute for Nuclear Technology in Karlsruhe and later industrially further developed by STEAG microParts. The physics behind this process is the production of very fine micro nozzles, which are "etched" in the so-called LIGA process. This is a micromechanical technique that allows the finest particles to be atomized even finer than was possible with previous inhalers. The company Boehringer Ingelheim found this technique exciting and bought STEAG microParts in Dortmund in 2004. The inhaler is manufactured in Dortmund and transported to Ingelheim, where it is filled and distributed.

Podophyllotoxin—Medicinal Weed

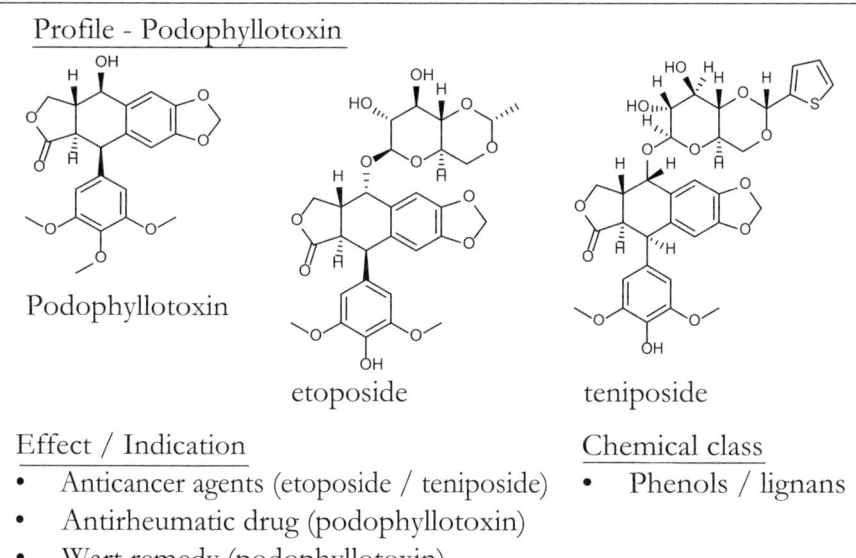

The study of the chemical class of lignans, which also includes podophyllotoxin, began early, and today the semi-synthetic derivatives are indispensable in the treatment of many types of cancer. Etoposide and teniposide, as the chemically modified natural substances are called in the clinic, make up about a third of the cancer drugs in our hospitals and even for students of pharmacy and chemistry, the natural substance is often unknown or underestimated.

From a chemical structure perspective, podophyllotoxin is not exciting, because lignans are among the oldest natural substances of evolution and probably emerged many millions of years ago when plants left the water and became land plants. Perhaps they were a byproduct when the plants needed a framework to be able to grow upwards and no longer had to stay on the ground. The invention of the shoot or stem in higher plants had the advantage of being able to use light for photosynthesis. In water, this was not important, as gravity was practically abolished and the plants swam like algae.

Jonathan Hartwell (1906–1991) studied *Podophyllum* species and their ingredients shortly after joining the National Institute of Health in 1939. He can be seen as the father of lignans in cancer therapy and also contributed significantly to the development of etoposide as the first semi-synthetic lignan and cancer drug. In the 1940s, he described the effect of *Podophyllum* extracts from *Podophyllum peltatum* (Fig. 15) against tumors, isolated podophyllotoxin along with 37 other lignans and related structures in 1950. The success was overwhelming and led to the Swiss company Sandoz starting to deal with podophyllotoxin as a drug against cancer in the 1950s. This was hard work at the time. From a wild mixture of substances, it was necessary to isolate a substance with primitive-looking devices and techniques, of which not even the chemical structure was known.

Plant extracts consist of several thousand substances, most of which are only present in very small amounts. A good estimate is that an extract consists of about 3500 to 5000 substances that the chemist has to separate. Fortunately, main components often found in extracts predominate and allow a first simple separation. What is achieved today by high-performance chromatography with specific separation materials in a separation column was still very laborious in the 1950s. The separation materials at that time were aluminum oxide, zeolites or sometimes just fine sand. Modern synthetic materials like reverse phases, where long carbon chains are bound to carriers, were only invented in the 1970s and introduced into analytics. Against this background, the achievements of natural product chemists at this time are remarkable.[180]

Fig. 15 *Podophyllum peltatum L.* (public domain, wikipedia)

Podophyllotoxin showed very poor solubility in water in all tests conducted by Sandoz. Poor solubility is always unfavorable for the pharmacist, because the substance must be in solution in the body to be able to work. Better solubility was to be achieved by a chemical change. The goal was to introduce a sugar building block, because sugars are considered water-soluble. However, they are also physiologically unstable, as there are many sugar-splitting enzymes in our body that would undo this chemical success. This difficulty could also be solved and the sugar could be protected in a further synthesis step. In the end, Etoposide (Vepesid®) and Teniposid (Vumon®) were approved as new cancer drugs by the FDA in the USA in 1983 and 1992 respectively.

"You know, the high demand for podophyllotoxin has led to an ecological disaster in the Himalayas." I was very surprised when my Indian colleague reported to me at a conference in Dehradun in 2005 that there are hardly any *Podophyllum* plants left in the high mountains. Many plant collectors set out and sold the roots to traders, who then resold them to extraction companies as intermediaries in the country. It seemed to be a lucrative business at the expense of the diversity of this sensitive ecosystem. I wondered how to

solve the problem, because in addition to the *Podophyllum* species with their high lignan content, there were many other plants that also contained lignans. I told Wim Quax, my then colleague at the University of Groningen in the Netherlands, about this amazing conversation, and he came up with the idea of investigating the native cow parsley *Anthriscus sylvestris* (Fig. 16), as lignans are also known here. Cow parsley grows wherever there is water nearby. This was the case in Friesland, there were many canals, and my team and I were able to collect and analyze many plants. Wim was right, we found lignans, but no podophyllotoxin. The biosynthesis stopped exactly one step before and we could only detect deoxypodophyllotoxin. This was worthless for a chemical process, because for glycosylation, as Sandoz carried out, we needed this hydroxylation in the molecule. Was the idea of elevating cow parsley from a weed to a medicinal plant therefore obsolete? No, because from the literature I knew about poisoning cases. The cause of the poisoning was not the deoxypodophyllotoxin, but the podophyllotoxin oxidized by a liver enzyme. Here was the solution to our problem. At that time in Groningen we had excellent equipment and a lot of experience in

Fig. 16 *Anthriscus sylvestris L.* (HOFFM.) (public domain, wikipedia)

the field of biotechnology, so the genetic modification of organisms was not a problem. The enzyme was known in the literature, we had the gene constructed, genetically modified the bacterium *Escherichia coli* in the first step and taught it to oxidize Deoxypodophyllotoxin. The bacterium did what it was supposed to do, and after cultivation we discovered our podophyllotoxin. The second step was not so easy, because the genetic modification of plants is difficult. For cow parsley, there was no gene taxi to bring the foreign gene into the plant, although the carrot, *Daucus carota*, belongs to the same family of umbellifers and had already been genetically modified.

To get a feel for the work, we built the gene for the liver enzyme into the tobacco plant. Tobacco is a grateful model plant that is not difficult to modify. The gene taxi was another bacterium that can cause plant diseases. These agrobacteria are attracted by an injury and penetrate into the plant. Once inside the plant cell, they build their own genes into the plant's genetic material and change the metabolism. We took advantage of these agrobacteria and instead of the disease-causing genes, we built in our liver enzyme gene. This trick also worked well with tobacco. We found the liver enzyme in the plant and it produced podophyllotoxin when we fed it with the precursor. My doctoral student from Indonesia was now supposed to introduce the gene into the cow parsley. This was certainly the most difficult task, because no one had tried anything like this before. It was not an easy task for her, because we had to bring the plant from the canals in beautiful Friesland into our laboratory.

We decided to cultivate the plant from seeds. We would take the young seedlings, incise them, infect them with agrobacteria, and try to make tissue cultures. Since the cow parsley is biennial, meaning it only blooms and forms seeds in the second year, this idea alone would take two years. If the doctoral thesis lasts four years, this is a real challenge. We tried it and were also successful in obtaining small seedlings. But the cow parsley did not like our lab. A plant that grows up to 1.8 m tall in the canals did not want to be larger than 50 cm in our lab, and these small plants produced very little deoxypodophyllotoxin. This was very frustrating and we learned that the plant lacked ecological stress such as cold, UV radiation, insect damage, and much more. Under such comfortable conditions as here in the lab, the cow parsley had no use for the lignans and the biosynthesis came to a standstill. We were able to confirm this phenomenon later when I asked colleagues in Europe to send us seeds and plant material for examination. It was clear that the samples from Iceland and Sweden contained much more lignans than those from the Balkans, where it was significantly warmer. My doctoral student managed to genetically modify the cow parsley and successfully complete her doctoral thesis. Unfortunately, we were not able to create a plant that produces enough podophyllotoxin to attract the interest of the industry. In addition, the author received a call to the Technical University of Dortmund at this time and the research remained in Groningen.

From Rubber to Medical Latex

Dear readers, in this chapter I would like to introduce you to a natural substance that is currently not or only limitedly used as a medicinal substance in the strict sense. The term is also incorrect and any medical lawyer would rightly point out that it is not an effective substance. We all know rubber as a natural latex product, which can be seen on the hands of every doctor in so many hospital soaps. The latex glove is as firmly associated with the image of medical personnel as the white coat or the stethoscope, but legally speaking, the glove is an interesting special case. According to European and international law, the latex glove is considered a medical device and is clearly distinguished from the active ingredients of medicinal products. But what makes the difference?

If we look into the Medicinal Products Act, we find right in the first paragraph the definition of the medicinal substance, which is intended to cure, alleviate or prevent diseases in humans or animals. We clearly read that the medicinal substance should have a positive biological or pharmacological effect for us. However, this does not explicitly apply to the medical product. These are only supposed to support and not have their own pharmacological effect. They should and often have to enable or enhance the effect of a medicinal product. Typical medical devices are, as a look into the Medical Devices Act shows, implants, infusion solutions, medical instruments, laboratory diagnostics, but also software, latex gloves and condoms. Yes, also condoms, because in addition to preventing pregnancy, they very effectively protect against transmissible sexually transmitted diseases such as chlamydia infections, gonorrhea, also called clap syphilis, HIV/AIDS, HPV viruses and many others.

What is rubber? Chemically speaking, rubber is a natural terpene polymer that is obtained from the latex of the rubber tree (*Hevea brasiliensis,* Fig. 17). With the discovery of the rubber tree by the Spaniards around 1500, the natural rubber obtained from it, known as wild rubber, came to Europe.

The first football in world history was probably also made of natural rubber, long before the first football game in history entered the annals in a Scottish monastery garden in 1497. On April 11, 1497, King James IV paid

Fig. 17   *Hevea brasiliensis* (public domain, wikipedia)

two shillings for a pair of fut ballis, and seven years later Queen Mary of the Scots is said to have indulged in the pleasure of a football game. So was Scotland the land of football? Doubts are appropriate, because none other than Christopher Columbus witnessed the natives' football game with a strange rubber ball on his second voyage to America (1493–1494) in Haiti. There were even stadiums in Latin America. Hernando Cortez mentioned in his 1519 records ball houses where the Aztecs chased this rubber ball. The Spaniards were interested in this ball and wanted to know what material it was made of. Hernando Cortez sent samples of the rubber from Quito in Ecuador to the Academy of Sciences in Paris with the request to investigate what kind of rubber it was. He provided the scholars with a brief description of how this rubber-like substance was obtained: *"In the province of Esmeralda grows a tree called "Heve" by the natives, which exudes a wise, milky sap when its bark is cut. The latter gradually solidifies in the air and takes on a dark color. In the province of Quito, this mass is used to coat fabrics and make them waterproof. The same tree grows on the banks of the Amazon, and the product obtained from them is called "cahutchu" by the natives."* An indigenous name derived from Caa (tears) and huchu for tree. This is very interesting because we learn that the name of the rubber plant, as well as the name rubber, comes from the indigenous language. The Spaniards had never seen this tree, but they brought natural rubber to Europe. At court, it attracted attention for a short time, but then fell into a Sleeping Beauty sleep for the next 200 years. The lack of enthusiasm was probably due to the strong smell and the physical decay of the rubber, which quickly darkened, stuck, became brittle and inelastic on the long journey. In warm weather, the rubber melted between the fingers and in the cold it became stiff and brittle. If the rubber was exposed to light, it quickly became unusable.

The first European who had the pleasure of seeing and exploring the rubber tree a little in 1743 was the French engineer François Fresneau (1703–1770). From France, he was sent to French Guiana in South America to develop the colony. He planned and built roads, mapped the colony, and developed a bellows to blow sulfur gases into anthills that were invading the cocoa and cassava plantations. He also took a liking to the local flora. While mapping the Approuague River, he discovered the rubber tree in 1747 and reported it to the Académie des Sciences in Paris. Rubber seemed to fascinate him, because on his return journey to France he took enough with him to research rubber in a small laboratory in his country house La Gataudière. He tried with little success to dissolve rubber in plant resins. After a few years, he then came up with the idea of turpentine, which wonderfully dissolved rubber. The idea was groundbreaking, because the dissolved rubber could be transported to France and Europe without damage. The solvent was evaporated and the rubber precipitated again and was usable.

When the Europeans discovered natural rubber, the indigenous peoples were already using it intensively and the Spaniards were amazed when they saw the Indians scratch the bark of the tree and a white sap came out. They were far ahead of the Europeans. The Indians not only knew how to impregnate fabrics with rubber to make them waterproof, but they also invented the first vessels or better bags made of rubber to transport and store water or perhaps also food. Humanity had to wait until 1839 for Charles Goodyear (1800–1860) to develop a method to transform the liquid milky sap or latex into a more durable, rubber-like mass through vulcanization. Charles Goodyear experimented in his kitchen with various additives like sulfur. In front of him were small strips of fabric that he coated with liquid rubber in which sulfur was suspended. Goodyear hung these strips of fabric on a line over the hot stove and didn't notice that a strip of fabric was not securely fastened and fell onto the hot stove plate. Unfortunately, he had to repeat the experiment, but when he took the slightly charred fabric from the stove plate, he noticed that the uncharred areas had excellent elasticity. The technical process of vulcanization, which others had tried before him, was invented.[181]

Charles Goodyear was not only the tinkerer who helped the car tire to break through. Few people know that he also made the first latex condoms in 1855. The first condoms had a wall thickness of 2 mm, and Goodyear made his first prototypes with side seams in the 1870s, with mass production beginning in 1895. They were probably not very sensitive, considering that today's condoms are between 0.06 and 0.07 mm thin. Perhaps Goodyear initially produced for household use or to earn urgently needed money. His financial situation was always serious. He died in debt and had no money to apply for a patent. This was obtained by the British Thomas Hancock (1786–1865), who took over his idea and received the patent for hot vulcanization in 1843.[182] He saw the potential of rubber and wrote to Joseph Hooker, the director of Kew Garden, in 1855 that he must ensure rubber plantations in India. Hooker took it kindly with a nod and promoted the project to English businessmen. Many waved it off, as they had not yet forgotten the laborious and loss-making business with cinchona bark. Should they now take such a risk again? India did not seem suitable to them. But Hooker did not give up, and another chapter in the history of biopiracy was to follow.

The Invention of the Disposable Glove

The idea of a protective glove in medicine was expressed early on by the surgeon Thomas Watson (1792–1882): *"In these days of ready invention, a glove, I think, might be devised, which should be impervious to fluids, and yet so thin and pliant as not to interfere materially with the delicate sense of touch required in these manipulations. One such glove, if such shall ever be fabricated and adopted, might well be sacrificed to the safety of the mother, in every labor."* [In these days of inventiveness, a glove could be invented that is impervious to fluids and yet so thin and pliable that it does not significantly interfere with the delicate sense of touch required in these manipulations (gynecological examinations and births). Such a glove, if it were ever made and accepted, could be sacrificed for the safety of the mother in every birth].[183] Watson recommended the glove especially for gynecology, because puerperal fever was rampant in hospitals and gynecologists often infected themselves with the papillomavirus on their skin. This is all the more interesting as he demanded this hygiene from his colleagues five years before Ignaz

Semmelweis' publication "Hoechst important experiences about the etiology of the epidemic puerperal fever in maternity hospitals" epidemic puerperal fever.[184] In 1870, disposable gloves for home use were introduced, which the surgeon Timothy L. Papin from St. Louis bought in street trade and put on in the operating room. After his studies in Paris, where he completed part of his training, he was completely convinced of the usefulness of gloves. Back home, he bought the simple gloves and brought the used ones home after the operation to the delight of his children, who blew them up as balloons. Regular use in medicine only began in 1894 after the introduction at Johns Hopkins Hospital in Baltimore. This happened on the advice and insistence of an operating room nurse, with whom a doctor fell in love.

She commissioned a rubber factory in Baltimore to manufacture disposable gloves for surgical purposes. The operating room nurse was supported by one of the most dazzling US surgeons of the time, who had taken a liking to the young lady. *"In the winter of 1889 and 1889, I cannot recall the month, the nurse in charge of my operating room complained that the solutions of mercuric chloroid produced a dermatitis of her arms and hands. As she was an unusually efficient woman, I gave the matter my consideration and one day in New York requested the Goodyear Rubber Company to make as an experiment two pair of thin rubber gloves with gauntlets. On trial these proved to be so satisfactory that additional gloves were ordered. In the autumn, on my return to town, the assistant who passed the instruments and threaded the needles was also provided with rubber gloves to wear at the operations"*[185] [In the winter of 1889 and 1889, I can no longer remember the month, the nurse responsible for my operating room complained that the solutions of mercuric chloride caused dermatitis on her arms and hands. Since she was an unusually capable woman, I dealt with the matter and one day in New York asked the Goodyear Rubber Company to manufacture two pairs of thin rubber gloves with gauntlets. During the trial, they proved to be so satisfactory that additional gloves were ordered. In the fall, upon my return to the city, the assistant who handed the instruments and threaded the needles was also equipped with rubber gloves to wear during the operations.], so argued one of the greatest surgeons of the USA William Stewart Halsted (1852–1922, Fig. 18). He was a surgeon from a wealthy New York family who was more interested in football than studying in his youth. It took several years before he became passionate about medicine and studied at the Columbia College of Physicians and Surgeons at the age of 22. He was a perfectionist and this showed at dinner parties he hosted at his house. Every turtle that went into the pot for the soup was personally selected by him. He took every coffee bean between his fingers and checked it. Every tablecloth had to be ironed

Fig. 18 William Stewart Halsted (1852–1922) (Release, Johns Hopkins University, USA)

so that it was wrinkle-free. He sent his white shirts to Paris for washing because they would become whiter there. However, this may be doubted, because on the way back he probably managed to smuggle his cocaine, but we will come to that later. In any case, his wife was completely annoyed by her husband. It is said that she suffered severe migraines because of her beloved William's behavior.[186]

Halsted was an extraordinary and ambivalent personality. He introduced new surgical techniques, invented the radical mastectomy for breast cancer, and was a role model for young surgeons. On the other hand, he had intimate relationships with nurses, first had a major cocaine problem and then an even bigger heroin problem. At the height of his addiction, he is said to have consumed up to 10 g of cocaine daily. Morphine- and cocaine-addicted doctors were not uncommon at the time, as both natural substances were not prohibited and were socially accepted. Halsted became aware of cocaine through an article by Henry Drury Noyes in The Medical Record magazine, which reported on the local anesthetic effect on the eye of the

Viennese doctor Carl Koller. This article prompted Halsted and some of his students in New York to try cocaine in a self-experiment, and they quickly developed a strong addiction. Halsted's cocaine problem was so severe that he was only able to work to a limited extent and was on withdrawal for years. Only a sailing trip to the Windward Islands, far away from all cocaine, made him abstinent. However, he never got his second addiction to morphine under control until his death. New York was too wild for Halsted and he looked for a quieter place to work. He went to Johns Hopkins Hospital in Baltimore, near the capital, which was a rather sleepy nest at the time. In 1904, the city was almost razed to the ground in a major fire and a few years later deindustrialization followed, which almost dealt the city a death blow. Halsted moved away from New York high society with his family and started his hitherto unsuccessful career as a surgeon for the second time.

As a pedant, he advocated absolute asepsis, hygiene, avoidance of excessive blood loss, and clean suturing of wounds. In this context, the introduction of the medical disposable glove is probably also to be understood. At that time, surgeons worked with bare, disinfected hands in the wound and Halsted introduced the glove because of antisepsis. Disinfection was not taken seriously and was backward in hospitals at the time (Fig. 19). For example, there was an orange handle at Johns Hopkins Hospital with which the surgeons cleaned their fingernails before the operation. They then dipped their hands and forearms in disinfecting solutions, containing potassium permanganateand mercury. The potassium permanganate stained the skin brown, decolorization was achieved by a bath in oxalic acid. For many surgeons and operating room nurses, skin diseases were quite normal occupational diseases.[187]

Despite Professor Sauerbruch in Vienna, hygiene was not yet a big issue. Surgeons wore black surgical gowns, on which the blood and pus stains of the last patients were still visible, as if medals were proudly testifying to their heroic deeds. Halsted wanted to change this. He was the surgeon who introduced short-sleeved white shirts that, like gowns, had to be changed regularly. He also banned the wearing of street shoes in the operating room and urged his grumbling colleagues to at least wear white tennis shoes, which we still know in hospitals today. Surgical instruments such as scissors and scalpels were already boiled, sterilized, and placed in a carbolic acid bath. Carbolic acid is chemically phenol, which microbiologist Joseph Lister introduced as a disinfectant in the 1860s. It is an extract of coal tar, obtained by distillation at 170 to 230 °C. At that time, it was the only disinfectant that did not corrode the metal of the surgical instruments. The new operating room at Johns Hopkins Hospital was to be modern, and Halsted was

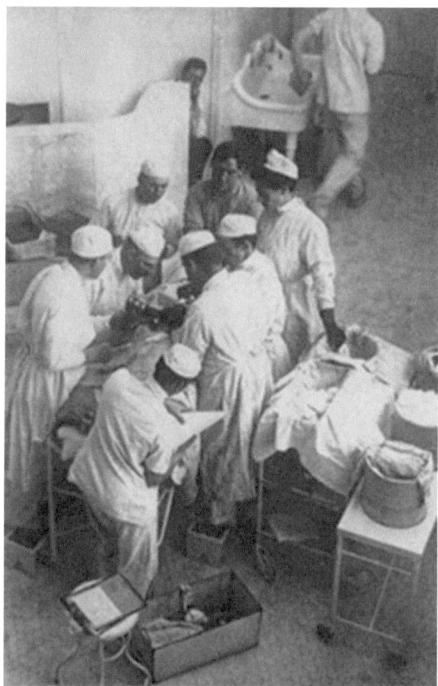

Fig. 19 W. S. Halsted (1852–1922) in the operating room at Johns Hopkins Hospital, around 1900 (Release, Johns Hopkins University, USA)

hard to dissuade from the idea of setting up a sterile tent in which the operating table stood. In New York, his patients still lay on a wooden table from the Prussian-French War of 1871.

The use of gloves in surgery is certainly not due to the desire to work aseptically. Rather, it is assumed that personal tissue injuries to the hand were to be avoided, especially in gynecology, where things were robust during childbirth anyway. Another reason was that doctors and nurses got skin rashes from chemicals like mercury solutions, ethanol, iodine, perchloric acid, and caustic lime, which were widespread in surgery at the time. Another amusing anecdote is that Halsted had rubber gloves with cuffs made for his operating nurse at the Goodyear Rubber Company in New York in 1890 to protect her arms from the chemicals. Because he wanted to marry her. A wife with cancer-red arms at the end of each working day, he did not like that. The gloves were made, but they had to be used multiple times. The surgical assistants boiled them after each operation to sterilize them. Before each operation, the freshly boiled and hot gloves were pulled over the washed hands of the surgeon. Werner Zoege von Manteuffel

(1857–1926), a surgeon in Estonian Tartu who introduced gloves in Europe, once described his work as follows: *"Whoever wears boiled gloves must work with boiled hands."*[188]

Most of his assistant doctors wore the gloves voluntarily, and one of his protégés, Dr. Joseph Bloodgood, later reported that wearing the gloves had reduced the risk of infection in hernia operations by almost 100%. Guest surgeons like Johann von Mikulicz (1850–1905) from Breslau were also impressed and von Mikulicz advocated their use in the OR after his return to Germany. Halsted was both thrilled and thoughtful, as it is reported that he wondered how he could be so *"blind not to recognize the necessity of constant wearing on the operating table."* Of course, there were also incorrigible doctors at Johns Hopkins who argued that the gloves were too rough, too uncomfortable, and too insensitive to operate well.[189] They were simply unsuitable. Halsted realized that he was not getting anywhere with good arguments and used a trick to convince his stubborn colleagues. He asked three blind ladies to put on medical gloves and use their fingers to feel and read Braille. The three ladies felt the small dots in the gloves and proved that there was no impairment of the sense of touch and thus of the highly tactile abilities of the surgeons. The last two surgeons seen operating without gloves in the USA were in the 1960s.

The Alternatives to Natural Rubber

Bayer AG recognized the potential of natural rubber. Waterproof clothing, rubber gloves, seals, hoses, and air-filled tires were the first practical applications that promised the company a lucrative business. However, there were also concerns about whether and how the market could be adequately supplied with natural rubber. The board of Bayer AG offered a research prize of 20,000 Reichsmark to the employee who invented a new process for the production of rubber or a fully-fledged substitute. Fritz Hofmann (1866–1956) did not hesitate and discovered the first process for the production of synthetic rubberfrom dimethylbutadiene.[182, 190] On September 12, 1909, the Imperial Patent Office granted Bayer AG Patent No. 250690 for the production of synthetic rubber. Even Kaiser Wilhelm II was thrilled and was *"highly satisfied"* with the synthetic rubber tires that moved his car.

After natural rubber was displaced in medicine by synthetic elastomers, it also fell into oblivion due to its strong allergenic effect. However, in recent years there has been a revival of natural rubber, this time from the dandelion. Each of us as a child has once plucked a dandelion and sent the

Fig. 20 *Taraxacum kok-saghyz* (public domain, wikipedia)

beautiful white seeds, which hang like a little man on a parachute, on their journey with a strong puff. It was noticed that a white milky sap ball formed at the torn-off spot of the stem. Chemically and botanically, this is also rubber or latex.

During the Second World War, the demand for natural rubber from tropical countries was high, but it could not be met. In the former Soviet Union, it was known since the 1930s that rubber could be made from rabbit food. At peak times, the Soviet Union cultivated 67,000 ha of Russian dandelion (*Taraxacum kok-saghyz,* Fig. 20) which covered about a third of the national demand. Sanctions were imposed against Nazi Germany and the import of natural rubber was no longer possible. The early interest in the Russian alternative was therefore very high. Its significance is shown by the fact that the National Socialists did not hesitate to set up a nursery as a research station for dandelion gum in Auschwitz concentration camp from 1942, employing 150 to 200 forced laborers. Nazi Germany lost the war and had access to rubber again during the economic miracle. Rubber from dandelion was too expensive and lost its importance. In the 2000s, the biologist Dirk Prüfer from the University of Münster came up with the idea of biotechnically obtaining rubber from dandelion. Dandelion unfortunately provides a very expensive latex sap, but it has the advantage of low allergenicity, which makes it interesting for medicine. So far, unfortunately, only a green bicycle tire is known as a prototype, but research takes time.

From Parisians and Fromsers

Even if the condom does not fit properly into this book, its importance in medicine is undeniable. Not only as a means of family planning, but also for the prevention and avoidance of sexually transmitted diseases, which have culturally shaped humanity and caused suffering to both women and men. In the chapter on sildenafil, we have already met the man's penis as an object of pharmacological care when it no longer functions properly. Fortunately, the majority of men have an erection, but besides the prevention of sexually transmitted diseases like syphilis in the past and HIV and papillomaviruses today, pregnancy prevention is the second reason for the use of condoms in German called "Pariser" (Parisians). The penis is certainly a physiologically extraordinary organ of the man. Let's start with its external appearance. It has no subcutaneous fat tissue, no matter how fat its owner is. If you get too fat, you have to buy new pants or shirts that are a few sizes larger, but the size of the penis remains the same. Let's move on to function. Penetration into the woman's vagina is only possible in the erect state, but 99.9% of the man's lifetime it remains flaccid. It would also be very uncomfortable and evolutionarily disadvantageous if the man were always to cultivate his land, sit in the saddle, wage wars or give speeches with an erect penis. Unlike many other mammals, humans do not have a penis bone. This is surprising, as many of our closest relatives such as chimpanzees and gorillas still have it. Why we lost it is unclear. One theory is that it could be related to our monogamous lifestyle.

The history of the condom is highly interesting and reflects very well the morality over the course of our cultural development. As so often, the first condoms in their simplest form can be found among the ancient Egyptians, and their constant improvement with various materials such as linen, animal intestines, to the modern latex condom, brings a smile to the reader's face. Many a book has been written about the history of the condom. I do not want to start here with the early prehistory in antiquity in order to be able to grasp the already large amount of information and anecdotes at all. Let's stay in the early past and see that the first mention of the modern condom can already be found in Meyer's Conversational Lexicon from 1851: *"Condom, a sheath made of goldbeater's skin (= the outer membrane of the cecum), which libertines pull over the male member before sexual intercourse to prevent fertilization and presumably also to prevent venereal infection. One of the ingenious inventions to which these corrupt times have led, although morally upright people despise them and hardly know their name anymore."* The moral finger was already raised. It is interesting that the material was mentioned in the definition. Goldbeater's skin is the outer skin of bovine ceca, which are washed and dried after the cow is cut open. A bovine cecum can be up to 30 cm long, is very tear-resistant and has a wall thickness of only 0.1 mm. The name comes from craftsmen who made gold leaf and divided the bovine ceca into 100 x 25 cm natural foils. These were natural protective foils, between which the wafer-thin gold leaf was placed. The tear resistance and the excellent impermeability to gases and liquids also made the ceca interesting for the zeppelins. About 700,000 bovine ceca were glued in up to seven layers onto the outer skin of the gas envelope to make a zeppelin halfway gas-tight.

In addition to Charles Goodyear, another pioneer of condom manufacturing should be mentioned, whose condom is still associated with his name today: Julius Fromm (1883–1945, Fig. 21). The Fromm family was an immigrant family from the city of Konin, about 120 km east of Poznan in today's Poland. During the imperial era, Poznan was still part of the German Empire and Konin was a city in tsarist Russia, which, after the Congress of Vienna, belonged to Congress Poland, but was administered by Russia in personal union. Konin was a city where a third of the population were Jews, and the Fromms were among the many who found work here more or less badly. In the years 1880/81 there were pogroms in Russia and the liberal and freer Germany became a refuge. The richer Jewish Russians moved to Charlottenburg, which the Berliners soon called Charlottograd. The poor settled in the Berlin Scheunenviertel, the catchment area of the Berlin poor, the left behind and migrants.

Fig. 21   Julius Fromm (1883–1945) (public domain, wikipedia)

With the industrially rising Berlin, just under 400 km away, many immigrants associated great hopes, and so Bartusz and Rigfa Fromm decided to seek their fortune in the big city. They knew it would be a tough start, but they wanted to take the risk and moved to the Scheunenviertel. Here, in imperial Berlin, was the slum of the city with run-down houses that were still half-timbered buildings. Simple and poor people lived here next to prostitutes, criminals, alcoholics, and families with many children. These were usually accommodated in a single room of an apartment. Here, their own rules applied and the criminal associations of Berlin, which called themselves ring clubsor savings clubs and bore names like "German Oak" or "Always Faithful", provided more security than the Blues, as the Berliners called the Prussian police.

The seven-member Fromm family arrived in the capital in 1893 and lived at Mulackstraße 9. They were in the eye of the storm, because only a few meters away was the Mulackritze, a pub where the ring clubs had their regulars' table and ran a brothel on a few bug-infested bunks under the roof. But celebrities like Gustaf Gründgens and Marlene Dietrich are also said to have been seen here. The Fromm family settled in, they had a good start, the father made a good living as a small trader of furs and with some money they moved to a larger apartment around the corner at Kleine Rosenthaler Straße 12.

The decision to establish a condom factory was bold in the prudish German society of the Weimar Republic. This was evidenced by the reputable business of the "Bookstore for Sexual Science", at whose shop windows the Berliners pressed their noses flat and which was often searched by the criminal police for indecent publications. The owner was regularly threatened by the state and in 1932 he had to go to prison for eight months *"for distributing indecent writings"*. Advertising for condoms was very restrained in the 1920s. The advertising message was directed at the German man: *"Protect yourself from sexually transmitted diseases!"* Contraception was not allowed to be advertised at all, as the state and church were suspicious when soldiers and believers were absent. Only from 1932 did Fromm dare more and praised the realistic wearing, the transparency of the "Fromm rubber", the pleasant smell and the silky softness. Germany became more relaxed, but still, at drugstores and hairdressers, slips of paper were handed over the counter unsolicited with each sale, on which the next sale was to be read: *"Please give me three pieces of Fromm's rubber discreetly."*

After the conservative and prudish years of the imperial era, people lived out their lust and sexuality (Fig. 22). Around Alexanderplatz there were 300 brothelsor hourly hotels, and anyone who unexpectedly found nothing here

Fig. 22 The girls of the Haller Revue in their Quadriga figure at the Admiralspalast on Friedrichstraße hold up a condom (permission bpk Berlin, DE)

went to Bülowstraße, Friedrichstraße or Wittenbergplatz, where during the inflation period the blowjob was offered for 2 trillion Reichsmark. A good 10,000 control girls, as the Berlin police called the registered prostitutes, and an unknown number of artificial silk girls, as the Berliners called the hobby whores, who were actually secretaries, stenotypists or saleswomen, offered their services. The situation can only be guessed at in all its explosiveness for Berlin when a letter from the Baden-Württemberg Ministry of the Interior from 1908 speaks of 20 to 30 prostitutes in Stuttgart. The Swabians already saw a big problem in this small number.[191]

Sexually transmitted diseases such as gonorrhea or syphilis were so widespread in Berlin in the 1920s that everyone knew someone who had the infectious disease. What to do if you get caught? Syphilis or hard chancre is a bacterial infectious disease caused by the bacterium *Treponema pallidum*. It is transmitted through sexual intercourse and causes ulcers on the mucous membranes and swelling of the lymph nodes in a fresh infection. If the infection is not treated and becomes chronic, it can lead to skin and organ involvement as well as disturbances of the central nervous system. In the Middle Ages, syphilis was still considered a punishment from God. The paintings of this time by Matthias Grünewald (1480–1530) of the little man with syphilitic bumps in the picture "Temptation of Saint Anthony" of the Isenheim Triptych or Dürer's depiction of a syphilitic are very well known.

Famous sufferers of syphilis were Catherine II, known as the Great (1729–1796), Wolfgang A. Mozart (1756–1791), Ludwig van Beethoven (1770–1827), Heinrich Heine (1797–1856), Charles Baudelaire (1821–1867), Oscar Wilde (1854–1900) and the doctor Ignaz P. Semmelweis (1818–1865). The latter went down in medical history as a hygienist who achieved great successes in the fight against puerperal fever.[192] He too took a liking to prostitutes, contracted the disease and was admitted to the Vienna mental hospital Döblin due to emerging nervous diseases, which were misinterpreted in the literature as dementia or depression, where the guards mistreated him and he died under unexplained circumstances at the age of 47.[193]

In the 1920s, syphilis was considered incurable because penicillin had not yet been discovered. However, there was hope at the beginning of the 20th century: Paul Ehrlich (1854–1915, Fig. 24) had already succeeded in developing the unfortunately highly toxic drug Arsphenamin (Fig. 23) in 1910. The name was programmatic: He called the drug Salvarsan®, composed of the two Latin words salvare for save and sanus for healthy. A very clever combination, because the expert also reads the word arsenic, as the drug contained arsenic, which is unusual today, but was in fashion in medical chemistry at the time.[194] Paul Ehrlich's idea was to tame and defuse the highly toxic arsenic. Apart from its narrow pharmacological range and the very high risk of dangerous side effects, it was so expensive at a price of 16,000 Reichsmark that only a few people could afford it. But it also brought in a lot, so it was worth transporting the drug to the USA on a commercial submarine during the First World War. The submarine was necessary to break through the British sea blockade. Salvarsan® was and remained scarce. Not only during the First World War, but well into the 1920s. It was the first effective chemotherapeutic agent and in demand worldwide, as syphilis was spreading rapidly across the globe.

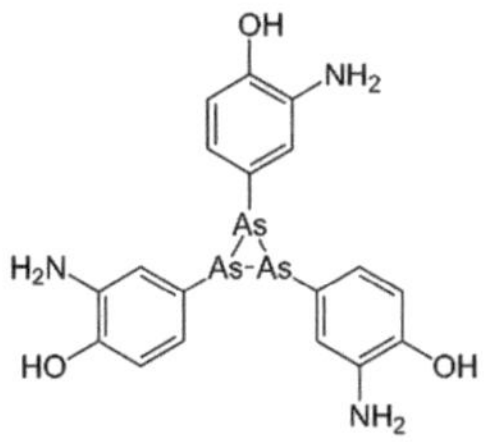

Fig. 23 Arsphenamin or Arsenobenzol (Salvarsan®) (work of the author)

Fig. 24 Paul Ehrlich (l) and Hata Sahachirō (r) (public domain, wikipedia)

In the development of Arsphenamin, Paul Ehrlich started from Arsanilic acid (Atoxyl®), another arsenic-containing drug against tropical sleeping sickness, with which the great Robert Koch experimented. Building on the results of the Frenchman Antoine Béchamp (1816–1908), who reacted arsenic acid with aniline, Ehrlich and his colleagues applied a rational drug search strategy for the first time in medical history to systematically track down an active ingredient. His collaborator Alfred Bertheim (1879–1914) synthesized more than 600 substances. The Arsenobenzol was the 6th compound in the series 6 with the number 606, which was further developed into Salvarsan®, which was launched by Hoechst in 1910.

A Japanese scientist, who came to Germany in 1876 as part of an exchange program of the Japanese government, contributed to this research. Hata Sahachirō (1873–1938, Fig. 24), who had already worked with the Japanese Robert Koch named Kitasato Shibasaburō (1853–1931) and had good knowledge in microbiology, deserves the actual recognition for the compound 606. He had already worked with Robert Koch and that was reason enough to send him to Paul Ehrlich. He assigned him the task of researching number 606. Hata experimented with a rabbit model and found the substance very active against syphilis and reported this to his boss. Ehrlich was thrilled and sent the substance to his colleague Albert Neisser (1855–1916) in Breslau, who successfully conducted the first human trials. As a dermatologist and bacteriologist, he already had experience with sexually transmitted diseases. As early as 1879, he made the groundbreaking discovery of the pathogen of gonorrhea, which was later named after him—*Neisseria gonorrhoeae*. Less groundbreaking, but the first major German medical scandal was his deliberate infection of healthy people with the syphilis pathogen. From the serum of the infected, he believed he could develop a vaccine. The project failed and in Prussia there were heated debates in the state parliament about this medical misconduct. The scandal did not stop human experiments, as Konrad Alt (1862–1922) in Uchtspringe near Magdeburg began new series of experiments in the fall of 1909. The only difference: Men with syphilitic paralysis, who could no longer move due to spinal cord damage caused by advanced syphilis, were to be treated with Salvarsan®. He published the very positive results from his point of view in 1910 in the Münchener Medizinische Wochenschrift and in the Berliner Klinische Wochenschrift, which prompted the Hoechst company to consider production.[160]

The use of arsenobenzene for the treatment of syphilis was not the original goal of Ehrlich's research.[195] The idea came to him through a publication by FritzSchaudinn, who was researching the syphilis pathogen at the Berlin Charité. Fritz Schaudinn (1871–1906) was not a physician, but a zoologist and discovered the pathogen in a syphilis ulcer as *"an extremely fine, very pale and only visible at very close inspection screw, which moved lively"*. His discovery in 1905 received international recognition, but not in Berlin. The prophet is worth nothing in his own country. The Berlin Medical Society also did not want to talk about the syphilis pathogen, even though other colleagues had found it in ulcers. The discussion was abruptly cut off by the chairman Ernst von Bergmann (1836–1907), who briefly noted after a lecture by Schaudinn: *"I close the discussion until the discovery of the next syphilis pathogen."* After 25 alleged syphilis pathogens had already been found in his career, he lacked faith. Schaudinn was not impressed and countered at dinner: *"I am not at all affected by this. What I have found is correct!"* Recognition was definitely assured for him and at the Dermatological Conference in Lisbon 1906 he was celebrated like a rock star. *"The dermatologists are paying me a tribute today that I will probably not experience again. (…) As soon as I opened my mouth, the guys clapped like crazy. I quickly picked up a few French phrases, grand honneur, merci beaucoup, again they stomped and clapped like mad. In short, it was wild!"* He would not live to see the discovery of

Salvarsan® three years later. After his return, he had to undergo emergency surgery for stomach ulcers caused by the syphilis pathogen. He had infected himself in a self-experiment and paid for it with a short life of only 34 years.

Syphilis was the AIDS epidemic of the 1920s in Berlin. There was a lack of an effective remedy, and the only one available was toxic and expensive. The idea of contraception and protection against sexually transmitted diseases fell on this fertile ground and the demand for Fromm's rubbers was high. But Julius Fromm did not come up with the idea through societal circumstances, but through his difficult family situation. His parents died early. The father was only 42 years old, the mother had just reached her sixth decade of life at 52 years. He did not want to be a cigarette seller, but to escape misery. His brother Salomon emigrated to England, Julius looked for an alternative and attended evening courses in chemistry, especially in rubber chemistry. He may have read about Goodyear and his condoms, so the idea might have come to him to become a condom manufacturer, as his son Edgar remembered later.

Shortly before the First World War, Julius Fromm founded the company "Israel Fromm, Manufacturing and Sales Business for Perfumes and Rubber Goods" in a backyard at the then Lephener Straße 23 in Berlin-Köpenick in 1912. Fromm's condom was better in shape and quality than Goodyear's original from the start. While the latter was coarse, still with seams and opaque, Fromm's condoms were seamless, thin and almost transparent. The reason was his better manufacturing method, because Julius Fromm made his condoms elegantly with a new dipping process. He had glass bodies resembling a penis made and dipped them twice in a rubber bath. He called them "Fromms Act", but no one knows exactly why this Anglicism crept in at the time. Julius Fromm was a perfectionist with a pronounced business sense. He knew that quality would prevail. No one would buy his condoms if they broke or did not allow a real feeling experience. He tinkered with the best method that would make him and his customers happy. The First World War was both a test and advertising at the same time.

The World War raged and many men were without a woman at the front for too long. The establishment of field brothels was no secret and was supported by the highest military authority. The soldiers were supposed to vent sexually after the stress at the front and the long abstinence. Separated by officers, who had their own brothels with an entrance fee of 3 Reichsmark, and the field brothels with 1 Reichsmark entrance fee for the troops, there was a condom obligation. A regulation for the operation of a field brothel even stipulated that condoms always had to be ready. A sign on the door pointed out to the soldiers their patriotic duty to wear their latex steel helmet over their best piece. The number of sexually transmitted diseases, especially syphilis, increased nevertheless, but with a condom only by 25% compared to 100% in unprotected sexual intercourse in other units. Most soldiers took the condoms, got to know and appreciate them, and that was the best advertising for the Fromms at the front.

Julius Fromm expanded his business and moved in 1917 from Friedrichshain to Elisabethstraße 17 in Berlin-Mitte, then to Landsberger Allee 73 directly at Alexanderplatz, but even here it soon became too cramped and he wanted to take the big leap. In Berlin-Köpenick, Julius Fromm bought a 16,000 m^2 plot of land. In the middle-class Friedrichshagen, no quarter of an hour's walk from Köpenick Town Hall, he wanted to have his condom factory built. He commissioned two young Jewish architects who wanted to place a modern factory and office building in the Köpenick green in the architectural style of New Objectivity. The plan was to produce 150,000 condoms daily. Was the location in Köpenick a good choice? What angered the residents was the aesthetics (Fig. 25). A modern building with a flat roof? That doesn't fit at all with the pointed gable roofs in the neighborhood. There were also problems with

Fig. 25 Julius Fromm's condom factory, Berlin-Köpenick, circa 1922

the authorities. The building department found that it was a business in a residential area. They would only approve the construction if there were no odor and noise disturbances. These occurred and the residents complained. And then this sulfur smell! Everywhere it smelled of vulcanization with sulfur. Added to this was the heavy traffic of delivery vehicles, the nightly rumbling of the rolling mills, and the smell of benzene, chloroform and gasoline, in which the rubber was dissolved. The factory remained, but it was destroyed in the Second World War. Today, a hardware store stands in its place.

Until 1960, condoms were made using the gasoline process, which was only late replaced by the latex process invented in the 1920s. This technique has remained almost unchanged to this day. The manufacture of condoms was anything but pleasant work. To prevent ripening and aging, the delivered rubber was mixed with ammonia in the exporting countries. The rubber bales, which mostly came from Malaysia, were crushed, washed, and left to swell for ten days with secret chemical additives. Only then could it be rolled, mixed, and dissolved in gasoline. In his 140 m long and 35 m wide factory, thousands of glass penises stood, which were pulled through the milky latex suspension. On each of the glass penises, a 0.03 mm thin, almost transparent film remained. The glass bulb went through a drying chamber and a second dip followed. When the raw condoms were dry again, they were rolled up with brushes. For separation and as a lubricant, Fromm initially used powders such as denatured cornmeal, carob bean meal, or Indian tragacanth. The disadvantage: rubber swelled and the rubbers could no longer be unrolled. Fromm replaced these additives with talcum and immediately patented this idea and improvement. Only in 1960 was the solid powder replaced by a liquid silicone lubricant. The modern spermicidal additives, such as modern nonoxynol-9, followed in 1968. Julius Fromm placed great emphasis on high quality and was one of the first to systematically take samples from a production batch and examine them. His modern quality control stipulated that a condom, filled with 18 l of air, should not burst. Most even withstood a pressure of 50 l. This simple test method had already been used by Giacomo Casanova (1725–1798), who inflated his condoms prepared from animal intestines and searched for holes (Fig. 26). To maintain the quality and motivation of the testers, there was an extra bonus for each found damaged condom.

Medical gloves are still made today using the same principle as condoms. A hand-shaped ceramic form is dipped several times into the liquid latex until the desired thickness of the rubber is reached. This blank is dried while constantly turning and then air-dried, removed from the ceramic form, and

Fig. 26 Casanova (l) testing the tightness of a condom (Permission Landesarchiv Berlin, Photographer Henri Fromm)

separated. To prevent condensation water from damaging the glove, the glove must cool down before packaging. After cleaning the exposed ceramic form, the next glove can be made.

For the conditions of the time, the production was enormous. In 1926, 24 million condoms were to be shipped from the Köpenick factory. The condoms promised a high quality that was missed in the USA. Here, the probability that they would burst or tear was over 50%. Late, but better than never, the American drug agency issued quality regulations for the US market in 1938, which the Fromms had already adhered to for over ten years. The condom empire flourished and grew. By the mid-1930s, the value of the company was 8 million Reichsmark, which would correspond to 120 million EUR today. This aroused desires and base instincts, if the owner was Jewish. His assimilation and his German citizenship were a hurdle, but the Nazis wanted to take them in order to Aryanize Fromm's work, as they mockingly called it. Fromm was to disappear as a Jew, his work and his fortune were to be given up, according to the idea of the National Socialists. Julius Fromm was not naive, but he did not believe that the National Socialists would go so far. He was mistaken. Before it hit him personally, he sent his children abroad in 1934. He tried to sell and agreed to a price of 4 million Reichsmark. Finally, he signed a notarial forced contract and sold for 200,000 Reichsmark to Hermann Göring's godmother Elisabeth Edle. In return, the Reich Air Minister received the castles Veldenstein in Bavaria and Mauterndorf in Austria from his godmother.

Julius Fromm emigrated with his family to London. He was determined to wait for the end of the war and return to Germany after the victory of the Allies. This was no longer granted to him, for he died a few days after the end of the war on May 12 at the age of 62. His son Siegmund tried to reclaim the company. It was now in the Soviet sector, and the Soviets were not interested in handing it over to a Jew and capitalist. This negative attitude was reinforced by the German communists, who saw in Julius Fromm a capitalist and exploiter. Another lie was that Julius Fromm had voluntarily sold to the National Socialists. The factory became state-owned, but at least Siegmund Fromm wanted to have the naming rights back from a cousin of Göring. This one smelled a business and with the downfall of the Third Reich his finances had also exploded and burned. The cousin allegedly did not want to sell and saw himself in the right. The purchase contract of 1934 was not a forced act, Julius Fromm had sold voluntarily. The Allied Command could not determine the opposite, there was a tug-of-war between the heirs and the cousin, who drove the price for the repurchase to 176,000 DM. For the conditions of the time, a gigantic sum. The Fromms paid and wanted to start the business in Lower Saxony's Zeven as quickly as possible. They were still lucky, because the former production manager was looking for work and they met at a fair in Leipzig. The British occupiers were also interested, because after the war sexually transmitted diseases were spreading again. A condom factory was quickly approved and they produced for post-war Germany.

The term "Frommser", "Fromms Act" or just "Fromm" was a brand name that stood for all condoms until the 1960s, but has disappeared from our brand consciousness today. We now generally speak of condoms and are not ashamed to buy, use and talk about them. On the contrary: condom stores can be found in almost every major city and condoms are advertised in all colors, shapes and flavors. With the advent of HIV infection, condoms have become more important than ever and sexually transmitted diseases such as syphilis, gonorrhea, chlamydia infections, genital herpes, papillomaviruses or trichomonads have been on the rise for years. Statistically speaking, every sexually active person has had at least one sexually transmitted infection in their life, even though many of them are asymptomatic. It cannot be proven, but condoms are still the best protection against sexually transmitted diseases and have saved more lives and prevented more diseases than many an antibiotic.

What remains of the rubber? Historians suspect that the cultivation of the rubber tree and the extraction of the rubber is the most profitable plant in world history, before coffee, tea and coca. Kew Garden and its affiliated gardens were very efficient in research and development and they led Brazil to the economic abyss in 1913. In return, it made Malaysia rich when it was still a British colony. Without natural rubber, the industrial development in the automotive industry would not have happened. Even today, car tires are produced with a high proportion of natural rubber mixed with synthetic rubber. Natural rubber has elasticity properties that have not yet been achieved by synthetic rubbers. In medicine, rubber has lost its leading role due to its allergenicity and is replaced by synthetic rubbers. However, rubber will still be found in our world for a long time.

Modern Drug Research

Modern drug research follows strict rational principles, which we highly value as empirically based science. No drug receives approval for use on and in humans unless its mechanism of action, its metabolism, and its whereabouts in the body are clarified. Approval without a solid understanding, as in the case of Thalidomide (Contergan®), would be unthinkable today. But what is pharmaceutical research or pharmacology as a science that has brought so much quality of life? Science can be defined as the systematic search for knowledge about the world itself and that, *"which holds it together at its core,"* to freely quote Goethe's Dr. Faustus. We make observations through experiments, formulate hypotheses, and discard them. What we find, we describe and document as accurately and precisely as possible. Only in this way can other scientists repeat them and come to the same conclusions.

Compared to mathematics or theoretical physics, the science of drug research is not an absolute, but an evidence-based science. This means that in medicine and biology, cells are often worked with that do not always deliver the same results. Results can fluctuate because researchers do not always feed the cells with exactly the same medium, because the temperature fluctuates due to handling, or because cells age over time and become limited in their function. We in the lab also know that cells in a warm lab grow faster in the summer than in the cold winter. Some plant cell cultures also go through the seasons in the incubator, even though they have never seen the sun. Much depends on the standardization of experimental conditions to detect deviations, not to misinterpret results, and to provide evidence (empirical proofs). In modern drug research, a variety of biological test methods such as animal experiments, cell culture experiments, but also enzyme or receptor models are used. Since the 1990s, the colleague computer with its programs from bio- and chemoinformatics has been used and with the upcoming Artificial Intelligence (AI), the simulation of drug activity will be indispensable. Many researchers hope that AI will drastically reduce the number of animal experiments in research and development. But if we are honest, the use of all strictly rational techniques is only part of the truth, because as in real life, chance and intuition play an equally important role. How high is the probability that the colleague chance or the colleague intuition will gain groundbreaking insights? Up to here, you could already read about some significant coincidences. How would the discovery of today's highly valued drugs have proceeded if chance had not existed? Some examples are intended to explain its importance with a wink.

Kairos as the fortunate moment of discovering a drug is quite present in history and still amazes today's scientists when they hear about the story behind the story. However, luck is to be distinguished from pure chance, which opens up a whole new chapter through incorrect handling or through the unforeseen product or result of an experiment. Random events have already been described in this book. Perhaps you remember William Perkin, who was supposed to synthesize quinine in 1856 and instead discovered the new dye mauveine. This accidental discovery was the starting signal for the chemical industry with Bayer (1862), Ciba (1859), and Sandoz (1882), which later became global companies (Table 2).

Let's try to capture chance statistically: Just over six percent of the drugs that are on the market today owe their discovery to chance. Of this small number, a third were accidental finds in the lab, the remaining drugs were discovered by chance during clinical trials or through reports from test subjects. Another 18.3% of the active ingredients were known, but additional effects were recorded through reports from professional staff. If we also consider the drug candidates that dropped out of clinical trials, about a quarter of the active ingredients were discovered by chance or intuition. Table 1 very impressively lists examples of drugs that were found by a lucky chance. Such a high discovery rate is something no pharmaceutical company can claim for itself.[196]

Some discoveries are so interesting that it is worth briefly reporting: Chloral or trichloroacetaldehyde was already discovered in 1832 by Justus Liebig, but only 37 years later, in 1869, Otto Liebreich (1839–1908), who was active in the field of pharmacology in Berlin, claimed that he saw a crystalline chloroform in the structure. He believed that chloral hydrate would

Table 1 Drugs found by chance or intuition

Drug	Year	False Assumption	Current Indication	Discoverer (Person or Company)
Potassium Bromide[386]	1857	Sedative	None, used as a remedy for epilepsy until 1860	Charles Lockock
Chloral Hydrate[386]	1869	none	Sedative	Otto Liebreich
Lithium Chloride[397]	1855	Gout treatment	Bipolar manic depression	Carl Lange
Methylphenidate	1955	Blood pressure increase	Attention deficit/ hyperactivity disorder	Charles Bradley
Paracetamol	1886	Wrong delivery	Fever reduction	Arnold Cahn Paul Hepp
Warfarin	1920	Discovered as poison in cows	Anticoagulant	Harold Campbell
Penicillin	1928	none	Antibiotic	Alexander Fleming
Nitroglycerin	1879	Originally explosive	Blood pressure reducer	William Murrell
Lysergic Acid Diethylamide (LSD)	1938	Gynecology	Psychoanalysis, Psychomimetic (approved until 1966)	Albert Hofmann
Streptomycin	1944	Antituberculosis agent	Antibiotic	Selman Waksman
Meprobamate	1945	Antibiotic	Anxiety reliever (in the EU market until 2012)	British Drug House
Chlorpromazine	1950	Anthelmintic	Antipsychotic	Paul Charpentier (Rhone Poulenc)
Imipramine	1950s	Schizophrenia	Endogenous Depression	Roland Kuhn (Ciba-Geigy)
Iproniazid	1952	Tuberculosis	Antidepressant	Selikoff Robitzek Orenstein
Benzodiazepines	1957	Synthesis error and sloppiness	Sedative	Leo Sternbach (Roche)
Cisplatin	1960	No Indication	Cancer Drug	Barnett Rosenberg
Mustine	1961	Originally a warfare agent in World War II	Cancer Drug	Hans-Joachim Kruse
Vinblastine	1950	Originally tested as an antidiabetic agent	Cancer Drug	Rober Noble

(continued)

Table 1 (continued)

Drug	Year	False Assumption	Current Indication	Discoverer (Person or Company)
Methotrexate	1961	Originally tested as a cancer drug	Rheumatism Drug	Pierre-Yves Degos
Potassium Bromide	1826	Sedative	Depression	Antoine Jérôme Balard
Sildenafil	1998	Blood Pressure Lowering Drug	Potency Drug	Solomon Snyder (Johns Hopkins)
Theophylline[142]	1859	Enjoyment of Coffee	Asthma Drug	Herny Salter
Aspartame	1965	Peptide Hormone for Gastric Ulcers	Sweetener	James M. Schlatter
Levamisole[408]	1966	No effect in mammalian model	Metabolite isolated from chicken feces (Thiazothielit) is effective	AHM Raeymaekers
Propranolol[419]	2008	Blood pressure reducer	Hemangioma	Christine Léauté-Labrèze
Flibanserin[420]	2002	Antidepressant	Potency agent	unknown

split into chloroform in the body, which was not true. The animal experiments were indeed a confirmation of his hypothesis, but the mechanism was different. Let's stick with the drugs that affect the brain. The inorganic salt lithium carbonate was administered to patients in 1855 under the now erroneous assumption that it dissolves kidney stones. The results seemed to satisfy Robert Bunsen (1811–1889) and Augustus Mathiessen (1831–1870), as the pain of kidney stone sufferers subsided. Four years later, they recommended their colleague Alfred Barring Garrod (1819–1907) to try lithium carbonate for the treatment of gout. The clinical picture in the eyes of the doctors in the second half of the 19th century was a deposition of uric acid in the body, which also led to heart and mental diseases. Gout patients show little enthusiasm for their pain and are understandably not in a good mood. This negative mood gave William Hammond (1828–1900) from Bellevue Hospital in New York the idea that mood disorders could be a central gout. He began treating manic-depressive patients with lithium salts. The Danish doctor Carl Lange (1834–1900) picked up these studies and treated hundreds of mentally ill patients in 1896, and even today lithium salts are strictly administered for manic depression and schizophrenia. The medical application was forgotten. After the Second World War, it was John Cade (1912–1980) who successfully tested lithium salts again in the psychiatric department of the Bundoora Repatriation Hospital in Melbourne. To this day, lithium salts are used in manic-depressive patients.

A semi-synthetic natural substance that is no longer legal as a drug today is lysergic acid diethylamide (LSD), which the Swiss chemist Albert Hofmann (1906–2008) discovered in the early 1940s. Hofmann was looking for new circulatory stabilizers from the ergot alkaloids, as he vividly describes in his book "LSD My Problem Child". When he performed the synthesis in his laboratory, Hofmann suddenly felt a strange feeling. He felt drunk, then a boundless energy flowed through him, followed by hallucinations and a reversal of his sensory perceptions. We will learn more about this very special moment later. His brain played a trick on him for two hours. Then everything worked normally again. Albert Hofmann must have accidentally poisoned himself in the laboratory. This was not uncommon in a laboratory for a long time, as safety regulations were rather lax. In the past, people used to eat or smoke in the lab. We also owe this bad habit to the discovery of aspartame (Fig. 1) in 1965. The smoker James M. Schlatter was supposed to synthesize short peptides for the pharmaceutical company G.D. Searle&Company that could have been used to treat gastric ulcers by inhibiting the peptide hormone gastrin. Between two syntheses, Schlatter put a cigarette in his mouth and noticed that it tasted sweet. Aspartame became one of the most famous sweeteners under the brand name Nutrasweet®. The discovery of penicillin, the potency agent Viagra® and the benzodiazepines can be put in the category "disorder and luck again". The first two active ingredients have already been described in this book, which is why the special feature of the benzodiazepines will be discussed in conclusion.

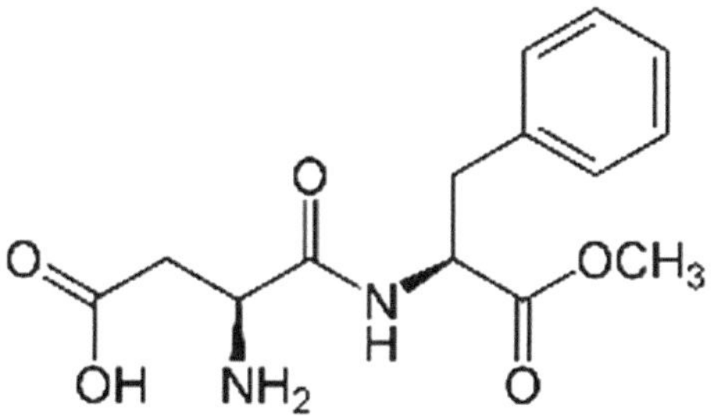

Fig. 1 Aspartame (work of the author)

Leo Sternbach and the Benzodiazepines

Leo Sternbach (1908–2005, Fig. 2) was born in Abbazia, now Opatija in Croatia, as the son of a Polish-Jewish pharmacist and a Hungarian Jew. Formally, he was an Austrian in the k.u.k. Monarchy, went to high school in Villach and then moved to Krakow to study pharmacy. Why Krakow? The economic conditions deteriorated after the First World War, antisemitism increased, his father ran a ghetto pharmacy in Krakow and little Leo had to go with him. Although he was Jewish, he was able to enroll at the university after learning Polish and passing his Matura. A scholarship allowed him to move to Vienna later, but the hostility towards Jews did not make life as a scientist any easier. A highlight was the lecture by Leopold Ružička in Vienna, whom he asked for a postdoc position. His quick yes prompted Leo Sternbach to move again to Zurich in 1937.[197]

After numerous stations as a medical chemist and persecuted Jew in Europe during the time of National Socialism, he emigrated to the USA

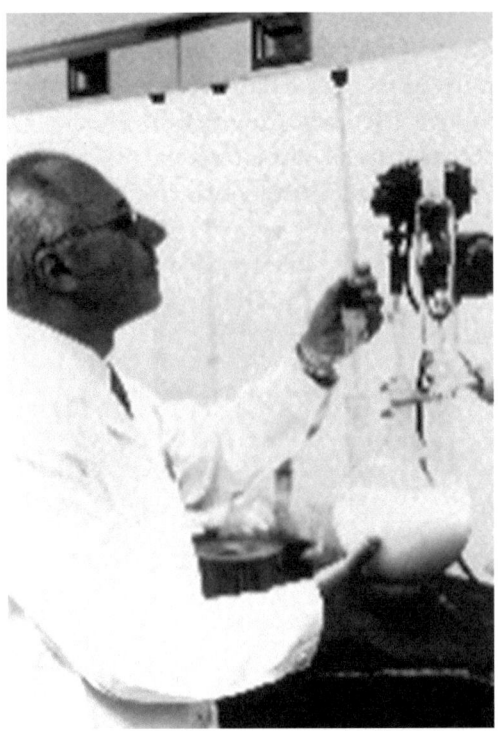

Fig. 2 Leo Sternbach (1908–2005), 1991 (Permission Roche, Basel, CH)

and worked for Hoffmann-La Roche in Nutley, New Jersey. The move to the USA benefited the company, as it pursued a pro-Semitic personnel policy. Sternbach was supposed to build up the American subsidiary in Nutley. As a chemist, he was tasked in the 1930s with synthesizing Heptoxdiazine, which were to serve as new dyes. The structures found did not convince as dyes and he put the work on ice, especially since Europe was overrun by the Wehrmacht and he felt very insecure as a Jew in this new world. Sternbach resumed the syntheses at the end of the 1940s, after reports about Chlorpromazine as a new antipsychotic aroused the company's interest. Äschlimann was Sternbach's boss and visited him in his office: "Leo, have you read about the new antipsychotics chlorpromazine or thorazine?" Sternbach nodded. "Can you imagine that we can synthesize similar substances to have something as well?" Sternbach thought of the old heptoxdiazines and suggested them to Äschlimann. "Good idea, start and then we'll see if we succeed." He let Sternbach go ahead. He rummaged in his old synthesis protocols and synthesized forty of the heptoxdiazines that he had developed at Professor Dziewonski in Krakow. They were hardly known and structurally different from chlorpromazine, but he thought it might be worth it. To his surprise, however, Sternbach found that the presumed structures he had synthesized at the time were incorrect. The structure elucidation revealed benzoheptoxdiazines, which were also chemically very unstable. The new candidates were tested and all disappointed in the animal model. Was it because this class of substances was not effective per se or because of the chemical instability? Sternbach gave it another chance and carried out the syntheses with methylamine to stabilize the structures. Some were tested again, but this time too he was unable to detect an effect. Clearly annoyed, his second supervisor, Dr. Goldberg, said it was time to stop this research. It would not bring anything. Sternbach followed his boss's instructions, cleaned up the lab, labeled one of the stable substances with the code Ro 5–0690, put it on the shelf and wanted to forget it. But the structures remained in his head and Sternbach found the topic too exciting not to occasionally deal with the benzodiazepines again. A few years later in 1957, Sternbach and his colleague Earl Reeder (1924–2003) stood in front of a full shelf in his lab. "Now we have to clean up here." Too many chemicals had accumulated over the years, and some had to go, because the shelves were too full with the many colorful containers with partially illegible labels. Reeder came with two containers and showed them to him. "We still have these two substances from these benzoheptoxdiazines. Should I throw them away? One is the base, the other is the hydrochloride. Boss, is it worth testing them?" Sternbach took the hydrochloride Ro 5–0690 in his hand and blew the dust off the lid. Yes, he remembered, his benzoheptoxdiazine, which he wanted to stabilize. He had succeeded, but had he ever tested it? He looked at the lab reports and flipped through the years. Here, Ro 5–0690, never tested, no structure elucidation. Before he wanted to throw the container in the trash, he had to check it one more time, even if he had little hope. "No matter, I'll do it. I'll bring the substance to Goldberg and fill out the paperwork." He put the container in his pocket and looked through the endless rows of other containers.

"This is really the last compound from you guys that I'm going to test!" Goldberg looked at him incredulously. "Yes, really the last one," Sternbach promised, but knew that he might have to break this promise. "Give me the forms. Did you fill everything out correctly? I hate this," Goldberg grumbled back. "Yes, but the structure elucidation is missing. We'll do that if it's active. Promise." Goldberg shook his head and put the paper and Ro 5–0690 in the envelope for Lowell Randall, who conducted the pharmacological tests. "Leo, that's it." He knew himself that Leo would eventually show up again in his office with new substances, but here and now limits had to be set.

One morning, a thick envelope lay on Sternbach's desk. He briefly looked inside and saw that it was the studies on Ro 5–0690. File it away immediately, he thought, but he wanted to look at everything closely. He flipped through the report. Randall had obviously been diligent, he had tested the substance on lions, leopards, and tigers at the San Diego Zoo. He frowned. That was unusual. He skimmed the first few pages. Human studies, Ro 5–0690 showed no antipsychotic effect, but he read that it was convincing as an anxiety reliever, and absolutely convincing at that. There was another note from Randall: "*Make more of this*". Sternbach sat down and thought. This could get interesting. Randall's request "Make more of this stuff" went through his head, because they didn't know what substance it was. Earl had to break down Ro 5–0690 and determine the structure from the fragments like a puzzle. Only then could they make more. He had to talk to Goldberg, that he couldn't keep his promise. He grumbled, but saw the great opportunity. "Go ahead, Leo. I won't let you go anyway." Ro 5–0690 was a complete success. The preparation exceeded all expectations and was launched on the market under the name Librum® in 1960. It was the breakthrough as an anxiolytic and was the starting shot for a series of benzodiazepines such as Valium® (1963), Mogadon® (1965) or Rohypnol® (1975), with which Hoffmann-La Roche made billions.

These happy coincidences have been scientifically studied for some time and are referred to with the technical term serendipity. This term was first mentioned by the Briton Haroce Walpole (1717–1794) in reference to the Persian fairy tale "The Three Princes of Serendip", where three princes make unexpected discoveries. Serendip is a real place in Sri Lanka. However, the term serendipity only became known worldwide after the Second World War through the sociologist Robert K. Merton (1910–2003) in his work "The Travels and Adventures of Serendipity". The author of this book can only confirm the Serendipity principle, as he repeatedly came across very useful information during literature research for his research, as well as for this book, which he had not even searched for. Fortes fortuna adiuvat! Fortune favors the brave!

Paul A.J. Janssen—the Self-made Pharma Entrepreneur

This book chapter would be completely incomplete if the name of a great medicinal chemist was not mentioned. Paul B. Janssen (1923–2003) was, in my opinion, the most successful industrial chemist in the field of medicinal chemistry and the second greatest Belgian before Eddy Merckx and after Father Damien. Although no subway station was named after him in Brussels, as was the case with Eddy Merckx, his achievements with 41 approved drugs, six of which are on the WHO list of essential medicines today, are outstanding. Janssen is certainly the prototype of a Steve Jobs of the pharmaceutical industry. Even during his PhD in 1956 at the University of Ghent on the pharmacology of a series of propylamines with Corneille Heymans(1892–1968), the Nobel laureate for medicine in 1938, he founded his own company. With the company Janssen Pharmaceutics, his childhood dream was to come true. He always wanted to research independently and the research should finance his life. It was not the desire for big money that drove him, but the will to be able to decide for himself when he could do what and how. He knew that his new active ingredients had to be protected by patents. From today's perspective, it seems naive, if not downright crazy, that a 27-year-old aspiring scientist founded his own pharmaceutical company without a concrete idea, which he also financed himself. His father supported the son in founding his start-up, as one would say today. He provided him with the third floor of his company, where Paul set up his first laboratory. His father ran a small pharmaceutical company in Beerse, northeast of Antwerp, that sold classic medicines like vitamins and organ extracts. A grant of 50,000 Belgian francs, equivalent to about

1,000 € today, and off it could go. A secretary was hired, who stayed by his side for over 40 years, an accountant and four scientists came on board, and the team started with the synthesis of the very first structures. He did not require his first scientists to have a degree, Janssen expected hard work, common sense, and a passion for the cause from them. Janssen had faith in them, if a good idea was put forward, it was supported and pursued.

An anecdote may illustrate this. The Belgians also had colonies like Belgian Congo (today Democratic Republic of Congo) and Ruanda-Urundi (today Rwanda and Burundi) under League of Nations mandate or as a UN trust territory. The handover of Belgian Congo took place in 1960, the return of the trust territory in 1962. After the brutal exploitation of the Congo under King Leopold II and the equally brutal administration by the Belgian state from 1908 to the end, Zaire, later Democratic Republic of Congo, did not find peace. Most Belgians, many of whom had worked in health and veterinary services for decades, returned to Belgium. Paul Janssen was looking for capable employees for his expanding company and invited them to join. Robert Marsboom, an enthusiastic parasitologist, was one of them. He had a passion for his subject and his knowledge impressed Janssen. Not to use his expertise for a good cause like antiparasitics would have been a shame. However, in the company, Janssen was advised against research in this field of parasites. There would be no market and it would not be economical. But Jansen had confidence in the man and liked the project. He instructed his secretary to procure a laboratory, a table, two chairs, test tubes, a microscope, and an assistant. The parasitologist was not a chemist and had no idea about organic synthesis, but that was not his task. Janssen's proposal was the development of a simple test system for testing the estimated ten thousand compounds in the library. In addition to his opioids, Janssen had a passion for parasitology and he often visited the new parasitological laboratory to get informed about the state of affairs. He was eager to learn parasitology. In the following years, they managed to find with Mebendazole an antiparasitic agent that proved to be highly active and later made it onto the list of essential medicines of the World Health Organization(WHO). Janssen was thrilled and after this success hired more scientists with special talents, even if there were officially no vacancies. He held the very modern view that the laboratory should adapt to the new scientists and not vice versa. The company could develop like an organic organism and not like an old rigid organization. Janssen was not the authoritarian boss in the lab who gave instructions from above. He expected his employees to come to him with their own new ideas and present them to him. *"Hello, what's new today?"*, he asked every day on his tour. You met him for discussions in the labs and usually not in the office or in meetings, as is almost the rule today. But what did you tell the boss when there was nothing new? The employees became inventive and only told him half the truth, so there was always something left for the lean times and the boss was not disappointed.

Just one year after its founding, his team had synthesized 500 new chemical compounds, from which seven approved drugs were later developed. What was the key to success? It is not known, but it was certainly his intuition to imagine how to chemically modify natural substances to make them more active. His chemists were already dealing with the so-called structure-activity relationship early on without having a high-performance computer or knowing the structure of the opioid receptors. In this research, the three-dimensional structure of the active ingredient is calculated and projected into the spatial arrangement of the amino acids in a receptor. This methodical approach from the gut without a computer would be unthinkable today. The first major success was in 1954 Isopropamide against stomach pain and stomach ulcers. A real blockbuster for the time, because with the emerging prosperity and the good food after World War II, people were no longer eating, but feasting, and stomach complaints were increasing. Isopropamide sold like hot cakes and led to the first expansion of the lab. All expansions and new projects were to be financed by old successes. This was the credo of Paul Jansen, who built a very solid company with this strategy, which was sold to the American company Jansen&Jansen in 1963. But that was not the end of it, the company had to be built up and organized. His ideas of management with flat hierarchies must have been revolutionary in the 1950s. Today they are quite normal in hip global corporations like Google, Facebook and Apple but also in many medium-sized companies. If a Krupp, a Bayer or an Akzo were organized according to the pyramid model, then here the communication levels were as flat as the Belgian hinterland.

A substance that was supposed to revolutionize psychiatry, was Haloperidol. Bert Hermans was looking for a new opioid at at Janssen Pharmaceutics and synthesized an unusual substance in 1958. Hermans injected a mouse with Haloperidol to test its opioid effect in the hot-plate test. He tilted his head, watched the little mouse, and wondered why it didn't try to escape. Mice under the influence of strong opiates tend to leave the hot plate quickly, almost in a flight-like manner. This mouse here didn't seem to mind the heat, it seemed indifferent to everything. Bertram reported to his boss, who saw the potential less for anesthesia, but much more for psychiatry. Janssen commissioned Bertram with the toxicology, which later showed that Haloperidol was non-toxic and well tolerated.

Under the then licensing conditions, there was no prohibition on conducting experiments on higher animals or directly on humans. For Janssen, there seemed to be no reason not to leave Haloperidol to a friendly psychiatrist: He could think about whether there might be an application in his clinical routine. Some time passed, but the opportunity arose when a young man with a severe psychosis was admitted. The young patient was very aggressive and suffered from strong hallucinations. He was hard to calm down, and the treating doctor remembered the substance that Paul Janssen had sent him. He dissolved a few milligrams and injected them into the psychotic young man, who then calmed down. What surprised him most was the fast and strong effect, which he did not know from the antipsychotics used so far. Formally, it was the first clinical study with Haloperidol on a patient. According to today's definition of clinical studies, however, it was illegal and, if anything, a simple medical experiment on humans.

What was missing was a really good animal experiment that was able to determine the antipsychotic effect in a rational and quantifiable way. It was the great time of cycling races in Belgium and Eddie Merckx was already being talked about. People were enthusiastic about their cyclists, and it was the riders with professional medical care who prevailed in the cycling scene. At that time, doping in cycling was not yet an issue and many athletes took amphetamines. These misused drugs will be discussed elsewhere. Janssen knew a doctor in his circle of acquaintances who looked after these cyclists. He often spoke of psychoses in the cyclists he looked after when there was too much amphetamine in the blood. The cyclists had to be stopped or were almost torn from their bikes when crossing the finish line because the performance-enhancing amphetamines were so stimulating. After the race, they fell into a mental hole and had delusions, as they also occur in chronic psychoses. These are typical side effects and risks that can also be seen today with amphetamine-like stimulants like Speed or Pep. Amphetamines as doping substances were banned in 1972. One of the last professional cyclists who was proven to have taken amphetamines illegally was Jan Ullrich. He is said to have taken these during his rehabilitation phase. Janssen tried administering it to volunteer cyclists without success. The group was too heterogeneous and ethically very problematic. But can amphetamine symptoms be transferred to an animal under standardized conditions? No, even the hamster in its running wheel, which might seem similar to the human cyclist, was simply not usable.

To this day, there is no reliable animal experiment that can mimic such a complex disease as depression or anxiety disorder in humans. This is certainly due to the multifaceted neurophysiology of humans, but also to their personal life history and genetics. In the laboratory under study conditions, all animals are the same age, the same size, the same weight, or equally sick. In human reality, this is not the case, each depression is as individual as the person affected. Nevertheless, I would like to briefly introduce an animal experiment that is frequently used in depression research. It is the Porsolt-swim test, which was developed for rats in 1977 and later for mice. It is considered the gold standard for testing the antidepressant effect of drugs. The principle is simple, but no less stressful for the animals that have to swim after being administered a drug. The rat or mouse is placed in a cylindrical container filled with water for 15 minutes on the first day. The saving edge cannot be reached because the mouse cannot grasp it and keeps slipping off. This experiment is repeated after 24 hours, with the animal only staying in the water for 5 minutes. The aim of the experiment is to measure the despair of the animals, who cannot find a saving way out of their situation. The animals in the water realize that there is no way out and despair sets in, which is why this animal experiment is internally also called despair test or "despair test". After some time, the animals give up and let themselves drift. This passive behavior is interpreted as a sign of depression or despair. The aim of the experimenter is to stop the time in which the animal is active without and with the active substance and tries to swim or climb. If a substance reduces the time in which the animal lets itself drift, this is interpreted as an antidepressant property. Substances that inhibit the uptake of serotonin into brain cells enhance swimming movements. Selective inhibitors of noradrenaline uptake can be detected by increased climbing movements. One problem with this animal experiment is, among other things, that it involves animals that are not depressed. This raises the question of how meaningful the results are when healthy animals receive antidepressant substances. Depressed people are rightly not sent swimming to see if and how severely they are depressed. In contrast to the swimming, healthy mouse, a score is determined for a depressed patient based on a questionnaire and a detailed medical history using the Hamilton scale, which indicates the severity of the depression.

Fashionable Diseases and Lifestyle Drugs

Fashionable diseases and lifestyle drugs are widespread in today's time.[198,199] You now know Viagra®, but also Propecia® against baldness in men? Sildenafil was the drug that was to become the hit of Pfizer AG and brought billions of dollars into the cash register, because impotence or erectile dysfunction was elevated to a real disease. We have discussed the discovery and development of caffeine in this book in detail. This inconspicuous substance was also the midwife of another active ingredient for a second problem area of men. None other than Julius H. Comroe Jr., the president of the American Physiological Society, put it so aptly when he defined luck: *"Happy coincidence is searching for a needle in a haystack and discovering the farmer's daughter."* This is exactly what the pharmaceutical company MSD achieved with Propecia®.

Propecia® is known to few readers and rather to the readers, because the drug contains the active ingredient finasteride (Fig. 3) for the treatment of androgenetic alopecia, i.e. male hair loss. When men come into their prime, the deforestation begins on the head. First, there are receding hairlines, later large football fields, which can lead to psychological problems in some men. What is the reason? The testosterone degradation product dihydrotestosterone gnaws at the hair roots and makes them shrink. Finasteride tackles the problem literally at the root and inhibits the effect of the culprit dihydrotestosterone. Finasteride itself is not a hormone, but blocks an enzyme that forms dihydrotestosterone. From this vanity, MSD has developed a very lucrative drug.

Hysteria can also be seen as a fashionable disease in women at the end of the 19th century. Today, the term is completely outdated and the disease, which was diagnosed in women, is no longer a recognized disease. It was seen as a mental illness that originated from the uterus. The French doctor Jean-Martin Charcot (1825–1893) worked in the psychiatric hospital Hôpital de la Salpêtrière in Paris as a neurologist and researched hysteria. His charisma was great and also attracted the young Sigmund Freud to Paris.

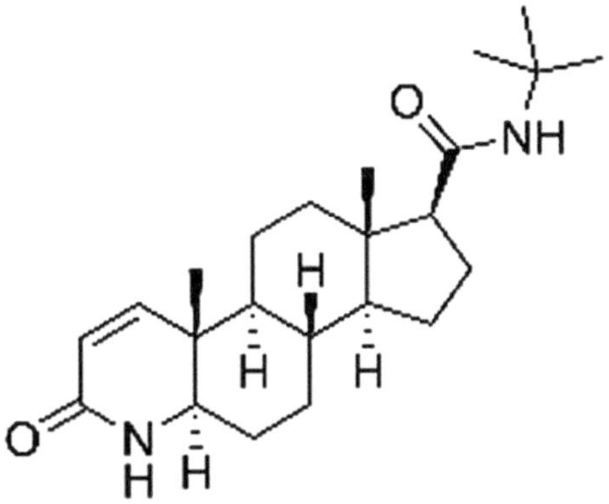

Fig. 3 Finasteride (work of the author)

His works "Studies on Hysteria", which were published in 1895, led him to psychoanalysis. Hysteria also made it into literature. Thomas Mann attributes the disease to Christian Buddenbrock in his work "The Buddenbrocks", who is diagnosed with hysteria by a doctor. Another work by Vera Buck is "Runa", which very vividly tells the situation of the young Runa in the Salpêtrière.

Last examples of fashionable diseases that were mistaken for real diseases. As described in the first chapter, neurasthenia was one of the most frequently diagnosed diseases before the First World War and was interpreted as general exhaustion, as we know it as burn-out. Are bore-out, leisure sickness and burn-out also fashionable diseases today or is there more to it? But there were also fashionable diseases that were completely mistaken for real diseases at their time. After the First World War, the Spanish flu was considered a cold. The war tremor of the returning soldiers in the First World War was dismissed as a character weakness and not as a psychogenic tremor of a war trauma. The consequences were devastating, as many soldiers turned to morphine and cocaine in their distress.

The Birth of the Pharmaceutical Industry

The pharmaceutical industry is one of the middle-aged industrial sectors in human history, which slowly grew since the beginning of the 19th century and took on a legal and economic form since the 1860s, as we know it from a modern chemical or pharmaceutical company. In Germany, or in the German Empire founded in 1871, a pharmaceutical industry emerged that can indeed be described as revolutionary. The traditional structures of the small pharmacy laboratories, which had started with extractions 60 years earlier, were left behind, and many well-known giants like Bayer AG no longer emerged from pharmacies, but from the coal tar industry. The tar was the precursor of petroleum as the source of all chemicals that made the industry flourish.[85].

But let's look back at the beginnings, when the pharmacies were the germ cells for today's well-known pharmaceutical companies like a Merck AG, Boehringer KG or the then Schering AG (Table2). These very successful companies had adopted and perfected their pharmacy tradition and the prevailing principles and ideas such as distribution structures, manufacturing practices, and technical procedures. Unlike the simple rural pharmacy, only a few of the emerging pharmacies, which later became large companies, produced their medicines or galenic products for their customers who

Table 2 Chemical or pharmaceutical companies founded by pharmacists in Germany (*not emerged from a pharmacy laboratory)

Year	Location	Pharmacist	Product
1800	Bremen	W.E. Stüve	Plant extracts Essential oils
Around 1800	Langensalza	J.U.C. Wiegleb	Pharmacy preparations
Around 1800	Erlangen	E.W. Martius	Pharmacy preparations
1811	Bürgel near Jena	J. Martiny	Pharmacy preparations Chemicals
1813	Teudnitz/Erfurt	J.B. Trommsdorf	Pharmacy preparations Chemicals
1814	Berlin	J.D. Riedel sen	Pharmacy preparations Chemicals
1820	Stuttgart	C.F. Boehringer*	Pharmacy preparations
1824	Oppenheim	F.L. Koch	Pharmacy preparations
1825	Berlin	J.E. Simon	Pharmacy Preparations Chemicals
1827/31	Darmstadt	H.E. Merck*	Pharmacy Preparations Alkaloids
1827	Berlin	J.D. Riedel jun.* E. de Haen	Quinine
1851	Berlin	E.F.C. Schering*	Salicylic Acid
1860	Rostock	C. Brunnengräber	Pharmaceutical Preparations
1869	Helfenberg	E. Dietrich	Galenic Preparations
1872	Frankfurt	K. Engelhard	Galenic Preparations
1894	Dresden	C. Stephan	Galenic Preparations

came to the pharmacy with a prescription from a doctor. They were already thinking bigger and saw that there was a regional need that they wanted to meet. In their own small distribution network, many pharmacies in the region and later in the empire were supplied. Significant quantities of medicines were produced in stock for shipping. This was significantly cheaper for the pharmacies to be supplied, but it also already placed demands on the quality of the manufacturers. In the patchwork of the Holy Roman Empire of the German Nation, quality and compliance with the pharmacopoeias were naturally not uniform. What was completely sufficient in the Kingdom of Prussia did not have to be enough in the Kingdom of Bavaria, and the Hessians or Westphalians had completely different ideas. For the first pharmacies that wanted to ship their medicines, it was a burden and already caused headaches for the Prussian medical bureaucracy in 1820 as to how to monitor drug production. Attempts were repeatedly made to standardize production in the countries of the empire, but this never succeeded due to particular interests until the founding of the empire. Somehow these problems seem very familiar in our present Federal Republic with its 16 federal states. Perhaps this fragmentation was the reason for the sluggish development of the German pharmaceutical industry, because nobody put money into getting the "shop" really going and investing in extractors or synthesis apparatuses. They were too big for their own region, but too small for the empire or Europe. The economic risk was high and so the pharmacy founders initially stuck to secret and specialty brands. Occasionally, price lists or catalogs were also distributed, but they preferred to stay among themselves.

The pharmaceutical industry found its origins in natural substances and especially in quinine. The now very well-known alkaloids such as morphine, codeine, scopolamine, and many others were discovered in the 19th century (Table 3). However, the cradle of the pharmaceutical industry can be seen in France, where very systematic research on plants was already being conducted in the last third of the 17th century. The Académie Royale des Sciences was pioneering, even if the methods such as pressing out, boiling with rainwater, or distilling were still very simple. What we describe today as comparative science was already the goal of the great plan of the Academy to compare extracts and their properties from different plants

The good extraction with ethanol was recognized by the pharmacist Simon Bouldoc, but only slowly gained acceptance. Reasons could be the high price, the lack of availability of pure ethanol, or the scientific assumption that salts in plants are also well soluble in water. The availability of pure alcohol was certainly a problem, which nicely explains the invention of the Bach flower extracts. The well-known Bach flower therapy is a method of alternative medicine invented by the British surgeon Edward Bach with a great love of nature. One might wonder why the alcohol content is unusually at 36%, which is due to the fact that Edward Bach (1886–1936) had no pure alcohol for extraction and only brandy available.

Germany's researchers only became interested in natural substances in the 18th century, and they threw themselves into the knowledge of their French colleagues. However, there was doubt as to whether pure substances actually existed, but Andreas Sigismut Marggraf (1709–1782) taught them with the isolation of sucrose in 1747 as the first pure representation of a natural substance ever, that pure substances can indeed be isolated from extracts. Carl Wilhelm Scheele followed with citric, wine, apple, and oxalic acid a few years later.

The first medically important natural substances were isolated in France at the beginning of the 19th century, and you already know the well-known gentlemen Pelletier and Desrone from the two chapters on cocaine and quinine. It was obvious that France could have become the motherland of the

Table 3 Discovery of important alkaloids

Year	Alkaloid	Plant	Discoverer	Effect
1806	Morphine	Papaver somniferum	Sertürner	Analgesic
1817	Emetine	Ipecacuanha Root	Pelletier & Magendie	Antiparasitic
1818	Strychnine	Strychnos nux vomica	Pellentier & Caventou	Cardiac tonic
1819	Atropine	Atropa belladonna	Runge	Mydriatic
1819	Colchicine	Colchicum autumnale	Pelletier & Caventou	Gout remedy
1819	Piperine	Piper nigrum	Orsted	Spice
1820	Quinine Cinchonine	Cinchona pubescens	Pelletier & Caventou	Malaria remedy
1821	Caffeine	Coffea arabica	Runge	Stimulant
1827	Coniine	Conium maculatum	Giseke	Neurotoxin
1828	Nicotine	Nicotiana tabacum	Posslet & Reimann	Insecticide
1832	Codeine Thebaine	Papaver somniferum	Robiquet Thiboumery & Pelletier	Cough remedy Starting material for opiate synthesis
1833	Colchicine Hyoscyamine Aconitine	Colchicum autumnale Atropa belladonna	Geiger & Hesse	Gout remedy
1839	Lobeline	Lobelia inflata	Reinsch	No application
1840	Chelidonin	Chelidonium majus	Maximilian & Probst	Wart remedy
1848	Papaverine	Papaver somniferum	G. Merck	Muscle relaxant
1860	Cocaine	Erythroxylon coca	Niemann	Local anesthetic
1875	Pilocarpine	Pilocarpus jaborandi	A.W. Gerrard E. Hardy	Diuretic Glaucoma
1885	Ephedrine	Ephedra sinica	N. Nagai	Cardiac tonic Asthma
1880	Scopolamine	Scopolia carniolica Hyoscyamus niger Atropa belladonna Brugmanisia spp.	A. Ladenburg	Cardiac tonic Asthma
1888	Arecoline	Areca catechu	Jahns	Stimulating natural substance
1888	Mescaline	Lophophora williamsii	Lewin	Hallucinogen

pharmaceutical industry, but laws surprisingly prevented this. According to historians, this was due to the "cumul" practice in France, which allowed established chemists to obtain a multitude of part-time positions, which in turn made it difficult for newcomers to gain a foothold in the chemical sector and diminished the competitiveness of the French chemical industry.

An additional factor that contributed to the stagnation of the French pharmaceutical industry in the 19th century was the legislation of 1803, which granted pharmacists a quasi-monopoly in the production of medicines. The German Empire and England, with the emerging tar chemistry, caught up and overtook France when the demand for effective natural substances exploded. Natural substances play an important role in the discovery of new active substances. Even today, in the era of biotechnology, we find a return to a natural substance structure in about 40% of approved drugs (see Table 4).

An interesting anecdote is that of the Oppenheimer quinine factory, which was founded in 1830 by Carl Koch (1786–1865) in the same city.[200] This factory was the first quinine factory in Germany and supplied the surrounding area with quinine until 1888, as malaria was prevalent in the Rhine meadows and the adjacent swamps. Quinine and cinchona bark are said to have been the beginning of the pharmaceutical industry in Germany. The founder Friedrich Koch (1786–1865, Fig. 4) built a thriving business, whose clients also included the British army. His son Heinrich (1833–1910) sourced the bark from India from the East India Company, but was able to withstand the impending price drop, as the Dutch brought cinchona bark to the market at low battle prices. He promptly converted the factory and yard back into a vineyard. The rise of the first pharmaceutical company was short-lived and with the oversupply in 1890, the company went bankrupt. But there were also other reasons: The pharmaceutical industry brought new fever reducers to the market with phenazone, antipyrine, phenacetin and acetylsalicylic acid, and there were too many producers and chemical factories that engaged in a brutal price war, which many like the Oppenheimer quinine factory did not survive.

Table 4 Drugs derived from natural substances

Natural substance	Drug	Indication
Adrenaline	Propranolol and other β-blockers	Hypertension
Quinine	Chloroquine	Malaria drug
Ephedrine	Salbutamol	Bronchodilator
Caffeine	Sildenafil	Potency drug
Coumarin	Dicoumarol	Anticoagulant
Coumarin	Chromoglycic acid	Mast cell stabilizer
Histamine	Omeprazole	Acid blocker
Cocaine	Lidocaine	Local anesthetic
Mevastatin	Statins	Cholesterol reducer
Snake venom	Captopril	Blood pressure reducer
Scopolamine	Thiotropium Bromide	Bronchodilator

Fig. 4 Friedrich Koch (1786–1865) (Permission CK Koch Winery, Oppenheim, DE)

The leading company in Germany was founded by Emil Merck in Darmstadt, who sat as a clever pharmacist and shrewd businessman in his father's Angel Pharmacy and read about the emerging industry in England. As early as 1823, he started the isolation and sale of morphine salts and four years later followed quinine. Already in 1827, he published the first catalog for his customers under the name Pharmaceutical-Chemical Novelties Cabinet, which offered 16 alkaloids and their salts (Fig. 5). Merck was the oldest company that produced and sold alkaloids of high quality. Not without reason, the French Société de Pharmacie in Paris awarded the golden Medal d' Encouragement for outstanding quality as early as 1830.

After the founding of the Reich, the economic structure in the country changed rapidly. The first chemical industry companies like Hoechst and Friedrich Bayer no longer emerged from pharmacies. The tar dye industry, which flourished in the 1860s, was the driving force of the chemical

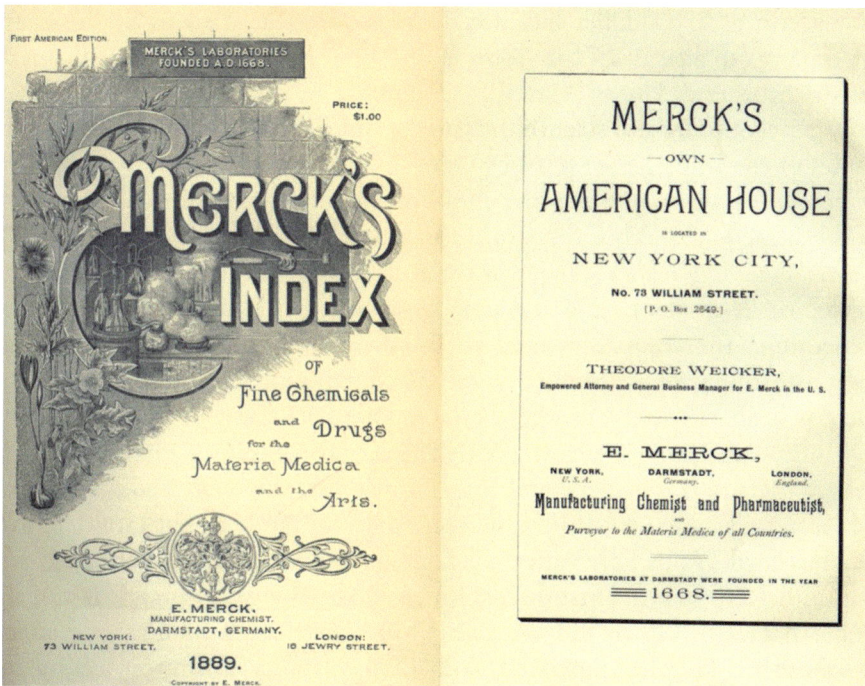

Fig. 5 Merck's Pharmaceutical-Chemical Novelties Cabinet, English, 1889 (Permission Merck AG, Darmstadt, DE)

industry on the Rhine, Ruhr, and Main. It was no longer pharmacists, but also chemists who laid the foundation for entire empires with dyes like fuchsine and later drugs. Finally, the long-awaited investments came, even if financed from the French reparations of the won war of 1870/71. It was supposed to be a boom, but just two years later the dye business collapsed when the prices for basic chemicals and aniline dyes plummeted. We have already read in the chapter on morphine that this recession accelerated the switch from dyes to drugs. From the 1880s onwards, the first research and development laboratories for drugs were established in the large chemical companies. These new establishments were supposed to offset the losses of the tar dye chemistry. The success came, the companies grew quickly, became joint-stock companies, hired more employees and for the first time also scientists. The special trick, however, was that the waste products of tar chemistry were used as starting materials for drugs. An example is para-nitrophenol. It served as a starting material for phenacetin and had until then been an unused by-product in the production of the dye Benzoazurin G.

Phenacetin as a painkiller has already been introduced to us in the chapter on aspirin. It was sold in 1888 as the first drug of Bayer AG, formerly Friedrich Bayer & Co.

What economic and scientific strategies did the first pharmaceutical companies of the Imperial era pursue? Were they driven to alleviate the suffering of people, as we might assume? A submission from the Hoechst company to the Imperial Health Office from the year 1884 is very sobering. *"There is always something useful,"* it says in the justification for registering their drug substance. Even if one did not exactly know what it was, one would find something. The companies tried to list drug substances and medicines in their assortment that promised good sales for minor or fashionable diseases such as headache, nervousness, women's ailments, fever, or insomnia. Even the intermediation via pharmacies was seen as inhibiting, they wanted to deliver their pills, powders, and drops directly to the retail trade. The word side effect was also avoided by the companies at the Imperial Health Office like the devil avoids holy water. Their medicines were simply perfect. It was not until 1890 that the first discussion took place between the officials of the Imperial Health Office and the companies about issues such as side effects and health risks. The Imperial Health Office saw its task in protecting the population, but it also saw itself as committed to promoting the German economy in the world. The world was to recover not only from the German essence but also with German medicines.

The State Intervenes—Drug Approval

The state control and supervision of drug approval was an early issue, even though it was not legally anchored in the Federal Republic of Germany and the former GDR until 1961 (FRG) and 1964 (GDR) respectively. There were no legal regulations from the time of the Weimar Republic or the Nazi era until 1961. With the Treaty of Rome in 1957, the birth certificate of the later European Union, even if it was not yet called that, Germany was the only accession country that had neither a national drug law nor a Ministry of Health. The assumption that Germany got a drug law because of the thalidomide scandal is wrong. The Ministry of Health was established on November 14, 1961, and the Drug Law was passed later that same year, but the thalidomide scandal, which sadly originated from the damage to fetuses caused by taking thalidomide in early pregnancy, only broke out with a letter to the editor from Scottish doctor Leslie Florence to the British Medical Journal on New Year's Eve 1961. Until this point, it was suspected, as reported by Der

Spiegel in its issue of August 16, 1961, that not the then frequent nuclear weapons tests, but the drug thalidomide was the cause. The scandal, however, undoubtedly led to an improvement in drug safety, which from 1971 and at the latest with the revision of 1976 demanded principles for the pharmacological-toxicological and clinical testing of drug specialties, which were subsequently implemented in the European Medicines Agency (EMA).

With the Drug Law of 1961, pharmaceutical manufacturers could live well, because it did not yet contain any obligation to test the effectiveness and safety of drugs. We have often dealt with the question in this book of how substances work, but the law makes some fine, legal distinctions. It can be safely said that every substance has an effect. The effect refers to any desired or dangerous interaction between a substance and a biological organism. It does not matter whether it is an active ingredient, a vitamin, or a toxin. The lawyer refers to this biological or pharmacological effect as effectiveness when it is a drug substance with a desired effect on humans. A drug substance is therefore only present when it is an active ingredient that is legally and officially recognized as effective, i.e., approved. This drug authority is the Federal Institute for Drugs and Medical Devices (BfArM) in Germany and in Europe the European Medicines Agency(European Medicines Agency—EMA).

The first Drug Law did not yet provide for approval. Drugs were registered with the then Federal Health Office and registered. If the effectiveness of the new substance was not yet generally known, a report on the type and extent of the observed adverse effects had to be attached. This was to ensure that the international competitiveness of German pharmaceutical companies was not restricted. Medical examinations were required, but not the conduct of clinical trials (Fig. 6). From 1964, political action had to be taken under the impression of the thalidomide scandal, and meaningful written assurances were required that the drugs had been tested in preclinical and clinical studies and that the current state of science had been adequately and carefully considered. Like all laws, the Drug Law is not static and has been amended over the decades. Here are some examples: The law was revised to meet the European requirements for a central approval and to enable market access for drugs from EU member states. Since the 12th amendment, documents for pediatric and adolescent trials must be submitted, as children are no longer considered small adults. But also smaller changes like the indication of an expiration date and the inclusion of professional information were prescribed with the 2nd amendment in 1986. It is important to know that drugs do not have a best-before date like food, but an expiration date, which guarantees safe use until then. With the 12th amendment in 2004,

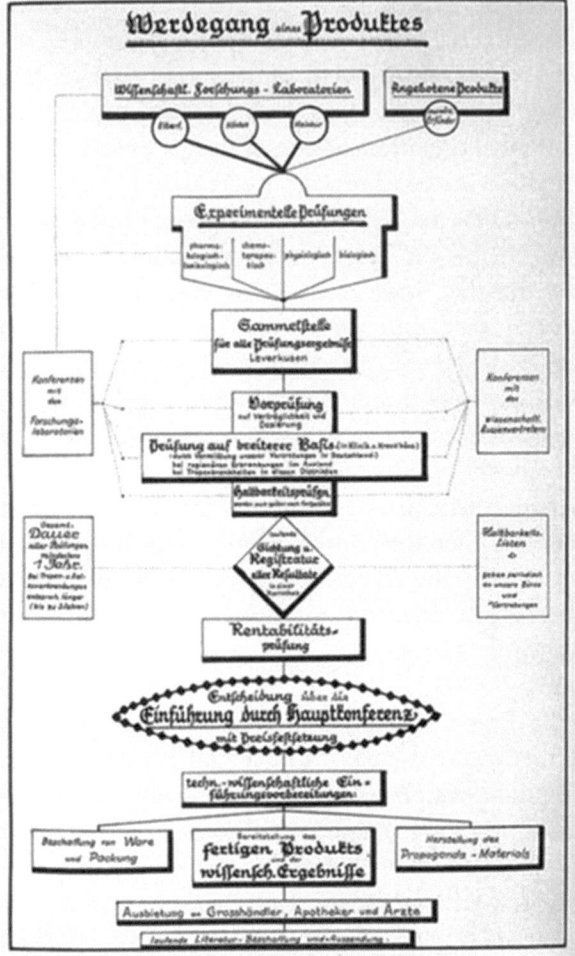

Fig. 6 Development of a product, phases of drug approval, around 1920 (public domain, wikipedia)1

mail order received its legal basis and with the same amendment it was also stipulated that the drug name must be printed in Braille.

Of Rats, Mice, and Humans—Animal Testing in Drug Research

A topic that polarizes people is the use of animals in pharmaceutical research. The negative reputation of the pharmaceutical industry is allegedly reflected in hundreds of thousands of animal experiments per year, which must be carried out in the supposed or officially required interest of risk assessment. In contrast to research institutions such as universities, the Max Planck or Leibniz Institutes, animal experiments in the pharmaceutical industry are legally required. The belief, called a misconception by opponents of animal testing, lies in the assumption that the mechanisms of action or the toxicity of active substances can be better understood with the help of animals such as rats, mice, dogs, or fish—in vivo veritas, as some colleagues say with embarrassment. But not only in physiological and pharmacological research are animal experiments used, but also in psychology. This begins with Ivan Petrovich Pavlov (1849–1936) in Leningrad, who classically conditioned dogs, goes on to important psychological experiments of the 20th century by B.F. Skinner (1904–1990) on operant conditioning, and ends with addiction research on rats by Peter Millner (1919–2018). Together with Bruce K. Alexander (*1939), he showed for the first time in 1981 in rats that addiction should not be considered a pharmacological problem, but a multifactorial event.[201]

Animal experiments have been conducted since the beginning of pharmaceutical research in the young companies of the 19th century, as already reported in this book with reference to Aspirin® and Heroin®. The reasons for animal experiments were trivial: there were no tissue cultures and out of ignorance, it was believed that the physiology and biochemistry of the dog or rat must be similar to those of humans. Above all, the philosophy of Richard Descartes (1596–1650), who was also a mathematician and natural scientist, shaped the image of the animal as a soulless automaton that follows a mechanism. He also saw humans as automatons or limb machines, and he could not be accused of any particular religiosity. With rational medicine, the heyday of physiology at the end of the 19th century, the optimism among doctors was so great that they had no ethical qualms about trying out drugs in animals, on themselves or on their patients without asking them. *"I have also noticed with ever-increasing astonishment that individual doctors, obsessed with a kind of research mania, disregard the areas of law and morality in the most worrying way. For them, the freedom of research goes so far that they break through any consideration for others. The boundary between human and animal is blurred for them."* The one who complained so bitterly about his colleagues was Albert Moll(1862–1939), who described 600 cases of *"human experiments"* in his book "Medical Ethics", as he called the practice of his colleagues. Albert Moll's book is extraordinary because it can be read as a guide for the medical profession, in which he sets out rules for the correct handling of health insurance, for the correct behavior at the dissecting table as a student, and also for good manners in the private life of the doctor.[202].

Animal experiments, also called vivisection, are older than the pharmaceutical research of the penultimate century and can be traced back to the philosophers and doctors Aristotle (384–322 BC) and Erasistratos (304–250 BC). Aristotle can be seen as the founder of biology, who dissected the first animals to understand anatomy. Erasistratos was an anatomist and surgeon and is possibly considered the first animal experimenter who worked with living animals.[203] He cut open pigs' trachea to find out how the lung worked. 300 years later, Galen of Pergamon, shortly Galen (130–200), conducted the first studies on dead pigs, monkeys, and other animals to practice comparative anatomy and physiology. His goal was to find the "truth", which for him was more the knowledge of what the animals looked like inside. Only in this way, according to Galen, could scientific progress be advanced. His anatomical studies on corpses in Alexandria, where it was allowed to dissect humans, gave him a great understanding that was beneficial to him in his later practice as a sports doctor and gladiator doctor in Rome.

In Europe, no scientific breakthrough in anatomy was expected during the Middle Ages, and it was not until the 15th century that Andreas Vesalius (1514–1564), the founder of modern anatomy, dissected and drew dogs, pigs, and humans, recording what he saw. With a few exceptions, such as Stephen Hales' determination of blood pressure in horses in England or William Harvey's description of the living and flowing blood circulation in 1628, animal experiments were not conducted on a larger scale until after the French Revolution in 1789.[204] Louis Pasteur studied infections in silkworms (Pébrine disease), dogs (rabies) and sheep (anthrax).[205] The reason for the great bacteriologist's interest in Pébrine disease may seem surprising, but the French silk production was threatened by a single-celled organism that infected the silkworms. The infection spread rapidly and almost drove the southern French silk industry to ruin. Pasteur identified the pathogen, clarified the infection cycle, and recommended not mixing infected butterfly eggs with healthy ones. Strict separation and clean work saved the silk industry.

Not only Pasteur, but many other scientists from all research directions conducted animal experiments, and the number of animals used increased rapidly. Of the 118 Nobel Prizes for Medicine and Physiology awarded between 1902 and 2020, 96 went to laureates who had also achieved their results with the help of animal experiments. With the amount of animal testing, the suffering of the animals was quickly brought to public attention. As early as the mid-19th century, animal experiments were considered an ethical problem, and the first animal protection societies were founded, such as the Society for the Prevention of Cruelty to Animals (SPCA) in England in 1860 and later the American SPCA with branches in Philadelphia, New York, and Massachusetts.

Resistance was necessary from their point of view, as the groundbreaking book by the young Englishman Charles Darwin "On the Origin of Species" from 1859 showed how close we humans are to animals, especially to apes.[206] They and Darwin were right, as modern genetics confirms. Humans differ in their genes by 7% from macaques and rhesus monkeys. The difference shrinks to chimpanzees to 1.23% of the genes and to only 0.23% of the genes that code a protein as a tool in our common cells. These few genes make a difference in embryonic development and our immune defense.

The consideration of our closest species has helped little, because in the age of rational medicine, people loved new discoveries in medicine and technology, but the well-being of animals was of secondary interest.[207,208] This is the only way to explain that during this time, physiology developed with 16 laboratories and chairs in the German Reich between 1870–1890. Driven by the idea of bringing biology under the laws of chemistry and physics, the laboratory animal was to develop into a model animal. Especially in bacteriology, all isolated bacteria were tested on animals. Many of the experiments were brutal, the animals were sedated with alcohol or opium because anesthesia had not yet been invented, and muscle twitching was suppressed with curare. By the end of the 19th century, scientists and doctors had tried almost everything on animals. This included surgical techniques, bone fractures, drug tests, simple chemicals, disinfection methods, and physiological experiments with severed nerves or amputated body parts. As a statement by Alexis Carel, who received the Nobel Prize for Medicine in 1912, shows, the enthusiasm for the suffering of animals was growing: "[…] *You have* […] *once again shown that the development of applied sciences follows the lessons from animal experiments.* […] *Animal experiments are the fundamental approach to scientific methods in biology, medicine, and teaching.*"

The idea of using animals for scientific work and the testing of active substances is not originally due to the optimism for the future of the industrial revolution from the mid-19th century, but the enthusiasm for rationality and technology certainly contributed to it. The advent of measuring instruments gave physiology a tremendous boost and a name. Precise measurement technology and a better understanding of chemistry enabled scientists to describe the life functions of humans and animals. Especially in the German Reich, physiology was outstanding and a driver of progress in medicine. In the land of thinkers and tinkerers, artificial ventilation was invented using an Otto engine and a mercury manometer, which initially benefited not humans, but dogs and rabbits in the animal laboratory. Carl Ludwig (1816–1895), who operated the first physiological laboratory in Germany in Leipzig, used his ventilator to maintain vital functions in animals treated with curare. He spoke of hybrid beings, as he saw an inseparable unity of experimental animal and technology. His technique was to revolutionize physiology, as living animals with opened bodies became accessible for many investigations.

In the absence of today's common receptor models, tissue cultures, and bioinformatics, researchers turned to the animal, which became second nature, as Amann described it in 1994.[209] The animal was connected to the known world, but this world became controllable and understandable, which was not the case for the wilderness outside the lab. This idea was further developed in the USA in the 20th century. The animal was not only to be a singular being, but to become the standard animal, from which the model animal could then be developed.[210] This was now the driving force to start breeding programs for rats, mice, and also other animal species such as the beagle or the guinea pig. The reduction of species diversity and genetic diversity, about which not much was known at the time, was the goal of the new animal breeding.

The use of experimental animals became more and more common, and especially in the USA, the number of animal labs increased significantly. The model was the German physiological institutes, and American students crossed the Atlantic to learn here. The American universities and their affiliated hospitals, which originally treated patients, recognized the value of medical research and established numerous animal holdings from 1910 to 1930 to breed animals and use them directly for experiments. On September 15, 1915, Simon D. Brimhall (1863–1941) was the first veterinarian to be hired as head of the experimental animal department at the Mayo Clinic in Rochester, Minnesota. His task was to care for the health of the animals, but also to successfully manage the breeding and conservation program. The animals were intended for experiments in surgery and pathology.

Between 1910 and 1930, breeding became heavily commercialized and breeders offered live animals of all kinds for research purposes. The most famous are the albino Wistar rats of the eponymous Wistar Institute and the BALB/c mice of the Jackson Memorial Laboratories. These animals were offered as standard animals, which were supposed to deliver reliable results under uniform conditions in the laboratories. The term "standard animal" was directly related to the newly developing world of work in the USA. Workers had to work hard in confined spaces under similar conditions as the experimental animals. Milton Jay Greenman (1866–1937, Fig. 7) was one of two directors of the Wistar Institute, which was originally an anatomy museum, and was interested in the theories and writings of Frederick Winslow Taylor. The engineer, founder of work sciences, had developed the basic principles of rationalization and division of labor in the young American industry. Greenman read his works "Shop Management" (1911)[211] and "The Principles of Scientific Management" (1911) and wondered if these ideas could not also be applied to rats and mice. They were reduced to technofactors or *"modern techorganic constructions"*.[209] The standardization of science was his guiding idea, which not only he, but thousands of science managers after him tried with dubious success, as the chapter on Dr. Chance shows. For Greenman, modern medical research included a *"standard or unit animal within accepted variation limits,"* as quoted by Bonnie Tocher Clause, who was close to him as an employee. An article by

Fig. 7 Milton J. Greenman (1866–1937)

Clause gives a deep insight into the ways of thinking and feeling about laboratory animals that are no longer comprehensible to us today.[212] Clause was enthusiastic and it was a stroke of luck for her that they could breed and standardize Wistar rats like machine screw threads or pure chemicals, just as the American industry needed technical parts. Rats were *"a living analogue to the pure chemicals that legitimize experimental science."* For Greenman, his rats were biological-technical hybrids that finally enabled the desired precision in the lab. The second director at the Wistar Institute, Henry Donaldson (1857–1938), showed in his book and table work *"The rat: reference tables and data for the albino rat (Mus norvegicus albinus) and the Norway rat (Mus norvegicus)"* how technically several thousand data should capture, standardize and quantify the rat anatomically, physiologically and biologically.[212,213] This could not work and it was criticized that his data were not as reliable as claimed. In the second edition of 1924, he backpedaled and admitted that there were fluctuations and deviations that had to be taken into account: *"In the very nature of the case such accuracy and constancy is unattainable, for all animals at all times are in a state of flux"*. The great difference between wild rat and lab rat was described by a rat historian with a fitting, albeit somewhat exaggerated analogy: *"The lab rat is the Hippocrates of the rat world; lab rats are to wild rats what Gandhi is to Hitler—they are a separate rat race of Kochs or Pasteurs or Salks or Madame Curies."*[214] Whether the lab rat is the heroine compared to the nasty house rat in our homes and sewers? Perhaps they are the Robin Hoods and Che Guevaras!

The name Wistar is derived from the German Wüster, the family that emigrated to British North America as German glassmakers as early as the 17th century.[213,215] Caspar Wüster (1996–1752), who soon renamed himself Wistar, was the founding father of the German-American family and gene line, who became wealthy less through his learned trade than through land speculation. He was probably the first German to make a fortune as a self-made entrepreneur in the British colonies. Only two generations later, his grandson General Isaac Wistar, whose marriage had remained childless, would endow a building to the University of Pennsylvania, which was initially intended as an anatomical teaching museum. It was named in honor of his father, who taught and researched as a doctor and anatomist at the university. The Wistar Institute, founded in 1892, was intended to become the leading anatomical institute in the USA according to Greenman's vision. Success came with the appointment of Henry Donaldson as an ambitious director and equally enthusiastic researcher. The breeding program began in

1906 and was a great economic success. The institute published five journals and the standard work on the rat (*Rattus norvegicus*) by Henry Donaldson was the best advertisement.[216] More and more experimental animals were requested and sent throughout the country. The rat was so coveted that the institute protected its trademark with the name WISTARAT®. Unusual for the turn of the century were the cooled laboratories in the summer and the hygiene. The reasons lay in the desired perfection, as described by Frederick Taylor in his manual, because at least 11,000 animals were to be bred annually, as Bonnie Tocher Clause describes in his excellent overview article about 50 years of breeding.[212] The beginning was difficult, but from 2,135 rats in 1912, the breeding program grew to 3,000 a year later, with a rising trend. Without the help and research drive of a woman, the Wistar Institute would not have achieved the significance it has today. Helen Dean King (1869–1955) came to the University of Pennsylvania as a biologist, where she assisted in anatomy and switched to the Wistar Institute in 1909. With great meticulousness, she dealt with the inbreeding program and the selection of characteristics of certain rats, which she sought to enhance through crossbreeding. Controlled pairings of siblings over years led to the famous Wistar rats, which can still be found in laboratories today. She was also the one who bred rats that became almost genetically identical clones, which prompted Henry Donaldson (1857–1938) to write to the board in 1915: "[The Institute] … *should be able to furnish perfect animals* [(The Institute) … should be able to provide perfect animals]". How did the wandering rat in the lab get a white coat? As written, various rats were crossbred, and among them were albinos, whose genes prevailed in later generations in terms of fur color. This is nothing unusual, think of the white-brown fox terriers, whose light color was deliberately bred in England to distinguish them from the dark game during hunting.

Unfortunately, experimental animals were legally considered as a thing or a kind of living measuring instrument that could be disposed of after use. This attitude has changed in recent years, the ethical value is not only anchored in the animal experiment laws, it was included as a state goal in the German Basic Law in 2002. Interestingly, animals are still treated as things in the Civil Code (90). What changed from 2002? With the constitutional amendment, animal experiments in pharmaceutical laboratories were not fundamentally prohibited. Nor were slaughterhouses and chicken farms closed or zoo animals released into the wild. The constitutional change gives the new state goal of the human-animal relationship finally a new direction, how humans can become more aware of their animal ethics. The philosopher Richard David Precht (*1964) has presented a magnificent work on the question of responsible dealing with our fellow creatures in this shared world with his book "Animals Think", the reading of which is highly recommended. But the question remains, how do we weigh the well-being of animals against the desire for safe medications for us humans?

How Should We Deal With Animal Testing?

There is an increased focus on the social structure of laboratory animals. Rodents such as mice and rats live in social communities, and it is their instinct to hide. In nature, rats and mice can move in an area of 3,000 to 8,000 m^2(rat) or 350 to 2,000 m^2 (mouse), while in the laboratory they are kept in small cages of 0.1 to 0.18 m^2/animal (rat) or 0.06 m^2/animal (mouse). Although the cages have become increasingly generous, the animals were kept in groups, they were given toys and their hiding places back, but for experiments they were isolated again. If they were lucky, they were anaesthetised before the experiment and did not wake up again. This extra effort is also worthwhile for the researchers. It has been proven that the normal behaviour of rats and mice cannot be reproduced for pharmaceutical research in a reduced and artificial prison world. Many studies show that not only is the cortisol level in the blood significantly higher than in wild comparison animals, but the fears in the standard cage are also more pronounced. In an experiment where mice can choose between normal drinking water and an alternative with an anxiety-relieving agent, the preference was clearly for the anxiety-relieving benzodiazepine.

What animal experiments are carried out in pharmaceutical research today, which are prescribed and which animal species are used? To anticipate it: Chimpanzees and related great apes are no longer allowed in pharmaceutical research. Other monkeys can be used and are legally referred to as non-human primates. Whatever animal is used, it must, like its parents, come from a breeding. This means that no animals from the wild or even stray animals from the street come into the laboratories for experimental purposes. Apart from the fact that taking from the wild violates ethical principles, the reasons for breeding also lie in the fact that all animals should have the same genetic background, as the Wistar rats do. It is of interest to have animals that show the same biochemical or physiological reactions, which can be explained by a uniform genetics of the breeding lines. A second example is the BALB/c mouse, which has been adjusted by inbreeding crosses since the beginning of the last century so that all animals are almost clones of themselves. The mice that founded this breed were brought together at the Memorial Hospital in New York by the mouse dealer Halsey J. Bagg (1889–1947) and the resulting sisters and brothers were genetically unified over 26 generations within 15 years.[217,218] The white albino mice with the red button eyes were passed on to various laboratories in the USA, where they were again mated with each other and the number of generations continues to increase to this day. Since 1920 to today, more than 200 known generations have been bred. If one assumes that in humans, who are incidentally a breeding line, the generation time is 25–30 years, then one could trace back humanity over 6000 years to the time of Noah's Ark with 221 generations (as of 2007).

In Germany, approximately 610,000 experimental animals are tested in the pharmaceutical industry (as of 2020), which is about 1,600 animals per day. The vast majority of experimental animals in German pharmaceutical laboratories are mice (58%), followed by rats (27.5%), rabbits (7.4%), guinea pigs (2.8%) and fish (1%). The proportion of dogs and monkeys is 0.16% and 0.17% respectively, which does not sound like much, but in absolute numbers this is still around 1,000 dogs and monkeys (as of 2013). This is still an alarmingly high number of animals, which must make us think even more about alternatives to animal experiments and tissue cultures. But will we really be able to do without animal experiments in the near future, as the Netherlands is aiming for from 2025? A waiver is not to be expected, as various laws such as the Medicines Act and the Chemicals Act, to name just two of the six laws that prescribe animal experiments, would have to be changed.[219, 220]

Mice, rats and other rodents have their biological limits and show clear differences to humans, which is due to their nature. This circumstance also allows justified criticism of the meaningfulness of animal experiments and the validity of the results with regard to the differences between humans and animals. Results from rodent studies on inflammation, stroke, Alzheimer's, diabetes, Parkinson's are doubted and some researchers fear that billions of dollars, euros or pounds have been spent in vain. Since the decoding of their genomes in 2003, we know that the mouse and rat share 99% of their genes with us. Nevertheless, scientists today are trying to humanise genes using genetic engineering so that the chimeric mice can mimic human diseases even better. An estimated 8,000 genetically modified mice are available, all of which are supposed to represent all diseases relevant to humans from cancer to immune deficiency and cystic fibrosis to obesity and diabetes. A genetically modified mouse, also called a transgenic mouse, that has made history in pharmaceutical research, is the Oncomouse. Its story will be briefly told here.[221]

Oncomouse—Probably the Most Famous Mouse in Science

Besides Mickey Mouse, the cancer mouse or oncomouse is probably the most famous mouse in science and society. On April 12, 1988, the U.S. Patent and Trademark Office granted the Harvard Medical School the patent US 4.736.866 for a genetically modified mouse that develops cancer after a chemical stimulus.[222] The two scientists Philip Leder (1934–2020) and Timothy A. Stewart (*1952) created the transgenic cancer mouse by introducing cancer genes into the embryo of a previously fertilized mouse with an ultrafine needle. Since the genes were not only introduced into normal body cells, but into the genetic material of every cell, including the sperm and egg cells of future generations, a species was created that could theoretically produce new oncomice forever. This prompted animal rights activists and other conservation groups like Greenpeace to take action against the patent, arguing that there should be no patents on life. Before we consider this question, let's first look at how the technology of this *in-vitro* fertilization developed.

As early as the beginning of the 1980s, a new technology for genetically modifying animals was developed. Although it had been possible for many years to isolate genes and specifically introduce them into gene taxis, so-called plasmids, this had so far only been possible with bacteria, fungi, and perhaps plants in a simple way. With animals, it was much more difficult. Because how could one make a human or a mouse from a simple tissue cell, for example, a HeLa cell? Impossible. Another method had to be found, which the geneticists borrowed from gynecology. There, a technical procedure for in-vitro fertilization in a test tube has been in use for many years, for which the Briton Robert Edwards (1925–2015) received the Nobel Prize for Medicine in 2010. The technique is simple, although strenuous in execution, which is why it is rarely used today due to new methods in genetic engineering. In gynecology, a very thin pipette is filled with a single sperm and pumped into the egg cell. The sluggish sperm is helped to overcome the wall to fertilize the egg, which it cannot do for biological reasons. The geneticists wondered whether this micromanipulator, as the device is called in gynecology, might also be suitable for introducing foreign genes into a fertilized egg cell of their experimental animals. What could be well applied in humans could easily be transferred to laboratory animals. There was only one problem, which is why the method of micromanipulation was heavily criticized by animal rights activists.

In the production of transgenic animals, a female mouse and a male mouse are first united, because only then can fertilization of the egg cell occur, which is then removed. The fertilized egg cells are surgically removed, unfortunately killing the egg donor. This is the dilemma of the animal rights activists. From the obtained egg cell, which divides and is called a blastocyst in a multiple cell stage, individual stem cells are removed and genetically modified. The foreign gene is transferred and the genetically modified stem cells are returned to the fertilized egg cell, which swam in a petri dish during these works. If the stem cells are well accepted, there is nothing to prevent reimplantation as an embryo into a surrogate mother. The implantation takes place in a pseudo-pregnant mouse, which previously copulated with a sterile male and was treated with hormones. Depending on the constitution of the mouse, it gives birth to 5 to 8 young. Rarely, due to the hormones, multiple births with up to 15 young are seen. The surrogate mother accepts the eggs and carries her genetic adoptive children. A disadvantage of this method is that not all genes are incorporated into the genetic material and the geneticist has to check which animal is a genetically modified animal. Today, a blood sample is necessary for this, in the past it was the cutting off of a small piece of the tail tip of the newborn mouse child. It is important to know that the development of the oncomouse preceded the development of in vitro fertilization technology, which will be briefly introduced.[223, 224]

In the fall of 1980, Richard Palmiter (*1942) and Ralph Brinster (*1932, Fig. 8) spoke on the phone because Allen Senear, a scientist in Palmiter's lab, had developed hybrid genes. He combined a gene with a foreign promoter, which in genetics is the starting point for reading a gene. This gene produced an enzyme that transfers phosphate groups to other proteins. This is not really exciting at first, but it is clever for tracing biochemical reactions. Palmier was looking for a molecular biological model to test whether his gene also worked. The two spoke on the phone, and Brinster suggested trying it with eggs from mice he had fertilized in his lab. Brinster welcomed the suggestion, as he had no gene but the model, and he wanted to see whether the introduced genes would remain in the egg cells or be destroyed. The experiment was successful, and Brinster was able to demonstrate in the following generation of mice that the introduced gene produced a protein that was active in the animals' liver. Was it just a lucky coincidence? He repeated the experiment successfully with a growth factor and obtained the "Big Mouse," as they called the oversized mouse. In the next experiment, they introduced parts of the SV40 virus genome, but these experiments did not go as planned. SV40 is a virus that can cause tumors in monkeys. The virus was isolated in 1960 by Ben Sweet and Maurice Hillemann (1919–2005)

Fig. 8 Ralph Brinster (*1932) (public domain, wikipedia)

from macaques and rhesus monkeys. Kidney cells from rhesus monkeys were used from 1955 to 1963 to produce vaccines against polio, and it was discussed at the time whether millions of people could have been infected with the SV40 virus through vaccination. Whether the virus is dangerous for humans is unlikely and not proven. Today, it is discussed whether and how similar viruses are able to excessively activate so-called oncogenes in our body, which can then lead to cancer. Oncogenes are activated proto-oncogenes that no longer allow gene regulation to function and promote cancer. In the very complicated development of cancer, normal and harmless proto-oncogenes can mutate through chemicals and the resulting oncogenes allow the excessive reading of genes in emerging cancer cells.

So what went wrong in the experiments? Brinster and his colleague Howard Chen observed that many mice suddenly died. They assumed that the mice were resistant to SV40, but the autopsy of the mice showed that brain tumors had formed. Both were not real experts in the field of oncology and could not classify the results. Brinster spoke with Terry van Dyke and Arnold Levinean from the State University of New York and with Abee Messingan from the University of Pennsylvania, who had more experience with brain tumors. All three looked over the data, examined the results, and found the data to be sufficient but not adequate, as they considered the triggering of the tumor in the mice to be too weak. Since the data situation was too thin for publication, the project would have almost ended without publication. They did not give up and looked for further promoters that should specifically trigger a tumor. One direction of their thinking was to check the role of SV40 and other promoters in pancreatic cancer. The connection was successfully detectable and they found a connection with the then still not properly understood regulatory networks. As so often, it was work in intellectual fog. Terry van Dyke continuously developed the SV40 model for brain tumors, without exactly knowing what he thought he would find. Palmiter and his colleagues found themselves in a dead end and lost interest in the work until the year 1984 came.

In 1984, Palmiter flew to Lorne, Australia to Lorne, Australia, for a scientific conference to give a lecture on his findings in kinase research. He also mentioned his work on SV40 in passing, but did not go into detail about these half-baked results. However, two listeners were interested and approached him. Jerry Adams (*1940) and Suzanne Cory (*1942) from the Walter and Eliza Hall Medical Institute (WEHI) in Australia suggested that he should not look at the level of oncogenes, but one step before. "Start with the proto-oncogenes." Palmiter shrugged. Why not, he thought, he had nothing to lose. They agreed to introduce the MYC gene into the egg cells to stimulate the oncogenes. The MYC encodes a protein that enhances the reading of other genes. This reading booster can promote tumors like Burkitt's lymphoma when it is mutated. However, this experiment did not bring the hoped-for success. He was able to prove that proto-oncogenes were transferred, but they were still far from causing a tumor in the animal model. The good news was that other researchers used these results as a basis for further groundbreaking discoveries. Their colleagues at WEHI, David Vaux and Andreas Strasser, found that the suppression of apoptosis can lead to the development of cancer—a real breakthrough in the field of cancer biology. Anton Berns (*1945) from the Netherlands Cancer Center (NCC) discovered regulatory genes that play a role in the development of cancer, and Scott Lowe (*1963) from the Cold Spring Harbor Laboratory recognized the role of interference RNA, which at that time showed a completely new mechanism in mouse models and in cancer. Everyone knew that sooner or later a breakthrough in applied genetics had to succeed, but a solid cancer model in the animal would be very helpful.

Two new scientists, who have already been introduced above, entered the stage of molecular biology and were to bring about the breakthrough: Phil Leder (1934–2020) and Timothy Stewart (*1952). Stewart is a New Zealander and worked as a postdoctoral fellow in Philadelphia with Beatrice Mintz (1921–2022, Fig. 9). Beatrice Mintz was the daughter of Jewish emigrants from Galicia, which belonged to Austria before the First World War. Due to anti-Semitic admission restrictions, she could not study in her hometown of New York or on the East Coast and went to the University of Iowa. At Mintz, Stewart learned all the important methods to genetically fuse embryos of mice of different origins. Mintz was also the first to create transgenic mice by introducing viral genes using retroviruses, and she developed the first animal model for skin cancer. These were the best prerequisites for the creation of the oncogene mouse.

Postdoc positions are short-lived in science, and it is normal for the money from the sponsor or funding organization to run out after one or two years. This also happened to Stewart, who moved to Harvard University to work with Phil Leder, who had made a name for himself by deciphering the triplet code in protein translation. Leder was a pioneer in the field of molecular genetics and structural analysis of mammalian genes. He wanted to develop a mouse model for breast cancer with Stewart and took him aside: "My old buddy Gordon Hager at the NIH has found a retrovirus that causes breast cancer in mice and he has brought the regulatory region of the virus together with an oncogene. Everything is squeaky clean and he was

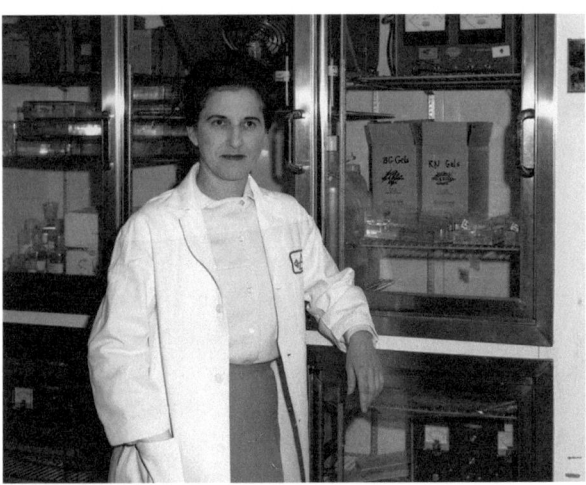

Fig. 9 Beatrice Mintz (1921–2022) (Permission from Smithsonian Institute Archives, Washington DC, USA)

able to regulate it with glucocorticoids." He nodded paternally to Stewart. Stewart looked at him incredulously. "I don't think it can be that hard to transfer this model to an animal. A half-year project—nothing more. You've learned everything from Beatrice in Philly!" Timothy Stewart nodded. "Why breast cancer?" he asked back curiously. "Well, it has something to do with the oncogenes. The highest rate of active organs is found in the epithelial cells of the breast," Leder replied briefly. "If everything works out, then you have a mouse where you can simply turn on cancer." Stewart had no idea about all the cloning and remembered his colleague Erwin Wagner (*1950) in the lab. The Austrian always talked about oncogenes and how they could be activated in mice by microinjection. That should work here too.

It took four times half a year until the MMTV-MYC mouse model was ready. As suggested by Erwin Wagner, he had built an oncogene into Gordon Hager's construct. After two years, he met with Phil Leder to report: "Hi Phil, the model is ready and the mice are alive. You can see the MMTV-MYC mice down in the stable." Leder looked up from his computer screen. "MMTV-MYC … what? No one can pronounce that. Let's just call it cancer mouse or oncogene mouse." Why not. From that day on, the lab mouse that could be reprogrammed into a cancer mouse with a chemical was called that. Stewart set about publishing,[298] but Leder came up with a new idea.

The biotechnological revolution in the USA prompted the US Congress to pass the Bayh-Dole Act in 1980. This law not only allowed, but demanded that American universities patent their research results funded by taxpayers. In contrast to today, this scientific policy change was something completely new for the country. Slowly, the will to patent and thus the greed to make money with research seeped into the universities. Harvard University quickly seized this new legal freedom and filed the results of the Oncomouse by Leder and Stewart as a patent application on June 22, 1984. However, there was not only a patent on the Oncomouse, Harvard University filed two more patents in 1992 and 1999 that referred to the original patent from 1988. The university granted licenses of the patents to the company Dupont, which now financed research and secured privileges to all new findings due to the promotion of industry. However, Dupont had no interest in using the patents themselves. The company bought the licenses for a lot of money, it received the right to its own exploitation and it could massively restrict the marketing by Harvard itself. The Oncomouse was a business model, serious research was not in demand. Each license granted by Dupont obligated the licensee not to share the technology with others without asking Dupont. This went so far that Dupont wanted the right to read every manuscript before it was submitted for publication in a journal. This led to great discontent in society and among researchers. The original idea of commercializing the Oncomouse probably also fell short of expectations because of this.

The anger about Dupont, the restrictive handling, and the obvious profit-seeking may have raised the political question of whether life can be patented. Harvard had drafted the patent so broadly that research by third parties was prohibited not only on mice but fundamentally on all rodents. The Nobel laureate (1989) Harold E. Vamus (*1939) was very annoyed, organized the dissatisfied scientists, and held a spontaneous meeting of opponents at a Cold Spring Harbor Meeting on mouse genetics. *"We went to war,"* he recalls in the journal Science. The echo was not long in coming and the National Academy of Sciences of the USA took up the topic. In 1994, the National Institute of Health (NIH) published a statement in which it demanded the freedom of research with transgenic mice. The institute went a step further and a newly founded institution distributed the transgenic mice freely in the USA. This could only be in the interest of Harold Vamus, who also became director of the NIH a short time later. During his time (1993–1996), he commissioned the Office of Technology to develop new guidelines that, contrary to the Bayle-Dole-Act, called on all research institutions not to patent and not to interpret the law as the *"eleventh commandment"*. The benefit to research from free access was higher than marketing an idea. We know that he could not prevail.

Animal rights activists and conservation groups in Europe protested against the granting of a patent to Harvard University by the European Patent Office. The Patent Office did not make it easy for itself, the long time until a decision in 2004, almost 25 years after the decision of the US Patent Office, shows the complexity. The patent opponents argued that under Article 53(a) of the EU Patent Law, patenting would not be possible due to a violation of morality. When drafting this paragraph, the legislator at the time did not think of the Oncomouse or biotechnology, but it was about the possibility of rejecting patent applications if they violated societal morality or public safety. What are patents that violate morality, in the legal sense of the word? To give some examples: The patenting of tools with which one can effectively break into other people's homes, or the creation of new, even stronger drugs should be prevented in this way. The European Patent Office faced a problem. Is it morally or ethically justifiable to genetically modify only bacteria? There have been countless patents on this for years. The second argument of the plaintiffs was that genetic modification could create new animal species? Was that really the case? A mouse remains a mouse, even if an additional gene has been inserted. Bacteria had not turned into other species either. A decision had to be found that was fair to the matter and not to the plaintiffs or defendants. The European Patent Office developed a purpose-oriented decision matrix. The morality, the benefit to humans, and the negative effects on the environment, animal suffering, and public perception should be considered and weighed. In the case of the Oncomouse, the patent judges found that the benefit for the development of cancer drugs was greater and that research with the Oncomouse, despite the suffering of the mice, would not lead to public disapproval. No new animal species would be created either. They granted the patent and the creation of transgenic animals was legally correct.

What has the Oncomouse moved in our society? Not much, say animal rights activists. Although it has triggered a boom in transgenic animals, which has brought enormous capitalization to the pharmaceutical industry, the scientific gain in knowledge has been rather small. Critics attribute the latter to Dupont's restrictive attitude. Only the massive intervention of Harold E. Vamus as NIH director with political pressure brought about a change. Dupont only gave in under the high pressure of the NIH when he not only threatened all researchers who were researching with Dupont's Oncomouse, but actually withdrew the funds. Free research, no insight into manuscripts by the industry, free exchange of animals between scientific institutions, but approval for industrial third-party projects. That was the deal.

Is Man Just a Mouse Too?

How useful are animal experiments in pharmaceutical research? It is undisputed that there is no animal model that can replicate the biological or physiological situation as it is in humans. There are no depressed mice or schizophrenic dogs, and the immune system differs significantly between humans, rats, and mice. We see this very clearly in the case of Corona. The mouse is not susceptible to the SARS-Corona Virus and it does not lead to infections in them. The same is true for dogs and cats, as we have observed in our pets. To develop a suitable mouse model, knockout mice are created, which can only become ill through genetic modification. However, we are still far from a human immune defense in rodents. This situation is precisely the motivation for using genetic engineering to modify animals so that a disease or defect can be better understood.

What are experimental animals used for in Germany? A look at the statistics of the Association of Research-Based Pharmaceutical Companies (VfA) shows that 80% of rats and mice are used for safety testing. If we compare today's pharmaceutical industry with the industry from the 1960s and 1970s, the transition from a research industry to a producing and selling industry is also evident in the type and number of animal experiments.[219] Research with scientific breakthroughs mainly comes from universities and research institutions. Here, the basic results are still tested using animal experiments and substances are validated that are suitable for elucidating a disease mechanism and its treatment. Only when promising results are to be brought to market readiness through patents does the researcher seek proximity to the industry. From this point on, standardized procedures and animal experiments are carried out to ensure the highest level of safety in three clinical phases. Although animal experiments are also carried out here, the majority of animal experiments have already been carried out at universities and research institutions.

The safety of animal experiments can be deceptive, and many researchers think too schematically in models, which is why they often rely unjustifiably on the results in animals. An example of the failure of animal models was discussed for Contergan®.[225] The change in germ cells after fertilization in humans is very sensitive to thalidomide in Contergan® in the early phase of human development. No changes could be observed in the tested rodents, mice, and rats. The human geneticist Widukind Lenz (1919–1995) discovered in 1961 that thalidomide damaged human fetuses. He called the research director Heinrich Mückter (1914–1987) on November 15 of that year. The pharmaceutical company Grünenthal vehemently denied this accusation, as their own investigations on rodents had not shown this. This was true, but the animal model was not sensitive enough to demonstrate the damage. The realization was that rabbits were prescribed as another sensitive animal species.[226]

The series of misinterpreted results could be continued. For example, American scientists reported in 1953 that the odor and flavor substance coumarin leads to liver damage. Coumarin is a good example that toxicity can vary greatly between animal species. Dogs can tolerate a maximum of 10 mg of coumarin in the short term, mice over a longer period 50 to 100 mg per kg body weight. A metabolism study in rats clearly demonstrated liver damage. However, the results of the rat study were misinterpreted, as the breakdown product in rats is indeed liver-damaging, but practically does not occur in human metabolism. This would have been bitter for us, as we appreciate coumarin in Christmas pastries, in green woodruff jelly, in the tonka bean but also in our May punch and would not like to do without it. Today it is assumed that moderate consumption of coumarin-containing foods is harmless for humans.[227] Most liver damage from the May punch is more likely due to alcohol than to coumarins. The belief that coumarins are harmful persists to this day.

So how do we get out of this dilemma of animal experiments? The pharmaceutical industry wants legal certainty that it will not be sued for omitting certain tests. However, we do not want patients to be exposed to a high risk of permanent damage or even death from new drugs. Our state has a duty of care to protect the physical integrity of its citizens as best as possible and thought early about drug approval (Fig. 6). We have already learned from Albert Moll that experiments on patients were already an ethical dilemma in 1902 and that administration for research purposes at a very early stage of drug development was not acceptable.[202] This is not a realization of the morally better modern constitutional state, because as early as 1900, the administration to children and those incapable of doing business was prohibited by a decree of the Prussian Minister of Culture. His colleague Rudolf Kobert (1854–1918), who was a respected pharmacologist, called on his colleagues at the 72nd Meeting of German Natural Scientists and Doctors to see ethical considerations as an indispensable prerequisite for every drug test on humans: *"Medical use or testing at the bedside of new remedies, which have not been sufficiently pre-tested in animal experiments, is animal cruelty to humans, or can at least easily become so, is therefore inhuman and violates the Codex ethicus medicorum."*[228]

What should be the best possible for humans is difficult to define. There is no absolute certainty, only different degrees of uncertainty. If the state, represented by its regulatory bodies, were to unconditionally prescribe animal testing, it would also lose sight of animal welfare, which is enshrined in our Basic Law in Article 20. There is no way out, and the demand to test a non-animal alternative in the cell culture model for every animal experiment and finally to test it on a few animals in different species remains a wish. Many different cell cultures are established today and are widely used in drug testing. This results in a wealth of different data, which can make reliable predictions with the help of statistical methods and computer models, but no cell experiment and no computer science can ultimately replace the animal.

What can we agree on if we as a society have to tolerate animal testing as a necessary evil? Possibly, animal tests may need to be conducted to rule out the toxicity of a substance, and pharmacokinetic studies must be provided to determine a dose for humans. But even finding the dose can be problematic. A sad story may illustrate this: Animal tests showed that the active ingredient BIA 10–2474 is non-toxic and harmless. In the dose finding in a further clinical study, this assumption led to a disaster. The Portuguese company Bial Porela developed BIA 10–2474 as an inhibitor to treat anxiety and improve mood. This substance interacts with the human endocannabinoid system, which is still incompletely understood today in times of cannabis hype. In a clinical study in 2015 in Rennes, France, there were severe neurological side effects and one death. What happened? After an initial phase with animal experiments and cell cultures, the inhibitor 90 was administered once to 90 healthy men. They received BIA 10–2474 and showed no side effects. A year later in January 2016, the study was to be repeated with six healthy men who received multiple doses of BIA 10–2474 on consecutive days. Neurological disorders occurred after just two days. The doctors initially suspected a stroke and took the subjects to the University Hospital Rennes, where brain death was diagnosed in a 49-year-old subject, who died a day later. The deceased was the artist Guillaume M., father of four children. His brother Laurent reported that Guillaume stepped in for a participant who did not come. How did the surviving subjects fare? Pierre-Gilles Edan, head of the neurological department, reported bleeding and necrosis in the brains of other subjects, from which they are unlikely to recover. Whether it was actually a question of dosage or, as was later suspected, a possible contamination or a misunderstanding of the pharmacology of the endocannabinoids, will no longer be clarified. The criticism of the associations and the French Ministry of Health was great, because the study design had been too ambitious and too streamlined. Further clinical studies with similar substances were discontinued, and it is unlikely that the substance class will be further developed in the future.[229,230]

Cell Cultures—the Immortality of Henrietta Lacks

They had been sitting in the Buick for a while. It was drizzling lightly on this January morning. "Babe, everything will be fine." He looked at her. She glanced briefly into his eyes and knew that nothing would be fine. She had a feeling. Maybe she had waited too long, but she didn't want to go to a hospital because she understood too little of what the doctors told her. She had already been to her small local hospital because of her children and she found it terrible. Now she was here at John Hopkins Hospital or whatever it was called. The only hospital that offered her free treatment that she could never have afforded. Where she grew up was poverty, bitter poverty.

Everyone lived on the fields and grew tobacco, like their ancestors who had to work hard as slaves. What was the difference between her and her family today? She didn't know.[231]

"Henniee, should I come with you?" he asked cautiously, but already knew her answer. "Dave, no, that's not possible, that's for women. What do you think will happen?" He nodded and searched for the cigarette pack in his jacket pocket. He found it, pulled it out. Three left. He took one and wanted to light it. "I'm going now, will you wait for me?" "Yeah sure, here." His cigarette bobbed between his lower and upper lip. She opened the door, light rain came towards her. Without looking at him, she walked up the stairs to the entrance. The white man at the top of the entrance held his arms immovably wide open to receive her. She recognized him, Jesus made of Italian marble. She walked past.

"Mam, my name is Henrietta Lacks, I have an appointment for my lower abdominal pain with Dr. Jones, I think." The white nurse looked at her sheet and slid her finger down the rows. "Yes, that's correct. Dr. Jones is in gynecology on the second floor. You should take the elevator. Can we take biopsy samples from you?" She nodded and didn't really understand what she wanted and what she was consenting to. She was one of the few in the Pleasant family who could read and write at all. She was taken out of school early and as long as she could remember, the sun of Virginia burned on her back. "Can I go to the toilet beforehand? Where can I find it?" "Then go straight to the second floor, that's the department for blacks, there you will also find the examination room for blacks."

"Mrs. Henrietta Lacks, is that correct?" She just nodded. Dr. Howard Jones (1910–2015), a white gynecologist, was the attending physician and sat in front of her at the desk with a cigarette in his hand, looking at the documents she had brought with her. The family doctor believed she had syphilis, he read in the doctor's letter. He continued to go through the letter: *"Middle school education. Housewife and mother of five children. Breathing problems since her childhood due to persistent respiratory infections and septum deviation in the patient's nose. Recommend surgical correction, patient declines. Patient has had toothache for five years. Tooth should be extracted along with others. Cooperative, drinks occasionally, nine siblings, unexplained bleeding since the second pregnancy, test for sickle cell anemia declined. Monogamous with husband for 15 years, mild neurosyphilis. Syphilis therapy declined. Last examination two months before fifth birth. Evidence of gonorrhea. Patient did not return to clinic for treatment."*

"Come in and undress. You can then sit on the chair." He pointed with the cigarette in the direction of the gynecological chair. I'll take a look at this. You say you have pain in your lower abdomen. "Have you had these for a long time and can you tell me when they first occurred?" "I don't know, they were just dull pains, I always had blood in my underwear and now they have become stronger. I felt with my finger and there was the lump." Dr. Howard spread her thighs a little apart and tried to see what was causing the pain. He had seen thousands of tumors in his life here at Johns Hopkins, but this one was different. It felt very soft, like a grape or red jelly bean, like he could buy downstairs in the cafeteria. Even simple touches caused the tumor to bleed. He cut the tumor from the wall. For comparison, a little cervical material from the opposite side as well and put both in the metallic kidney dish. He knew he had to send both to George immediately. Jones looked again at her documents and leafed through her recorded personal data. Admitted already on September 19, 1950. He lit another Lucky Strike. The colleagues saw nothing unusual. Either they had messed up, which he did not believe, or this cell mass was growing at a furious speed. He feared the worst for Mrs. Lacks.

Epidermoid cervical carcinoma stage 1, was written in the pathologist's examination report. Uncle Dick, as they called the chief physician Richard W. Te Linde (1894–1989), and Howard had informed him that he should carry out the standard therapy on Mrs. Lacks. This was the insertion of a radium implant into the uterus. She could no longer have children. He wondered, had anyone already told her this? He leafed through the medical record. Yes, there it was. He could save himself this difficult conversation. "Hello Mrs. Lacks, please come in." Henrietta Lacks, who was waiting in the hallway and was led by the nurse into the treatment room, felt insecure. What was happening now? She saw the wooden table on which she would probably lie soon. Dr. Jones had called her in early February and told her that it was very serious and the lump was a tumor. She had considered not coming, but the pain in her lower abdomen was getting bigger. She went alone to *"John Hopkin"*. She had not told Dave or anyone else in the family about the call. They should not worry. She told everyone, the doctor just wanted to do a few more tests. Now she was standing in the treatment room. Where was Jones? There was another doctor here!

"Mrs. Lacks, my name is Dr. Wharton, I will insert a small glass ampoule into your lower abdomen". This is a radium implant that will fight your cancer. Before I start, you need to sign this statement. He pulled a sheet of paper from the drawer and handed it to her: *"I hereby authorize the staff of Johns Hopkins Hospital to perform all surgical procedures and any type of anesthesia, either local or general anesthesia, which they deem necessary for proper surgical treatment"* and here on the line she should sign. She did and there followed many medical examinations and preparations for the operation the following day.²³¹ A taxi brought the glass radium tubes from the radiation clinic at the other end of Baltimore. Johns Hopkins had a lot of experience with this therapy since an older colleague, Howard Atwood Kelly (1858–1943), got to know radium in France with Marie and Pierre Curie. He often brought radium samples in his shirt pocket to the USA and developed a new procedure that replaced the inaccurate surgery. Kelly underestimated like many the great danger of radioactivity and later died of radiation-induced cancer as one of the great four of Johns Hopkins Hospital and is considered the founder of modern radiation therapy.

Lawrence Wharton Jr. sat between the legs of Henrietta Lacks, positioning the small glass vial at the correct spot on the cervix wall. He also took additional samples to give to Otto Gey (1899–1970, Fig. 10), who was trying to cultivate cells in the basement. This freak, he thought, this guy wants to succeed in his self-built lab? He wished him good luck. A nurse took the samples to the basement. *Cell culture—George Otto Gey,* was written on the door. He had been here many times and had always found only this Otto Gey, his wife Margret, and the assistant Mary in front of hundreds of vials

Fig. 10 George Otto (1899–1970) and Margaret Gey, circa 1950 (Permission Johns Hopkins Hospital, Baltimore, USA)

and bottles. It gave him a shudder each time, as it seemed surreal to him, a bit like the Frankenstein movie with the wonderful actress Mae Clarke, which he had seen in the cinema years ago. He would have loved to have his opinion confirmed that Gey was some kind of Frankenstein doctor, for he knew that all the freezers were filled with blood, placentas, tumors, dead mice, and a dead duck that Gey had brought back from hunting 20 years ago and that didn't fit in the home freezer.

He looked for George and glanced at the walls, which were filled with cages containing mice, guinea pigs, rabbits, and rats that stared at anyone who entered. "Hey George, I have something for you again. Where should I put it?" he called into the room. He looked around, there was no free space. George came over to him. "No problem, hand it over. I'll give it to Mary right away. Thanks, buddy." He took the glass bottles and went to Mary. The door behind him closed. "Mary, work for you." He put Henrietta's samples in the isolator. Mary was his assistant and looked at him while chewing. She was eating her lunch and more samples lay on the desk between coffee cups, sugar, and powdered milk. The animals in the cages looked at her pleadingly, hoping she would give them some of her tuna sandwich. "Ah, thanks," she said briefly. She acted as if it didn't concern her. It can wait, it's break time now. She knew that speed was always crucial with all cells. But she was tired and had been cutting tissue, transferring cultures, and evaluating old cultures all morning.

The basement was supposed to be a lab, but everything was improvised. Mary Kubicek had only recently been in Gey's lab and didn't know whether to be happy or to run away. She stayed because she could work with Margret and experienced something new every day. Margret was an operating room nurse and was supposed to help Gey with his cell cultures. Gey and Margret had set up the lab here in the catacombs (Fig. 11). The two had accomplished an amazing team performance, even though the hospital provided almost no money. He, the skilled craftsman, built a rolling bioreactor and an aseptic isolator from the old incubators. She, who was well versed in sterile technique and lab work, complemented him perfectly. The isolator, in which Gey placed the samples, was such a self-construction. It was just 150 × 150 cm in size, equipped with an air filter and a water vaporizer that sterilized the isolator before entering. When she was in this thing, she closed the door, then let in water to flood the floor and protect it from the intrusion of germs from the air, then she started flaming all the glass and metal equipment with the Bunsen burner. She had learned sterile work from Margret, and she knew her stuff and was the most important person down here. She almost never had a contamination and if so, it was Minnie who had rinsed

Fig. 11 G. O. Gey's lab at Johns Hopkins Hospital, circa 1950 (Permission Johns Hopkins Hospital, Baltimore, USA)

incorrectly again. Margret had hired Minnie for the dishwashing kitchen and explained everything to her in detail. This went as far as the Gold Dust Twins soap, which Margret had bought in large quantities and now filled an entire wall in the back part of the house when she heard that the company was going bankrupt. Margret was the drill sergeant who rarely smiled. She patrolled the lab with her arms crossed. If she found a mistake, she could become very unpleasant.

Mary put on a new white coat, put on her cap and mouth guard, and went to the isolator. She started the sterilization procedure for the box in which she would work for the next two hours. She took the new tissue, cut it into small cubes each one millimeter in edge length, picked it up with the pipette, and transferred the small cubes into tubes where solid blood medium had already been prepared. A few drops of the nutrient medium were given to the small cubes and the tubes were sealed with a rubber stopper. The tubes were briefly flamed at the edge and labeled. She took the first two letters of the first and last name. "HeLa" was written in large letters on each tube, and she put all of them in the incubator that Gey had built himself. A milestone in medical history was created and she was done for the day.

George Otto Gey was a tall, wiry man, born in Pittsburgh to German immigrant parents. He came from humble beginnings, growing up in a shack next to a steel mill. The family cooked whatever the garden provided, and young George dug a small pit and collected coal for his family and neighbors to cook and heat their homes in winter. He financed his biology studies at the University of Pittsburgh by working as a carpenter and painter. He was skilled with his hands and built a microscope camera from parts he found at the junkyard, which he used to capture slow-motion images of dividing cells and play them back on film. Because he had to finance his tuition with jobs, his studies took eight years. Gey was a visionary and tinkerer who quickly became excited about problems and spent nights jotting down ideas and concepts on napkins at his desk, kitchen, and pub. In addition to his work on cultivating HeLa cells, his second invention is less well-known. However, it can still be found in many labs today. It is the roller bottle (Fig. 12), which slowly rotates to keep the cultures on the wall or in the medium moving or to better supply them with oxygen. He invented this culture technique because he believed that blood and tissue

Fig. 12 Roller bottles in G. O. Gey's lab, circa 1950 (Permission Johns Hopkins Hospital, Baltimore, USA)

fluid are always in motion and that this movement can only be beneficial for a cell culture. He joined Johns Hopkins Hospital in 1937, and his task was to examine biopsy material. There was no suitable culture medium available. He believed that blood would be ideal, as it surrounds the tissues in the body. Working with human blood was very difficult at the time for ethical and technical reasons. He saw an alternative in the use of animal blood, especially from chickens. When blood was scarce, he would go to a chicken farm, catch a chicken, and lay it on its back. With one hand, he held the feet and pressed his elbow on the neck. With his free hand, he disinfected the chest and inserted a syringe into the heart to draw blood. If the poor chicken survived, it was allowed to go back; if it died from stress, he brought it to Margret for dinner

Henrietta's cells held up remarkably well, as Gey found. Every morning, Mary checked them and transferred the cell clumps to the roller bottles, which moved at about two rotations per hour. There wasn't much to see, except that they didn't die, but Gey thought that was already a success. He warned everyone not to celebrate too early. However, Henrietta's cells grew incredibly fast. Every morning, Mary had to get new roller bottles and prepare new vessels, because the small cubes from the first day had become large, visible clumps. Although Gey was still cautiously optimistic, Mary was convinced that these cells were completely different. They grew twenty times faster than Henrietta's normal cells, and as long as they received heat and food, they didn't want to stop.

Three weeks after the radiation therapy that Henrietta Lacks had to undergo on April 10, 1950, George Gey sat in a television studio of WAAM Television in Baltimore. He had been invited to talk about successes in the fight against cancer. Using a pointer, he explained to the viewers on a board what cancer cells are, how quickly they divide, and what they look like. During the lecture, he pulled a small bottle out of his jacket pocket and held it up to the camera so that viewers could see a floating lump. "Let me show you a bottle that contains a large amount of cancer cells." He weighed the bottle in his hand and everyone could see it. Now he tried to impress the public and convince them that the fight against cancer could be taken up with these cells. Gey wanted to be the selfless front-line fighter against cancer and sent HeLa cells to anyone who asked for them. Shipping by mail was not common, and Gey sent his cells by plane, sometimes asking pilots and flight attendants to carry a tube on their bodies. A few drops of culture medium and body heat ensured safe transport. The cells made their way in saddlebags on donkeys over the Andes to Chile and on bicycles to Holland. Once there, they were passed on to friendly colleagues. Now all investigations were possible that were unthinkable with patients. On October 4, 1951, Henrietta Lacks died and her body was transferred to the cooler for Black people. Unlike her cells, she was to make one last journey to Clover, Virginia. Her cells were to go into the world and never die.

Henrietta Lacks (1920–1951) was a boon for modern cell culture technology, even though, to her regret and with her early death at only 31 years old, no further help for healing could be given. We can posthumously be grateful to her today, even though much went wrong in the loving and empathetic treatment of her. In the 1950s, which in many states of the USA resembled apartheid in South Africa, Henrietta Lacks was a second-class black person. The alleged consent to the implantation of the implant that made her infertile was a misunderstanding. She would never have agreed, as she said shortly before her death, because she wanted to have more children. The bioethical rules for the removal and handling of one's own cells were practically non-existent and did not correspond to today's internationally recognized rules. The Lacks family did not know for years that their mother's tumor cells, known as HeLa cells, were metastasizing worldwide through shipping and culture. Today, we owe medical breakthroughs in the fields of cancer research, genome sequencing, new methods in genetic cell manipulation such as interference RNA, development of the polio vaccine against polio, and they were shot into space with Sputnik 6 in 1960, where they multiplied even faster. Without HeLa cells, there would have been significantly fewer cancer drugs and no Nobel Prizes for Harald Zur Hausen (2008) and Elizabeth Blackburn, Dr. Carol Greider, and Dr. Jack Szostak (2009).[232]

Even though HeLa cells were the first to be cultivated according to our ideas, the science of animal cell culture is of course older. The exact date can perhaps be located with the discovery of the cell as the smallest viable biological unit in the 16th century, when the Dutch optician Zaccarias Janssen from Middelburg saw a wonderful division in a leaf with a precursor of the microscope. Antoni van Leuwenhook (1632–1723) from Delft discovered bustling creatures in water and on his oral mucosa in 1683, as well as red blood cells, without knowing what to make of it. It was not until 1838 that the German anatomist and doctor Theodor Schwann (1810–1882), together with his colleague Matthias Schleiden (1804–1881), further developed the concept of the cell from plants to animals and humans as the "cell theory". Now science understood the importance of animal tissue for therapy and attempts were made to establish cultures from living biopsy and autopsy material. Encouraged by the ease of their colleagues from botany, who could easily create plant tissue cultures, this should also succeed for animal and human cultures, but it was a fallacy. Tissue could indeed be easily taken and brought into culture, but the cells did not divide or only a few times and then perished.[233, 234] If it was possible to dilute the dividing cell cultures up to five times and transfer them to other vessels, i.e. to passage, one was satisfied. 20 passages were already considered extraordinary at that time, 50 passages as visionary. Cancer and tumor diseases were known, and none other than Rudolf Virchow (1821–1902) was the first to classify tumors into types and lines. Wilhelm Roux (1850–1924) was the first to deal with the culture of cells. He came from a Huguenot family in Jena, where he also studied medicine. In addition to his interesting work on blood flow and blood pressure in the liver, he was fascinated by embryonic development and cell division. In a hanging drop, he was able to show the cell division of frog eggs and later the division and cultivation of cells from the chicken embryo in a salt solution for a few days. At least, Roux had succeeded in proving that cells can also develop and divide outside their organism.

The American Ross Granville Harrison (1870–1959) studied medicine in Bonn in 1892/1893 and fell in love with the German Ida Lange, whom he married in beautiful Altona near Hamburg in 1896. He is rightly considered the father of animal cell culture and should have received the Nobel Prize for his work on the cultivation of nerve cells. He was nominated 15 times, but never awarded.[235] Because of this marriage and his closeness to the Germans, which also showed in some visits to Bonn after his call to Yale University, he got into trouble at the outbreak of the First World War. German scientists were researching in his laboratory, who worked and slept in the Osborn Tower of the university. This tower was relatively close to the coast and could be seen from the sea under favorable conditions. The Germans left the light on at night, which seemed suspicious to the American counter-espionage, as they assumed that secret position lights were being set for the German submarines already sighted on the coast.

In the laboratory, Harrison took up the old work of Roux and became one of the leading embryologists. He was interested in how nerves grow and whether they connect in a special way between cells. Roux's idea of culture in a hanging drop of water on an inverted microscope slide seemed perfect for him, as he could observe the development of nerves in the spinal cord and brain in tadpoles. For the first time, he was able to keep tadpoles and later nerve fibers from chicken embryos in culture for a short time. But it was all in vain, all cultures died after a short time. Harrison desperately wanted to extend the survival time and meticulously cleaned all equipment, boiled the scalpels, scissors and tweezers, and autoclaved all fabrics and papers. It should not be due to a lack of hygiene and he rightly suspected that it was due to the nutrient media and the lack of supply to the cells. His experiments aroused broader interest, as the potential of the tissue culture technique conceived by Harrison was immediately recognized and described as an *"amazing step in the history of biology"*.[236]

Inspired by Harrison's success, Montrose Burrows (1884–1947), who worked with Alexis Carrel (1873–1944) at Rockefeller on chicken embryos, later also studied guinea pig, cat, dog, and rabbit embryos. Burrows wanted to find out how cells could survive better, and his idea was that this depended less on the cells themselves and more on the nutrient medium. His visit to Yale University to see Harrison showed him that the tadpole experiments worked with water as a nutrient medium, but not with blood or lymph fluid. His visit came to an end, and he returned to New York with the idea and realization that there were more problems to solve for blood as a nutrient medium. Blood as an addition to nutrient media lasted surprisingly long, but today it is considered obsolete in biotechnical fermentation. Cell culture research was a bloody affair until the 1960s. We have already read that Gey, who first cultivated HeLa cells, devoted a lot of time to the question of the best composition of nutrient media, and he was, as reported, a frequent guest on chicken farms.

Burrows and Carrel introduced the term "tissue culture" into science in their joint publication in 1910. Carrel was a French surgeon and researched the transplantation of tissues and organs. In 1912, he described the first cell line ever developed for heart muscle cells. It was a simple fragment from embryonic chicken hearts and was passaged over a hundred times until a bacterial contamination of the culture almost ended it. Arthur Ebeling, who also worked at Rockefeller with both of them, took care of Carrel's valuable culture and washed it with nutrient solutions and consistently applied aseptic techniques. Antibiotics like penicillin were still unknown. Ebeling was successful and was able to save the culture, which could be kept alive until 1946, two years after Carrel's death. This cell line was phenomenal for scientists, as it showed that it was possible with the right nutrient media and the right hygiene. But that was only half the truth, as Francis Peyton Rous (1879–1970) was to show.[237] He developed a biochemical method in which cells could be better isolated by using an enzyme that we all have in our small intestine. The enzyme was supposed to be much better than any scalpel, no matter how fine. Trypsin is an enzyme that breaks down proteins into small peptides, which are important for digestion, absorption, and utilization as amino acids in our body. You put the tissue pieces in a trypsin bath and wait until the tissue piece dissolves. It is assumed that many of the isolated cells did not survive this process, but the basic technique of digestion was so convincing that it can still be found in laboratories today. Rous remained connected to tissue culture his whole life and dedicated his research to the treatment of virus-related tumors, for which he received the Nobel Prize in Medicine in 1966. It was probably the longest incubation period for a scientist, as he had already managed to produce cancer in other chickens by injecting a muscle tumor 50 years earlier, in 1916. There were two small miracles, as the rooster with the specific tumor as a model was brought by a farmer from near Baltimore. The second miracle was that the committee still wanted to remember his work 50 years later.[238]

The foundations of modern tissue culture were laid at the beginning of the 20th century. The preparation of tissue from living animals and humans was successful, the understanding of technical requirements such as oxygen supply, asepsis, and composition of nutrient media was present, only tissue that did not die so quickly was missing. An initial solution to this problem was the study of cancer cells. These tumor cells were quickly discovered as immortal tissue that needed to be cultivated. If so-called primary lines from original biopsy material, which are mortal, were first obtained, research quickly moved on to tumor lines and from the 1950s to immortal lines like the HeLa cells. Today, more than 50 cell lines can be found in research. Typical lines are the MRC-5, WI-38 or Vero lines, which are also important in vaccine production. Early vaccine production since the 1960s could not do without Vero cells, taken from the kidney of the green monkey, to breed viruses. They are indispensable for the production of vaccines, in which attenuated rabies or polio virus are bred on feeder cells. Viruses can only multiply in foreign cells that they infect, reprogram, and usually kill to create billions of new viruses. Feeder lines are needed only for this purpose. Unfortunately, millions of monkeys had to give their kidneys and their lives for our vaccines. These in vitro cell cultures in drug search and production are indispensable in pharmaceutical research. They enable the screening of several hundred thousand substances in the shortest possible time and have drastically reduced the number of animal experiments. Mammalian cells are also of great benefit to biotechnology, as they are used to produce recombinant therapeutic proteins such as blood clotting factors, follitropin or erythropoietin, to have drugs against cancer, stroke, blood clotting disorders and much more in therapy.

The Chimpanzee as a Pharmacist—Even Animals Know What's Good for Them

Do animals know the benefits of medicinal plants? Proud owners of dogs and cats know that both animals eat grass on some days to make vomiting easier. The reason is simple and somewhat unappetizing. The hairs swallowed during grooming accumulate in the stomach, are not digested, and must be vomited out. It is unlikely that puppies and cats learn to eat grass from their mothers. It is more likely that it is a relic from the early days of the wolf. The wolf and tiger are the genetic ancestors that we humans have domesticated as dogs and cats. That wolves and tigers eat grass was already described in 1944. As fecal examinations revealed, the reason was not too much hair in the stomach, but the excretion of tapeworms.

Is it only our pets that take grass as a kind of medicine, or is the intake of other plants known in more animal species and can we humans learn from it? Many healers in all rural areas of the world report sick animals, but also insects, that specifically seek out plants to find relief. The researcher Benito Ryes from Venezuela knows that animals eat the seeds of the Cabalonga tree, which is known for its antiparasitic effect. Perhaps the seeds are appreciated by the animals to get rid of intestinal parasites like worms or to reduce their number. Of course, one could also argue that it is a random combination of food and plants that contain an above-average proportion of pharmacologically active natural substances. This is the case with the cute koala bears in Australia, who feed on eucalyptus leaves. Eucalyptus contains eucalyptol in the essential oils that the bears eat. As a side effect, the animals smell good and their fur is free of parasites like lice or fleas, because they do not like the oil secreted through the skin. On the other side of the Pacific, in Costa Rica, the brown howler monkeys do not suffer from intestinal worms like their relatives, the spider monkeys, because they eat figs, which the spider monkeys do not like. Figs are known for their laxative effect.

Animals follow their instinct and use the available food sources such as plants, insects or other animals if they like the taste and smell. Unlike us humans, who go to the supermarket and can satisfy their hunger from an incredible range of at least 15,000 products in an average supermarket, in nature scarcity is the only cook. In Mahale Mountains National Park, it was observed that local chimpanzee groups do not have 15,000 plant species on offer, but specifically seek out and eat up to 28 plants. This is highly exciting, as these are centuries-old medicinal plants of the local Tongwe tribe. An interesting question is who copied from whom here. Eloy Rodriguez (1947) referred to the behavior of animals that specifically seek out and eat plants for the treatment of diseases, as zoopharmacognosy and established a very current field of research.[239]

Chimpanzees in Tanzania have a similar taste to us humans and like both sweet fruits and bitter roots and leaves. Above all *Vernonia amygdalina* is very popular. Sick chimpanzees pluck the leaves, bend the stem and suck out the very bitter juice, which they would normally despise. The plant is so effective that the monkeys are healthy again after 24 hours. Researchers wanted to know exactly and caught some of the animals to examine their feces. They isolated worm eggs and cultivated them with the extracts of *Vernonia amygdalina*. The experiments clearly showed that the extracts killed the worm eggs and freed the chimpanzees from their plague. The question arose why the animals removed the outer leaves and did not eat them right away. The assumption was that the concentrated parts of the plant were the juice inside, and that was also true. A surprise brought the analysis of the leaves. Here the steroid glucoside Vernonisid B1 was found, which is very toxic for chimpanzees. It is not surprising that the same plant is also considered very poisonous by the Tongwe tribe. The chimpanzees perceived the toxic effect as less strong, which might be due to a different metabolism. By the way, the clever animals pass on the knowledge of their usefulness against worms, amoebas, malaria and stomach problems from one chimpanzee generation to the next.[240]

But isn't that too simple? All plants that are bitter do not have to have a healing effect. We also know that animals try many plants and leave the bitter ones aside. This developed into a feeding protection as a defense mechanism in plants to survive. But why do animals eat these bitter plants when they are sick? We do not know and maybe we will find an answer when we look at ourselves. We humans have the expectation that a medicine is only good if it tastes bitter. A folk wisdom that we learned as children from our parents. We always remember that bitter medicine made us healthy. Perhaps it is for the same reason why chimpanzees specifically seek out the plants in their environment that are particularly bitter. By the way, this also applies to laboratory mice that have never experienced real nature. With malaria infected mice were allowed to choose between normal water and water with the bitter substance quinine. They preferred the bitter water to get healthy again.[364]

African porcupines eat a herb with the cumbersome botanical name *Aeschynomene cristata* when they have digestive problems. The porcupines dig in the ground and eat the root of this bitter-tasting plant. A healer of the WaTongwe tribe observed the little pigs and wanted to know why they did this. The plant was known to the WaTongwe as poisonous and no one would have thought of using it as a medicinal plant. He saw that some porcupines were losing blood with their stool. But just a few days after they had eaten the roots, the bleeding stopped and they seemed to have regained their natural zest for life. He decided to recommend the root to his tribe members for intestinal infections, but they responded with speechless and incomprehensible rejection. Did their shaman and healer have no idea about the toxicity of this plant? The healer demonstrated it to them by taking a small amount of the powder himself. His survival and the absence of poisoning symptoms convinced them. In the laboratory, the plant was tested and not only against intestinal infections, but also against infections, especially gonorrhea, the extracts showed great success.[239,240]

A final example is the consumption of insects as medicine. In traditional Chinese medicine, caterpillars, which are mistakenly referred to as worms, are eaten against infertility and erectile dysfunction. This winter worm-summer grass (冬蟲夏草) is a dead caterpillar that wanted to develop into a moth. This is infected by a fungus in winter, which kills or paralyzes it in summer. Fungal hyphae grow from the dead head, producing medically interesting natural substances. In China, Nepal, and Bhutan, the dead caterpillars are collected, dried, and offered as powder on the markets. These caterpillar powders are popular as an aphrodisiac, but also against almost all other diseases, from harmless colds to diabetes to cancer. By now, the powders are so sought after that they are even offered in convenience stores like Seven-Eleven in Taipei, Taiwan.

Not only mammals know about the importance of medicinal plants. Birds also seem to take advantage of the benefits when building nests. In our latitudes, it is the starling that is looking for aromatic herbs when building its nest and weaves them into the structure of the nest. Its American counterparts choose yarrow, goldenrod, or flea herb because they know that the essential oil kills or keeps away ectoparasites like lice, bugs and other vermin. The sparrow in Spain performs a special feat. Researchers have observed some birds collecting cigarette butts and incorporating them into their nest.[241] Is it the lack of suitable soft nesting material that poses new challenges for the birds in the city? The filter certainly makes it comfortable for the young to rest, but it is probably the nicotine in the butts that has a strong insecticidal effect. Who doesn't know the story by Wilhelm Busch about the delousing of the dog in tobacco suds?

Into a barrel full of tobacco suds
They dunk him, skin and hair,
Even though he strongly resists
And is totally against it
Then in a stable
He is interned with caution,
Until, what one finds to criticize,
Gradually disappears.

Healing Earth from Pharmaceutical Mining

It doesn't always have to be plants that humans or animals use for healing or alleviating diseases. Often, animals are observed eating dirt, small stones, parts of termite mounds, or simply just soil. We humans also eat soil, and this has not only been observed among indigenous peoples such as the Aborigines in Australia and peoples in China and Africa.[242–244] No, even here in Germany, quite a few people swear by healing earth, obtained from ice age loess deposits, when they have stomach problems. The bookseller and later lay natural medicine practitioner Adolf Just (1859–1936, Fig. 13) suffered from a nervous disorder 100 years ago and the then still simple conventional medicine brought him no relief. He decided to completely turn away from civilization, became a vegetarian and moved to the forest near Braunschweig, where he lived solely on what the forest provided. He lived in a simple hut without electricity and water and slept on the uncomfortable wooden floor. Whether it was the back pain from this simple bed or the awareness of training his body remains his secret to this day. He threw all conventions overboard and performed his physical exercises naked in the forest, which prompted some walkers to call the police. Just wanted to live holistically and saw himself as part of the emerging nature, which also included the recognition of the healing power of the earth. After some time, however, he also left the forest and took on the personal concern and task of founding the Jungborn in 1896. With the purchase of 24 acres of land, it was the largest natural healing bath in Europe in the middle of the Harz Mountains. Up to 350 guests came to take cures, among them Franz Kafka (1883–1924). Many guests were treated with healing earth, which Adolf Just mined in the Harz Mountains. He founded the Luvos-Heilerde company, which initially found new loess deposits in the Harz and after the division of Germany in the Taunus. Today, several thousand tons of healing earth are mined and obtained there every year. Luvos is probably the only pharmaceutical mining company in the world (Fig. 14)

Fig. 13 Adolf Just (1859–1936) (Permission Luvos-Heilerde, Friedrichsdorf, DE)

Why animals eat soil may again have to do with stomach and intestinal problems. Soil or clay bind toxins in the stomach or intestine that cause discomfort when released by the wrong food or by parasites. Clay and soil consist, among other things, of bentonite, which can normalize the acidic pH value in the stomach and helps with diarrhea. For our domestic cows in the pasture, soil is a normal part of the diet. Soil can make up to 20% of the ingested food and cows like to eat torn out grass tufts with remnants of soil still attached. If you can, observe what the cow in the pasture is enjoying on your next walk. You will be surprised.

High in the Virunga Mountains, which are located in the African country of Rwanda on the border with the Democratic Republic of Congo, there are eight volcanoes. This area is also home to mountain gorillas and it is the spot where Dian Fossey (1932–1985) is buried. Fossey was found murdered in her hut on December 27, 1985. The death was never solved, but as for the motive, it is clear that it had to do with her fight against the economic exploitation of mountain gorillas by tourists. Her concern was to protect her gorillas and to research their behavior. Her favorites had a tendency to eat yellow volcanic stone. The strong animals traveled to the slopes

Fig. 14 Historical Luvos Healing Earth Production in Blankenburg, Harz (Permission Luvos-Heilerde, Friedrichsdorf, DE)

of the volcanoes, broke off pieces of the stone with their hands or teeth, and ground them in their large leathery hands. The finely ground powder was especially eaten during the dry season, when the diet had to be changed. When it became dry, the animals ate bamboo, lobelias, and ragwort. Ragwort in particular contains toxic alkaloids, which were bound and rendered harmless by the fine stone powder.

These insights, in a field almost unknown even to science, are already being actively used today. New Zealand and Australian livestock farmers like to let their deer and stags graze on pastures with plants such as *Hedysarum coronarium*, *Lotus corniculatus* and *L. pedunculatus* with high tannin content because they have to administer fewer deworming agents. How much we humans can learn from the behavior of animals, however, is questionable. Certainly, earth and loess will not help us cure complex diseases like cancer or diabetes. But we should remain curious, as many plants that are eaten by animals are still unknown to us and could provide an effective natural substance.

Biopiracy—the Curse of Colonization

European countries set out to turn African, Asian, and American countries into colonial empires. We know the British Empire, the French world empire in Africa, and the South American countries that belonged to the Spanish crown. In addition to soldiers and priests, scientists always went along, who were interested in the animals and plants in the new colonies. They not only brought back plants that they found beautiful and useful, like potatoes or corn, but also spice and medicinal plants, about which the indigenous population told them. The colonial masters made use of this traditional knowledge. They collected and dried plants or put them alive in flower pots to bring them to European greenhouses. From these plants, natural substances were later isolated, which meant enormous progress in the development of active substances. Morphine, cocaine, and quinine have already been written about.

What is so bad about us using these plants and natural substances, even if they are not native to our latitudes? The injustice lies in the fact that individuals and later companies enrich themselves with this knowledge and the indigenous communities, who generously share the traditional knowledge, do not receive fair compensation.[245] In the past, it was the knowledge about a certain plant that helps with infections or pain, passed down orally from generation to generation over centuries. Today, it is plants, microorganisms, animals, and their genes that live in a national flora and fauna and are of interest for the discovery of new drugs. At the time of colonial conquest, no one would have thought of their own behavior as wrong or even illegal. This theft of plants or traditional knowledge is now called biopiracy. What does the world understand by this term today? Simplified, this behavior can be described as unauthorized removal of living organisms of all kinds from foreign ecosystems or as unauthorized use of traditional knowledge, mostly from developing countries.[246] In both cases, it is important to know that biopiracy includes patents or commercial activities that are not compensated for by third parties from these countries. To protect these countries and the local population, who have the knowledge and the plants, animals, or microorganisms, international agreements were concluded early on, of which the Nagoya Protocol is the most well-known today. This protocol is an international supplementary agreement to the UN Convention on Biological Diversity and is intended to protect developing countries and enable a profit balance when commercial successes are achieved with biological material from their countries.[247] According to the current definition, which you have read above, all imported plants and natural substances from which we have obtained modern medicines would be a case of biopiracy. Legally, this is not the case, because the Nagoya Protocol sets a deadline until October 10, 2014. All plants that were already in laboratories or botanical gardens outside the donor countries before this date are no longer covered by the protocol.

Kew Garden—the Marigote Bay of Biopirates?

The island had seen wilder times before Johnny Depp filmed "Pirates of the Caribbean" here. The old pirate bay Marigote Bay reminds us of the time when pirates made the Caribbean unsafe and especially taught the Spaniards to fear. The times of the buccaneers are long gone and the small Caribbean state of St. Lucia lives today from bananas and tourists who invade like pirates and make bars and hotels unsafe.

After piracy with royal permission became insignificant for the English crown, the island no longer played a role and became more of a problem case in 1730, as England focused more on secure sea trade than on sinking enemy ships. The Caribbean and the tropical zone became an economic area where sugar, coffee, cocoa, rubber, sisal, and other plants were grown. They wanted to achieve greater things in the Empire and saw botany in a tentatively beginning industrialization as the starter of a "New Economy". The chemical industry was not yet advanced enough with its technical possibilities for mass production, and botany with huge plantations was supposed to help out. For the project of botanical plundering of the colonies, whose blueprint was the very lucrative triangular trade with slaves, tobacco, sugar, and cloths between Africa, North America, and the motherland, an inventory of the green treasures of all colonies was needed. It was supposed to be botanists and taxonomists who determined the exciting plants that English captains and local governors sent home.

In southwest London, Princess Augusta founded a botanical garden in 1759 on a piece of land in Kew, a former royal estate in the London borough of Richmond upon Thames. It was more of a pastime and a bit of a collecting passion for Augusta, who cultivated exotic plants here. It was fashionable at the European royal courts to plant plants from the colonies or from scientists they sent, if they ever came back and did not die of epidemics. Well-known names such as Alexander von Humboldt or Joseph Banks, but also companies like the Dutch VOC, which supplied a Swede named Linneaus with plants in Leiden, contributed to the ever-increasing species richness in Europe's botanical gardens. The collecting frenzy was huge and nobody asked the population in the colonies if they agreed with the theft.

But were the gentlemen of Humboldt and Joseph Banks and later Georg Forster also the first biopirates? No, botanical souvenirs can be found in history even earlier among all travelers. The most famous route of the first possible "biopirates" according to our current definition was the spice route, which Marco Polo traveled to China in the 13th century. This was followed by the Atlantic route of Columbus, who brought tobacco and other interesting plants to Spain on his first voyages. Everyone took what they found and filled the greenhouses and cabinets of curiosities of their noble patrons. Even today, you can find plants and trees from this time in the greenhouses of Berlin, in the Kew Royal Garden, in Paris, in the Missouri Botanical Garden and in the New York Botanical Gardens. Although the researchers of that time could be seen as biopirates according to today's legal understanding, research, the gain of knowledge was the focus of their travels and not their material enrichment. Biopiracy today is the deliberate theft of traditional knowledge or genetic resources in order to have an economic advantage without compensation.

As described in the previous chapters, the prohibition of the export of the cinchona tree from Colombia and Peru was punishable by death. Taking the silkworm out of the country was also punishable by death in China. The Brazilians were not squeamish either and threatened the death penalty for smuggling the plant or seeds of the rubber tree. Rubber is a drastic example that first brought a city and then an entire country to the economic brink. In the biopiracy crime story, Kew Garden played a significant role as a representative of imperial interests. The economic loss for the countries with a rich biodiversity is still enormous today. According to current estimates, about 29,000 plants have been taken from countries with high biodiversity in Africa, Asia and America by about a thousand companies and individuals with an estimated value of 60 billion US dollars without profit sharing.[245]

Even though in this chapter the Kew Garden is used as an example for collecting, exhibiting and economically exploiting, almost all botanical gardens in Europe and America were not fundamentally different and also thought about how their plants could be turned into green gold. Tragic was the role of the Jardin Colonial de Laeken near Brussels, which was established by King Leopold with the clear purpose of researching the plants of the Congo. The costs were transferred to the government of the Congo, the profits remained in Belgium. Every colonial power and each of their colonies saw the economic importance of plants, which is why botanical gardens were founded on then insignificant islands like Fiji (1886), Barbados (1886),

Jamaica (1886), St. Lucia (1889) or British Honduras (1894). The goal was to research economically interesting plants as systematically as possible.

The colonial botanical gardens had a certain autonomy within the British Empire, which depended on certain criteria. The "First Class Establishments" had to meet Kew criteria and each garden had to have a museum, herbarium, laboratory and library and be run by English directors. These gardens included those in Calcutta, Ceylon, Mauritius and Jamaica. In the ranking followed "Second Class Gardens" and "Botanical Stations", the latter were also run by native gardeners. This colonial system was a great advantage for Kew Garden. At the end of the 19th century, 700 botanists were active in the Empire, who were trained in Kew Garden. The garden was comprehensively informed by its offshoots, received all rights and had no costs to bear, which had to be taken over by the regional governments. The disadvantage was that they did not exercise direct control and did not know exactly what was going on in the gardens.[248]

The Kew Botanical Garden, today one of the most species-rich and beautiful gardens in the world, played a prominent role as it served the interests of a rising colonial power seeking to break monopolies in countries outside its empire. Agriculture, as the English politician Bruce wrote in 1910, was the real goal of the *"struggle of the nations of the northern temperate zone for control of the tropics,"* and so Kew Garden became political. Kew Garden played a role in research and agricultural sciences in the national interest, this research was called "Economic Botany". It was to become the hub for the desires from South America that were to be cultivated in the colonies of Asia. In the commemorative publication for the 100th anniversary of the garden, the purpose is clearly highlighted: *"One of Kew's tasks is to send plants of economic and horticultural value to all parts of the dominions and colonies where the conditions for their cultivation are suitable."*[151] With Joseph Hooker, who himself brought 7,000 plant species into the garden, where Joseph Banks had already archived 3,400 species from his travels with James Cook, commercial interest was paramount. The development of high-performance varieties for their own colonies was a state goal. The new imperial view of botany was also a springboard for rising to high political offices and for scientific careers. Today, genetic engineers and biotechnologists are in demand in the industry. 150 years ago, it was botanists and natural product chemists. Service for the crown in the colony as a botanist was very lucrative and promised interesting jobs with high decision-making responsibility, far from London bureaucracy, which were usually directly subordinate to the governor. The jobs were not without danger, as shipwreck was almost a daily occurrence and the reports of Kew Garden precisely recorded which plants from the colonies went down with the ships and which botanists were missing. Everyone knows the film "Mutiny on the Bounty", which transported breadfruit trees. These were to be brought from Asia to London to Kew Garden. A shipwrecked botanist was also reported because he saved himself in a lifeboat to Batavia in Indonesia and was involved here in the founding of the Dutch botanical garden Buitenzorg in 1817.

What Hooker did as director in Kew was the benchmark for the other European countries. Especially the Germans were very impressed, they imitated the English model and even surpassed it in the end. After a tough struggle with the colonial department of the Foreign Office and with modest means from the Reichstag, in 1891 the director of the Berlin Botanical Garden, Adolf Engler (1844–1930, Fig. 15) founded the Botanical Central Office for the Colonies in Berlin.[249] Engler was a thoroughbred Prussian, born in today's Polish Zagan, he studied biology in Breslau (today Wroclaw, Poland). Appointed to Berlin, he was the first director and founder of the Royal Botanical Garden and Museum in Schöneberg for four decades, and he initiated its move to Berlin-Dahlem in 1900. The German botanical gardens, Professor Straßburger wrote in 1893, *"are now called upon to promote colonial interests, and have as a model the magnificent achievements that the Botanical Garden in Kew near London can rightly be proud of."*

Fig. 15 Adolf Engler (1844–1930) (public domain, wikipedia)2

The "Notizblatt des Königlichen Botanischen Gartens und Museums zu Berlin", essentially a Kew Bulletin for the German Reich, appeared from 1895 and reported diligently on the colonial significance and especially on East African plants. Two houses of the Berlin Garden were dedicated to colonial affairs and corresponded with experimental stations in Africa. Although Engler was only in East Africa for a few days in 1902 and 1905, he dared his great opus with the title "Die Pflanzenwelt Ost-Afrikas und der Nachbargebiete".[250] From our current perspective, this study would be unthinkable, as Engler only had poor black-and-white photos, dried herbarium specimens, and a few plants in his new glass tropical house.

The new Institute for Agriculture in Amani in German East Africa with its team of botanists, zoologists, and chemists was already referred to as the future "Kew of tropical Africa" by a British visitor in 1904. Indeed, the institute founded in 1896 was the most modern agricultural research facility in Africa. It was to become the most species-rich and research-intensive botanical garden in the world. African poisonous plants were researched as well as insects and their control. The first African plantations with cinchona trees were grown here as well as eucalyptus trees from Australia or hemp from India. The German biopirates circumvented prohibitions like their English and French colleagues by growing the export-prohibited sisal plant from Mexico. The Botanical Central Office in Berlin-Dahlem not only collected and archived seeds and plants from the German colonies. It also sent German vegetables to German East Africa, today's Tanzania, in return. The Imperial Biological Experimental Institute Amani in the Usambara Mountains supplied the country and German colonialists with tomatoes, onions, carrots, and watermelons. The climate at an altitude of 800 to 2200 m with sufficient rainfall was well suited for this. But also the supply of the homeland with coffee, cocoa from Tanzania, and rubber from Cameroon were important research areas of the German scientists.[251]

After the establishment of the gardens in Kew, the area and first small greenhouses were used to store the collection that Joseph Banks, as a former botanist and companion of James Cook, brought back from their joint sea voyages to London. He is said to have grown and propagated the plants he brought with him in Kew. With so much expertise, he was immediately appointed the first director. He was given the honor and the task of organizing and building the garden, which took a good 30 years. After Banks, William Jackson Hooker (1785–1865, directorate from 1841–1865) and his son Joseph Dalton (1817–1911, directorate from 1865–1885) followed, who saw the garden as a place for the collection and exhibition of various plants, but also gave it an economic and political function with the cultivation of industrially useful plants such as tea, coffee, rubber, or coca. The import of rubber from Peru and Brazil seemed to explode, as the Empire paid 300,000 pounds in 1854 and already 1.3 million pounds twenty years later, which reduced the currency reserves. There was a good success story of biopiracy with Cinchona, and why should this not be repeated with rubber. Father and son Hooker gave Kew Garden a new task in the time of competing economies on the European continent. Their garden was to be the central hub for the identification and research of still unknown plant species that were sent to London, and they took on the role of mediator between the local growers and "Plant Hunters", as the biopirates called themselves.[151]

For the rubber plants *Hevea brasiliensis*, it becomes clear how Kew Garden functioned. All botanists of the Empire knew that an alternative for South American rubber had to be found. British governors in Zanzibar, Niger, Ghana, and India had rubber-bearing plants like *Landolphia owariensis* (Ghana) or *Funtumia elastica* (Zanzibar) searched for and sent seeds to London. The seeds were grown, the content determined, and if advantageous, the plant was propagated. Seeds from these local sprouts were again sent to colonial botanical gardens, which were all organized according to the model of Kew Gardens. Kew Garden sat like a spider in the web and controlled and advised the cooperating gardens. This gave rise to the "New Botany", as Hooker junior or "Botanical Imperialism", as historians called the exploitation of colonial crops. Hooker was just as interested in medically relevant plants and new food crops like cane sugar, which was to be grown in Queensland, Australia. He promoted the cultivation of kapok in Burma as a cotton substitute, plantations of palm oil plants in Malaysia, and those of the tea plant in Ceylon.

With the shipped plants, harmful fungi, parasites also arrived as blind passengers from one continent to another, and the damage was huge for some regions. Known are spreads and economic setbacks for Sri Lanka, Indonesia, Malaysia, and even Fiji, where attempts were made to grow coffee. Much worse was the fact that the introduced pests could spread massively, as they lacked natural enemies. "*There is no doubt that this pest was introduced into the Botanical Gardens on imported plants—whether from Kew, or some other country remains an open question [...] All our most serious insect pests are imported ones.*", so complained the Ceylon Observer.

The King of Biopirates

The feather collector and adventurer Henry Wickham (1846–1928) was a daredevil, and that didn't keep him in his English homeland for long. At the age of 20, he set off for Nicaragua to explore nature. He became known as the feather collector because he found the colorful world of birds very fascinating and it was the latest craze in London to stick one of these colorful feathers on a lady's hat.[252] Fashions come and go and his feathers were soon no longer in demand. He was not short of money, as he had a job as a forestry official in the British colony of Honduras, but he wanted more. He quit, booked a ship passage to Brazil and tried unsuccessfully to become a rubber farmer on the Rio Tapajos, about 500 km east of Manaus, the city of the rubber barons. Wickham was a passionate writer and wrote travel reports that made it to the desk of Joseph Dalton Hooker, who came up with the idea of rubber cultivation in a British colony in Asia over a cup of tea. The two got in touch and Hooker asked Wickham to smuggle rubber seeds from Brazil to London for money. A life-threatening assignment, because if the smuggling was discovered, there was no one to literally save his neck. Was he really that daring? Later, after the successful coup, Wickham told that he had stored the seeds in a gunboat directly behind the bow below the waterline next to the ammunition. If the smuggling had been discovered, he would have blown up the boat. Underwater, the cargo would no longer have been findable. Wickham falsely declared his cargo as orchid seeds and smuggled 70,000 seeds to London. Hooker and his commissioned smuggler were lucky, as only eight seeds sprouted and thrived splendidly in the greenhouse.

Nevertheless, another 20 years of research in Kew Garden would pass before the seedlings could be brought into the world. The eight plants formed the basis for 2,700 seedlings, of which 1,900 went to the Botanical Garden of Colombo in present-day Sri Lanka, 18 to the Botanical Garden of Bogor in Indonesia, and 50 to Singapore. The original eight seeds from Kew Garden that were brought to Britain still form the genetic basis of all rubber trees outside Brazil today. The breeding was continued very successfully in the Heneratgoda Botanical Garden in Colombo and developed into a trading place for young rubber trees throughout Asia.

The trees thrived splendidly in Asia, as they had no natural enemies or pests and yielded a higher yield due to denser planting. At the end of the colonial period leading up to World War II, only 14,000 tons came from Brazil, compared to a massive amount of 1.1 million tons from Asian cultivation areas. Similar dense plantations could not be grown in Brazil and large distances of up to a hundred meters had to be maintained due to pests, which was very bad for business.

The rubber barons were rich. Exotic animals were kept, there were private yachts, racehorses were doused with champagne, and one baron's horse stable was so opulent that he moved in himself. Money was no object. That's why the decadent image of lighting a cigar with a banknote originated not in New York, but here in Manaus. When rubber from Malaysia hit the world market, the Brazilian rubber barons panicked. Brazil's Paris was on the verge of economic collapse. Manaus was the city of trade, which brought a lot of money into the Paris of the Tropics. It was the only city on the continent that had continuous electric light and a functioning drinking and sewage system available. The Opera House was built with Italian marble and tiles from Alsace. Even the opera singer Enrico Caruso (1873–1921) performed here, despite his panic fear of malaria. The opening took place on New Year's Eve 1896 and the ladies arrived in fur despite the tropical heat. Around the opera, paving stones made of a rubber-sand mixture were laid to avoid disturbing the guests with horse-drawn carriages. The splendor of the opera house faded with the abrupt end of the rubber boom in 1912. The last performance took place in 1917 and the house was left to itself. The climate and termites turned the opera house into a ruin, but it was saved. After a comprehensive renovation, the opera house shines again in new splendor. For the second opening in 1990, the tenor Placido Domingo was invited, who was not afraid of malaria.

Biopiracy Today

At least 40% of our current treasure trove of medicines can be traced back to natural substances. Even though we live in the age of biotechnology and computer chemistry, interest in natural substances remains high. Their extraordinary chemical structures allow the pharmaceutical industry to test with a higher hit rate than comparable synthetic substances.[253] Statistically, a pharmacologically active structure is found in every 125th plant. Despite the good successes, interest in plants is declining, because many of the estimated 250,000 higher plants are related and their metabolic pathways are well known. Interest is now focusing on microorganisms and, of course, genes. Building gene databases is a gold mine, and selling or licensing sequenced genes is a strong business model. When a gene is sequenced, a machine analyzes the sequence of the four bases of DNA. It is the code of our genetic material. In this Eldorado, entrepreneur Craig Venter has emerged as a pioneer. His company collects organisms from ecological niches such as sulfur springs, geysers, or the deep sea and looks at which special genes exist. The Galapagos Islands, which belong to Ecuador, were also on the wish list. There was a memorandum of understanding between Venter and the Ecuadorian government about access and collection rights, but it was never signed. Nevertheless, numerous samples were taken and illegally brought to the USA for sequencing. Ecuadorian environmental groups accuse Venter and his company of biopiracy and demand their return.[254]

Rare and special gene samples also concern us humans. Pharmaceutical companies and universities have a great interest in obtaining "pure" genomes from indigenous peoples. The goal is to understand genes that are involved in our civilization diseases such as asthma, obesity, diabetes, or depression. Indigenous peoples, who have little contact with Western civilization, its diet, and medicine, are therefore of particular interest. The Australian genome company Autogen signed a contract with the government of the state of Tonga in 2000 to gain exclusive access to the genetic database of the inhabitants. Two years later, the Australian company already held 41 patents related to obesity and diabetes and claims to have licenses. Autogen is willing to pay licenses according to its own statements, but resistance arose in the population because they did not see their privacy sufficiently protected. A similar attempt was also made in Iceland and with the Yanomami, where the genome of the population is similarly homogeneous.

The behavior of the Chinese state towards its own population was also outstanding. In 2002, the genomes of 200 million Chinese in the north of the country were collected, who had not, incorrectly, or incompletely consented by today's standards. However, in all the cases described of sequencing human genomes, it is not so much about biopiracy, but about bioethics and the question of how to deal with such highly sensitive information from people. Spiritual questions also need to be clarified, as the Yanomami demand the blood samples of their relatives from the freezers of American companies and universities in order to bury them in their earth. Of course, there are many more interesting cases that would have to be described here. However, it would not be helpful to explain all of them here, as many patterns such as the theft of the plant, the insufficient documentation of genes, or neglect of indigenous peoples repeat. Therefore, I would like to highlight five cases from recent years that have a clear reference to natural substances.

Hoodia—the Thirst Quencher from the Dessert

For the San in the Kalahari Desert in hot Namibia, the *Hoodia* cactus was highly valued when they roamed and hunted in the desert for days and felt thirst and hunger. They dug up the cactus, which botanically is not a cactus but belongs to the dogbane family, ate the roots, and thus quenched their hunger and thirst. The South African research institute CSIR began research with the aim of identifying and marketing the thirst and appetite-suppressing extract or active ingredient. The researchers found the effective natural substance P57 and had it patented. The project was sold to the American company Pfizer for further development. Whether this project was too small or too expensive is not known, but the buyer Pfizer sold it to the English pharmaceutical company Phytopharm without considering the license fees for the San. When pointed out this mistake, the spokesman for Pfizer said he believed the San were extinct. Later, the Dutch company Unilever bought all rights to further development and negotiated a contract with the San for participation in success. The market would have been huge with an estimated turnover of at least two billion US dollars. However, since the steroid glycoside P57 turned out to be hepatotoxic, the project was discontinued.

Healing Curry

We love *Turmeric* when we visit an Indian restaurant and have curry and other delicious dishes prepared for us. In curry powder is *Curcuma longa* a component and the ingredients of this Indian plant are responsible for the yellow color and the pleasant taste. But turmeric can do even more: In 1995, the Indian (!) researchers Suman K. Das and Hari P. Cohly from the University of Mississippi filed a patent with the number 5.401.504 to market the invention of wound healing with turmeric extracts. The Indian Council for Scientific and Industrial Research (CSIR) objected and sued against the granting of the patent. The plaintiffs argued that the effect of wound healing had already been described in ancient Sanskrit scriptures. They were granted the right.

The Potion of the Gods from the Home Garden

Ayuhasca is a potion from the Amazon region, which is used in shamanic healing rituals. It is a psychedelic brew that can be cooked only from the vine *Banisteriopsis caapi* or together with the leaves of *Psychotria viridis*. The psychedelic ingredients are dimethyltryptamine alkaloids. Its use is probably known among indigenous people in the Amazon basin of Brazil, Venezuela, Bolivia, and Peru for centuries. However, this did not stop Loren Miller from patenting the Ayuhasca preparation of a supposedly newly discovered subspecies of the vine *Banisteriopsis caapi* with the new name "Da Vine" in 1986. In the patent, he protected the therapeutic areas of cancer, Parkinson's, and psychotherapy. His claim that he had found the supposedly unknown vine in his home garden in the Amazon region was audacious. The patent was revoked in 1999 after the umbrella organization of indigenous organizations in the Amazon basin (COICA) informed the public and international protests and lawsuits against the patent took place for four years. Everyone was satisfied with their success and could not believe that the patent was granted again two years later. Loren Miller now believed that his vine was composed quite differently in the analytical examination than the pressed sample specimens he had received from museums. The question is interesting why the International Patent Office PTO never asked whether the plant was patentable at all against the background of traditional knowledge. They finally decided in 2000 that Loren Miller had the right to use the patent. The patent expired in 2003 and can no longer be extended.[255]

Umcka What? The Strange Story of a Geranium

Dear readers, a very well-known and popular phytopharmaceutical in Germany was also accused of being an act of biopiracy. It is probably the medicine with the most unpronounceable name in the pharmacy range. The plant comes from South Africa and Lesotho and is mistakenly referred to as a geranium again and again, although botanically it belongs to the genus of pelargoniums. It was brought to Europe by an Englishman at the beginning of the 20th century and from this point on its story has developed like a crime novel. Unfortunately, the historical records are not very meaningful, as many documents are no longer complete, and were even classified as "confidential" by the British government until 2089 and are not accessible to the public. This phytotherapeutic with the name Umcklaobao® made it into the British Parliament. But make up your own mind about the biopiracy of the plant *Pelargonium sidoides* (Fig. 16), whose root extract you can buy as Umckaloabo® for colds.[256]

Fig. 16 *Pelargonium sidoides* DC (public domain, Wikipedia)

Out of Africa

Kagaitse looked at the frail man standing so thin before him that two additional holes had been punched in his belt. He came from a distant land on the other side of Africa, a place Kagaitse had never seen, but he had already met the English after the Boers. He heard the wheezing and coughing from the man's throat, who was asking him for help. He knew this lung disease, which many of his brothers and sisters suffered from. They had to toil for the whites in the soil of his Zulu land to satisfy their greed for sparkling diamonds and gold. Should he help him? His brothers would accuse him of disgrace for helping a white man who oppressed his people. But he was a human being, and the gaunt Englishman could not be blamed for the misery in his country.

"Englishman, I know what disease plagues you, and I can help you." The pitiful creature was named Charles Henry Stevens (1880–1942), born in 1880 in Birmingham and stood at seventeen years old before the mighty shaman of the Zulu, looking at him expectantly. "UmKhulane," Kagaitse simply said, and Stevens understood nothing. "UmKhulane," he repeated. "You suffer from a lung disease, you have a cough, fever, and shortness of breath. Look at yourself, you eat nothing anymore and are too thin. Do you also suffer from uHlabo?" Stevens became increasingly nervous. Perhaps he should not have listened to the South African doctor Mike Chichitse in Cape Town. He had advised him to come to this godforsaken place. He had already administered the reddish-brown broth that made Stevens vomit. But Mike swore by his whiskey bottle that the plant extract of this medicine man could cure him of consumption. How much he had to suffer on the long sea journey from England and his doctor at home said he had to go to a mild climate, away from the rain and cold of England. Only in this way could he fight his consumption. Unfortunately, there were no medications. "UHlabo?" He had never heard the word and shrugged his shoulders. "How do you say that in your language?" Kagaitse asked. He pointed to his chest, coughed, and grimaced in pain. Stevens understood immediately. "Yes, I also have chest pain, yes it hurts," he spoke the last words softly. "I can help you, but you have to stay with me for a long time. Do you have money?" Stevens nodded and pulled out a few pounds and shillings from the leather pouch in his dirty trousers. Kagaitse weighed the money bag in his hand. "Good, you will live behind my house. You are lucky, it is the dry season." Stevens went into the small drafty hut and chased away the chickens.[257]

He stayed with him for three months. They traveled into the vastness of Basutoland in search of this inconspicuous wine-red plant in the African grass steppe. It was unbelievable to him that he could look to the horizon but saw nothing and no one. In England, he always saw a house, a farmer, or smoke from a chimney. There was no hut here, no hill, just a few birds in the sky, the eternal horizon, and then this inconspicuous plant. It had dark green, kidney-shaped leaves and a velvety hair. This plant was easy to overlook and looked delicate. But when you dug up the roots, they were as thick as the carrots in Birmingham and as dark brown as a Guinness. Sometimes the medicine man took the large roots between his hands, which were as big as frying pans, and with his fingers, he broke the root open so that the marrow glowed fire red. He collected the roots and learned that the large roots were up to twenty years old and full of healing substances. He should put the plants with the young, thin roots back into the ground so they could regrow. Back in the kraal, they boiled the roots. He observed that the women boiled the roots with the milk of their goats. There was nothing else at the end of the world. When a bottle of whiskey came to the village once, the medicine man said, water is for the goats, whiskey for the people. Then he nodded, took a big gulp, and passed him the bottle.

Out of England

Stevens recovered from the alleged tuberculosis, which is why he was sent to South Africa. He must have been happy and surprised, for in Europe there was no medicine and this small plant, which later received the scientific name *Pelargonium sidoides*, had helped him. He named the plant Umckaloabo, packed his things and a few kilograms of the roots, and set off for home. Things were not going well in England, and he returned to South Africa in 1904 to fight as a volunteer in the Boer War with the Scottish Yeomanry and then with the Cape Mounted Police. When the war was over, he wanted to stay and opened a workshop where he repaired motorcycles. A fire destroyed everything. Nothing was insured and there was no money for a new start. What was left for him, how could he make money? He turned to an industry he didn't know much about, but he knew how to procure and organize things. Stevens produced the *Pelargonium* extracts that the medicine man had once shown him. Under the names Lungsava and later Saccom, he sold the medicine in a country where many tuberculosis patients ensured good sales. He did not become rich, but it was enough to live as long as he got roots. That changed, the root deliveries stalled or almost dried up. Frustrated, he gave up and returned to England for the second time in 1907. In cold London, he was asked why he had returned from the beautiful colony. Stevens did not want to admit that he was a poor pharmacist or an untalented businessman. He had given a lot of money and many of his medicines to the poor. He had been the victim of a nasty extortion or he lied that he had been arrested and fined for distributing alcohol to the population.

He did not give up and, with the support of some doctors, founded the CH Stevens Company in London in 1907. Since tuberculosis was considered incurable on the continent, he already saw himself as a wealthy man. His company sold the root extract as Stevens' Consumption Cure.[258] According to his statements, the recipe contained 80 grains of Umckaloabo root and 13 1/2 grains of Chichitse per ounce. Business was going well until the British Medical Association took notice of him. It was this persistent Mrs. Miller who wanted her 10 pounds back because she believed that the extract had not helped her husband, as he had died suddenly. She insisted on her money, as Stevens advertised his plant extract with the guarantee "no cure, no pay", and she was very dissatisfied. If Stevens had paid, the story of his herbal medicine would have turned out differently, but he did not want to pay, and the story took its course. The lady turned to the British Medical Association (BMA), which is still the country's medical organization today, and they checked whether Mr. Stevens was a doctor at all. This was not the case and the Association wanted to know more. They took up the case because they believed it was a case of drug counterfeiting.

At the beginning of the 20th century, the kingdom was flooded with many so-called medicines that had no effect. While many were harmless and promised healing for colds or simply served to strengthen, some were problematic because they fraudulently advertised with the promise of healing for cancer or tuberculosis. Was this extract from Mr. C.H. Stevens of 204 Worple Road in Wimbledon effective against pulmonary tuberculosis at all? What was this plant from the South African colony that this Charles H. Stevens advertised in the Daily Telegraph and the Manchester Guardian? This had to be investigated and the chemist Harrison was commissioned to check whether it had been prepared according to the methods of the British Pharmacopoeia. The chemist commissioned by the British Medical Association came to the following conclusion that *"the medicine extract was a clear red liquid, and the analysis showed that it contained 23.1 parts of alcohol, 1.8 parts of glycerin, and 4 parts of solids, about 1 part of tannin and 0.2 parts of ash in 100 liquid parts. The solid substance agreed in every respect with the solid components of the decoction of Krameria or a mixture of this decoction with a little kino. The formula seems to be: Rectified spirit… 23.7 parts, glycerin 1.8 parts, decoction of Krameria (1 to 3) to 100 parts. Or it can be made with a tincture of Krameria… estimated cost for 2 fl oz-1½ d'."* The unpleasant consequence for Stevens was that his Umckaloabo extract was included in the book "Secret Remedies". This book contained all the obscure and ineffective medicines of the kingdom and was a blacklist for doctors and patients of what they should not prescribe or take. The book was sent to all doctors in England and of course also distributed to patients who asked.

Stevens was taken to court and the judge found the evidence of the BMA and the commissioned chemist very convincing.[259] He ruled in favor of the plaintiff and the BMA and convicted Stevens. In the oral reasoning of the judgment, the judge described him as a fraudster and accused him of quackery. Stevens had exploited the situation of poor people who had only seen the newspaper advertisement. He had informed them incorrectly or not at all. It was scandalous that he had exploited the situation of the poor in this way. He concluded with the words: *"I will now say what I have so far refrained from saying. I think this is a deliberate and well-thought-out fraud. It is scandalous that poor people are made to give their money for the hope of healing for these quack remedies without value."*[260]

Stevens appealed, as he saw himself wronged. An appellate court upheld the appeal, and on October 24, 1912, the parties met for the first of two hearings. Stevens wanted his rights and compensation for his business loss and sued the British Medical Association. He was the only plaintiff ever to resist the inclusion of his person and his medicine in the book "Secret Remedies". Since his listing in 1910, he could no longer sell Umckalobo in the USA. The US Post fished all his deliveries out of traffic, and that hurt. The jury consisted of 42 jurors and a difficult trial was expected, as both sides wanted to bring in many experts to clarify the case.

Stevens' lawyer, Mr Tindal Atkinson, rose and made his plea: *"Honorable court, my client has only used plants that are not listed in either the British or South African Pharmacopoeia, so the analysis cannot be correct. One cannot find something if the plants are unknown here in England. Yes, if the chemist's analysis had been correct, then his client would indeed have committed a fraud with cheap plants, but he would have risked losing his reputation at any time."* The honorable lawyer Atkinson told the court the story of the miraculous healing in Basutoland, today's Lesotho, and how he had come across the South African plant. The African medicine man had given his client the roots of the plant with instructions to boil and drink the infusion. After just two months, he no longer felt any symptoms of tuberculosis and upon his return to England, his doctor declared him healthy. After his service in the South African army and police, he was very successful in producing an extract, not only with the substance of the African healer, but also with the addition of Chijitse. This addition was only important to make the extract more interesting in pronunciation. The Zulus like these click sounds and he wanted to take advantage of that. But, and this is important, he can prove that in South Africa many hundreds of people have been cured of tuberculosis. If necessary, he would have witnesses and doctors summoned before the honorable court. It is clear that the first trial caused considerable financial damage to Mr. Stevens.[261]

The court called Mr. Harrison to the witness stand, who as a chemist had conducted the first examination. He now admitted that it was not Krameria as a plant, but he really did not know what it was. He retracted the accusation that Stevens was a swindler, but stuck to his opinion that everything was quackery. A second expert was consulted, who was familiar with the histology and anatomy of plants. Edward Morell Holmes took on the case.

He examined a root and the plant that Stevens had given him under the microscope and described it as a plant from the geranium family. Edward Morell Holmes concluded that it was impossible to identify the material as Umckaloabo. He would say that it is a plant that occurs everywhere in the world. Further experts were consulted, who disagreed with Mr. Holmes' opinion. They agreed that it was neither Krameria nor cinchona bark. In the end, only one thing was certain: They did not know what kind of plant it was, but they knew that it did not come from Europe. In his closing argument, Stevens rose and delivered a fiery speech against the BMA, which he described as defamatory, and pointed to the sacks of roots standing next to him. The jurors could be grateful that he did not call several hundred very satisfied patients as witnesses. The jurors retired for deliberation and returned after three quarters of an hour to inform the surprised court that they could not agree on a verdict, no matter how long they deliberated.

The case ended in a draw in court and Stevens could not enforce his claim for damages. However, he stuck to his lawsuit, filed an appeal, and appeared in court again on July 16, 1914. All accusations and expert opinions from the first trial were brought up again. This second lawsuit was dismissed and he was sentenced to a fine of 2,000 pounds, including the costs of the proceedings, which Stevens is said to have never paid. The trial lasted only fifteen minutes and the jurors immediately believed the BMA. Why was this trial apparently so clear and quickly decided? The jurors had previously read an 891-page report from the BMA, which was presented to the British government, and they had received a two-page explanation, which mainly complained about the problem of cancer and consumption drugs. The doctors' organization had prepared well and convincingly argued that all these substances could not be found in the British Pharmacopoeia.

Stevens did not like this report at all. A trump card to refute it was supposed to be the African doctor Dr. Arthur Alfred Henry Bennett, who worked as a ship's doctor and surgeon in Liberia and was called as a witness by Stevens. He declared with conviction that Umckaloabo had also been taken in Liberia, where he had practiced like in South Africa for 22 years. He described Umckaloabo as a decoction of the leaves with water and salt and he believed that it contained a plant parasite that was a kind of antidote to cough.[262] The court initially believed him as a medical witness, but then the dark side of his person came to light. When asked about his professional career, the witness stated that he had worked as a surgeon, but also as a commissioner in Liberia. It got even more confusing, because the honorable doctor had no academic title, served three prison sentences and the real Dr. Bennett was in Australia. The fake Dr. Bennett was previously arrested on January 24, 1914 in Gravesend for forgery and pretending to be a doctor.[263] He was arrested. He acknowledged his arrest with the words *"Perjury! That's a nice charge."* He was tried in 1915 at the Old Bailey, a criminal court in London, because he had no residence in Gravesend. Unfortunately, no further information is available on how the trial was conducted and what his sentence was. Stevens had found the worst possible witness, who now also destroyed his reputation.

His second witness was the company's own bacteriologist Dr. Lord, who was employed for five shillings a week to write very personal letters with instructions for the patients, which were included with the medicine bottles. Dr. Lord lived in a Salvation Army home for alcoholics and his claim that Umckaloabo could cure consumption in 15 minutes was refuted by the BMA through their own investigations. These showed that it was not possible to cure people within 48 hours of taking Umckaloabo. The question arises why Stevens cited witnesses of such dubious reputation, or whether he completely underestimated the BMA.

He did not want to give up and went to court for the third time in 1915, but he was denied another lawsuit. Interestingly, economically, it did not affect him that his Umckalobao medicine was listed in the "Secret Remedies". Yes, he had to accept losses in sales, but people quickly forgot about the trial and the report. The BMA's poison list of false or ineffective medicines remained virtually unknown to the public.[264] Only the Times and the New Statesman reported, the others remained silent, as they did not want to destroy a business that brought in very high advertising revenues for questionable medicines annually. The hope for healing was too great, and, as the judge correctly pointed out, many of the poor patients could not read or could read very little. They understood newspapers, but a specialist book probably exceeded their intellectual abilities at the time. Nevertheless, Stevens became a wealthy man and the 2,000 pound fine, which was a lot of money at the time, did not really affect him. He sold his Umckaloabo extracts for about two pounds, which was equivalent to a worker's weekly wage. He had 50 employees who made extracts, capsules, and lozenges in Wimbledon. When he died in 1942, he left his family a considerable sum of 140,525 pounds.

Out of Germany

But how did Umckaloabo come to Germany? Did it simply become popular on the British Isles despite the ban? The missionary doctor Adrien

Sechehaye from Geneva played a significant role in its introduction, even though Stevens orchestrated much of it behind the scenes. The Swiss doctor learned about Stevens' remedy in 1920 and treated around 800 patients in his homeland with moderate success until 1930. He described 64 cases in detail and presented his findings to the Geneva Medical Society. The Berlin Charité became aware of the study and his work when he compiled it in 1929 and published it a year later. Sechehaye wrote the report in French and translated it into German, Romanian, and English. The English translation was quickly produced and published by Fraser&Company. Interestingly, the publisher's address is blacked out in these books and in the accompanying advertisements. What could this mean? Sechehaye claims never to have met Stevens, which is credible, as he did not want to be suspected of favoritism or collusion. If you look closely at the location of Fraser&Co.'s publishing house, it is located near Worple Street in Wimbledon, and no other books seem to have been published there. One might suspect foul play.

The German translation arrived in Berlin in 1931, where tuberculosis was a major problem. The further path of the Umckalobo medicine in the 1930s is largely unknown and requires scientific research in the archives. It is certain that supply from England ceased with the onset of World War II. It is also known that Stevens had major supply difficulties from Lesotho and South Africa during the same period. The Second World War brought business to a standstill. His son took over the business for a few years and sold the company to the German JSO-Werke, which had been founded by the pharmacist Johannes Sonntag (1863–1945). He bought Umckaloabo because it was an interesting finished pharmaceutical product that was easy to use. Later, the company was renamed ISO-Arzneimittel. It is located near Ettlingen and is now part of the Wilmar Schwabe GmbH.

Why was the Schwabe company attacked and accused of biopiracy? The company is alleged to have stolen indigenous knowledge of the Zulu and Xhosa and profited from it. The criticism in the 2000s was directed against the procurement of the Pelargonium roots and the granting of patents, which gave Schwabe the right to use and market ethanolic extracts. The plaintiffs argued that an ethanolic extract was not a novelty and therefore not an invention. Such techniques have been known for at least 100 years and are used with other plants. The patent would effectively result in a monopoly on extraction, and the indigenous population would no longer have the right to extract their own plants. But who was the indigenous population? The case became even more complicated. Because the Xhosa, who sued Schwabe, had migrated centuries ago into the area of the Pelargonium, where the San tribe had previously lived. They had taken over the knowledge about the Pelargonium from the San. Another valid point of criticism was the payment of the plant collectors in southern Africa. Collecting Pelargonium roots in South Africa is difficult and expensive, but the local collectors did not even earn the South African minimum wage. They had to dig out the deep roots in laborious and time-consuming work and sold them to middlemen who paid them about €1.20 per kilogram (as of 2006). These middlemen sold the kilo for €80 to Schwabe or other companies worldwide. Another reason was that the German company Schwabe did not have a permit to collect wild plants. The Karlsruhe-based company argued that it obtained the roots from a South African dealer who had a permit. In 2008 and 2009, public pressure became too great, and the company withdrew this and other patents to settle the dispute. Schwabe founded the Umckaloabo Foundation in 2010 to ensure that profits from indigenous knowledge are distributed fairly.

In addition to the issue of biopiracy, there is also the problem of sustainable procurement of the roots. The demand for Pelargonium roots is still very high, and it is estimated that up to 150 tons of roots are harvested annually. This puts a lot of pressure on the Pelargonium stocks. Lesotho already decided in 2004 to protect *Pelargonium sidoides*, which only intensified illegal wild collection. The plants were brought to South Africa and mixed with legal collections. Today's stocks are also protected in South Africa, and the company is trying to cultivate Pelargonium in South Africa, Mexico, and Kenya. In addition to finding the right climate, which is ideal in Lesotho and the Eastern Cape, the long growth time of the roots is also crucial. Pelargonium with the desired large and strong roots have to stay in the ground for up to twenty years to become perfect.

Pelargonium sidoides is representative of many other medicinal plants that are collected worldwide and marketed primarily in Europe and the USA as phytopharmaceuticals. According to WWF calculations, up to 80% of the traded quantities of medicinal plants come from wild collections, the origin of which is not clearly clarified. Only eight of the economically important medicinal plants are currently produced according to the rules of pharmaceutical cultivation. That biodiversity and sustainability are important issues in the plant-processing pharmaceutical industry does not need to be mentioned separately.

Are Drugs Only Bad?

Drugs harm humans because they can lead to physical and psychological dependence, which severely restricts or makes one's own actions impossible. Brutal, but honest. The massive dependence of people who regularly and compulsively consume cocaine, cannabis, alcohol, or medication often poses a great burden on relatives, friends, and of course society. However, the term addiction falls short, because in our modern world we also know the addiction to television, sugar, sports, work, or sex. Somewhat simplified, one could say that all substances or experiences that can stimulate our reward system in the brain are potential addiction factors. At this point, one could write an entire chapter or even a book about the societal significance and the neurobiological backgrounds of addiction, but that is not the aim of this work. Rather, I would like to describe the history of addiction using some natural substances such as cocaine and opium, as was already hinted at in the preceding chapters.

Are only we humans in nature so obsessed with drugs and intoxicating natural substances, if we also consider alcohol as a popular drug? With great cheerfulness, we remember the stories of Michel from Lönneberga, in whose house at least one drug party took place. The pets staggered drunkenly across the yard after eating spoiled cherries, because the fruits were sweet and the animals did not know what fermentation and alcohol is. Is the amusing description of the drunken animals the exception in the animal world or reality? Do animals get intoxicated and drunk when the opportunity arises? The Karlsruhe biologist Mario Ludwig (*1957) reports that hedgehogs in Great Britain have a massive alcohol problem. The reason: hobby gardeners set up beer traps to catch snails. This is perfect for the hedgehog, as the beer is served along with the snack. In the end, they lie drunk in the flower beds. Meerkats seem to have a high addiction potential, as they attack hotel complexes individually or in hordes on the Caribbean island of St. Kitts and steal high-proof alcohol from hotel guests or drink it straight away if they feel unobserved. By the afternoon, the meerkats are lying on the beach next to the drunken hotel guests, sleeping off their intoxication in sociable rounds side by side. The monkeys were brought to the Caribbean as pets with slaves 300 years ago, where they spread due to a lack of enemies. They got a taste for it because rotting sugarcane fermented and formed alcohol. Since sugarcane is no longer grown and tourism in the all-inclusive clubs offered an alternative, they got their stuff here. A study shows that about 15% of the monkeys on St. Kitts have a real alcohol problem and another 5% are heavily alcohol-dependent. But heavyweights also like alcohol. The fact that elephants are not averse to alcohol is shown by the film "The Desert Lives" and the scientific report by Ronald Siegel and his colleagues.[265] But what about real drugs? Are these also known in the animal world?

The Sami, an indigenous people in northern Sweden and Finland, have observed that reindeer like to eat fly agaric mushrooms and then wander disoriented through the swamps, making strange noises. The same phenomenon is observed on the other side of the Atlantic in Canada with wild caribou. The Khanty, who live in northern Siberia, are not sure why brown bears there feed on fly agaric mushrooms. They believe it is a stimulant when they have to present themselves during the mating season and the muscarine possibly boosts self-esteem. In any case, it encouraged this people to try the mushrooms and sent them into ecstasy. Today it has become a good, but also strange tradition. The mushrooms are rare and the ibotenic acid, which is a hallucinogen and is contained in them, is excreted through the urine. Depending on rank and hierarchy, the urine is collected in bowls and drunk again, which should be possible up to four or five times until the dilution no longer allows any effect. The shamans of the Khanty try to use the intoxication to communicate with the spirit world. There is no evidence, but drunken reindeer, intoxicated shamans, and sleds in the snow that can fly are supposed to be the reason for our Christmas idea that Santa Claus flies with the sled.

When we move from the cold regions of the earth to the warmer latitudes, we learn that other animals also enjoy drugs. In Tasmania, Australia, kangaroos deliberately invade poppy fields, eat the capsules of the poppy plant, and lie down in groups with their peers in the fields to sleep off their intoxication. In the Amazon region, the preparation Ayahuasca, which induces hallucinogenic states of consciousness, has long been described by local shamans. The plant *Banisteriopsis caapi* was studied and DMT was identified as a potent natural substance. Nature observers have filmed jaguars eating the roots and leaves of the plant and then rolling on the ground in a state of intoxication. Goats also like to get a boost. The idea of drinking coffee and being stimulated by caffeine, we owe to goats that ate the red fruits of the coffee bush *Coffea arabica* and became restless in the herd. According to legend, an attentive shepherd named Kaldi somewhere in the highlands of southwestern Ethiopia became curious and reported this to the abbot of the local monastery. He had the red berries brought to him, roasted them, and brewed the first coffee in world history. He named the drink Kahveh, which means stimulating or invigorating, from which the commonly used names Cafe or Coffee are derived. The monks of the monastery liked to drink the new black beverage because they hoped it would invigorate their nocturnal religious ceremonies. Islam, which prohibits alcohol, brought coffee to the Arab world, where it spread from Yemen. Coffee only came to us very late, supposedly with the Turkish invasion in Europe in 1683. More on this in the previous chapter about caffeine.

Excesses with alcohol and drugs in the form of plant extracts have been described since ancient times. However, a look at our time shows that the offers and addiction possibilities were rather limited 300 to 400 years ago. In Europe, the common people did not know plants from South America or Asia that had an intoxicating effect. This knowledge was reserved for the nobility or some scientists who lived in countries with colonies. It is likely that alcoholism was widespread among Europeans, and some may have turned to belladonna or henbane. However, the effect must have been too intense for prolonged use to be assumed. It was rather healing women, so-called witches, whom people believed took such devil's stuff and therefore talked nonsense in delirium. Witches had an undeserved bad reputation. They were the first healers, perhaps also pharmacists, because they had a little knowledge of the effect of medicinal plants. Overdose and ignorance sometimes led to rituals like the ride on the broom. We still tell our children about these rituals today and explain why Bibi Blocksberg can fly on a broom. It is likely that riding on the broom or moving the broomstick between the thighs had more the character of masturbation and self-satisfaction, which was triggered by the consumption of the extracts taken.[266]

Which new drugs have reached Europe that are very popular and have the potential to become addictive? Many now think of cocaine and morphine, but it was rather everyday substances that probably everyone has in their cupboard. I'm thinking of sugar and chocolate from South America, coffee from Ethiopia, tea from India or China, and tobacco from North America. The English quickly fell for sugar and tea. Queen Victoria is reported to have been an early victim of high sugar consumption and had not a single tooth left in her mouth at her death because caries had completely eaten them away. August the Strong in Saxony was not only strong but also obese and died of diabetes because he appreciated sugar. Coffee and tea became popular and their popularity as stimulating pleasures remains unbroken to this day.

It is reported of the Incas of South America that they chewed coca leaves and stuffed them into their mouths to slowly let the cocaine pass from the plant pulp into the bloodstream. Clever, because the tongue is a well-blooded muscle and absorbs cocaine very well. If cocaine is swallowed, enzymes in the stomach and intestines are active to split it and make it ineffective. The Indians did not chew with the intention of getting high, as doctors, then artists, and later the partying bourgeoisie did from the end of the 19th century. No, chewing coca leaves was to maintain performance at work, especially in the higher mountain regions. *"You take three pretty leaves from your coca bag and hold them together like a cloverleaf. This is called Quintu. You blow over the leaves, first for the most important Apu, the Ausangate,"* said journalist Sven Weniger for Swiss Radio in Peru in 2014.[267] "Apu" means divine and gives the spiritual aspect when it comes to climbing the 6384 m high Ausangate, explains mountain guide Diego Nishiyama. Because the Indians know about the performance-enhancing effect, they also give coca leaves to the llamas, who as ruminants chew the leaves for hours and serve as pack animals. The Indians could not imagine that the pale man would isolate a pure white substance from the plant, which was first welcome as a local anesthetic and then brought misery to the streets of the western world.

At the turn of the 19th to the 20th century, only a few people experimented with cocaine. We have met the most famous scientist and doctor—Sigmund Freud. He himself experimented with cocaine and administered it to his patients to better excavate their childhood memories, which he later called complexes, from the unconscious. Freud began to use cocaine in 1884 and emerged from his melancholy, which today can be scientifically interpreted in literature as mild depression. The cocaine gave him a pleasant feeling that promised him relief and drove away his worries and fatigue. Interestingly, he is said to have written his essay "On Coca" (1884) under the influence of the drug, in which he spoke very positively about the effect and harmlessness of cocaine. Of course, there was also justified criticism. One year after the essay was published, Prof. Erlenmeyer, an authority in the field of morphine addiction, spoke out against the use of cocaine. For him, cocaine was as dangerous as heroin, and he described cocaine as addictive as heroin. The impacts were getting closer. His friend Obersteiner claimed in 1886 that cocaine had driven some of his patients into dependency. The young Freud had published too uncritically and too quickly, concluded Freud biographer Ernest Jones (1879–1958), who also knew him very well as a friend since 1908.

Before the so-called drugs got their bad reputation, they usually started their careers as research substances in the pharmaceutical industry. Born innocent, but made a steep gangster career? That was certainly not the way of the classic drugs. At this point, the term drug should also be avoided, even if it is provocatively in the title. It is better to refer to these types of pharmacological substances as psychotropic substances. This meets the requirement that all active ingredient candidates of the pharmaceutical industry should become a drug that could have helped mankind. The term drug is borrowed from Old German, as it originally meant dried, or "drugged" plants. This is why you can legally buy drugs in the pharmacy today, but the pharmacist hands you a tea bag with chamomile or peppermint. Therefore, it makes more sense to speak and write about psychoactive substances when talking about consciousness-expanding substances.

We must rid ourselves of the notion that these prohibited psychotropic substances such as LSD, Ecstasy, cannabis, heroin immediately lead every person into severe physical and mental dependency and that life as a junkie is predetermined. Please don't misunderstand me and don't think that I, as a pharmacist, want to release drugs too liberally and downplay the dangers. But we need to change our view of these substances. We live in a society that is getting older and older. Women and men born from 2020 onwards have a very good chance of living to be 90 years old and older. The quality of life in old age will be good, but diseases or simply the decay of mind and intellect will shape us. Our parents and previous generations died early from cancer, heart attack, or the effects of hard work, so diseases like Alzheimer's, dementia and other neurodegenerative diseases did not even occur due to lack of lifetime. Even today, we see that mental and intellectual performance is declining in the group of people aged 70 and older. The justified desire to have drugs that slow down or stop this mental decline is justified and entitles the pharmaceutical industry to research. Many of the psychotropic substances certainly misused today have the potential to produce a consciousness-expanding effect in a positive geriatric sense.

Now one can argue about the term consciousness expansion. What is this expansion? Apart from perception, which is usually directly associated with drugs, the spectrum of possibilities is broad and does not necessarily have to include chemical substances. For Zen Buddhists, silent sitting is already an expansion of consciousness,[268] for others bungee jumping, wild dancing, holotropic breathing, sex or a marathon run that whips the endorphins up. For the friends of chemistry probably alcohol, nicotine or some of the synthetic or plant substances already mentioned in the closer selection. What I want to say, and I hope we can agree on the smallest common denominator, is that everyone has to find their own way to expand consciousness. I define this from the perspective of natural substances and their use as a means to enable clear, empathetic and perceptive thinking again in old age.[269] Unfortunately, the real world looks a bit different for mostly young people.

It is often overlooked that these people have probably not taken a therapeutic dose, but a toxic one, which inevitably leads to these mental and intellectual derailments. The physician also refers to this as an overdose, which was deliberately smoked, snorted, or injected. But what happens when we use these psychotropic substances in much smaller quantities, like only a tenth of the known usual dose? This concept, known under the term "microdosing", shows a completely different reality, and the physiological effects are impressive. As will be shown below for lysergic acid diethylamide (LSD), volunteers have reported good effects with microdosing. Shouldn't we rethink and give ourselves the chance to reassess the benefit-risk ratio of psychotropic substances in a new and unbiased way? What speaks against releasing these psychotropic substances like LSD, Ecstasy or Psilocybin for people over 75 years of age under medical supervision? Do we believe that these seniors will handle their lives irresponsibly, all become addicted and rob banks out of financial need with their walkers? Very few seniors are heavy alcoholics or excessive smokers. They want to improve their mental, emotional, and psychological fitness and have learned to handle psychotropic substances. It's about being able to concentrate better, lift the mood, improve friendliness, emotional perception, and not about diving into a parallel world because it was so cool as a student in the shared flat. Isn't it time to at least initiate such a discussion in our society?

Amphetamines—Drug or Medicine for the Fidgety Phil?

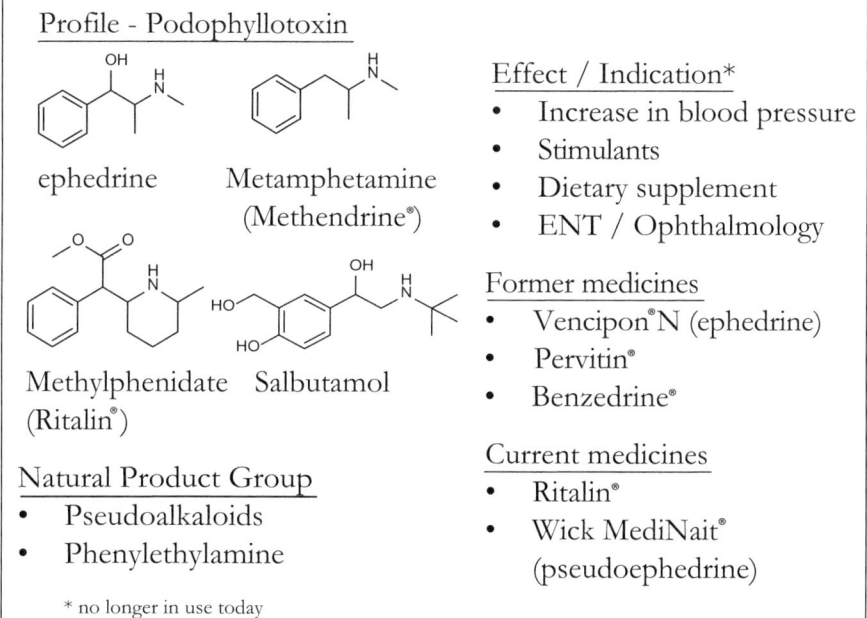

Asthma is one of the terrible diseases where you feel like you're suffocating because not enough air gets into your lungs. Readers who suffer from asthma know this disease all too well. To give the fortunate healthy ones a sense of what it's like for an asthmatic, try breathing through a straw for a minute. It will be very uncomfortable.

Bronchial asthma is usually an allergic, often chronically inflammatory narrowing of the bronchi. It leads to shortness of breath, coughing, and wheezing, which are referred to as asthma in Greek and have given the disease its name. Asthma diseases have been known since antiquity and almost all known doctors of antiquity have given treatment recommendations or tried to find the cause. Treatments with plant extracts, salts, or metals were not very successful and reflected the helplessness. Towards the end of the 19th century, when Henry Hyde Salter (1823–1871, Fig. 1) discovered that allergens such as cat and dog hair as well as hay can trigger asthma attacks, theophylline from the coffee plant and ephedrine from the ephedra herb were used as the first active ingredients. To the credit of cats and dogs: it's

Fig. 1 Henry Hyde Slater (1823–1871) (public domain, Wikipedia)1

not their hair that triggers an allergy, but proteins in their saliva that are transferred to humans when they lick their fur.

In Chinese folk medicine, also called Traditional Chinese Medicine (TCM), the plant Mahuang (麻黄) was considered very effective in the treatment of asthma. Botanically, the folk name *Ephedra sinica* and the active ingredient Ephedrine are hidden behind it. Although the discovery of ephedrine in 1885 immediately led to a therapy with the pure natural substance, it took another 20 years until Robert Leper Doig reported in February 1905 in a publication about adrenaline and ephedrine for the treatment of asthma. Both natural substances have side effects. They do not act selectively in the lung, adrenaline cannot be administered orally, ephedrine euphorizes and inspired the development of amphetamines. The breakthrough in modern asthma therapy was only achieved with Salbutamol and its modern successors. How long the journey was, how fiercely scientists fought for fame and money, and how quickly synthetic chemistry displaced the natural substances again, is shown by the exciting history of these natural substances.

Ephedrine

The discovery of ephedrine (Fig. 2) is attributed to the Japanese chemist and pharmacist Nagayoshi Nagai (1844–1929, Fig. 3), who studied medicine at the Dutch University Nagasaki, now a faculty of Nagasaki University, from 1866 onwards. In 1887, he isolated ephedrine and a chemically related structure, pseudoephedrine, from *Ephedra vulgaris*. The *Ephedra* plant is native to Asia, especially India and Pakistan, where it is known under the name *Ephedra sinica*. Outside of Asia, there are *Ephedra* species on all

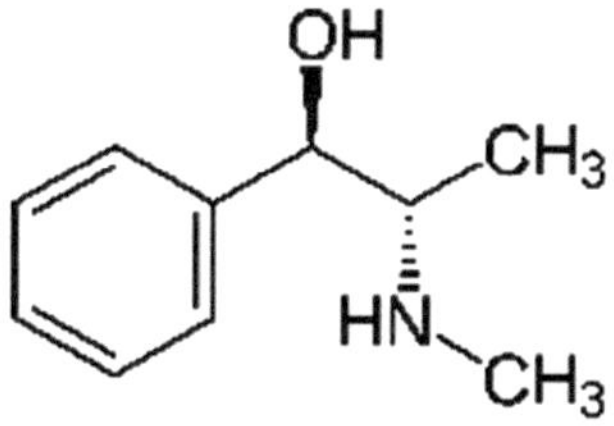

Fig. 2 L-(-)-Ephedrine (work of the author)

Fig. 3 Nagayoshi Nagai (1844–1929) (public domain, wikipedia)2

continents, but they contain fewer ingredients than the Asian species. In Europe, *E. distachya* (Fig. 4) is known for its very low alkaloid content.[270]

Nagayoshi Nagai was born to a Samurai family and grew up in the province of Awa on the island of Shikoku, which was considered the poorhouse of Japan until the Second World War. His father, Rinsho Nagai, was a herbalist and naturopath in the Western sense. Nagai lost his mother at an early age and his father felt guilty because he could not help her. As a passionate follower of Japanese folk medicine "Kampo", he turned to Western medicine and encouraged his son to learn Chinese and Dutch in order to send him to Nagasaki. Nagasaki was the center of knowledge transfer from Europe and America in Japan. The feudal lord Narihiro of Tokushima sent the young Nagayoshi to the city with a scholarship to learn about Western medicine: *"On November 25, 1863, the students visited the great Shinto shrine of Ise to*

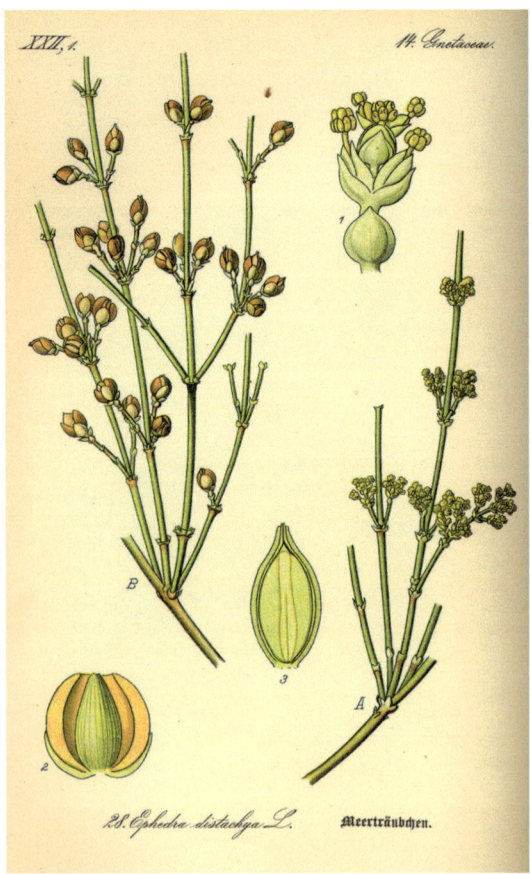

Fig. 4 *Ephedra distachya* C.A.MEY (public domain, wikipedia)1

pray for protection during their journey. Then at 8 a.m., Nagai set off with his fellow students Tachiki, Igi, Hayami, Yamamoto, Takeda and his servant Hanshichi from Tokushima. On the day of departure, he wore an undershirt, a shirt, a warm underjacket, an Obi, Hakama trousers, a black Haori jacket, shin guards, indigo blue Tabi socks and Zori sandals. He also had a long and a short sword, a lunch box, a pen, tobacco, and a pocket watch with him. He said goodbye to his friend Takarashi with a sake drink. He also said goodbye to his friend Izuki. So he left the village of Akui with regret ... The journey through wind and rain lasted 24 days," reports Ikuta Chiaki about the life of Japanese students in the late 19th century.[271]

His studies were heavily influenced by natural sciences and in addition to medicine with Prof. C. G. van Mansveldt, he took a liking to chemistry with Anthonius Franciscus Bauduin (1820–1885). Under the influence of Hikoma Ueno, with whom he lived in Nagasaki, he increasingly turned to chemistry, which attracted him more and more. Ueno later went down in history as the founder of Japanese photography. Nagai did not complete his studies in Nagasaki but in Tokyo, where he followed his mentor and teacher Koenraad Wolter Gratama (1831–1888), who awarded him the doctorate in pharmacy in 1871 with the topic "Contribution to the knowledge of eugenol". At the end of the 19th century, Japan opened up in great strides to catch up with the industrialization of the Western world. Having isolated itself from the rest of the world for centuries, only allowing Portuguese and Dutch through the port of Nagasaki, it abandoned this self-isolation at the end of the Edo period in 1867.[272] Japan wanted to connect to European knowledge and launched an educational offensive. It sent eleven scholars to Europe and Nagayoshi Nagai was allowed to officially travel to the German Reich as one of the eleven selected students. He received a postdoctoral scholarship in Prussia.[273] At the Friedrich-Wilhelms-University, he worked for 13 years, spending five of them with August Wilhelm von Hofmann before he moved to London. Officially sent to Berlin as a medical student, he had to study botany as a minor, which bored him and led to a break with medicine. His father had already taught him everything in botany, he did not want to continue medicine and secretly studied chemistry under the guise of being a medical student. He immediately stood out in chemistry and Hofmann and his assistant Franz Mylius (1854–1936) gladly took him in and took care of their only Japanese in the semester. In the lectures and in the lab, he received preferential treatment. *"This Japanese was quick with his fingers and enjoyed admiration,"* as Bernhard Lepsius (1854–1934) said about him: *"We often admired Nagai's manual dexterity. Thanks to the Japanese habit of eating with chopsticks, it was easy for him to pick up small objects from a beaker filled with liquid using two glass rods held in one hand."*

August Wilhelm Hofmann wanted to keep Nagai in Germany at all costs, introduced him to German women and encouraged him to marry one of them. Without Nagai's knowledge, he asked his subtenant Marie von Lagerström, who lived at Artilleriestraße 108, to act as a mediator. Many Japanese lived with Mrs. von Lagerström and she was called "Aunt Japan" by the young men.[272] The lady made a great effort and also found the suitable partner. Therese Schumacher (1862–1924) from Andernach was on her way back to Frankfurt and visiting Berlin. Marie von Lagerström paired the two over tea and Therese Schumacher was to become Nagai Therese (Fig. 5). But

Fig. 5 Nagai couple, right Therese Nagai née Schumacher, 1896 (public domain, copyright expired) 1

that took a while, because Nagai had to return to Japan for official duties in 1883 and on this occasion he asked his parents if he could marry. An outrageous request in early 20th century Japan, but they agreed. He happily returned to Germany in 1896 and married his love in Andernach. A carpet is said to have been rolled out from the Schumachers' house to the church. The two led a very equal marriage, raised their children Japanese and despite his love for Germany, Nagai could not get used to German food. An anecdote describes his love for home cooking. During a vacation in Germany in 1907, he fell ill and only Japanese Umeboshi plums helped him get back on his feet.[271]

As a bachelor, Nagai initially lived with the two years older diplomat Shūzō Aoki in Berlin, who also studied medicine in Nagasaki. Nagai poured out his heart and shared with him his thoughts that medicine might not be his great passion. Shūzō Aoki could very well empathize with his friend's situation, as he too had dropped out of his medical studies in Berlin and completed a very successful study of economics and politics. The switch from medicine to chemistry could not be hidden for long. The father in distant Japan was not pleased when he learned that his son would not become a doctor after all. He had placed very high hopes in him. Shūzō Aoki had to mediate and achieved that Nagai was allowed to finish his chemistry studies with the father's blessing. Shūzō Aoki himself took a liking to foreign policy and advised a Japanese government delegation that visited the German Reich on their world tour in 1871 under the leadership of Iwakura Tomomi (1825–1883). He must have done a good job, as he was not only the Japanese envoy in Berlin, but later also the foreign minister of his country.

Although the Nagais only lived in Germany for a short time, their interest in this country lasted a lifetime. The Japanese government had high hopes in him that he could organize the establishment of the pharmaceutical industry in Japan, and brought him and Therese back to Japan in 1893. In addition to his professorship in pharmacy and chemistry at the University of Tokyo, he was the chief engineer of the company founded in 1897, Dainippon Pharmaceutical, today's Sumitomo Pharma. Medicines were imported from China as Traditional Chinese Medicine (TCM) and only a few western medicines came from Europe. The imported products were of poor quality, as the manufacturers knew that there were no quality controls in Japan. Nagai was now faced with the great task of building an infrastructure to produce medicines in his own country and to allow a good supply of the Japanese population at a high quality standard.

Therese also remained in the academic world and took up a professorship in German language at the Japanese Women's University founded in 1901.

The spread of German cuisine and culture was a particular concern to her. Her husband supported her and promoted women at the university. He had seen in Berlin and at other German universities that from 1902 at least professors' daughters were studying there. The first women who learned chemistry and pharmacy in Japan and later also received their doctorates came from that women's university, where he built the Kosetsu chemistry faculty with financial support from Denzaburo Fujita and also taught himself. The Women's University was very successful. The former director, Mrs. Konodo, reported on Nagai's commitment: *"Professor Nagai's commitment to women's education was quite impressive. He always tried to achieve recognition and equality for women. And looking back from the current state, I am very grateful to Mr. Nagai. At that time, everyone said that a woman only needs to become a good housewife and a good mother. A woman's career was not welcomed. Mr. Nagai said: 'Enough with this restraint of women!' and always advocated for higher education and good training for women. He used to say that with better education they would also achieve better quality of life and a higher standard of living. He was very progressive and enterprising. What used to be phrases has become reality bit by bit. This includes the equality of men and women, and in this area Mr. Nagai was truly groundbreaking."*

As a pharmacist, Nagai had a brilliant career in science and society: He was the director of a high school in Tokyo, where German was taught as a foreign language for the first time in Japan, he became president of the Pharmaceutical Society of Japan, the Japanese-German Society, which he founded together with former Prime Minister Taro Katsura and his old Berlin acquaintance Shūzō Aoki, and he promoted the establishment of pharmaceutical institutes in Tokushima and Kumamoto. In short, he was the father of Japanese pharmacy. His work continues to have an impact in Germany today. In 1990, the Study Foundation for German-Japanese Cultural Exchange in NRW, based in Düsseldorf, was established by his grandson Teigi Nagai with a generous foundation. The aim of the foundation is to promote the exchange of young people from both countries and to deepen mutual understanding.

Let's return to ephedrine (Fig. 2) with Nagai. In 1883, he returned home to Japan by ship and arrived in Tokyo, where he was tasked with setting up the chemical-pharmaceutical institute. There, G. Yamanashi from Osada University had already isolated a substance from *Ephedra vulgaris* in 1885. However, due to his early death in 1887, he was unable to determine the structure, which is why Nagai took over the work. Shortly before his death, Yamanashi showed Nagai a black extract of *Ephedra sinica* in his office at Dainippon Pharmaceutical. On the surface, some white crystals could be seen, which Nagai immediately recognized. He had learned the tools of structural elucidation in Berlin, and it didn't take long for him to solve the puzzle of the structure in 1885, which he published in the German Pharmaceutical Journal in 1887. Since ephedrine has two spatial centers and therefore tends to form four different stereoisomers that can also interconvert, it is not a simple chemical substance. Strictly speaking, ephedrine is not a true alkaloid, as the nitrogen is not arranged in the ring as in morphine or quinine. In chemistry, we speak of pseudoalkaloids. If the two stereocenters are aligned the same way, both are left- or right-rotating in polarized light, it is ephedrine. But if one center is right-rotating and the other is left-rotating or vice versa, it is pseudoephedrine, which has a weak pharmacological effect. In contrast to Yamanashi, Nagai found these four ephedrine stereoisomers and understood their chemistry (Fig. 6). Once purified, he gave the ephedrine to his colleague Osawa Kenji and Mr. Kinnosuke Miura from the Medical Faculty in Tokyo, who were to determine the effect in dogs. More horrified than surprised, Miura reported to Nagai that his ephedrine was far too toxic to be administered orally to a human. Perhaps it could be used in ophthalmology to dilate the pupils and examine the back of the eye. Nagai concluded his work with the synthesis of ephedrine and put everything in the drawer of oblivion. There were no established research networks between research groups in chemistry and medicine in Japan, which is why he could not continue working with ephedrine. The collaboration between Nagai and Miura was the absolute exception. Japanese doctors did not trust research in their own country and wanted to continue imports. Since medical research in Japan was underdeveloped, the dusty drawer was not opened again until 40 years later.

Europe showed interest in this fantastic substance. The Darmstadt company Merck started an industrial process for the extraction and sale of ephedrine from 1886. However, this ephedrine was not pure and the two German chemists Ladenburg and Oelschläger discovered as early as 1889 that it must be pseudoephedrine. It was not until 1902 that the Englishman Miller succeeded in cleanly isolating pseudoephedrine from the European *Ephedra* species. He suspected that the Chinese *Ephedra sinica* preferred ephedrine and the European species *Ephedra vulgaris* and *E. helvetica* preferred pseudoephedrine.[274] Ernst Späth (1886–1946) and Rudolf Göhring from the University of Vienna synthesized ephedrine and all stereoisomers in 1920. With pure ephedrine, pharmacological research could be conducted much more effectively and the American-Chinese pharmacologist Ko Kuei Chen (1898–1988) and Carl Frederic Schmidt (1893–1988) resumed animal testing at the Peking Union Medical School in 1923.[275] Ko Kuei Chen

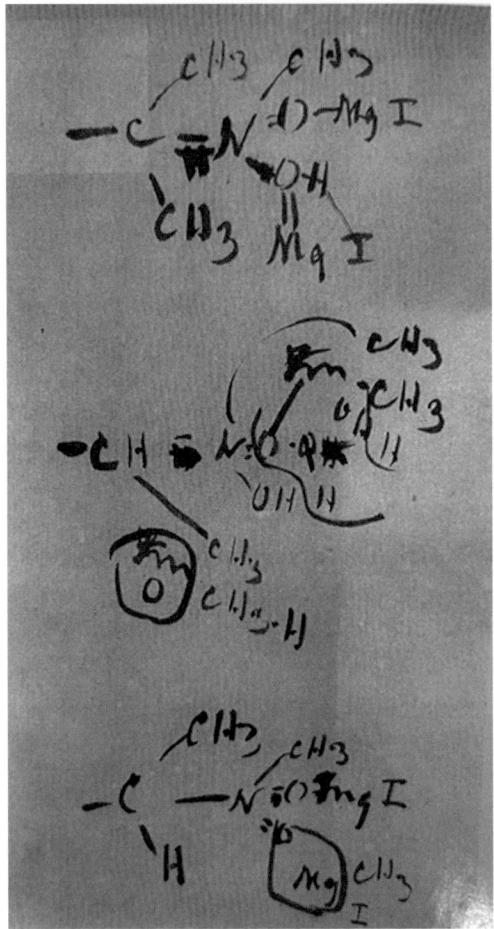

Fig. 6 Drawing of ephedrine by Nagayoshi Nagai in his lab book (public domain, copyright expired) 1

returned to Beijing in 1923 and met there the American Carl Frederic Schmidt, whose father had previously emigrated from Germany to the USA. Carl Frederic Schmidt taught at the Peking Union Medical School, which had been founded by American and British missionaries before the First World War and expanded with funds from the Rockefeller Foundation and academic support from Johns Hopkins University.

Ko Kuei Chen and Carl Frederic Schmidt extracted *Ephedra sinica*, which they received from northern China. They used 60% alcohol and sodium carbonate as a base and precipitated the alkaloid. This was dissolved again in hydrochloric acid with water as a salt, precipitated again with alcohol and purified. The physical principle of precipitation as a base and re-dissolution as a salt in a diluted acid is a well-known and proven principle in chemical process engineering. But not only in technology, but also among drug dealers, it is a known purpose for the production of the drug Crack. Cocaine is mixed with dissolved soda and the highly soluble salt of cocaine hydrochloride is precipitated as a base, collected and dried. The cocaine base is pure cocaine, very fat-loving and is now smoked. Since there is always a crackling and popping sound when heating in the crack pipe, called cracking in English, the consumers named the drug Crack. Smoking the base, which becomes noticeable in the brain after just a few seconds, gives the consumer an additional very intense kick. Unfortunately, a pronounced dependency can already occur after the first crack pipe.

Ko Kuei Chen and Carl Frederic Schmidt injected dogs with ephedrine and observed an increased heart rate and vascular contractions in the kidneys. The two were specialists in adrenaline (Fig. 7) and found in animal experiments that ephedrine and adrenaline have similar effects. They were very surprised that ephedrine not only worked for two hours like adrenaline, but almost twelve hours. Both immediately recognized the potential and sent the isolated substance to Leonard George Rowntree (1883–1859) in the USA at the Mayo Clinic. His clinical results confirmed the results of Chen and Schmidt, and Rowntree proposed to the American Medical Association in 1926 to approve ephedrine as an asthma medication. The first publications triggered an ephedrine rush and researchers as well as patients rushed to the new asthma remedy.

While only a few kilograms were needed at the beginning of the 1920s, demand increased to tons in less than ten years. In order to gain a market monopoly, companies tried to buy up the entire harvest of a year in China. This did not succeed, as the Chinese producers recognized their chances and the pharmaceutical industry very successfully produced synthetic ephedrine and other new asthma remedies. Later in the 1930s, patients were dissatisfied with the effect of ephedrine and pharmaceutical companies combined

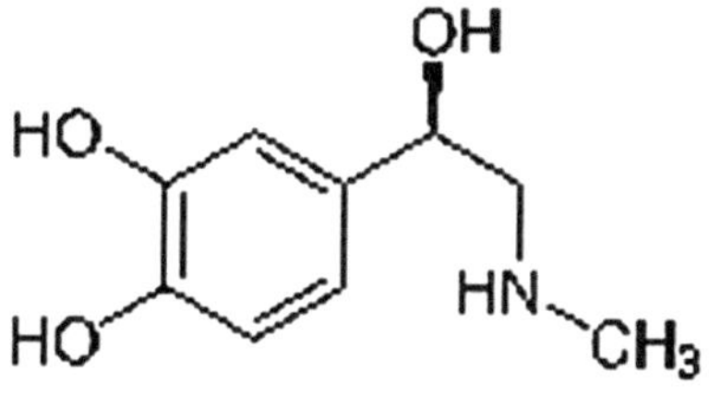

Fig. 7 L-Adrenaline (work of the author)

ephedrine with adrenaline in tablets. One reason for the dissatisfaction were the side effects after systemic administration, as doctors call oral intake. As an alternative, physicians considered whether it would not be better to inhale the active ingredient through the lungs. But from the towel over the head of the heated extract in the pot to modern inhalation systems, it was still a long way.

Giving an active ingredient or an essential oil directly into the lungs is probably as old as smoking, which mostly served cultic purposes. Later, smoking became a fashion and developed as a habit in many societies. In a world where smoking had become fashionable, there were no health concerns. The fact that smoking and cigarettes can also trigger lung cancer was only in the consciousness of most people since the 1950s. In Europe, the USA and China, people smoked merrily. Around 1550 BC, inhalation was already used for medical purposes in Egypt. The ancient Egyptians knew the inhalation of essential oils and used an apparatus that looks like a can and was found between the legs of a mummy in Thebes. It is also known what the deceased Egyptian inhaled: It was henbane *(Hyoscymus niger)*, which had a high content of tropane alkaloids and he probably took this to treat asthma. At first glance, one might consider the can to be a very simple and early precursor of the modern inhaler, filled with a very effective tropane alkaloid that has survived in a modified form to this day.

The inhalation of aerosols with tropane alkaloids was popular on the Indian subcontinent around 600 BC. Indian doctors recommended the therapy to their patients and the knowledge forgotten in the Orient diffused back and was thus rediscovered. Why inhalation was no longer practiced in the Arab world in the meantime is not known, but it may be related to inadequately functioning applicators that were not popular with patients. The famous doctor Avicenna (980–1037, Fig. 8), whose full name is ʿAlī al-Ḥusayn bin ʿAbdullāh ibn al-Ḥasan bin ʿAlī bin Sīnā al-Balkhi al-Bukhari, took up inhalation therapy and also described it in his Canon of Medicine (Fig. 9). If this rather unmemorable name of the Persian physician does not mean much to you, you may remember the book and the film "The Physician", in which he is not mentioned, but the doctor comes very close to the historical model.

The first vaporizer, reminiscent of the function of a modern inhaler, was a kind of teapot with a spout that was taken into the mouth to inhale the steam. It was conceived by Hippocrates, further developed by Galen of Pergamon, and used in China to vaporize *Ephedra*. Surprisingly, there was almost no further development in this field until the beginning of the industrial revolution, and smoking cigarettes with lung-expanding plant drugs

Fig. 8 Avicenna (980–1037) (public domain, wikipedia)

was the standard. A better understanding of materials, but also the plight of people, especially children, who quickly developed asthma in cities with high air pollution, may have been reasons for the now beginning development of inhalers. The English doctor Philip Stern described in his work "Medical Advice to the Consumptive and Asthmatic Peoples of England" the first recipes and mechanical inhalers that should be prescribed by his colleagues. But it was John Mudge (1721–1793), who continued Stern's findings and recommended inhaling opium juice for coughs, who first used the term inhaler in his book "Radical and Expeditious Cure for a Recent Catarrhous Cough".[628] He also had inhalers made that looked like a coffee pot, but had a small tube on the top of the lid that served as a mouthpiece. A hot liquid was poured in, the mouthpiece was enclosed with the lips, and the water aerosols, which carried active ingredients or essential oils, were inhaled. According to today's state of scientific knowledge, the penetration depth of the water particles into the lung was not sufficient. Particles or water droplets larger than 250 µm at most reach the upper branches of the lung. They condense on the oral or nasal mucosa, are swallowed, absorbed

Fig. 9 The first page with invocation of Allah from a manuscript of Avicenna's Canon from the 16th century (public domain, wikipedia) 1

in the duodenum, and act systemically via the bloodstream. The production of the right particle size is a high art in pharmaceutical technology, if the particles or droplets are smaller than 5 μm, they are exhaled with the breath stream. Somewhere in the middle lies the perfect particle size, which not only brings the drug to the finest branches of the lung, but also deposits it there and allows it to unfold its effect.

Pharmaceutical technology in the 18th century was not so advanced as to offer such a delicate technique to patients.[276] Not a few technicians were inventive and designed various inhalation devices made of metal and ceramic, and doctors believed they had found a way to treat tuberculosis and laryngitis by vaporizing chlorides, iodine and Cresol. But it was not until Helbing and Pertch presented an aerosol generator with a propeller in 1899 that was supposed to bring the fine droplets into the upper bronchi. This was the birth of sprays, which are still found in their basic physical form in modern inhalers today. Most inhalers were not ready for series production at the turn of the century and came as prototypes into hospitals, which is why the majority of patients still had to resort to asthma cigarettes.

A real advancement was the invention of the so-called Air-Jet nebulizers by the American company Riker from the 1950s onwards. They achieved a breakthrough in inhalation technology. The principle is that the droplets, when they flow through a nozzle, as we know it from a jet engine, are greatly accelerated and torn apart. These now very fine droplets penetrate deep into the bronchi and transport modern drugs such as corticosteroids, chromoglycin, and beta-sympathomimetics. For the drugs that are not water-soluble anyway, water is no longer the suitable transport medium. Organic solvents took their place, as they could dissolve the active ingredient well, were microbiologically clean, and physiologically harmless. These propellants consisted of fluorine and chlorine and were called fluorocarbon hydrocarbons, or FCKW for short. Their common property was that they destroyed the ozone layer in the stratosphere after being released into the environment. Chlorinated and fluorinated hydrocarbons are banned from spray cans and are now only permitted for medical purposes. They moderately affect our climate in relatively small amounts. But here too, there are further advancements and the FCKWs are being replaced by the chlorine-free HFKWs.

Chlorine is the culprit, as it causes the actual ozone depletion. Light splits the FCKWs and chlorine radicals are formed. These bind to an oxygen atom of the ozone, oxygen is formed and the two chlorine radicals combine to form chlorine. Gaseous chlorine always exists in pairs as Cl-Cl (Cl_2), which is split into two radicals in the stratosphere due to light. It becomes clear that even small amounts of chlorine are problematic in our environment over a longer period of time due to the constantly repeating splitting reaction. Worldwide, an estimated 600 million inhalation devices are sold annually. Each of these devices used to be filled with 10 g of FCKW. By replacing FCKW with HFKW in these inhalers, the release of chlorine into the atmosphere is reduced by about 4000 t.

We have already introduced a very modern inhaler in the chapter on scopolamine. I don't want to withhold a second one from you and explain it in the brevity of this book. For the comminution of particles, solvent and propellant-free dry inhalers like the Turbohaler® were developed, which crush the particles with the help of a piezoelectric vibrator. Piezoelectricity means that a solid substance, when under pressure, builds up an electrical voltage. The two brothers Jacques and Pierre Curie, the latter was the husband of Marie Curie, noticed that when tourmaline crystals were deformed, an electrical charge was created on their surface. When chemical substances are further crushed, they are physically very small particles, but they tend to cluster together because there are also charges on their surfaces that attract each other. The pharmacist speaks of agglomeration. You know this phenomenon from your own kitchen. When you bake a bundt cake, you sprinkle it with powdered sugar and not with granulated sugar from the sugar bowl. Chemically, both types of sugar are identical, they only differ in their degree of comminution. The very fine powdered sugar sticks together and it not only loves to stick to its own kind, but also to surfaces like the bundt cake. Very fine medicines are no different. To separate them again, the pharmacist then speaks of deagglomeration, he takes very fine piezo crystals or piezoelectric ceramics, applies an electrical voltage and sets the crystal or ceramic into very fast vibrations. This very fine shaking or shaking separates the tiny particles from each other again.

Ephedrine and adrenaline, both used in the 1920s for the treatment of asthma, were not in competition with each other.[143] The adrenaline from Parke, Davis & Co. and Hoechst AG was not orally bioavailable and had to be either injected or administered with a primitive inhaler. The inhaler was too expensive and the injections were unpleasant for the patients, which is why ephedrine was the favorite. Demand increased and the need could hardly be met. Ephedra plants for extraction were grown regionally limited in China, but when Japan occupied Manchuria in 1937, exports collapsed and other countries like India and Pakistan took over. Ephedra chinensis was now grown, harvested, and extracted on a larger scale. The Mediterranean variant with a lower content of right-handed ephedrine was found in Spain and up to 2.5 kg per day were isolated by Zeltia Laboratories SA in Porrino. At the same time, organic chemists developed syntheses for the production of ephedrine. Of particular interest is that biotechnology made contributions early on to solve the procurement problem. If yeasts are fed with benzaldehyde, they are given a biosynthetic precursor that transforms benzaldehyde into the desired ephedrine in a simple biochemical step. With the chemical and biological syntheses, the ephedrine crisis for the USA and Europe was over and cultivation collapsed again. Now the farmers in India and Pakistan were the ones to suffer.

The end of the era came with significant side effects. From the 1950s onwards, unwanted effects such as insomnia, restlessness, and anxiety became a problem in public health care, and professional journals reported on patients who took daily doses of one to two grams. Other drugs such as glucocorticoids and theophylline, which were now new to the market, were supposed to replace ephedrine. At the latest with the introduction of salbutamol, ephedrine was given a place of honor in the history of medicine and in this book.

Adrenaline

In the wake of the discovery of ephedrine, two other substances came to the fore and ultimately became very important drugs in asthma therapy. One substance was adrenaline, also called epinephrine, it was discovered as a hormone and neurotransmitter and plays a major role in physiology and medicine today.[277] The second substance is not a natural product, it was synthesized more or less by the way, forgotten and took the place of a legal drug with tremendous force, which after its ban and numerous chemical modifications became a worldwide toxicological scourge—amphetamine.

Let's look at the three structures shown before. You can see that amphetamine is very similar to ephedrine because it does not have a hydroxyl group in the side chain. Adrenaline has two more hydroxyl groups on the aromatic ring. These chemical differences also lead to different effects. Amphetamines stimulate the brain and adrenaline tends to affect the body and leads to the well-known acute physical stress reactions such as rapid heartbeat, pupil constriction, sweating, and others.

Everyone knows adrenaline as the substance that is released at the nerve endings during acute stress reactions and is equated with stress. This is correct, but in addition to the immediately effective stress molecule, we also know cortisol, which is detectable in higher concentrations in the blood during prolonged stress. Adrenaline was mentioned in the chapter on cortisol, but the neurotransmitter from the adrenal medulla is worth being brought onto the biochemical stage and into the spotlight. Drum roll and here it is, the adrenaline, which seems to be a close chemical relative of ephedrine. Its effects were described even before its discovery, but no one could classify them medically. It was a substance that Alfred Vulpian (1826–1887) found in the adrenal gland and dyed blue with iron salt in 1856. He was a physiologist and cut open the medulla of the adrenal gland to examine these sections under the microscope. At that time, it was very common to dye tissue or cells with dyes or salts. He noticed that the adrenal medulla and not the adrenal gland was dyed green, which he found very unusual. Iron salts were very popular dye reagents that form blue to greenish complexes with phenols, depending on how many hydroxyl groups are present in aromatics.

John Jacob Abel (1857–1938) was born in Cleveland as the son of farmers George Abel and Mary Becker. His mother died when he was 15 years old, and from then on he was plagued by financial worries. He had to drop out of his studies in the third year and worked as a teacher in La Porte, Indiana, USA. The only good thing in this difficult time was his acquaintance with Mary Hinman, whom he later married and with whom he had three children. After three years at the school, he returned to the University of Michigan, resumed his medical studies in 1883, and his passion for physiology was rekindled. Abel was fascinated by Germany, the heyday of physiology and the emerging pharmacology. He studied for six years in Baltimore, Leipzig, Paris, Vienna, Heidelberg, and Strasbourg, where he received his doctorate in medicine in 1888. He explained his long study time with the supposed lack of knowledge that had always tormented him. He wanted to know everything about medicine. He took his first professorship in Ann Arbor, Michigan, where he founded the first chair of pharmacology at the University of Michigan in the USA. His old alma mater in Baltimore saw his potential and wanted to bring him back. Abel liked the idea of returning and became a professor of pharmacology and biological chemistry at Johns Hopkins University in Baltimore in 1893. The subject of biological chemistry was very modern even 120 years ago, because in the minds of most scientists there was either biology or chemistry. Abel had one worry when he went to Baltimore. Who should be his successor in Ann Arbor? It had to be a distinguished scientist, because he did not want to see his chair disappear into oblivion. He returned to Strasbourg and consulted his former mentor Oswald Schmiedeberg, who recommended his assistant Arthur Robertson Cushny (1866–1926) from Scotland. Cushny was his postdoc, who had been researching in Strasbourg for a year. Cushny was an excellent choice, because he was to discover the filtration-reabsorption theory for the concentration of urine in the kidney together with Carl Ludwig (1816–1895) the filtration-reabsorption theory.

Abel is considered one of the early pioneers of hormone research and discovered a precursor and later adrenaline in the adrenal gland in 1897, after a guest had stolen the idea from him. The thief came from the Far East and was named Takamine Jōkichi (1854–1922, Fig. 10) from Takaoka in Japan.[278] This Samurai and chemist, with this knowledge, succeeded in isolating adrenaline in 1900. Who was this Japanese man, unknown in the scientific world? An unknown chemist who did not collaborate with universities, let alone teach or research. He was a member of a group of eleven students who, like Nagai in 1879, were allowed to leave Japan after studying in Tokyo and earning a doctorate at the University of Strathclyde with

Fig. 10   Takamine Jōkichi (1854–1922) (public domain, wikipedia)2

a government scholarship. In 1883, he returned to Japan and took his first position in the newly established Ministry of Agriculture. As a civil servant, he was allowed to visit the cotton exhibition in New Orleans in 1884 and brought back two souvenirs from the USA, a lump of phosphate and the love for Caroline Hitch, whom he married a year later. His government sent him on another official trip to Great Britain in 1887 because they had discovered the benefits of phosphate as a fertilizer for agriculture, and Takamine Jōkichi was commissioned to buy more of this raw material, which was to be further processed in Japan. He took the steamer back to the USA to Louisiana. There, Caroline Hitch was still waiting. He was sure that she was the woman for life. He proposed to his 18-year-old love and married her on August 10 of the same year.

Takamine Jōkichi changed his job and accepted the offer to become deputy director of the Japanese Patent Office. But that didn't hold him for long, he quit and founded his own company. Tokyo Jinzo Hirjo, today Nissan Chemical Industries, produced fertilizers for Japanese farmers, who bought them only hesitantly for fear of contaminating their soils. The business was not going well, and he thought about a second business idea that would fundamentally change his life. But it was not yet the knowledge about adrenaline, but about the enzyme Takadiastase. He isolated it from the mold *Aspergillus oryzae* with the Japanese name Kohji, which is used in Japan for the production of soy sauce and miso. In the USA, where the production of whiskey and beer was booming, the industry was urgently looking for enzymes that could better break down starch. Because more sugar from malt starch meant more alcohol. His Takadiastase could break down the starch in a third of the usual time. It also split bran, which was particularly cheap to buy as a by-product in the food industry and was attractive for US breweries.

Business in the USA called, his wife and he packed their things and returned to the USA, this time for good. Takamine Jōkichi sold his knowledge for a lot of money to the American company Parke, Davis & Co. He was now a wealthy Japanese man who considered New York his new home, but he remained a businessman. In 1897, he founded another laboratory, first on East 109th Street in New York and later on the other side of the Hudson River in Clifton, New Jersey.[63] The so-called laboratory in New York was 13 square meters large and consisted of a former janitor's apartment without ventilation. At the beginning of this start-up, there were only two employees, Takamine Jōkichi himself and a laboratory assistant from Japan. Takamine Jōkichi always excused the tightness with a reference to Henry Ford in Detroit. He had also started his career as an engineer in a small workshop. With the increase in work with adrenal medulla from the Meat District, the complaints of the residents on the second floor about the smell and noise became louder and louder. He decided to move, and Parke, Davis & Co. in New Jersey helped. More on that shortly.

How did Takamine Jōkichi come to steal ideas and why did the samurai want to isolate adrenaline from adrenal medulla? He was visiting the laboratory of John J. Abel and the young professor of pharmacology told him all too freely about his problems with isolation. John J. Abel, who had insufficient experience in the field of chemistry, isolated a strange product with the empirical formula $C_{17}H_{15}NO_4$. Feel free to calculate and you will find that there are too many carbons for adrenaline. Takamine became suspicious. Because he knew that Otto von Fürth in Strasbourg had extracted 0.4 g of the same compound with the empirical formula $C_5H_9NO_2$ or $C_5H_7NO_2$ from 2000 pig adrenal glands. Both statements were wrong, because neither Abel nor von Fürth had any idea that they were not holding the correct adrenaline, but an oxidation product. Takamine calculated and came up with the empirical formula $C_9H_{13}NO_3$. This aroused his ambition, and in the summer of 1900 he brought the Japanese Keizo Uenaka into the laboratory, a student of Nagai, the discoverer of ephedrine. He and the great Nagai, with whom he had traveled to Europe as a student, had a lot in common. Takamine's father was also a doctor, he left his parents' house at the age of eleven in the direction of Nagasaki, studied at the Dutch school with the same teacher Koenraad Wolter Gratama and began medical studies, which he also later dropped out of.

The summer of 1900 was hot and on July 21 the temperature rose early to 30 °C. Mr. Uenaka locked the cellar door of the brick building between Central Park and Hudson River, only the small but spotlessly clean sign pointed the visitor to the Takamine Laboratories. Mr. Uenaka was back early after a long night. He had not seen much of this city when he came from his homeland in the mountains of Hyogo prefecture to this metropolis six months ago. He took one of the test tubes in his hand and held it against the sunlight that fell through the small windows. Something moved, there were crumbs, maybe even crystals, that broke the light. Could he be lucky? All his experiments had failed the night before. "That can't be," he muttered into his beard. He filtered the small pieces, which he did not think were crystals, washed them with water and ethanol. He looked around, where was the reagent Ferric chloride? He found it next to the sink and dropped a drop onto a crystal with a pipette, which immediately turned green. That was the color that Alfred Vulpian had seen about 50 years ago. Uenaka was amazed, but he still didn't trust the matter. A second proof had to be provided. He repeated the dye reaction with Iodine solution and the crystals turned carmine red, just as Vulpian had described. He was now sure to have isolated adrenaline. Takamine came by in the morning and was immediately shown the test tube with the crystals. A deep look into his eyes and a stern question whether he was sure he had worked correctly convinced Takamine. Excited, he instructed his lab assistant Parke, Davis & Co. in New Jersey to ask if they could deliver more bovine kidney medulla.

Uenaka ordered the adrenal glands and sat down to write down the protocol in illegible handwriting. Did he want to prevent anyone from stealing the knowledge? Or did he want to save his boss from disgrace in case everything turned out to be wrong? Uenaka had no intention of showing his lab books before his boss died, and kept them locked away. Even after Takamine's death in 1922, he kept them hidden. In 1960, six years after his own death, his son found them and handed them over to Prof. Dr. Aiko Yamashita, who deciphered the illegible handwriting.[279] Today they are located in the National Museum of Technology in Tokyo and two things stand out: The red headline "About Adrenaline" and the name of the lab assistant "Wooyenaka". What had happened here? The chemical name "adrenaline" was not common until it was introduced to the market in the USA, because to this day the term epinephrine is used in the USA. Uenaka must have introduced this term retrospectively into the protocol. But was it Uenaka at all? Yes, he Americanized his name because the Americans never pronounced it correctly, so in New York he was Mr. Wooyenaka.

The adrenal glands arrived on July 30. The container contained 8 kg, which Uenaka immediately processed, because despite being stored in ice, the biological material spoiled quickly in this hot summer. On August 4, Uenaka managed to isolate sufficient quantities for an animal experiment. Mice lived in the lab. Takamine and Uenaka caught three of them and dripped an ethanolic adrenaline solution into their eyes. The mucous membranes and tissue around the eye turned white instantly. From the literature, both knew that adrenaline extracts have exactly this vasoconstrictive effect. Now they could provide the proof with pure adrenaline. This did not prevent Uenaka from conducting a self-experiment. On October 13, he dripped a watery solution in a ratio of 1:1000 (0.1%) into his eye, looked at the white eye in the mirror and made notes in his lab book: *"A thousandfold dilution based on water is correct for practical use."* This concentration was to be the standard for the next ten years.

Takamine sat in the small lab, looking at his colleague. "Are we sure we have the active ingredient that Abel calls epinephrine and wants to isolate?" Uenaka shrugged his shoulders. "I don't know, but it can't be otherwise. All the results suggest that we are correct." Takamine remained silent for the time being. "You know he has published six papers and is a luminary in this field. We are just two chemists in a drafty lab!" Uenaka rolled up his white sleeves. "Boss, he never isolated adrenaline. He probably isolated a derivative that he obtained with benzyl chloride. That's not a natural substance." Takamine thought and drank his cup of green tea. "You're right. We should convert our substance with benzyl chloride and see what we get. Do you have everything here?" Uenaka nodded vigorously, jumped up, ran to the chemical shelf, and came back with benzyl chloride. Takamine stood next to him and they synthesized the substance that Abel thought was his adrenaline. They sent the new samples and the unchanged adrenaline to Parke & Davis. On October 30, they had the third proof that adrenaline must have been in the shell.

Abel had not expected Takamine to overtake him, but he did, and a few years later, Takamine demonstrated in an animal experiment that his adrenaline was 2000 times stronger than Abel's or even that from Strasbourg. Abel reluctantly acknowledged his competitor's success: *"The years of effort on my part in this once mysterious field of adrenal and medullary biochemistry, marked by many errors, eventually led to the isolation of the hormone not in the form of the free base, but in the form of its monobenzoyl derivative."* The isolation of adrenaline was a mammoth task at the time. Takamine managed to extract four grams of adrenaline from bovine adrenal glands. In the summer of 1900, he was able to see the painstakingly extracted and purified crystals. He immediately recognized the potential, as he was very familiar with the research of his colleague Nagai and saw the potential for asthma. He filed a patent that he wanted to sell to Parke, Davis & Co. The company was not exactly thrilled when the Japanese man stood at their door with a vessel. Their employee Thomas Aldrich was working on the same project. He was practically at the same stage as the friendly samurai. He came from Johns Hopkins University in 1898, where he had worked with John J. Abel. They were just a few weeks away from having the knowledge and the patent. Now the friendly Mr. Tamakine was one step ahead of them. They had the adrenaline given to them, Thomas Aldrich checked the identity and quality, and they tested the adrenaline on their own animals in Detroit, where the company was based. The results must have been convincing. Because in a publication in the Journal of American Physiological Society in 1901, Aldrich wrote about Takamine: *"I regard Takamine's adrenaline as a great discovery. Shortly after Takamine, I also received some crystals. A few months later, I was able to produce a sufficient amount to perform an elementary combustion analysis. I soon found that Takamine's crystals and my own were similar, and upon further investigation, I found that the two crystals matched in every respect and were identical."* They invited Takamine and Uenaka to come to Detroit and began discussing the brand name. The epinephrine was to be marketed as a hydrochloride salt and called Adrenaline®. The name is derived from "ad" to and "renal" kidney. The ending "in" indicates that it is a drug like penicillin or insulin. The brand name has prevailed worldwide to this day and has displaced the name epinephrine. This is now only used in the USA.

Abel heard about the patent and was beside himself. The patenting of the samurai was unacceptable to him, no, it was theft! He believed in industrial espionage. Despite his incorrect isolation, he considered himself the true discoverer of adrenaline and wanted the patent rights. Abel went to court and argued that his adrenaline, like all natural substances, was always contaminated with other accompanying substances. These impurities would have made purification difficult, but the adrenaline was always in the contaminated fraction. From his briefcase, he pulled out seven publications that he had hastily written since 1901 to prove that he was right. He bought adrenaline and repeated his experiments and analytical methods to demonstrate

Takamine's alleged errors in structure elucidation. Takamine never responded. Today we know that Takamine's adrenaline was not completely pure and contained traces of norepinephrine as a "sister compound", but this knowledge was not available to any of the protagonists at the time. The court dismissed the lawsuit and ruled in favor of Takamine. Disappointed, Abel wanted to state his position again after his retirement and published a comment in the science magazine Science: *"After I had completed the investigations described in the [journal] and while I was still striving to improve my procedure, I was visited one day in the fall of 1900 (if I remember correctly) by the Japanese chemist J. Takamine, who examined with great interest the various compounds and salts of epinephrine that were presented to him. He specifically asked whether I did not think it possible that my epinephrine salts could be produced by a simpler procedure than mine ... Takamine returned to his own laboratory, produced concentrated gland extracts, and obtained chain-like crystal clusters by adding ammonia."*

The history of the extraction and structural elucidation of adrenaline is not yet complete at this point. After the elucidation of the simple structure of adrenaline, it was quickly synthesized, and a veritable flood of research and publications has followed to this day. Parke, Davis & Co. was not the only company that recognized the economic potential of adrenaline. In Europe, Otto von Fürst cooperated with Hoechst AG. The German company launched 1900 Suprarenin as a drug on the market, and today it can be assumed that it was not pure adrenaline. The head of the chemical department wanted to forego extraction from animal adrenal glands and started a synthesis project. The goal was the complete synthesis of Suprarenin. The work was successful, and Hoechst AG was able to register three patents by 1905 to defend itself against the ever-increasing number of competitors. The knowledge from the USA did not worry anyone. Because the patent of the American samurai referred to the isolation, not the synthesis, which the Germans massively promoted. The substances synthesized in Frankfurt were tested in animal experiments by academic cooperation partners in Marburg by Hans Horst Meyer (1853–1939) and his assistant Otto Loewi (1873–1961). Meyer already described a possible synthesis, which was then taken over by Hoechst. However, one question remained unclear: What about optical rotation and effect? There was a publication by Hermann Pauly, who had described this phenomenon in 1903 on extracted adrenaline, but the data changed with each new attempt. The work of other colleagues, such as that of Hooper Jowett from the Wellcome Chemical Laboratories in England, did not contribute to the clarification.

Friedrich Stolz (1860–1936) and Henry Drysdale Dakin (1880–1952) were able to provide answers with their independently conducted syntheses, their work was of particular importance for the industry and the market launch. Now the great Bayer on the Rhine also became aware of adrenaline and got involved. A project was set up and with its own syntheses and three patents, a piece of the cake was brought to Leverkusen. All good things come in fours, thought Nagayoshi Nagai in Tokyo. He found the chemistry of adrenaline exciting and developed another synthesis, which he also patented.

Another three years passed before Franz Fächer, who got a job at Hoechst AG, delved into stereochemistry and came into contact with the publications of Pasteur on tartaric acid. It was worth it, because with the support of his doctoral supervisor Emil Abderhalden (1877–1950) at the University of Berlin and his assistant Franz Müller, Fächer was able to send the pure D- and L-isomers to the pharmacologist Robertson Cushny in Strasbourg just two months later. Cushny tested in animals and showed that the left-turning form is much more active, increasing blood pressure and urinary urge in animals, while the right-turning form shows no significant effect.

Today, adrenaline is only used in a few medical cases. One indication is insect stings, when the affected individuals are highly allergic. These endangered people carry ready-to-use syringes to inject themselves with adrenaline if the sting closes the glottis and shortness of breath threatens. Every second is important if the emergency doctor is not on site quickly enough. Adrenaline, which is injected directly into the heart, is recommended in emergency medicine for acute cardiac arrest. Highly diluted adrenaline is mixed with local anesthetics. This achieves a vasoconstriction that delays the transport of the painkiller. Adrenaline also plays an important role as an antidote in cases of beta-blocker poisoning.

The exciting story of adrenaline continues, and it was very surprising when L. J. Haynes, K. E. Magnus and P. C. Feng from the University of the West Indies in Jamaica found noradrenaline in the plant *Portulaca oleraceae*. The same plant was recently reported to be valued in Asian folk medicine for treating asthma. It is certainly exciting to describe all this, but in the end, the two natural substances paved the way for synthetic drugs such as blood pressure-lowering beta-blockers or salbutamol. It remains astonishing with what willpower and with how little support the samurai and chemist pursued his goal. He did not collaborate with universities, but financed his research himself. He was a self-taught bio-process engineer to understand how adrenaline is isolated from biomaterial.

Salbutamol

The Scottish pharmacologist David Jack (1924–2011) discovered the active ingredient Salbutamol (Fig. 11), derived from ephedrine and adrenaline, for asthmatics. With his discoveries, he was probably the scientist who achieved the most for asthmatics. The discovery of Salbutamol would have crowned his career, Jack also discovered the principle of beta-2 receptor agonists and the first inhalable corticosteroid (Beclomethasone) in 1972. Salbutamol, which he found in 1969, is still the best-selling asthma medication of all time. The active ingredient was a breakthrough in asthma therapy, as thousands of people died annually from an asthma attack in Europe and the USA.

In medical textbooks, Salbutamol is referred to as an adrenoceptor agonist that acts very selectively on beta-2 adrenoceptors with a 500-fold higher affinity than adrenaline. This important receptor type will be briefly explained. In our body, there are three types of beta receptors. Firstly, the beta-1 receptors in the heart, the beta-2 types, which are mainly found in the lungs, and less significant beta-3 receptors on brown fat cells, which are important for the breakdown of fat and heat production. For medicine, the beta-1 receptors in the heart are important, as modern beta-blockers for lowering blood pressure target these. The stimulation of the beta-2 receptors in the smooth muscle like the lung leads to relaxation and the cramps in asthma decrease.[280,281]

Jack David worked for the British pharmaceutical company Allen & Hanbury, which became a subsidiary of Glaxo in the 1960s, experiencing rapid development in post-war Britain. Glaxo was to evolve from a company that produced milk powder to one of the most innovative pharmaceutical companies in the world. Sir Jack David, who was knighted by Queen Elizabeth II in 1993, played a major role in this. He was the son of a miner from Markinch Fife and it is a coincidence that in the same year just a village away James Whyte Black (1924–2010) was born, who received the Nobel Prize for Medicine in 1988 for the discovery of beta-1 receptors.

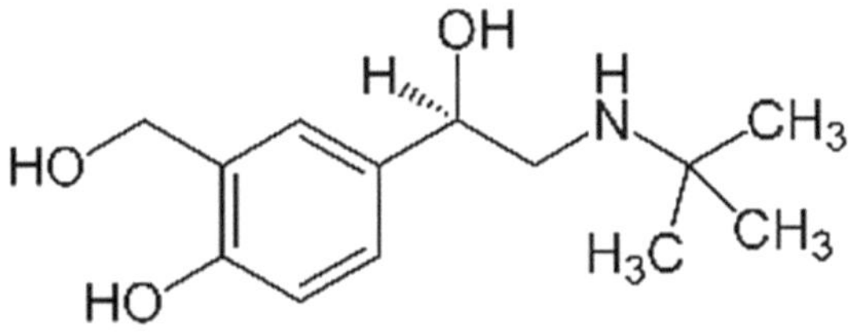

Fig. 11   R-Salbutamol (work of the author)

Jack completed an apprenticeship at Boots, studied pharmacy and pharmacology at the University of Glasgow and at the then Royal Technical College. He found his first job in 1951 at Glaxo, which he did not like. In 1953, he moved to the American company Smith, Kline and French (SK&F), where, by coincidence or not, James Black also researched from 1964 to 1973. There, Jack developed waxes that were used to coat medications. This must have been too boring for him as well, so he decided to do a part-time doctorate at Chelsea College of Technology, later University of London, with Professor Arnold Beckett. The qualities of the young man and scientist quickly became known and in 1961 he was offered the position of Director of Research and Development at Allen & Hanbury, whose laboratories were located in Bethnal Green, London. After his odyssey, he returned to Glaxo when Glaxo took over Allen & Hanbury.

David Jack was hired as a pharmacist, who until then had been making currant pastilles at Allen & Hanbury. Together with Paul Girolami, Glaxo's CFO, he focused on the development of common diseases such as respiratory, cardiovascular, and digestive diseases. Jack was very interested in the emerging research on the newly discovered receptors. As early as 1966, he was able to present Salbutamol, the first selective beta-2 adrenoceptor agonist, which Allen & Hanbury brought to market in 1969 under the name Ventolin®. It was a great success and the medication was gratefully accepted, as the asthma-related mortality rate in the UK was 2000 cases per year. Finally, there was a better alternative to ephedrine, theophylline, and isoprenaline. In particular, isoprenaline could be dispensed with, which as an early beta-2 agonist was not selective enough and had side effects on the heart. A major disadvantage of Salbutamol was its short duration of action of only a few hours. His team wanted to solve this problem and in 1969 developed Salmeterol, a new asthma medication with a duration of action of up to twelve hours.

Mothers Little Helpers and Methylphenidate

Before we conclude this chapter, we must address the chemical group of amphetamines, which are not natural substances and strictly speaking do not belong in this book. However, they are active substances that—like many other synthetic substances mentioned in this book—were developed from the natural substance ephedrine. Interest in ephedrine waned in the 1950s because an amphetamine, which had also been in a dormant state, reappeared out of nowhere and has significantly changed future society to this day—methamphetamine. This substance, which quickly embarked on a criminal career, achieved a macabre cult status with the series "Breaking Bad", which was first broadcast in 2008.[282] By the way: The blue methamphetamine crystals in the series come from a store in Albuquerque called "Candy Lady", the real crystals are colorless.

The young Romanian Lazăr Edeleanu (1861–1941) came to study in Berlin and researched in Hofmann's working group from 1883 to 1887, where Nagai Nagoyashi was a doctoral student a few years before him. Hofmann wanted his new protégé to deal with dyes. The exploration of amphetamines as a class of medical substances was never considered.[283] Thus, the discovery of amphetamines, which he called phenylisopropylamines, received little attention and disappeared after publication on the shelves of Humboldt University in Berlin, without the professional world taking an interest in it. It was no stranger who synthesized methamphetamine in the course of the structural elucidation of ephedrine. We know him, Nagayoshi Nagai, introduced the substance in 1893, but his later colleague Akira Ogata (1887–1978) from the University of Tokyo took care of the structural elucidation. He found crystals and filed a patent in 1921 because of its stimulating effect. Ogata was a pharmacologist with good chemical knowledge and immediately recognized the potential for Japanese society, whose work culture is hierarchical, collectivist and probably deeply rooted in the traditions of samurai culture. The word Karoshi means death by overwork and has its origin not coincidentally in Japanese society. In such a performance society, the potential of methamphetamines is great and the pharmaceutical company Dainippon Seiyaku distributed it under the brand name Philopon (Japanese ヒロポン, Hiropon). The name is composed of two words: "fatigue" (hiro) and "at once" (pon). Another explanation is that Philopon was borrowed from the Greek word philoponos (diligent). This would definitely fit the Japanese work culture.

Lazăr Edeleanu did not become known because of the amphetamine. He had certainly long forgotten this chemical substance after his doctorate with Hofmann. With 212 patents and a revolutionary method for processing crude oil, he was an extremely productive technical chemist in the field of petrochemistry. Edeleanu had a difficult childhood, youth, and school time. The parents were poor, the father a simple turner and there was no money for the young Lazăr to go to school. He lived in damp cellars and earned his school fees with tutoring. After graduating from high school in 1883, the drudgery continued to save money for a chemistry degree in Berlin. Five years later, he received his doctorate on the topic "On some derivatives of phenylmethacrylic acid and phenylisobutyric acid".[283] No colleague was interested in the field, but 20 years later American scientists picked up the research topic because they were looking for a better asthma remedy than ephedrine or adrenaline.

The allergist George Pines and the biochemist Gordon Alles recognized the development potential of amphetamines for the treatment of asthma and saw a cheap alternative to ephedrine. They tried amphetamine on themselves and found that it had a significant effect on the nervous system and the bronchi. The study was expanded to more patients and the physically stimulating effect did not escape George Pines either. It was Pines who eventually gave the class of substances its name. The compound Alpha Methyl PHenyl EThyl AMINe is summarized, the word amphetamine was born and is as familiar to us today as the trade names Coca-Cola®, Persil® or Aspirin®. Gordon Alles patented his discovery and approached companies, which were very reserved. Probably because Smith, Kline & French had long since developed Benzedrine®, which contained an amphetamine. Gordon Alles found out about it, was angry and accused the company of industrial espionage and idea theft. Smith, Kline & French gave in and involved him with five percent of the turnover to quickly remove the sand in the gears of the future money machine. The amphetamine was convincing and was brought to market in 1929 in the midst of the global economic crisis by Eli Lilly and Smith, Kline & French. The company management believed they were facing an economic disaster, because who was supposed to buy amphetamines in a time of falling stock prices and record numbers of unemployed? But they were greatly mistaken, because Benzedrine® sold like hotcakes. Were there so many asthmatics? No, despite a lack of money, the inhaler was bought, probably to bring a little happiness into the bleak life of the unemployed and new poor. After ten million sold packages, Smith, Kline & French wrote *"Ten million buyers who can't be wrong"* on the package, but perhaps they were also millions of addicts.

Smith, Kline & French noticed that amphetamine suppresses hunger and thirst, prevents fatigue, and activates human energy reserves. The issue of hunger became interesting because Smith, Kline & French discovered a new, unknown target group: the modern woman. Being slim and no longer adhering to the beauty ideal of Peter Paul Rubens became the norm. The "Empire-Line", little bust, tall and slim was the ideal for women from the 1920s onwards. This ideal was far-fetched, but it found role models in the starlets emerging from the new dream factory near Los Angeles. The 15-year-old Judy Garland, who would have turned 100 in 2022, was a victim of this beauty ideal. She took high doses because, in the eyes of the film bosses, she was too old and her femininity was not marketable. She desperately wanted to play the childlike Dorothy in the 1939 film "The Wizard of Oz". She got the role of her life. The price was filming for up to 72 hours at a stretch and to help her calm down in the evening, doctors gave her barbiturates as sleeping pills. She persevered, her breasts remained flat, and she remained the childlike idol. She sang the song "Somewhere Over the Rainbow" fantastically and it took on a second, special meaning. For Judy, it became normal to be stimulated. She had her first breakdown at the age of 19 during filming. Instead of rest, the doctor gave her more Benzedrine® for her "amphetamine cure". During the filming of "Summer Stock", she completely lost control and was a nervous wreck. The film company MGM fired their pill-addicted diva in 1950, whom they had made dependent, with the terse explanation that she never or too late came to the shooting days. She was the original entertainer who unfortunately found a tragic end to her career at the age of 47 due to an overdose.

Not only in American society was methamphetamine popular, as it promised artificial paradises in an emerging performance society. In the 1920s, methamphetamine was legal and available without a prescription. In addition to amphetamine, drugs and alcohol have always been a part of the societal avant-garde. Without guaranteeing completeness, some figures from politics, science, and culture who also took drugs for mental stimulation are mentioned in Table 1.

In Germany, many people took Benzedrine® or Pervitin®. The night watchman did not fall asleep, the nurse on night duty kept going, the exhausted housewife and mother could better look after her children, who also swallowed the stimulants to optimally prepare for their school exams. Berlin, the modern Babylon, was allegedly flooded with amphetamines because the seemingly inexhaustible cocaine supplies of the German Wehrmacht were running low after the First World War. It is true that the supply from South America was faltering because the Allies blocked it. The price for cocaine had exploded and a chemical alternative had to be found, because Germany no longer had any colonies that could supply the stimulant for the head. The pharmaceutical industry in the German Reich had to become creative with its possibilities and the available raw materials. The workshop of the world became again the pharmacy and later the dealer of the world. Methamphetamine was part of this strange success story.

Amphetamine and methamphetamine were so deeply rooted in society until the 1950s that the former airline PanAm listed the Benzedrine inhaler in the onboard menu, and among young adults, the "B-Bomb" was completely normal when the inhaler was unscrewed and the amphetamine cotton ball was dipped in coffee. In the USA and the Anglo-Saxon countries,

Table 1 Famous personalities and their rarely or often highly valued plant drugs, amphetamine, and alcohol

Person	Alcohol and plant drug	Note
Adolf Hitler (1889–1945)	Amphetamines[19] Cocaine	Polytoxicomane
Aldous Huxley (1894–1963)	Mescaline	
Charles Baudelaire (1821–1867)	Absinthe Cannabis[284,285] Cocaine[284,285]	
Truman Capote (1924–1984)	Alcohol[284]	
Judy Garland (1922–1969)	Amphetamine[284,286]	
Hermann Göring (1893–1946)	Morphine[19]	
Barack Obama (*1961)	Cocaine[287]	
Hans Fallada (1893–1947)	Morphine[284,285]	
Sigmund Freud (1856–1939)	Cocaine[24,25]	
Paul Gauguin	Absinthe	
Gustaf Gründgens (1899–1963)	Morphine[284]	He needed morphine, first for migraines, then because he was addicted
Vincent van Gogh	Absinthe (Green Fairy)	He painted the "green fairy" in his work "Café table with absinthe" (1887)
Ernest Hemingway (1899–1961)	Alcohol[284]	
Friedrich von Hardenberg, Novalis (1772–1801)	Cannabis[284] Opium[286]	Praised and lauded opium in his poem "Hymns to the Night"
Heinrich Heine (1797–1856)	Opium	For pain treatment due to myatrophic lateral sclerosis
E. T. A. Hofmann (1776–1822)	Alcohol Opium[286]	
Ernst Jünger (1895–1998)	Mescaline	He also found LSD very exciting and processed his trips in literature
Gustav Klimt (1862–1918)	Cannabis[284,288]	
Wolfgang Neuss (1923–1989)	Cannabis Alcohol	LSD is also said to have been involved
Klaus Mann (1906–1949)	Morphine Heroin	
Sigmund Freud (1856–1939)	Cocaine[24,25]	Wrote the first scientific treatise on the pure substance cocaine
Edgar Allan Poe (1809–1849)	Opium[286] Alcohol	His last words were: "Lord, help my poor soul"
Elvis Presley (1935–1977)	Amphetamine	He took an estimated 10,000 tablets per year and supported Nixon in the war against drugs

(continued)

Table 1 (continued)

Person	Alcohol and plant drug	Note
Rio Reiser (1950–1996)	Cannabis[288]	
	Alcohol	
Jean-Paul Sartre (1905–1980)	Mescaline	He found amphetamine at least as good
	Amphetamine[286]	
Friedrich Schlegel (1772–1829)	Opium	
Georg Trakl (1887–1914)	Cocaine[286]	Died of a cocaine overdose
	Opium[286]	
Henri de Toulouse-Lautrec	Absinthe	
Rainer Werner Fassbinder (1945–1982)	Cocaine[289]	
Konstantin Wecker (*1947)	Cocaine	

Benzedrine® and amphetamine were mainly consumed. The prescription and intake of amphetamines reached epidemic proportions in the 1960s. In 1970, an amphetamine was on eight percent of all medical prescriptions and public awareness of their risks and side effects was growing. The dispensing by doctors and pharmacists was increasingly regulated and by the end of the decade, amphetamines were considered narcotics and could no longer be dispensed.

The result was the production of methamphetamine in illegal laboratories and the sale on the black market. In Western societies, but also in Africa and Asia, well-known drugs like Ecstasy, Tic or Speed are widespread. Their synthesis is very simple, and a first-year chemistry student is capable of performing the synthesis according to the Pervitin method. The synthesis also made it into film history with Walter White from the series "Breaking Bad", which as a tragicomedy shows the whole misery of organized crime and abuse. While for the chemist Walter White the problem was getting the strictly controlled precursor chemicals, today Mexican drug chemists are targeting the herb *Ephedra*. The imported goods from China are extracted and the ephedrine is converted into methamphetamine (Crystal Meth) in a chemical step.

Much has already been written about the role of amphetamines in the 1920s and during the Second World War in Nazi Germany and among the Allies. Norman Ohler has very vividly and grippingly described in his book "Blitzed: Drugs in Nazi Germany" how the Nazis swallowed the methamphetamine in Pervitin®. The blitzkrieg was on speed and Hitler was high in the Führerbunker. *"Hitler's body was trembling and was repeatedly powdered up with the many pills and narcotics. He had often wondered whether the prophylactic injections, … were not a kind of exploitation of Hitler's health and life, the catastrophic consequences of which were now becoming apparent,"* Joseph Goebbels confided in his diary. Who gave the preventive injections? It was Hitler's personal physician, Dr. Theodor Morell (1886–1948), a chubby Hessian, specialist for skin and venereal diseases, who in the dying Weimar Republic became a fashion doctor for non-existent diseases. A doctor who enjoyed the dubious trust of the stars, starlets and celebrities of Berlin and liked to sell amphetamine injections with vitamins as wellness injections. It was said that he set the injection so well that one did not feel the puncture.[19]

Germany was supplied with Pervitin® by the Temmler-Werke in Berlin, which contained the active ingredient methamphetamine. In contrast to amphetamine, attaching an additional methyl group to the nitrogen resulted in faster absorption into the brain, as the methamphetamine was more fat-soluble and also showed a significantly stronger effect. Pervitin® was often abused. Hermann Buhl, who participated in the German-Austrian expedition to the 8126 m high Nanga Parbat in the Himalayas, had methamphetamine in his blood, which served as a model for subsequent mountaineers.[284] The rumor that American athletes had won their medals at the 1936 Olympic Games with Benzedrine® was the impetus for Fritz Hauschild of the Temmler-Werke in Berlin to manufacture and sell methamphetamine under the name Pervitin®. The company, which was based in Berlin-Adlershof, produced millions of the small white tablets that were delivered to the people, the leader, and the Wehrmacht. Almost 800,000 tablets could be pressed daily, the Wehrmacht had placed large orders for 35 million tablets. Even though the number of 800,000 tablets per day seems high, modern tablet presses can easily produce 500,000 tablets per hour for larger ibuprofen tablets and up to five million mini tablets for the "pill".

The search for new applications for amphetamine continued. A new market was to be tapped. At the beginning was the not entirely serious question of whether amphetamine could help in the treatment of narcolepsy. Because in Europe, millions of people had fallen ill with the so-called sleeping sickness, encephalitis lethargica, by 1927. Narcolepsy is a disease where patients suddenly fall asleep during the day. After the First World War, there was great puzzlement among doctors. Was it a mental illness or a viral infection? Within a few years, a third of the five million acutely ill people died. The sick fell asleep while eating. After waking up, they suffered from nausea, fever, and headaches. They could be awakened again, but in severe cases, death followed. In Germany today, about 4,000 people are affected, the causes are not yet fully researched. It is suspected that the gene for the neuropeptide orexin, which regulates the sleep-wake rhythm, could be mutated. Anyone who has seen the movie "Awakenings" with Robert De Niro and Robin Williams knows the disease. The film is based on the true story of the recently deceased doctor Oliver Sacks (1933–2015), who as a young doctor in New York worked with the victims of the American epidemic. The young Oliver Sacks experimented with L-DOPA, which is similar to adrenaline and amphetamine. He had modest successes, as after a short time his patients became resistant to L-DOPA, which was broken down faster and faster in the body and became ineffective.

There is a second amphetamine and a specific indication that should be addressed. Methylphenidate is an amphetamine and an approved drug that is still used medically today and whose discovery is due to a coincidence. The neurologist Charles Bradley first described the effect of amphetamine on highly active children with attention deficit, now known as ADHD, which he administered to children with neurological problems in 1937. He was not thinking of therapy, but was testing a diagnostic procedure with the cumbersome name pneumoencephalography, in which cerebrospinal fluid was drained and replaced by air. The children suffered from severe headaches, which he tried to alleviate with amphetamines. Some of the children, who might be diagnosed with ADHD today, became calm and later showed better performance in school. It seems contradictory that an amphetamine, which we know to be stimulating and contributes to being able to dance in the disco from Friday to Sunday morning, should have a calming effect on children. Why a stimulant has a dampening effect in the child's brain was inexplicable. It can be somewhat explained by the effect of cocaine, although less strong and without high addiction potential. Methylphenidate inhibits the reuptake of dopamine and noradrenaline into the synapse of nerve cells. The synapse is the end of a nerve where an electrical nerve impulse is converted into a chemical impulse. In the so-called synapse at the end of the nerve, there are many transport containers filled with various chemical transmitters. When the electrical signal arrives, they merge with the cell wall and release the substance, which can trigger an electrical impulse on the opposite nerve. Each cell deals sparingly with these important transmitters and would like to recycle them. Transport proteins in the membrane actively bring the substance back into the nerve cell and fill up the new transport container. If cocaine or methylphenidate is present, the transporter is blocked and the dopamine must remain in the synaptic cleft between the two nerve cells. With cocaine, this leads to a long-lasting kick, with methylphenidate it seems to be different. Although the dopamine also remains in the gap here, the dopamine effect in other brain areas is significantly increased. The result: You can concentrate better on your thoughts and are less distracted by external stimuli.

Charles Bradley, a doctor at the Emma Pendleton Bradley Home, administered Benzedrine® to children and initiated a pilot study. He already suspected the side effects and warned early against improper administration in children. With Methylphenidate, approved in 1955 under the trade name Ritalin®, it was believed that the ideal drug had been found, whose stimulating effect should lie between amphetamine and caffeine. The studies were repeated, expanded, and extended to "problematic adolescents" in the USA in the 1960s. Only in the 1970s were German children first treated by the child and adolescent psychiatrist Gerhardt Nissen (1923–2014). His experiences with new findings are still valid today and even led some educators to consider administering Ritalin to poor students, as a significant improvement in academic performance with Ritalin could also be observed in healthy students. From the 1980s onwards, things became uncomfortable and with increasing awareness, parents questioned the administration of Ritalin to their children. The Church of Scientology got involved and saw Ritalin as a drug of evil. The spread of criticism was funded by an organization close to the Church of Scientology called "Citizens Commission on Human Rights". Another organization called "Commission for Violations of Psychiatry Against Human Rights e. V." was no less vehement in its criticism. The fronts were hardened and a factual discussion was no longer possible, which is why Ritalin supporters assumed that any criticism was anyway controlled by psycho-sects.

Children with excessive activity in their movement or activity do not seem to be a problem of our time and overactive children have already been described by Heinrich Hoffmann in his poem about Fidgety Philip:

"Will Philip be quiet today"
"at the table?"
Thus spoke in a serious tone
the father to his son,
and the mother looked silently
around the whole table.
But Philip did not listen,
to what his father was saying.
…

See! He swings too wildly,
until the chair falls backwards;
there is nothing more to hold him;
he grabs for the tablecloth, screams.
But what's the use? At the same time
plates, bottle, and bread fall."

The drug Methylphenidate was synthesized in the last months of the Second World War in 1944 by the chemist Leandro Panizzon, who was employed at the Swiss company Ciba, now Novartis. As already often mentioned in this book, self-experiments were not unusual and both he and his wife Marguerite took Methylphenidate. Dear Marguerite was a moderate tennis player and suffered from low blood pressure. Leandro was amazed when his wife turned into a Wimbledon star after taking the drug and confidently hit the balls over the net. Her friends called her Rita and Leandro found the nickname original to later name the drug "Ritalin".

The aim of his development of Methylphenidate was to find a better amphetamine with fewer side effects such as loss of appetite and lack of sleep. It was supposed to be taken for low blood pressure or mild depression. Panizzon and the marketing department were not yet in agreement. A remedy for attention deficit or hyperactivity was not considered because this mental problem in children was not perceived in post-war societies. Children played in the yard, romped over the fields that no longer exist today. Children also just went away, came back tired in the evening and did not have to check in with their mobile phone in between. With the growth of our cities and new media, our children are born into a new, narrower world. Physical exertion becomes a hobby for a few hours a week or in physical education. The author has to think about when he last saw children in his neighborhood playing or fooling around unsupervised outdoors. In the 1990s, Methylphenidate experienced a boom as a drug for treating attention deficit hyperactivity disorder, or ADHD for short, and was administered to hyperactive children, mostly boys. To this day, the number of small patients with actual or suspected ADHD continues to rise. In 2013, 2.4 billion doses were produced, a good 80% of them in the USA. This corresponds to a turnover of 31 billion dollars per year. While 2 tons of the active ingredient were produced in 1990, today it is 30 tons. The trend is rising.

LSD—Lucy in the Sky of Diamonds

> **Profile - Lysergic acid diethylamine (LSD)**
>
> **Effect**
> - Hallucinogen
>
> **Former medicines**
> - Methysergide®
> - Deseril®
>
> **Natural Product Group**
> - Alkaloid

1938 was the big year of Albert Hofmann (1906–2008, Fig. 12). It was the birth of one of today's most well-known substances. From the ergot alkaloid lysergic acid, to which he attached two ethyl groups at the nitrogen, he created at Sandoz AG in Basel Lysergic acid diethylamide 25 (LSD-25). More

Fig. 12 Albert Hofmann (1906–2008), 1993 (Permission Novartis, Basel, CH)

about the history of ergot alkaloids from ergot (*Claviceps purpurea*, Fig. 13) can be found in the next chapter. A little knowledge of the cultural history is necessary to understand how LSD was discovered as one of the most potent psychotropic substances. The Swiss pharmaceutical company was not looking for natural substances that alter our consciousness, they wanted drugs that could be used in obstetrics and for Parkinson's disease. Albert Hofmann was responsible as a chemist for the synthesis of derivatives from natural substances, in order to extract the effective component from the complex ergot alkaloids. Once again, the question in medicinal chemistry was which structural elements are important for the effect and which can be eliminated to avoid side effects such as psychosis and gangrene.

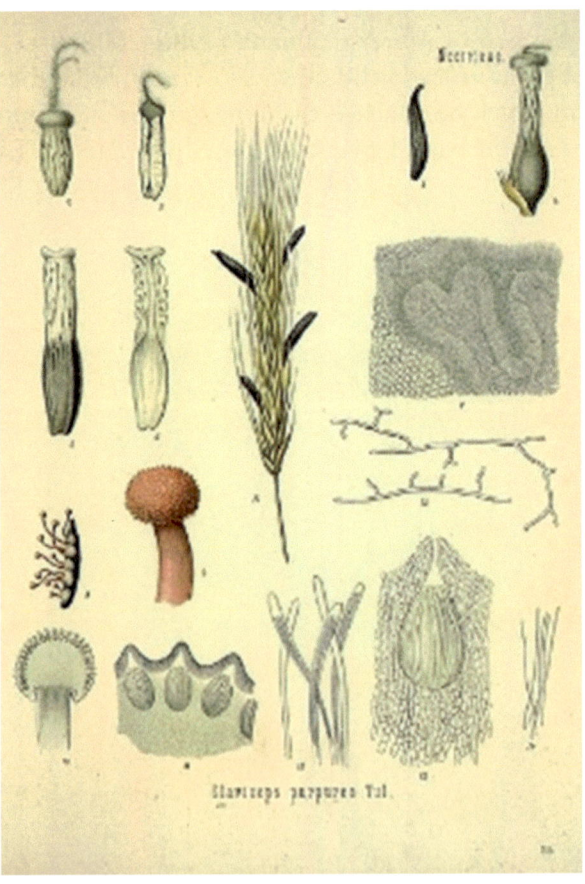

Fig. 13 *Claviceps purpurea* (public domain, wikipedia)1

Who was this Albert Hofmann, who joined the services of Sandoz AG as a chemist in 1929?[290] He always wanted to work with natural substances and therefore preferred to accept the offer from Sandoz. Other Swiss companies only offered positions in pure organic chemistry, which did not appeal to him. During his doctoral thesis with Paul Karrer (1889–1971), who researched polysaccharides, vitamins and flavins at ETH and in 1937, together with Walter Norman Haworth (1883–1950), received the Nobel Prize for the elucidation of the structure of carotenoids, flavins and vitamins A and B, he examined the gastric juice of the vineyard snail. This may not sound exciting at first, but the research was important to understand how chitin is broken down into amino sugars. Chitin is the hard shell material of insects and crabs. It is the second most common natural substance after cellulose and is found in fungi and insects, where it is part of the exoskeleton. Albert Hofmann was able to clarify the structure in his dissertation, which from a biological point of view is a greater achievement than the discovery of LSD.

The pharmaceutical department of Sandoz was quite manageable in 1929 with three PhD chemists. The department head Arthur Stoll assigned him the task of isolating heart-active substances from the sea onion, which he accomplished with flying colors and completed in 1935. Not that he would have been bored, but Hofmann was looking for a new, exciting field of research and asked Arthur Stoll if he could work on the ergot alkaloids. Here, Stoll had stopped in 1918 with the discovery of ergotamine, as the Americans and Brits had made considerable discoveries in this decade. Stoll agreed, but he gave him the following advice. *"I warn you of the difficulties you will encounter when working with ergot alkaloids. They are extremely sensitive, easily decomposable substances, quite different in stability from the compounds you have worked with in the field of cardiac glycosides."* Anyone else would have felt enormous respect and asked themselves again whether they were not risking their career with this research, but Hofmann felt a *"sense of anticipation of creative happiness."*[291]

His optimism was dampened when he filled out the order form for half a gram of ergotamine and wanted to have it signed by his boss. The latter came into his office and reprimanded him: *"Dear colleague, if you want to work with ergot alkaloids, you must familiarize yourself with the methods of microchemistry. It is not acceptable that you consume so much of my precious ergotamine for your experiments."* Albert Hofmann had to prepare his own ergotamine, which he needed to extract lysergic acid, for his microchemical work, i.e., work on a small scale. At that time, people were very frugal and only inferior goods from Portugal were available for research. Before the

Second World War, most sclerotia were collected and sold in Portugal and Spain. Here, in larger quantities, was found the ergotoxin, a mixture of three ergot alkaloids, which was amorphous. Hofmann seems to have had a difficult relationship with his boss. Stoll was very frugal with his natural substances, a new laboratory hood with a fan that sucked up the toxic fumes and expelled them outside was rejected with the justification that the old ones with a gas flame would do just as well *"and in the laboratory of his doctoral father Willstätter, this type of ventilation was also sufficient."* In discussions with his chemists, Arthur Stoll always referred to his work with his boss and the achievements of Willstätter. He was simply open to everything old.

His work was successful and Hofmann was soon able to isolate not only a very unstable lysergic acid, but also to clarify that ergotoxin was not a pure substance, but a mixture. By reacting with propylamine, he had synthesized the first semi-synthetic ergot alkaloid. In a series of experiments, he wanted to find more lysergic acid derivatives that differed from propylamine, and LSD-25 was born. A means to stimulate circulation and respiration, so Hofmann believed, and handed the substance over to Ernst Rothlin for animal testing.

In the oxytoxicity test, LSD-25 failed and only achieved 70% of the effect of ergobasin as a standard. This test examines the toxic effect on the uterus. Ernst Rothlin reported that the mice were restless under anesthesia. This was not unusual and the new substance LSD-25 aroused neither Rothlin nor the chemists' greater interest. Sandoz decided to abandon further development. The substance disappeared into a box on a shelf for five years, but Hofmann felt there had to be more. As he later wrote in his book "LSD—My Problem Child", he had a strange premonition.[292] On April 16, 1943, when Ramón Mercader was sentenced to death in Mexico for the murder of Leo Trotsky, he synthesized LSD-25 again to check its analeptic effect. In the process, he accidentally ingested some lysergic acid or LSD. When he wanted to rest at home, he noticed a strange effect.

Kaleidoscopic, colorful, and very intense illusions of his brain with unnaturally plastic representation of body shapes over hours completely captivated Albert Hofmann. In the following days, he immediately set to work to get to the bottom of the phenomenon. He repeated his self-experiment, reduced the dose from 250 to 25 micrograms, and also tried all other LSD derivatives synthesized up to that point. After 40 minutes, he no longer felt well. He was no longer able to speak properly and he felt that he was losing control of himself. Hofmann called for his assistant and asked her to drive him home. April 19th went down in the world history of the hippies as Bicycle Day, because Hofmann, accompanied by his assistant, rode home on a bicycle.

During the Second World War, few people drove cars, so the two set off on their bicycles. Hofmann felt as if he was pedaling in place and not moving forward, but his assistant thought they were going quite fast. When they arrived home, Hofmann asked her to get milk from the neighbor to get an unspecified antidote. He drank two liters, but it didn't get better. Fear rose in him that he might now die, leaving a widow and his children as orphans, and he regretted his reckless decision. Frightened, he asked the lab assistant to call his family doctor. The doctor shook his head and looked at Hofmann lying on the couch. He could not find anything that would be a reason for the impending death. The breathing, pulse, and blood pressure were normal, only the pupils were slightly dilated. He carried Hofmann to bed, sat next to him, and waited until his wife arrived from Lucerne. Mrs. Hofmann was informed that her husband had suffered a breakdown. She left the children with the parents and returned to Basel. By her late arrival, he was already better and the terrible hallucinations became pleasant, as he now perceived everything intensely and the sensory impressions changed. The sound of the door handle and the passing car were translated into color. He was grateful to have escaped the impending madness. He slept peacefully that night, had no hangover, and was refreshed and clear to his own surprise. Breakfast tasted good and he wrote a report for Stoll, in which he described the experiences of the last night:

"16:20 Ingestion of the substance"
"17:00 Onset of dizziness, feeling of fear, visual disturbances, paralysis, laughter. Cycling home."
"From 18 to about 20 o'clock severe crisis, see special report:"
"The last words I could only write down with great difficulty. [...] the changes and sensations were of the same kind [as yesterday], only much more profound. I could only speak intelligibly with the greatest effort, and asked my lab assistant, who was informed about the self-experiment, to accompany me home. Already on the way home by bicycle [...] my condition took on threatening forms. Everything in my field of vision swayed and was distorted as in a curved mirror. I also had the feeling of not moving from the spot with the bicycle. However, my assistant told me later that we had been going very fast. [Having arrived at home] dizziness and faintness became so strong at times that I could no longer stand upright and had to lie down on a sofa. My surroundings had now transformed in a frightening way. [...] the familiar objects took on grotesque, mostly threatening forms. They were in constant motion, as if animated, as if filled with inner restlessness. The neighbor's wife [...] was no longer Mrs. R., but a malicious, insidious witch with

a colored face etc. etc. Later during the fading of the intoxication: Now I gradually began to enjoy the incredible play of colors and shapes that continued behind my closed eyes. Kaleidoscopically changing, colorful fantastic structures pressed in on me, opening and closing in circles and spirals, spraying in fountains of color, rearranging and crossing, in constant flow. Particularly strange was how all acoustic perceptions, such as the sound of a door handle or a passing car, turned into optical sensations. Every sound produced a corresponding, lively changing image in shape and color."[292]

His story sounded too fantastic when he told his colleagues about LSD and its effects. Stoll and Rothlin didn't believe a word and pressed him with the question of whether he had not weighed himself. No psychotropic substance is known to date in this small amount. He repeatedly assured that he had weighed himself and done everything correctly. They remained skeptical, but Rothlin and two of his employees were brave and also took LSD in a self-experiment, however, only a third of the dose of Hoffmann, and they could confirm the effect. Today it is known that Hoffmann's dose was overdosed by a factor of 10,000. Already 15 to 25 nanograms, that is nine zeros after the comma, if you want to express the amount in grams, are enough to achieve a first effect.

In animal experiments with rodents, no comparable effects were observed as in humans and it appears that LSD can only produce similar effects in higher animals, such as great apes. In mice, movement disorders and changes in licking behavior are observed. In cats and dogs, behavioral changes are observed as in humans, but at significantly higher doses. So is LSD toxic? As always, everything depends on the dosage, but the amount of poison is very high in humans, LSD is well tolerated. If the lethal dose (LD_{50}) in mice is 50 to 60 milligrams per kilogram of injected substance, it is a third in rats, 0.3 milligrams in rabbits. An elephant, which was administered just under 0.3 g of LSD, died and the lethal dose was calculated at 0.06 milligrams per kilogram. From toxicological reports it is known that humans have survived the 100,000-fold lethal dose of mice. However, it can be summarized that the toxicity seems to increase with the size of the animal.[293] There are also known cases of death after LSD consumption, which, however, are due to suicide or states of confusion. Today it is known that LSD can lead to death by respiratory paralysis.

Hofmann and his colleague Franz Troxler synthesized further LSD derivatives after 1945, none of which achieved the effect of LSD-25. Not only the substituents were changed, but also the stereochemistry. Lacking meaningful animal models, both took the new substances themselves up to a maximum dose of half a milligram. It was professionally described whether the substances were more psychotropic or narcotic and how the strength was perceived compared to LSD. However, the two did not want to find an even stronger LSD, the idea behind the test was to find a lysergic acid that was not psychotropic, but could have a strong influence on serotonin receptors. This is how Hofmann describes it in his book. This is an interesting question, as knowledge about serotonin receptors was practically non-existent shortly after the Second World War. Troxler found an LSD with a bromine as a very active agent for the treatment of migraines. This BOL-148 was the chemical and conceptual precursor of Methysergide, which came onto the market as Deseril® for interval treatment of migraines.

While the dose for an LSD trip is about 100 to 250 micrograms, a research group led by Nadia Hutten at Maastricht University has shown that microdoses of 10 to 20 micrograms already have effects on a clear and controlled consciousness. This is confirmed by an anecdotal report from a friend of the author. After taking a small dose of 10 micrograms, he experienced a very intense walk barefoot on a meadow. The colors of the meadow and the flowers were very clear, he heard and distinguished the background noises of the birds and other animals in the garden much more clearly. Nadia Hutten was able to confirm these anecdotal reports that 20 micrograms improve cognitive perception, already 5 micrograms lead to a mood enhancement and all without a hangover. We should not take these scientific data as an occasion to trivialize or exaggerate LSD. But it is worth thinking objectively about LSD. In an aging society, where dementia and declining memory performance are becoming a common disease, drugs as psychoactive substances can certainly be a help. If microdosing helps elderly people to find more conscious moments in their lives, it must be discussed how drugs can be used legally to improve quality of life.[269] As I have already explained in the introductory part to this chapter, I am convinced that arguments such as procurement crime among seniors or addiction are wrong and do not contribute to a good discussion. While Germany does not even want to start this discussion, other countries are further along. We already see today in the USA and England that start-ups are dealing with the topic and the discussion in both countries is being conducted without prejudice.

Lysergic acids and ergot alkaloids are mentioned in the same breath as ergot, but these exciting natural substances are also found in many tropical bindweed plants. None other than Albert Hofmann discovered the ergot alkaloids in the Mexican magic bindweed *Turbina corymbosa* (Fig. 14) in his private research. This bindweed plant has been used for centuries by the Aztecs and Mayas in cultic and religious ceremonies and was

Fig. 14   *Turbina corymbosa* (public domain, wikipedia)1

called Ololiuhqui. On one of his private trips, he got to know the plant, tried it out and found similar, but much milder effects compared to LSD. Laboratory tests showed him that the extract was not LSD, but a chemically very similar substance, lysergic acid amide LSA, which lacked the two ethyl groups. The natives crushed the seeds of the bindweed and mixed them into their beer, which was made from agaves. The effect was a profound intoxication that awakened memories from Hofmann's earliest childhood. The Spanish conquerors, who came into contact with the magic bindweed for the first time in the 16th century, found all this too suspicious. They banned the natives from consuming it, but no one obeyed the ban. What alarmed the Spaniards was the trance and the observation that the shamans left their bodies after ingestion to communicate with their gods. But for the Spaniards there was only one God and to speak to him, one went to church or prayed. The conquerors did not like this. Hernando Ruiz de Alarcón was both fascinated and disgusted, but he wrote everything down to report to his lord in Spain. In Chapter VII of his anthropological work "Tratado de las supersticiones y costumbres gentilicias que hoy viven entre los indios naturales de esta Nueva España" [Treatise on the superstitions and customs that live today among the Indians in New Spain], de Alarcón was not only outraged about the drunkenness of the consumers, their shamans and seers, but also about the fact that some Spaniards had also taken this devil's stuff. The misuse as a truth drug in court proceedings went too far for the pious man of God. He concluded his report with the sober statement that neither he nor the church were able to ban the Ololiuhqui, as the natives were afraid of punishment if they committed *pampa àmo nechtlahueliz*, i.e. declared Ololiuhqui to be their enemy.

Although Hofmann referred to Lysergic acid amide (LSA) as a plant substance, researchers in the 2000s found out that it is endophytes that produce the LSA.[294] Endophytes are bacteria or fungi that live in plants. Similar to how we humans have a gut flora, microorganisms also live in plants, their number in a plant is estimated to be at least 20 and up to 100. The reasons why microorganisms live together in plants are diverse. They are advantageous for the plant in the uptake of nutrients from the soil. Conversely, the microorganisms benefit from the sugar from plant photosynthesis. Their support in defending against other invading pathogens is under discussion, as they can build up common biosyntheses with the plant to form defensive substances. It is a very exciting field of research whether certain "plant" natural substances are actually only formed by plants or whether it is a partnership. The biosynthesis of lysergic acids must be attributed to the microorganisms according to current knowledge. The seeds are already infected during their formation after the fertilization of the flowers and the endophytes are taken to the next generation, where they form the lysergic acids again. Their task is probably to see them as LSA producers in defense against predators. After eating the LSD plant, hallucinations occur, their protective and escape reflex no longer works and they become prey to their predators more quickly.

The medical application of LSD has been controversial since its discovery and after a dubious career in the hippie communes of the western world, a worldwide ban followed in 1966. Since this time, the exploration of positive effects has hardly been possible, even though the political attitude towards the over 50-year-old restrictive decision is changing somewhat today. Sandoz AG began serious research after the discovery and saw great potential as an antidepressant to treat alcoholism and schizophrenia. In the 1950s, LSD was made available to psychotherapists and psychiatrists for self-study. This was intended to enable them to better empathize with the neuroses and the world of their mentally ill patients. However, this did not succeed, as the mental problems of their patients were different. Intelligence agencies tested LSD as a truth serum and the British army as a chemical weapon on its own soldiers. When they saw the effect, they no longer thought anything of LSD. Because if the enemy soldiers were as unpredictable as their own, they might just do nonsense or even press the red button. That was not a good idea.

In a very painful disease, LSD has found an application as an illegal substance for cluster headache, an extremely painful migraine for which there is no effective therapy. The sufferers report hours of worst headaches, which they can only alleviate by inhaling pure oxygen. Patients bang their fists against walls or oxygen bottles and tear their hair out, so much do these people suffer. Science knows little about these headaches. There are no reliable medications and oxygen only helps in the short term. What it does in the head is unknown. Probably the blood vessels are narrowed, but that is not certain. On the website clusterbusters.org, patients report their suffering and the testimonials about LSD and Psylocibin impressively show that these substances deserve a second chance in medical research. How great is the therapeutic potential? BOL-148 was tested experimentally as a non-psychoactive substance. In a study at the Hannover Medical School, the pain disappeared in five out of six patients without any major side effects. But it is still a long way to a clinical study, because the US Food and Drug Administration (FDA), but also other approval authorities, are still very restrictive with lysergic acid derivatives. A small exception is Switzerland. Franz Vollenweider, emeritus professor of psychiatry at the Psychiatric University Hospital Zurich, wanted to know exactly in the 1990s: Can hallucinogens help with depression or alcohol addiction? The politicians believed the calm and sober scientist and gave him permission to conduct studies with LSD and Psilocybin. He was inspired in 1978 by a conversation with Albert Hofmann, who encouraged him to research with LSD and to study medicine, because only in this way could he legally work with the hallucinogen. He followed the advice and studied biochemistry and medicine. Later he switched to Psilocybin, which he considered to be better controllable in the human body. He organized two double-blind studies on the effect of Psilocybin and depression like alcohol addiction. After completing the studies, he sees chances, but also risks of hallucinogens and remains skeptical about microdoses, as he sees no evidence that minimal doses make more relaxed or creative. What remains is a big question mark. Whether LSD is a problem child or becomes a prodigy can only be answered when research is liberalized.

Cannabinoids

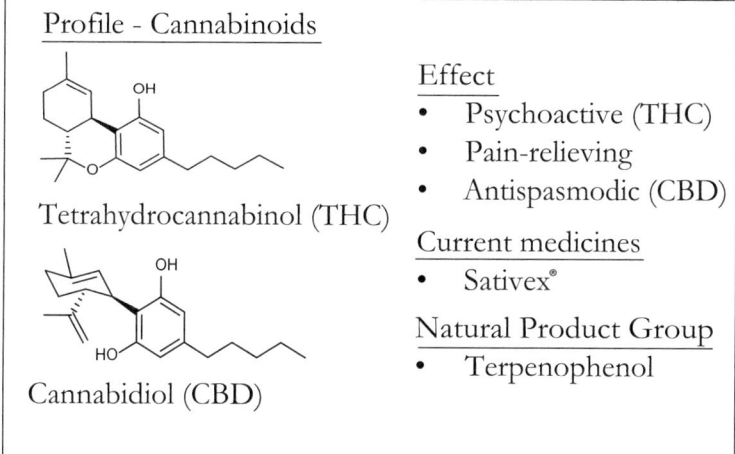

Cannabinoids and Cannabis have a history as turbulent as LSD. However, it seems more likely here that they will be legalized for medical purposes and later also for recreational use. In the last 1,000 years of human history, this plant has had a solid career as a medicinal plant. The oldest find of Cannabis or Marijuana was in a Chinese grave, as discovered in the excavations of the Yanghai tombs near Xinjiang. We do not know why the Chinese relatives put cannabis in their ancestors' graves, but from their Indian neighbors, we know very well today that cannabis flowers and their extracts were given for epilepsy and pain. Perhaps the mourners in China wanted the deceased to fare better in the afterlife if he could take his medicine with him.[295] Many books have been and are being written about the history of hemp, because besides the stories about the intoxication, this plant has written human history like no other. No expedition of the Royal Navy and no British Empire, where the sun never set, would have been possible without hemp ropes and hemp sails on the war and trade ships. Several tons of hemp, which had to be replaced on the ships every two years, testify to the high demand and the immense importance of this plant worldwide (Fig. 15).

Before the discovery of cotton as a softer fabric, the work clothes of many Europeans and Asians were made of hemp. The decline began with the industrial revolution when hemp was replaced by cotton in the spinning mills, steamships displaced the sailing ships, and hemp was no longer needed for the large sails. With the prohibition of the intoxicating cannabis plant in the 1920s, the low-THC industrial hemp was also heavily regulated

Fig. 15 *Cannabis sativa* L. (public domain, wikipedia)1

in the 1950s and 1960s, and cultivation was only allowed with permission. Cultivation took place on the smallest areas and technically no longer played a role.

The psychotropic active ingredient of the cannabis plant is the Tetrahydrocannabinol. Although another 150 cannabinoids are known, none has such a pronounced effect on our psyche as this one. Tetrahydrocannabinol, abbreviated as THC, is chemically a peculiarity, because in comparison to other intoxicating natural substances such as morphine or ergot alkaloids, there is no nitrogen in the molecular formula. Perhaps this was also one of the reasons why THC was overlooked and only isolated very late in 1964 by Yehiel Gaoni and Raphael Mechoulam (*1930–2023) at the Weizmann Institute of Science in Israel. He also clarified the structure of Cannabidiol (CBD), the second important cannabinoid, which was already isolated in 1940 by Roger Adams (1889–1971), who is considered the godfather of cannabinoids. Adams also made a structural proposal, but was wrong, which is why Raphael Mechoulam is known as the discoverer of Cannabidiol in history.[296]

Fig. 16 Raphael Mechoulam (1930–2023) (public domain, wikipedia)2

Raphael Mechoulam (1930–2023, Fig. 16) was born in Sofia and moved to Israel in 1949 to conduct research first in Haifa and then in Jerusalem. He recalls that the discovery of THC raised many questions, as it was not known how the substance would act in the body. It was not until 20 years later that Pfizer researchers discovered during the screening of synthetic cannabinoids that these reacted with an unusual receptor. These receptors belong to the family of GPCR proteins, of which there are about 1,500 different types in the human body. Using similar receptors in cattle, Allyn Howlett discovered the CB1 receptor in 1985, which mediates the feeling of intoxication in the brain. A few years later, in 1993, the second cannabinoid receptor, CB2, was discovered. This receptor is responsible for immunological and anti-inflammatory reactions outside the brain. The researcher Kenner C. Rice was able to locate these receptors in the brain and demonstrate them under the microscope with labeled synthetic cannabinoids. To make sure these were not ordinary receptors, his team made identical tissue sections from humans, dogs, monkeys, and rats. All animals exhibited the interesting receptors. Stephen J. Lolait and his team perfected the matter and were the first to map in the brain exactly where the cannabinoid receptors are located, and found that the receptors responsible for the intoxicating effect are almost exclusively found in the head. Today we know that there is a second group of cannabinoid receptors, few of which are in the head, but most are found in all other organs and have more to do with the function of our immune system. Many scientists wondered why there are receptors for cannabinoids in our body at all.[297, 298] The answer was found at the end of the 1990s: Similar to how endorphins act on opioid receptors, there are endogenous endocannabinoids that act via cannabinoid receptors. Little is known about the physiology and function of the cannabinoid receptors in our nervous system. It is discussed that the endocannabinoid system could play an important role in anxiety disorders, sleep, body temperature regulation, and inflammation. This is a very exciting area for drug research, where we will experience many surprises in the future.[299] But how did it all start and what therapeutic approaches have been developed since the discovery of cannabinoids in medicine?

The modern medical use of cannabis was initiated in 1838 by the Irish doctor William B. O'Shaugnessy (1809–1889), who experimented with hemp in the British colony of India. His research areas on hemp extracts for rheumatism, tetanus, cholera, pain, and epilepsy were very broad. He meticulously and conscientiously collected all information about possible reliefs and cures of patients who took hemp extracts in India. His profound knowledge led other doctors and scientists in distant Europe and the USA to trust him and start their own research with hemp extracts. The treatment of neuropathic pain conditions, the prophylaxis of migraines, and the treatment of menstrual complaints in Queen Victoria by her personal physician John Russell Reynolds are the best-known examples. That the menstrual complaints of the queen were always a burden and she was not averse to cocaine wine, we have already read.

Cannabis for the Treatment of Epilepsy

We owe not only the first scientifically substantiated reports on cannabis to William B. O'Shaugnessy, but also the invention of the intravenously administered isotonic saline solution. Not during his service as a doctor with the East India Company in Calcutta, but in the no less dangerous London, where one cholera epidemic chased the next, he came up with the idea of compensating for his patients' water loss in diarrheal diseases through infusions. Cholera is a dangerous infectious disease in which patients can lose several liters of water per day through diarrhea. Especially children dehydrate due to diarrhea and besides the administration of antibiotics, the supply of water is an important part of the therapy. O'Shaughnessy had a great interest in Indian folk medicine, which included also *Cannabis indica*. He was impressed by the analgesic and antispasmodic effect in rabies and tetanus. From this observation, the origin of epilepsy treatment emerged. If you consider all the miraculous effects of cannabis, probably two or three real indications remain: epilepsy, post-traumatic stress disorder, and pain in cancer therapy.

The little Charlotte Figi (2006–2020) had a hard life in the 13 years she was given. The little girl suffered up to 50 epileptic seizures a day and at the age of six, the classic and approved antiepileptic drugs no longer helped. She suffered from Dravet syndrome, a genetic disorder in which there is a mutation of the sodium channel. Cells pay close attention to a concentration gradient of salts such as potassium, magnesium, calcium, and sodium between outside and inside. If this gradient is disturbed, especially in nerve cells in the brainstem, it can lead to false stimuli and reactions, which in Charlotte led to severe forms of epilepsy.[300]

The first epileptic seizure occurred when Charlotte was three months old. The doctors diagnosed a gene mutation at position 2791 of the SCN1A gene, where the amino acid arginine is replaced by a cysteine. When she was five years old, the doctors gave her and her parents little hope. She was considered to be out of therapy, all common drugs would no longer work. The desperate parents heard about cannabidiol and consulted their doctor Sanjay Gupta. Out of desperation and because there were no good alternatives, Charlotte took CBD oils containing cannabidiol until her death. Joel Stanley, who breeds this low-THC and CBD-rich cannabis plant in Colorado, where cultivation and sale are legal, provided the special cannabis strain. His oils contain CBD and very little psychoactive THC. As a worker in the oil fields of Texas, he had no idea about the Wild West of cannabis. But he left one oil business and successfully set up his own with the other plant-based CBD oils. He sells his oils under the name Charlotte's Web, which is to remind of little Charlotte. In any case, the CBD oils surprised everyone. Because in combination with the prescribed medications, Charlotte was able to reduce her high seizure numbers to two to three per month.[300]

Loewe and Goodman discovered the antispasmodic effect in 1947, shortly after the Second World War, and found that not THC, but CBD is responsible for this positive effect. Although cannabis extracts had been used in China and the rest of Asia for a very long time,[301, 302] it was Mechoulam who proposed cannabidiol as an active ingredient against epilepsy in the 1980s. Unfortunately, the effect of CBD on humans could not be researched because cannabis was banned in the USA, even though the antispasmodic effect in mice and cats was already known in the early 1970s. It took another 20 years until a molecular biological effect for cannabidiol was actually found to reduce epileptic seizures, which to the surprise of many was not due to cannabinoid receptors. Other receptors seemed to be responsible, which act on dopamine, glycine, noradrenaline, and serotonin. The exact mechanism of action still needs to be researched. In two large studies from 2013 and 2014, it was shown that both in children and adults the frequency of seizures decreased by up to 80% by taking three milligrams of cannabidiol per day. A final large study confirmed the good effect of cannabidiol again and it is to be hoped that CBD will provide new ideas for future innovative drugs given epilepsy.[303]

Cannabis for the Treatment of Glaucoma

"The guy is using cannabis, I can see that right away," grumbled the emergency doctor when he saw the dazed patient in the emergency room of Los Angeles Medical. The hippie movement was over, but marijuana had not yet lost its place in the left-liberal student scene. Like a rabbit with red eyes, the young man stared at the doctor who was examining the pupils' reaction to light stimuli. Everyone knew that chronic stoners have red eyes because THC promotes blood circulation and lowers blood pressure.[304, 305] Ophthalmologists were interested in this phenomenon as they considered it a dangerous side effect of smoking pot. Then in the 1970s, a patient suffering from pigment glaucoma and seeing the characteristic halos in his eyes with increased intraocular pressure came along. He noticed that the halos disappeared when he smoked marijuana.[306] He applied to the court to be allowed to purchase and use marijuana for the treatment of his glaucoma. This permission was granted and he became the first legal user of cannabis for medical reasons in the USA. The eye disease with high internal pressure is called glaucoma. Glaucoma is a creeping disease of old age and can lead to blindness. The reason is the high sensitivity of the optic nerves at the back of the eye, which die off with increased intraocular pressure. We know that our societies are all aging and the number of glaucoma diseases will drastically increase in the next 20 years. New active ingredients like cannabinoids are therefore of high interest.

Studies on the effect of cannabinoids to lower intraocular pressure were difficult due to the ban. The first studies in the 1970s and 1980s were very small and did not meet today's standards as they only included a handful of healthy volunteers and no glaucoma patients. The results were not very meaningful. They only showed that cannabis lowered the eye pressure in two-thirds of the subjects by about 25% for two to four hours. In the early 1980s, further poorly controlled studies were conducted in which the volunteers quit smoking and took eye drops, which however showed no effect on intraocular pressure. The interested doctor Robert S. Hepler looked into stoners' eyes and later discovered that THC was responsible for the effect. But the fear of ophthalmologists in the following decades was that a prohibited substance was being administered and they wondered whether the effort was worth it when other approved drugs are on the market? Mahmoud ElSohly (*1946) from the University of Mississippi saw it differently and thought about chemical derivatives of THC and cannabinol, which he synthesized and tested. The effect was clearly demonstrable in animal experiments and the tested mice and rats showed a reduction in intraocular pressure due to THC, but not due to CBD in the animal model. However, rat and mouse experiments cannot simply be transferred to humans. The rodents are nocturnal animals and their eyes are adapted differently to the nocturnal environment than ours. Another interesting phenomenon is also that the mouse—unlike the rat—is not sensitive enough and does not show a significant reduction in intraocular pressure.

Another problem that had to be solved was solubility. Most cannabinoids like THC or CBD have very poor solubility in water, which is between 3 and 10 milligrams per liter depending on the publication. The production of eye drops is further complicated by the fact that poorly soluble drugs, which precipitate in the form of small microparticles, are perceived as unpleasant in the eye. Good eye drops should be a clear solution that does not contain any suspended particles and does not trigger tear flow. Tear flow is not desired at all, as the drug, which is only in contact with the eye for a few seconds, is quickly washed out again. It is therefore a high art to produce good eye drops. They must be more viscous than water and stay on the eye, but they must not stick together or even obstruct the view. The problem of solubility and the residence time in the eye should be solved by two excipients. The US company Emerald used hyaluronic acid to increase viscosity and cyclodextrins to dissolve THC and cannabinol.[307] Cyclodextrins are ring-shaped sugars consisting of five to seven glucose units. To this day, these eye drops are not available in the USA or Europe, but initial trials against placebos seem to be successful. Cyclodextrins have been known for many decades and are found in everyday products. Anyone who wants to bind unpleasant odors with a room spray unknowingly sprays cyclodextrins, which are dissolved in the propellant gas. In instant teas, flavorings are bound so that they do not evaporate, in cat litter they bind the smell of cat pee and in Japan in the 1990s cyclodextrins were woven into shirts so that the Japanese man does not smell so strong under the armpit.

The Declaration of War on Cannabis

How did the worldwide ban come about, even though the plant was valued for its fibers in technology and for its extracts in medicine? Historians believe in a coincidence between an official without a task in the Federal Bureau of Narcotics (FBN) and the emerging plastics industry in the USA. This official was Harry Jacob Anslinger (1892–1975, Fig. 17), who was also a diplomat of German-Swiss origin and a fervent advocate of cannabis prohibition.[308] His father left Switzerland because he did not want to be conscripted into military service, went to Pennsylvania in the USA, and worked there as a

Fig. 17 Harry Jacob Anslinger (1892–1975) (public domain, thanks to Pennstate University Library and Archive)2

barber. Harry Jacob was the eighth of ten children and his father sent him to the railroad as a teenager, which was laying tracks in Pennsylvania at the time, to earn some money. The boy was smart and urged his father to let him go to school at least in the mornings. The time on the construction site was formative, as he came into contact with Chinese, Blacks, and Latinos, who worked as simple construction workers. He also clashed with the Mafia, which extorted protection money from the Sicilian workers. Big Mouth Sam was the boss of the local Mafia and despite his big mouth, young Harry Jacob was not intimidated and threatened to kill him if he did not leave his people alone. Very brave for the young man, but Big Mouth Sam left his customers alone. We do not know the reasons. Since this time he hated the Mafia and in his later job as president of the FBN he should recognize the danger and pursue the Mafia before Edgar Hoover, who still downplayed the danger of organized crime for America as FBI director.

He sat at a large desk in the office of the FBN, the predecessor of today's Drug Enforcement Agency (DEA), and was involved in the enforcement and control of alcohol prohibition in the USA until 1933. At the age of 38, he was appointed head of the FBN and was the first, before Nancy Reagan with her "Just Say No" campaign and Richard Nixon, to declare the war on drugs (War on Drugs). His power and the harshness of the laws he enforced are still felt today in the USA, where 3.5% of Americans are in prison for violations of the Marihuana Tax Act of 1937. This law was less about the tax and more about the prohibition of the acquisition and consumption of cannabis. In his opinion, opium and morphine should also not be used medically. Nevertheless, in the 1930s he organized the storage of 300 tons of opium in the USA, as he saw the Second World War coming. Anslinger's opium was stored in the vaults of the US Treasury in Washington D.C., which were cleared of gold with the opening of Fort Knox. The background to his aversion to opium was the large number of Chinese who immigrated as refugees via San Francisco at that time and brought with them their preference for smoking opium. It was a time when crime and drug abuse were equated. A vivid example was alcohol prohibition and the spread of the Mafia, represented by Al Capone. Only through alcohol could the Mafia become big and only the Mafia could smuggle and sell alcohol illegally, according to Anslinger. With the repeal of alcohol prohibition in 1933, there were no more federal subsidies for his federal agency and he had to find another drug to justify subsidies and a reason for his agency. Cocaine and heroin were uninteresting to him, as there were too few consumers and no politician would have approved the subsidies that would have kept everyone in work and bread. Cannabis was perfect.

His crusade was to be against cannabis, which was legally brought into the USA by Mexican immigrants, because millions of people consumed cannabis. But he had to find good reasons, because cannabis did not cause serious crime, which would argue for a ban. As president of the FBN, he questioned doctors and asked police departments across the country to send him all reports of murders related to cannabis. He waited, but no reports reached his desk. But then came the long-awaited report from Florida, which would finally provide him with the reason for his campaign.

Victor Licata (1912–1950) from Tampa, Florida, is said to have murdered his family with an axe on October 17, 1933, under the influence of marijuana. Victor Licata denied ever having smoked cannabis, and a review of his medical records in 1998 showed that he never took marijuana and was mentally ill. His brother was diagnosed with schizophrenia, and two of his great-uncles were also mentally ill. Victor Licata was found not guilty due to schizophrenia and acute episodes and was never convicted. The report never stated that he was under the influence of drugs. However, the suspicion of the investigating police officer W. D. Bush was enough for Anslinger to finally get his alleged case that a teenager became a murderer because of cannabis. Bush claimed that Licata had been smoking cannabis cigarettes six months before the crime. The police were not sure, and the police chief backpedaled, but emphasized: *"Maybe the weed had a small indirect part in the alleged insanity of the youth, but I'm now declaring for all the time but the increasing use of the narcotic must stop and will be stopped."* [Perhaps the weed had a small indirect part in the alleged insanity of the youth, but I am now declaring for all time that the increasing use of the narcotic must be stopped and will be stopped]. Anslinger made use of the mass media and received a lot of support from William Randolph Hearst, the publisher of tabloid journalism, to turn the anti-marijuana sentiment from the state level into a national movement. He used the so-called "Gore Files," a compilation of quotes from police records, to dramatically depict the crimes committed by drug addicts. They were written in the concise and poignant language of a police report, as this one also described the case of Victor Licata:

"A twenty-one-year-old boy in FLORIDA killed his parents, two brothers and a sister while under the influence of a Marihuana 'dream' which he later described to law enforcement officials. He told rambling stories of being attacked in his bedroom by 'his uncle, a strange old woman and two men and two women,' whom he said hacked off his arms and otherwise mutilated him; later in the dream, he saw 'real blood' dripping from an axe." [A twenty-one-year-old boy in FLORIDA killed his parents, two brothers and a sister while under the influence of a marijuana "dream" that he later described to law

enforcement officials. He told rambling stories of being attacked in his bedroom by "his uncle, a strange old woman, and two men and two women," who he said hacked off his arms and otherwise mutilated him; later in the dream, he saw "real blood" dripping from an axe.]

The strong propaganda on the radio and in the newspapers was successful. In 1937, cannabis was criminalized by a federal law that Anslinger had co-written. Now the war against cannabis could be waged. The war chest was additionally filled by a large donation of 10,000 USD from the banker A. Mellon, who was a shareholder of the chemical company DuPont. In times of recession, this was an unimaginably large sum, but it was well invested, as the company wanted to sell nylon fiber and the natural hemp fiber interfered with the business. Allegedly, the Mafia also supported the FBN with money. The best deals are only made when a substance is banned and there is no free market.

Who did Anslinger pursue? The white boy in high school? No, he hated blacks and was known as a rabble-rouser who frequently used the N-word in his vocabulary. He despised the hated jazz scene in New Orleans, New York, and Chicago, where cannabis and other drugs were consumed. He believed that this strange music was only played because the musicians had no real sense of time after consuming cannabis and time was stretched. Anslinger's war was less a war against drugs than a war against culture, an attempt to suppress the radical freedom of jazz for people of color. *"There are a total of 100,000 marijuana smokers in the USA, and most are Negroes, Hispanics, Filipinos, and entertainers. Their satanic music, their jazz and their swing are the result of marijuana consumption. This marijuana causes white women to seek sexual relations with Negroes, entertainers, and others and if you smoke just one joint, you will probably kill your brother."*[309,310] The jazz singer Billie Holiday (1915–1959) became his particular object of hatred, which he had pursued by his agents. The drug-addicted singer was systematically destroyed. In a letter to her friend, she wrote: *"The witch hunt and the constant pressure made me think of the last resort, death."* Her tragic life, described by Johann Hari in his book "Drugs," ended in methadone withdrawal, handcuffed to the bed and two police officers at the door.[308] Anslinger did not let up even in death.

Anslinger's efforts were not only directed at artists and regular consumers, but also at scientists. Under strict observation by the FBN was Roger Adams, who was introduced as the first scientist to successfully isolate CBD. With some luck, he received a license in 1939, two years after the cannabis ban, to research cannabis, but Anslinger's officials watched his fingers and mouth very closely. In a social gathering, Adams must have spoken too positively about cannabis. His informants brought Anslinger the news of the *"pleasant effects of using this drug."* The public reprimand followed directly, for *"in his opinion, this drug is bad for human consumption and should therefore be eliminated."* The "Office for Strategic Services", the predecessor of today's CIA, became interested in cannabis as it was looking for a truth serum to quickly distinguish friend from foe during wartime. In the USA, the "Manhattan Project" began, named after the New York district where a secret office of the project was rented. Everything was subject to the strictest secrecy, as the USA wanted to build the atomic bomb before Nazi Germany. Unreliable employees and soldiers were not allowed to participate. Cannabis extracts were administered as a truth serum to find out if there was a German spy among them. Roger Adams was appointed to the "National Defense Research Committee" to support the war efforts. Hoover considered this appointment a very bad idea, as he suspected Adams of being a sympathizer of the communists. Adams received clearance with delay and presented himself as an anti-fascist in his public appearances. He was always suspected as a leftist and cannabis researcher of doing something against American society. Unjustly, but in the upcoming McCarthy era, the eponymous politician McCarthy (1908–1957), who incidentally had an alcohol and morphine problem, already considered un-American behavior sufficient for persecution. Adams remained steadfast, researched and published 27 studies on *cannabis* in the American Journal of Chemistry and since 1959 the Roger Adams Award has been awarded every two years in his honor by the American Chemical Society.[310]

In historical retrospect, Anslinger was a hate preacher against drugs, blacks, Latinos, or all combined.[308] The doctor Henry Smith Williams visited Anslinger in 1931 to ask him to reconsider the application and enforcement of the cannabis law, for no addict can be cured with harshness. *"Doctors,"* he said to him, *"cannot treat addicts, even if they wanted to."* He demanded *"tough judges who are not afraid to throw murderers in jail and throw away the key."* Although the Supreme US Federal Court demanded treatment of addicts before imprisonment in 1925, Anslinger commented on this with the words *"building clinics for heroin addicts would be like setting up department stores for kleptomaniacs."* His FBN ignored the federal court. Historian John C. McWilliams is convinced that today's prison industry in the USA is a result of Anslinger's harsh and inappropriate prison sentence, as he writes in his book "The Protector".[310]

The Cannabis Moss of the Maori

Cannabis sativa seemed for a long time to be the only plant that biosynthesizes cannabinoids, but recent research paints a different picture and a moss in distant New Zealand also contains cannabinoids in small amounts. Although this was already published by a Japanese research group in Tokushima at the end of the 1980s, nobody took the scientific publication seriously.[311] It slumbered in the archives for a full 30 years until the big hype about cannabis broke out. From the 2010s onwards, my research group from Dortmund became interested in it.[312] Together with colleagues from Auckland, we collected the moss and analyzed it again. As in James Cook's times, we searched in the fantastic Kauri forests of the North Island and found the moss *Radula marginata* on shady stones and on the bark of dead trees. An initial examination on site, in which the dried moss was smoked, showed a weak effect. However, the laboratory analysis clearly showed that the extract contained cannabinoids. The concentration is much lower at about two percent than that of the cannabinoids in cannabis. Further investigations showed that moss cannabinoids dock onto the good CB2 receptor and probably have a positive effect on the immune system. The exciting question, of course, is what ecological effect Perrittonin, as the most important cannabinoid in the moss is called, could have. We do not believe that it is a protection against grazing, it is more likely that it has an insecticidal effect. Perhaps this is also a good reason for the cannabis plant to protect itself against beetles, caterpillars or other six-legged grazing enemies.

Whether the Maori in New Zealand as indigenous people smoked moss is not securely documented. There are no written records in the archives that the author searched on site. There are no reports from English colonists or researchers who met the Maori and wrote about such a custom. Conversations with younger Maori also do not provide a clear picture, as smoking was unknown in their culture until the arrival of the Europeans. However, it cannot be ruled out that the knowledge of the slightly intoxicating effect of the cannabis moss was nevertheless passed on orally from generation to generation. Was it a coincidence that a Japanese scientist received the moss from a colleague who was on an excursion? Why not, Dr. Chance would not be the first time in the lab.

Cannabis—a Bright Future?

What does the medical future of cannabis and cannabinoids look like? Both will not have a common future, even though they seem inseparable. Cannabis flowers, briefly called cannabis, will be released from medical tolerance in the coming years. In a country that wants to ban smoking, it is incomprehensible that medical smoking is accepted. This reminds us of the time after the Second World War, when the *Ephedra* herb was sold in pharmacies as a smoking product for the treatment of asthma. At the turn of the century, Cigarros Indios made from *Cannabis indica* were freely sold in tobacco shops in Brazil as a sleep aid. These Indian cigarettes were offered by the Parisian company Grimault & Co., which added datura and belladonna to the hemp. From 1872 they were on the market in the German Reich and immediately became market leaders. The plant material for the German hemp cigarettes came from the company Bronchiol, which has been producing in Berlin since 1900, or from the pharmacist family Trnkoczy from Laibach. The future belongs to the extracts and the pure chemical cannabinoids, which are still largely unexplored today. The many so-called rare cannabinoids, which occur in very small quantities in the plant, are synthetically replicated and tested. Ingenious chemists will, as with all known natural substances, chemically alter the structures and adapt them so that their effect on the endocannabinoid receptors can more effectively cure diseases such as epilepsy, pain, post-traumatic stress disorders, etc.[313,314]

We still know too little about the physiology of the endocannabinoid receptors, so new insights for drug research and new therapies are expected in the near future. What are these receptors, whose function is so difficult to grasp? With the opioid receptors, it was very clear. It was about pain and its suppression, but should the endocannabinoid receptors only be responsible for the high in us? That makes no sense. To understand what these receptors mean for us mammals and us humans, we have to go far back in evolution. In the time of hunters and gatherers, the cannabinoid receptors were certainly important for the development of the intelligence of our species. They are not only found on a cell surface that transmits a stimulus into the cell interior, but they are closely integrated into a complex biochemical network of endocannabinoids. Don't worry, we humans do not produce our own cannabinoids, we have only named this network after them, they are polyunsaturated fatty acids that act metabolically converted. Well-known fatty acids such as arachidonic acid, which we already got to know in the chapter on aspirin®, react at the two receptors CB1 and CB2. Now it also becomes complicated for biochemists. Because suddenly we are dealing with endogenous substances that act as prostaglandins in pain and inflammation or as leukotrienes in immune responses in a cascade. As in a well-balanced mobile with many substances, enzymes and receptors, a plant cannabinoid now pulls on the string and changes everything. Like the baby who is excitedly watching the swinging and rotating movements, we do not understand the mobile either, but we are fascinated by the effects on our body.

Scientists wondered if there is a connection between high-fat diet, endocannabinoid receptors, and our eating behavior.[315] Cannabis consumers often report that smoking joints stimulates their appetite. This can only be demonstrated in the animal model in rats, but not in humans. The BMI as a measure of overweight is usually lower in consumers, which is explained by stress factors and the generally lower food intake in this group. However, it is known that white fat cells show increased CB1 receptors, which can be stimulated by THC to absorb energy in the form of fat and sugar. We humans as intelligent dry-nosed monkeys need a lot of energy for our brain, which accounts for about a fifth of our energy turnover. The size of the brain is not a criterion for intelligence, rather it is the networking of our nerves and nerve pathways that distinguishes us from many of our conspecifics. Did we need high-energy food like fat and meat to give us an evolutionary advantage over all other species? It is interesting that in the gorilla only three percent of the food consists of fat, the rest of fiber and carbohydrates.[828] In humans, the fat content in the diet can be up to 30%, as is the case with the Inuit. With the animal fats, we take up more highly unsaturated fatty acids, which are advertised as omega fatty acids. Once absorbed, not much happens with these special fatty acids that our body cannot produce itself. It stores them as a reserve in the cell membranes and converts them into prostaglandins and leukotrienes as needed. Perhaps the endocannabinoid system can be seen as a brilliant metabolism for the absorption and utilization of unsaturated fatty acids from our hunter-gatherer times? That is an exciting question. Because it is not unlikely that a reward system has developed in our brain that gives the endocannabinoid receptors a new reward function when meat, fat and thus unsaturated fatty acids come to the table. It is interesting that many other reward receptors, such as those for dopamine or glutamate, are closely linked to our taste receptors. These findings were an incentive for pharmaceutical companies to develop new active substances for smoking cessation, as slimming aids or as a drug against anxiety states.

We still know too little about the complex endocannabinoid system. We still have much to learn and only discover by chance what cannabinoids can do. When the world went into collective quarantine under SARS-Covid-19, newspapers reported that smoking cannabis could make the course of Corona diseases milder. Publications showed that cannabidiol attaches to the ACE receptors and blocks the virus's docking site, preventing it from being absorbed into the cell. An exciting insight that shows us that there is still much to expect.

Healing Poisons

It seems to be a great contradiction that poisons are supposed to heal. We understand a poison to be a substance or mixture capable of causing severe damage or even death in us humans—and also in animals. We know poisonous substances from classical chemistry like arsenic or cyanides, but the deadliest poisons are found in the realm of animals and plants. This does not sound nice, but we must and should be aware that the discovery of medicinal substances would not have been possible without poisons and drugs.

Our ancestors quickly learned, when they ate and tasted plants, that some species were not at all digestible or even deadly. In the clan or tribe, it was closely watched whether the neighbor, friend or relative was writhing on the ground, had to vomit, got muscle cramps or closed their eyes forever. From then on, these plants were avoided by the rest of the tribe and the knowledge was passed on to the collective knowledge over generations. Our ancestors learned in thousands of self-experiments and involuntary field studies what later became an antibiotic, chemotherapeutic or remedy for high blood pressure. These poisons were defused in the chemical laboratory and are now an important part of our treasure of medicines like methadone from opium poppy, captopril from snake venom or blood pressure reducers from the red foxglove.

Whether a substance is a poison or a good remedy depended, as Hippocrates (460–370 BC) wrote, fundamentally on the amount ingested. The ancient Greeks found this rough distinction inappropriate. The first clear distinction between good pharmakon and poison is made in Latin by Paracelsus. The term "Venum", which stands for the poison of plants, animals or minerals, is created. Other Roman authors such as Pliny the Elder (1st century AD), Quintus Serenus (2nd and 3rd century AD) and

Caelius Aurelianus (5th century AD) follow suit and distinguish it from the "Remedium" and "Medicamentum" for the good pharmakon.

The term Pharmakon, as we understand it today as a medicine, was in ancient Greece the generic term for remedies. Already in the Homeric epics of the 8th and 7th centuries BC, the term "pharmakon" is used, and the root word pharma probably means magic. The Greeks already knew that a pharmakon can have both healing and poisonous properties.

In ancient Greek, in addition to the Venum, there is also the term "Ios", which means animal poison. Other meanings are arrow and rust, patina or verdigris. Possibly the Greeks had the idea that the administration of an animal poison by a sting was the equivalent of an injury by an arrow. A connection to rust and patina is also conceivable, as bacterial infections such as tetanus can arise from contamination in wounds. In Aristotle's zoology (384–322 BC), a reference to scorpions and venomous snakes is made for the first time. Vague definitions of what poisonous animals are, we already find in early cultures. Here too, the Greeks were very interested in providing a classification. According to today's understanding, a subdivision was made according to transmission routes such as stinging, spitting, biting or unnoticed exhalation and exhalation. It is not surprising that in addition to the classic poisonous animals, which the Greeks called "Iobola", such as scorpions and snakes, fleas, bugs, mosquitoes as well as rabid dogs, biting bears, wolves, crocodiles and even humans were included in the list. The fact that cats, rats and fleas were counted among the poisonous animals may be due to the ignorance of the actual bacterial causes at the time. The ancient doctors did not know that the supposed poisonous animals were mother ships for the actual pathogen.

Poisonous animals exist on all continents, with Australia seeming to produce particularly poisonous animals. Why there are poisonous animals certainly has to do with the selection pressure in evolution. Every ecological niche or refined strategy to make prey, to get sexual partners, had to be exploited. Animal poisons play a special role in natural substances, as their chemical structure differs significantly between animal and plant toxins. While in plants it is mostly simple structures like alkaloids, in poisonous animals we often find peptides or proteins. These proteins are complex and only in the last 20 to 30 years have we gained access to their structural secrets due to improved analytics.

People poison themselves with animal poisons out of ignorance, they are accidents and less by intent. It is clear that few people in modern times are in constant contact with venomous snakes or are threatened. Even if the number of poisonings seems to be small compared to other dangers such as traffic or accidents in our societies, it is worth researching the poisons of animals. Today, four drugs are known that can be derived from snake venoms. But who as a researcher would like to work with such poisonous substances when the risk of dying is not insignificant? Today, researchers take a different approach to avoid being stung, bitten or poisoned. Catching an animal is enough to get a complete analysis of the poison from the genes, the blood or the poison glands. With the help of genetic engineering methods, the substance can be reproduced without having to hunt or kill more animals.

From Bee Stings and Frog Poisons

One of the most well-known poisons is bee venom. Perhaps you have been stung by a bee or wasp before and still remember the piercing pain. You can still clearly see the redness and swelling that lasted for hours and physiologically means a strong blood flow. Therefore, in the 1920s to 1940s of the last century, the venom of the bee *(Apis mellifica)* was used to treat rheumatism, as advocated by the German-American doctor Constantin Hering (1800–1880). He was also the one who presented the first comprehensive pharmacological study of bee venom in 1857. With the advent of painkillers like Ibuprofen, interest waned, but in the 1990s there were still two larger clinical studies, which however made contradictory statements about the benefits and were again forgotten.

In the poison dart frog *Phyllobates terribilis*, the strongest animal poison known to date has been found: Just 0.5 milligrams of the steroid alkaloid batrachotoxin (Fig. 1) can be lethal, as the half-maximal lethal dose for mice is already at 0.002 milligrams. However, it is doubtful whether the frog's

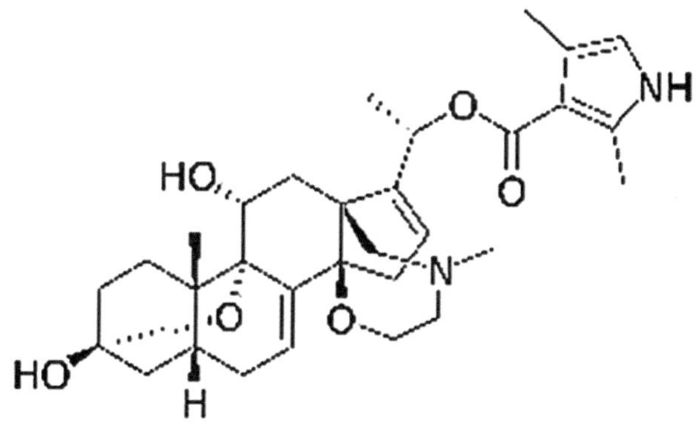

Fig. 1 Batrachotoxin (work of the author)

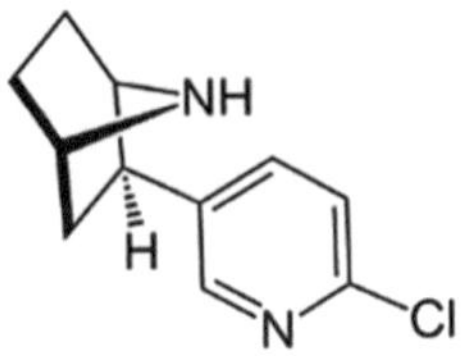

Fig. 2 Epibatidine (work of the author)

poison is a true animal poison. If you examine the frog's skin, you can isolate lice that presumably produce the poison from the frog's ingested food and thus protect it from predators. By keeping these beautiful and colorful frogs in terrariums and changing their diet, they quickly lose their poison. A frog poison that has made it into preclinical pain research is epibatidine (Fig. 2) of the tree frog *Epipedobates tricolor*, which John W. Daly and Charles W. Myers discovered in 1974 on the Río Jubenes in the southern rainforest of Ecuador near the Peruvian border. They collected 300 frog skins for their analysis at the NIH in Bethesda, USA, to find new painkillers.[316] But how could he discover the effect in the jungle and safely select more frog skins with the active ingredient? The answer is simple and takes us back to the chapter of self-experiments. He licked all the frogs, and when the tip of his tongue felt numb and bitter, he knew exactly that this poison frog had enough poison on its skin. With the extract, he performed the Straub test for opiate effect and many mice raised their tails. But he did not find the substance, it was somewhere in the extract, the amount was too small. He had to go back to Ecuador. Daly flew back to the South American country in 1976 and went to the same place where he found the frogs on his first visit. The shock was great, because where the frogs croaked, there were banana and cocoa plantations. The few frogs he still found were no longer poisonous. Not a single one could numb his tongue. He went further to the road that passed the plantations and a little further into the forest. Here he found his poisonous favorites, of which he collected and skinned 750. From these skins, he obtained a complexly composed extract of 60 milligrams. He isolated 21 milligrams of pumiliotoxin, which has been known since the 1960s from Panamanian poison frogs and South American ants, and another 500 micrograms of epibatidine. He performed the Straub test with the extract and saw that the mice's tails stood up. As a control, the mice were injected with naloxone, an opiate inhibitor, but to Daly's surprise, the tails stayed up and the mice did not lick their paws on the hot plate. Was the test faulty or did this epibatidine have a completely different, unexpected effect? What was this chemical structure in the frog's skin? Daly traveled to Ecuador again in 1979 and 1982 to collect more material and find out what the frogs eat. He found no useful clues and could not clarify the source. He rightly suspected that there must be a connection between the diet and the toxicity. To this day, this question has not been answered. The unusual alkaloid with a chlorine atom was structurally clarified only in 1992 and synthesized the following year.[317] Now there was enough substance available for all further investigations. Today it is known that it is 200 times stronger than morphine. However, it seems to have a different mechanism of action, as it more closely resembles nicotine and reacts with nicotine and muscarinic receptors, not with opiate receptors.

What happened to Epibatidin? The American pharmaceutical company Abott became aware of the substance and cooperated with Daly and the NIH. A synthesis program was initiated and the candidate ABT-954 was included in the clinical trial. However, the phase II studies were discontinued due to gastrointestinal side effects. The substance and its effect caused a great stir in the media and it was falsely claimed that Abott was clinically testing the natural substance Epibatidin, which was not the case. To this day, the favorite food of the frogs in Ecuador is still unknown and improved analysis techniques show that about 50 alkaloids are still waiting to be discovered in the skin of the tree frog.

The Viagra of the Renaissance

Casanova swore by it, Cardinal Richelieu, Madame du Barry and the mistresses of Louis XV. also took it. Half of Europe in the 19th century knew the deadly Viagra, obtained from a green metallic shimmering beetle—*Lytta vesicatoria* (Fig. 3). This insect is better known under the common name "Spanish Fly", although it is neither from Spain nor a fly. This insect is a type of beetle that occurs all over Europe, from Great Britain to Poland. The beetle became known as a potency enhancer because the alkaloid Cantharidin (Fig. 4) accumulates in the males. This causes a strong irritation of the urinary tract and can lead to an erection up to a permanent erection. Dear men, a permanent erection may sound tempting, but it is a medical emergency. If such an erection lasts longer than four hours, it can lead to serious problems, such as permanent erectile dysfunction, which can worsen to gangrene. In the case of low-flow priapism, as the hard permanent erection is called, pain treatment is carried out. In some cases, blood is taken from the corpus cavernosum and sometimes vasoconstrictive drugs like

Fig. 3 *Lytta vesicatoria*, Spanish Fly (public domain, wikipedia)

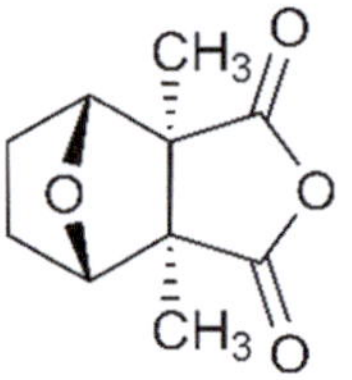

Fig. 4 Cantharidin (work of the author)

adrenaline are injected. The famous violinist Niccolò Paganini (1782–1840) suffered from priapism, which was triggered by the sight of a woman.[318] The poor man sought treatment from Samuel Hahnemann, the founder of homeopathy, but the treatment had to be discontinued unsuccessfully because Paganini fell in love with his wife and made advances that clearly displeased the great doctor.

Hand on heart, what do you think? Are there more plants or animals that produce toxins? The answer may surprise you, but animals and plants are balanced, perhaps there are fewer plants than animals that biosynthesize toxins. No one knows exactly how high the number is among higher plants, but according to Kingsbury in 1979, there are about 750 toxins in 1000 plants, and if we only consider those that cause a strong effect in humans, then the number is significantly lower. Plant and animal toxins seem to balance each other out.

Cases of poisoning by plants are rather rare in our latitudes. The most dangerous animal in Germany is the deer, as every year 30 people die in car accidents involving deer. The low number of poisoning cases is certainly due to the civilizational alienation of modern humans, who hardly come into contact with plants anymore, except when visiting a botanical garden. Their environment is characterized by monocultures from the hardware store and the eradication of subjective weeds in their own garden. Apart from consciousness-expanding drugs like cannabis or psychogenic mushrooms, poisonings in adults are rare. In emergency medicine, they are attributed to the collection of wild herbs for personal tea or mushrooms in late summer. The risk of poisoning is higher in children, as they follow their play instinct and curiosity and try and eat plants with tempting flowers or fruits. In cases of poisoning by ingestion, the yew *(Taxus baccata)* with its beautiful red pseudo-fruits is at the top of the list. This is followed by the cherry laurel *(Prunus laurocerasus)* and rowanberry (*Sorbus aucuparia*, Fig. 5). The latter is believed to be poisonous in folk belief, which is not true. Rich in vitamin C are the fruits of the rowanberry, which are processed into compote and jam. For intestinal and stomach complaints, the tannins and the parasorbic acid are probably responsible.

Fig. 5 *Sorbus aucuparia* L. (public domain, wikipedia)1

In recent years, a plant that has an effect on our skin has proven to be particularly problematic. Giant hogweed or the Heracleum contains phototoxic furanocoumarins, which can trigger meadow dermatitis under the influence of UV light outdoors. In gardeners or people who like to work in their garden and come into contact with the plant, it can lead to redness and even severe blistering, which can cause intense burning. Since the plant is long and hollow, children like to cut it off, hold it to their eye, and play with the supposed telescope. Severe blistering of the eye is known and must be treated by an ophthalmologist. However, the toxic ingredients of the plant have also advanced medicine a good deal. The furanocoumarin psoralen (Fig. 6) is used in PUVA therapy to treat psoriasis and neurodermatitis. PUVA is the abbreviation for **P** soralen **UV-A** therapy. For phototherapy, the doctor and dermatologist Niels Ryberg Finsen (1860–1904) received the Nobel Prize for Medicine in 1903. He is to this day the first and only Faroese, the inhabitants of the Faroe Islands, to have been awarded a Nobel Prize. Currently, the natural substance psoralen is no longer used and is replaced by the natural 8-methoxypsoralen (8-MOP). This 8-MOP is taken orally two hours before irradiation or applied to the skin as a cream and then irradiated with UV-A light. Since the UV light can only penetrate a few millimeters into the skin, the active ingredient is first activated in the skin and reacts in the upper layers of the skin. In recent years, PUVA therapy has been replaced by irradiation with UV-free blue LED light, which also has a growth-inhibiting effect on the excessive proliferation of skin cells and a reduction of inflammation.

Three more natural substances are particularly close to my heart. I have chosen a microbiological, an animal, and a plant poison, which either became drugs or were tested as candidates. Of course, this is a personal selection. You have already discovered other plants and natural substances in this book, which could have been listed in this chapter with equal justification. Other poisons like Curare, Digitoxin, Tropa-Alkaloids or Strychnine were elevated to drugs in the past, their very strong toxicity remained, which is why they were replaced by new and safe drugs and have disappeared again from the pharmacopoeia. The chapter on ergot may seem familiar to you, but besides LSD, there was also very serious research that resulted in drugs that have had a great significance in the last 70 years.

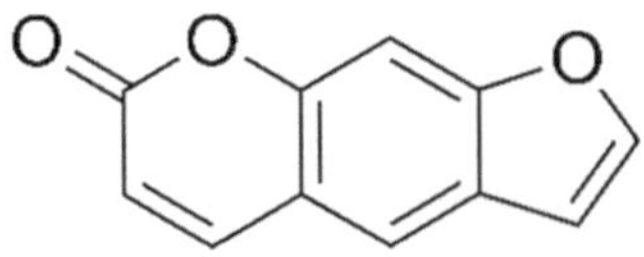

Fig. 6 Psoralen (work of the author)

Ergot—From Plague to Mr. Hofmann's Problem Child

Brief profile – Ergotamine

Structural formula:

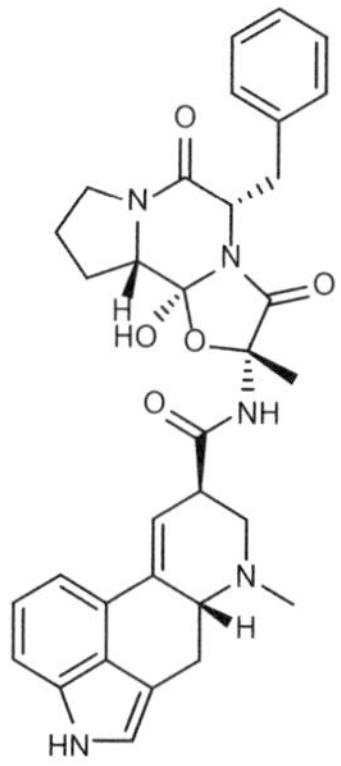

Natural substance group:
- Alkaloids
- Ergot alkaloids

Derived substances:
- Methylergometer
- LSD
- Methysergide
- Bromocriptine

Occurrence:
- Claviceps purpurea, ergot

Effect / Indication:
- Vascular contraction
- Contraction of smooth muscles

Possible medical significance:
- Obstetrics
- Migraine prophylactic

He saw the farmer lying in the dirt of his hut. He gave him the last rites. The young farmer moaned and repeatedly asked for water, for he was thirsty and felt the fire of hell within him. He raised his arms and opened his mouth with its rotten teeth and actually believed that the fire was now coming out of his throat, at least he could see it blazing and later testify to it before his Lord and God. The woman was already dead and curled up on the bed or whatever you might call the damp bundle of hay in the corner. Flies sat on her eyelids. Her newborn child, whom he had been able to baptize and thus snatch from the devil just a few days ago, had long since been transferred to the graveyard into the deadly earth. What had ended the short earthly existence of these two poor souls here and the other sinners in the village in the service of our Lord Jesus Christ? What had happened here? The feudal lord and the nobility seemed to have been almost spared, but here death knocked relentlessly on each of the shaky doors and they willingly opened them. He stood in the semi-darkness of the hut of the two and the piss of the pigs and other damp filth seeped through his poor leather shoes into his trousers. The two pigs became restless, the lost soul of the farmer could no longer feed them. There he lay on the ground and breathed confused prayers into the air, which he did not quite understand. He still held his calloused palms together, and the blue, brittle fingernails and fingers intertwined tremblingly. Suddenly the poor man let his hands sink and a short groan released the last breath from his chest. The Lord had taken him to Himself. The farmer fell face down into the mud. The monk looked around in the dirt of the hut, hoping to find a few belongings. What could he still turn into money? The oil had to be paid for and no craftsman would make a coffin for the two if a few coins did not jingle into his hands. Many of the farmers and craftsmen could no longer use their hands. The fingers were missing, they walked as cripples without toes or pushed themselves without legs begging through the mud of the streets, hoping that a coin or a piece of bread would fall into their open hands. He saw the two pigs. One for each expired soul and a wooden coffin. That would be enough, if he could find a carpenter at all in these godless times of plague. He looked out of the open door, it was raining and raining and had been for weeks.

The monk Sigebert of Gembloux must have been greatly affected by the raging disease in Lorraine when he reported in his chronicle in 1089: *"It was a year of plague, especially in the western part of Lorraine, where many, whose insides were consumed by the holy fire, rotted at their gnawed limbs, which turned black as coal. They either died miserably or continued an even more miserable life after the rotten hands and feet were cut off."* He was completely helpless and the doctors of his time could not help him. It would take until the 17th century from the first descriptions of this disease, which occurred endemically after the consumption of infected rye, until the cause was recognized. The church tried to provide comfort and advised the veneration of Saint Anthony, who lived in the Theban desert in Egypt in the 4th century and is said to have performed miraculous healings there. In his honor, the disease was called St. Anthony's fire, because the sick had the feeling of burning internally and felt an inner fire. Today this disease is called ergotism, derived from "ergot", which in French means ergot, because in France St. Anthony's fire raged particularly fiercely. Between the 9th and 11th centuries, there were regular epidemics in Germany, France, and Spain with up to 40,000 deaths. A very severe epidemic is documented for the year 857 in Xanten on the Lower Rhine. *"It swept people away through a terrible decay, so that body parts detached and fell off before death,"* as the "Annales Xantenses" report. In 954, ergotism is mentioned in Paris, 994 in Aquitaine in southwestern France and in Limousin. Not only the geographical assignment remains interesting, because in France ergotism was characterized by gangrene of the fingers and toes, while the chroniclers reported cramps in the German form, even if here the extremities fell off.

The church not only advised prayer. The Antonite Order, which belonged to it, made it its task to help the people who suffered from ergotism. In order to increase the importance and credibility of the order, they were not squeamish and transferred the bones of Saint Anthony from Constantinople to the province in the village of La-Motte-aux-Boix near Grenoble and founded a monastery. In the coming years, another 370 monasteries were established, which were used as hospitals. Up to 4,000 people suffering from ergotism were cared for.[319] The Antonites received their hospital order, which was confirmed by Pope Boniface VIII in 1247 and merged into the Order of Malta in 1777. The transport of the bones was a splendid idea and the basis for a comfortable living as a monastery. People flocked to the Saint Antoine monastery, where they could also purchase medicines such as the Antonius wine, the Antonius ointment for vasodilation. This wine had to help, because the relics were dipped into it to transfer the holy power.

We owe the Antonites not only their mercy and the care of the sick, but also a magnificent work of art that Matthias Grünewald (1480–1530) created between 1512 and 1516.[320] In Isenheim near Colmar in Alsace, he accepted the commission for a wall altar that still takes one's breath away today. Grünewald was an ascetic and great man who was very open to the political currents of his time. His career is interesting, because despite his brilliant art, which was certainly influenced by Albrecht Dürer and Lucas Cranach the Elder, he first worked as a water master on castles, which today might be a kind of water engineer. In his middle life, he became an artist and served as a court painter for the Archbishop of Mainz, but after completing his masterpiece, he left his service and earned his living as a soap maker in Frankfurt.

What kind of impression must the Isenheim Altarpiece have made on the common people and the monks? Look at the triptych and turn your gaze to the right picture of the altarpiece. It is called "The Temptations of Saint Anthony", but pay attention to the hooded man in the lower left. He raises his left hand, missing fingers. A clear sign of ergotism, in addition, many bumps can be seen, perhaps from the plague that raged just a few decades ago. The red genital area clearly indicates the then little-known syphilis, which probably came to the Old World with the discovery of South America. In the painting, in the middle, a white cloth with a lost foot can be seen. Beggars in the Middle Ages often carried their rotten limbs with them to reinforce their plea for a donation. On the right, a white man without his right leg can be seen, and the monk in the blue robe seems to be hallucinating. Of far greater interest is the multitude of plants at the bottom of the adjacent picture. Whether these were plants for the treatment of ergotism is questionable, but from the left you can see plantain, verbena and broadleaf plantain (in front of Anthony) as well as bulbous buttercup, couch grass, glandular root, spelt, wound clover, deadnettle, poppy, gentian, speedwell, swallowwort and cyperus grass. All were common medicinal plants at that time and belonged in every well-stocked apothecary garden. But Matthias Grünewald was not the only painter who immortalized ergotism in his pictures. The theme was a popular motif of that time. Before him, Hieronymus Bosch painted a group around Saint Anthony hallucinating from ergot in 1501. In the triptych with a similar title "The Temptation of Saint Anthony" you can see a mummified foot on a white cloth in the middle picture. Art historians also suspect a reference to ergotism here.

How did the poisonings feared by the population come about? In addition to the climatic conditions, which will be discussed later, the change in diet is probably worth mentioning. From antiquity to the early Middle Ages, wheat was on the menu, which is not affected by the ergot fungus. Rye, which was called "grain" in peasant language, prevailed because it was hardy and survived the winter. It was more resistant to frost, drought and soil requirements.[321] Rye was considered an ancient grain and the staple food of the people, which is why the nobility preferred wheat to distinguish themselves according to their status. This also had the advantage that it did not or rarely suffered from ergot. Wheat was not a grain of northern Europe. It came with the Roman legionaries, who ate it in the legion cities. It was imported along with olive oil from the south. Like the Roman legionaries, wheat also retreated behind the Alps with the collapse of the Roman Empire. The logistics broke down and the Germanic farmers had to resort to what grew well locally—rye and oats. With the invention of the iron sickle, it was no longer necessary to harvest close to the ground, and rye as a naked seeder willingly released the grains from the husk. However, the rye grains had to be ground, and so the many thousand water mills in Germany, France and England were created, which we can still romantically view today.

It would take another 600 years before the cause of the terrible St. Anthony's fire in humans was recognized. The French monk Robert Drumont was on the trail of the mystery and wrote in 1125 that *"the grain was mixed with a dark corrupted grain, as a result of which the flour became sanguinolent, discolored and poisonous."* However, his hypothesis was not compatible with the four humors theory and the prevailing doctrine of the time, which blamed a thickening of the bile juices for the fire. The farmers recognized that the black sclerotia were the cause of the terrible St. Anthony's fire and sorted them out of the harvest. From the 17th century onwards, ergotism became rare and there were only occasional poisonings. The last epidemics in Europe occurred in 1926/1927 in Sarapol in the former Soviet Union, where 11,000 people were poisoned, 1928 in Manchester and 1951 in Pont Saint-Esprit in France. In the 1980s, there was once again heavily contaminated rye in Germany because it was grown without fungicides when organic farming began. Today, dark ergot grains are automatically detected and sorted out using lasers. Cases of poisoning are no longer known.

The special effect of *Claviceps purpurea* was recognized in medicine and the discovery of its components aroused great interest in research.[322] As early as 1808, the American gynecologist John Stearns recommended administering ergot extracts to accelerate childbirth. Ergot alkaloids were used from 1824 in the USA and England to stop postpartum bleeding. Because of their wide distribution, they were included in the pharmacopoeia of the USA in 1820 and in the pharmacopoeia of London in 1836.

Fig. 7 Henry Hallet Dale (1875–1968) (public domain, wikipedia, creative commons 4.0) 2

British chemists at Wellcome were concerned with the question of which natural substances in the sclerotia could have an effect on the uterus. The biochemist, physician, and later Nobel laureate Henry Hallet Dale (1875–1968, Fig. 7) and the chemist George Barger isolated ergot alkaloids in the Burroughs Wellcome Research Laboratories in London, but could not assign a structure to the substances. Dale remembered well the ergot gruel, as he revealed in his speech on the occasion of the Nobel Prize ceremony in 1936 in Stockholm. His boss Henry Wellcome, who ran the Wellcome Laboratories, had brought Dale into his company in 1904 because he thought he was doing him a favor by entrusting him with ergot research. After all, there were still a few sacks of sclerotia in the basement and for the young Dale that would be a nice exercise. He just wasn't interested. It was the pharmacological Pandora's box, he complained. But it didn't turn out that bad, on the contrary, the research with the ergot alkaloids should show him the way to acetylcholine and its receptors, for which he received the Nobel Prize together with Otto Loewi (1873-1961) from Vienna.

Why couldn't Dale isolate the ergotamine, purify it, and cross the finish line before Arthur Stoll? Barger recalls in his memoirs that both made a fatal mistake. They found many crystals in the ergot lye before the Swiss and mistakenly assumed that it was ergotoxin and only a single substance. Only in 1943 did Arthur Stoll and Albert Hofmann clarify that it was a mixture of three similar alkaloids: ergocrinin, ergocristin, and ergocriptin. Now Barger and Dale understood why they kept getting non-reproducible results. The disappointment was great, even though they had simultaneously carried out further great work on acetylcholine, histamine, and tyramine, which were just as exciting.

It was Arthur Stoll (1887–1971) who discovered and purified ergotamine as the main alkaloid of ergot in 1918. Arthur Stoll was born in Schinzach-Dorf, Switzerland, in 1840 and, like Albert Einstein before him, attended the Aarau Cantonal School, which he completed with his Matura. Arthur Stoll initially studied botany and geology at ETH Zurich, but under the influence of his later mentor Richard Willstätter (1872–1942), he switched to chemistry. After his doctorate, he followed his doctoral supervisor Willstätter to Berlin at the Kaiser Wilhelm Institute for Chemistry. His research area was the chlorophylls, which he had already dealt with during his doctorate. He qualified as a professor in 1915, but he did not complete his habilitation in Berlin. Willstätter went to the Ludwig Maximilian University of Munich, his student followed him again and in gratitude he was appointed Royal Bavarian Professor of Chemistry in 1917. An offer from industry seemed more attractive to him. In the same year, he left Bavaria, returned to Switzerland to the pharmaceutical company Sandoz (now Novartis), and made an impressive career there. Together with his colleagues and the most famous researcher Albert Hofmann, who became famous as the discoverer of LSD, he succeeded in bringing ergotamine to the market under the name Gynergen® for obstetrics in 1921. In addition to ergotamine, the researchers at Sandoz found other ergot alkaloids like ergosin and ergocristin. The special thing about these plant alkaloids is that in addition to the nitrogen-containing building block typical for alkaloids, a small peptide residue is attached. This made it so difficult to clarify the discovered chemical structure. It was not until the 1950s that this question could finally be clarified. Stoll clarified the structure of ergotamine as the first ergot alkaloid in 1951 and ten years later Hofmann succeeded in synthesizing it. The structures found by Stoll and later chemically converted to LSD by Hofmann, as described in the previous chapter, are interesting.[322]

Walter A. Jacobs and Lyman C. Craig from the Rockefeller Institute for Medical Research in New York made a structural proposal for lysergic acid in 1936, which was confirmed almost 20 years later by the synthesis of dihydrolysergic acid.[323,324] These works represented an important step for the further understanding and in-depth research of the entirety of the ergot alkaloids.

Earlier works from the 1920s had shown that the vasoconstrictive effect and the benefit in migraine was great, but the known uterine constricting effect could not be attributed to any substance.[325] Dale simply named the substance Ergot-X, but that was to change with John Chassar Moir (1900–1977) who entered the scientific stage in the early 1930s. Moir had completed his studies at the medical faculty of the University of Edinburgh in 1922 and specialized in obstetrics. The young Scot was also interested in biology and the German language. With a Rockefeller scholarship, he went to Leipzig and Berlin in the 1930s and read fluent German medical literature. His interest in the alkaloids of ergot grew and the discovery of the hemostatic alkaloid for use in obstetrics was to become his research area. The young gynecologist worked in 1932 as an assistant to Professor Browne at the University College Hospital in London. He read works from the USA that showed that fresh ergot extracts had a particularly good effect on the uterus. Excited and with the idea in mind to find Ergot-X, he quit his job and went to Dale, who was enthusiastic about his project. Moir had developed a strange pharmacological model: It consisted of a rubber balloon that was inserted into the uterus of a puerpera. The balloon was connected to a water manometer. The pressure fluctuations in the manometer were recorded. He administered ergotoxin and ergotamine orally and intravenously with moderate success, as the balloon showed no pressure change. Moir was disappointed that he could not detect any extraordinary effects and wanted to try again with the extract as a reference. To his surprise, this showed a very strong contracting effect. The needle of his manometer shot up and he feared for the safety of his test subject, who was lying in front of him on the bed. However, she was fine, and on later visits he assured himself of her well-being, as he wanted to repeat the experiments.

He brought the records to Dale, who immediately recognized that Moir had found something special and called Harold Ward Dudley (1881–1935), who had been working as a chemist with Dale since 1932. He immediately set to work, which he was fortunately able to complete before his early death in 1935 at the age of only 48. Dudley isolated an alkaloid, which they called

Ergometrine. The name already indicates its use: Metrin stands for Uterus menstrus (womb) and was used for bleeding after childbirth. On October 2, 1935, the day of Dudley's death, the discovery of ergometrine was published. However, a continuation of the successful British work in the field of ergot alkaloids was no longer possible. Dudley and Barger died early, Dale was looking forward to his retirement in the 1940s, Moir took up a professorship in Oxford and did not want to research ergometrine anymore. The field was left open to the Swiss, and Stoll and Hofmann at Sandoz became the ergo-experts of the post-war period.

The pharmaceutical company Sandoz was the world leader in the production of ergot alkaloids until the 1950s. They were followed by Boehringer Ingelheim in Germany, Galena in the Czech Republic, and Gedeon Richter in Hungary, to name just a few. To prevent theft and illegal production of LSD, production took place at secret locations. Rye fields were deliberately inoculated with the fungus *Claviceps purpurea* and allowed to grow. The extraction process could be significantly simplified with a few biotechnological research efforts. When in late summer, in August, the black sclerotia fall to the ground to form spores in the next spring, which can infect new rye fields via the air, this spore formation was shortened in the laboratory. For this purpose, the sclerotia were collected, the surface was disinfected to avoid foreign germs, and the fungus was propagated in the laboratory on agar plates. This was followed by the transfer and propagation in a liquid nutrient medium to prepare for use in the field again. In the 1950s, workers still walked through the green rye fields with a sponge in their right hand and a blunt nail board in their left, penetrating the still immature rye ears with the spore suspension. In the 1980s, the infection routines were automated. At the end of August, the sclerotia could be collected manually or automatically using a laser in the early years. After the Second World War, Portugal and Spain were still the main growing countries, followed by cultivation for pharmaceutical purposes also in Germany and behind the Iron Curtain in the Czechoslovak People's Republic. At peak times, up to eight tons of ego-peptides and 15 t Lysergic acid were isolated worldwide.

Ergot alkaloids roughly consist of two parts. On the one hand, the lysergic acid, which is considered the actual alkaloid, and a small peptide of three amino acids, which is bound to the lysergic acid. The name lysergic acid comes from laboratory language, as it is the product after the lysis of the alkaloids. Interestingly, all ergot alkaloids have the same lysergic acid, which is biosynthetically derived from the amino acid tryptophan. The resulting peptide alkaloids, of which more than 80 are known today, make ergot alkaloids really interesting for medicine, as they are able to stimulate dopamine receptors and are simultaneously active at the serotonin receptor. The fact that ergot alkaloids can now mimic dopamine or serotonin explains their vasoconstrictive effect and their effects on the uterus, which is why they were later used in obstetrics. Their hallucinogenic effect is lower, as they cannot cross the blood-brain barrier due to their peptide portion. Only when this part is broken down in the body does the lysergic acid also reach the brain in small amounts.

The brief history of ergot alkaloids cannot be concluded without mentioning further chemical derivatives. Both chemical modifications were motivated by the desire to make the effect of the ergot alkaloids more specific and to reduce side effects. A chemical modification to LSD, which gave lysergic acid a completely different pharmacological direction, has already been mentioned. In fact, according to the accounts of Rudolf K. A. Giger and Günter Engel, the Novartis AG seriously considered discontinuing research due to the hallucinogenic effect. However, the story continued with Methylergometrine(Methergine®), Dihydroergometrine (Dihydergot®), Methysergide (Deseril®) and Bromocriptine (Parlodel®), which entered the clinic as further drug substances.

Albert Hofmann wanted to specifically eliminate the double bond in lysergic acid and succeeded in doing this in ergotoxin. These ergot alkaloids without a double bond are called ergolides and were later used to treat mild hypertension in older people and also for the treatment of dementia patients and stroke victims. However, recent studies have shown little success, which is why they have lost clinical significance in recent years. The Ergoloid Mesylate (Hydergin®) and Dihydroergometrine (Dihydergot®) have a stronger effect on the alpha-adrenergic receptors and release more norepinephrine, which leads to a reduction in blood pressure. Norepinephrine is the antagonist of the stress hormone adrenaline, which is released during acute stress. Hydrated ergotamine has also proven helpful for migraine patients. Migraine is a neurological disease that many people perceive as very strong, one-sided headaches. It can be accompanied by nausea, vomiting, and a high sensitivity to light and noise.

Bromocriptine was the last real innovation in ergot chemistry and the active ingredient with the bromine atom in the lysergic acid structure of Ergocryptine entered the clinic in 1975. It was to be a last great achievement, as bromocriptine mimicked dopamine. It simulated its effect at the receptor and was therefore used in the treatment of Parkinson's Disease and disorders of the pituitary gland (hypophysis). In people with Parkinson's disease, the lack of dopamine, which can no longer be produced in the atrophied substantia nigra, is the main cause. It does not release dopamine, but it is like knocking on the door of the cells in the brain under a false name. The symptoms can be alleviated in Parkinson's patients who suffer from severe hand tremors and rigidity of movement. In disorders of the pituitary gland (Hypophysis), bromocriptine inhibits the release of the hormone, so that excessive growth of the pituitary gland (acromegaly) or tumors can be treated.

What came after the ergot alkaloids? Was it the end of a highly interesting group of natural substances? The ergot alkaloids were taken off the market a few years ago because their side effects were too great compared to their benefits. However, new interesting drugs were developed from the lysergic acid framework. Similar to the morphines, chemists took a close look and rediscovered the indole structure. Although Albert Hoffmann had already recognized the importance of the indole structure when he discovered the psilocybin, it was not until the 1970s that Peter Stade brought the indole structure to the center of interest. Migraine patients could benefit from this, because in 1982 Peter Stadler from Albert Hofmann's group, managed to synthesize the Tropisetron as a prophylactic against migraines. Not for the treatment of migraines, but against vomiting during chemotherapy, it came to the patients in 1995 as Navoban®. The doctors recognized that a new group of serotonin receptor agonists had emerged, substances that act at the receptor as if they were the activator. Tegaserod the of both chemists R. Giger and H. Mattes came as Zelma® or Zelnorm® into the clinic. Without the knowledge of the ergot alkaloids, these structures would not have been conceivable.

Snakes—from Venom to Blood Pressure Reducer and the Birth of Computer Chemistry

Brief profile – Captopril
Structural formula:

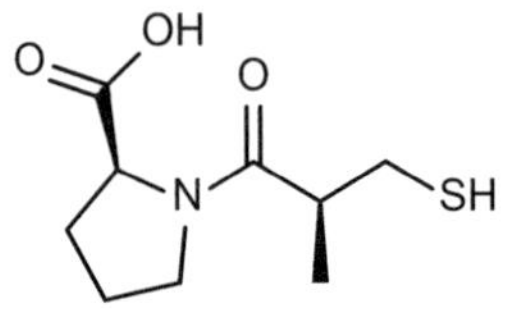

[Structure of Captopril]

Natural substance group:
- Amino acids derived from teprotide

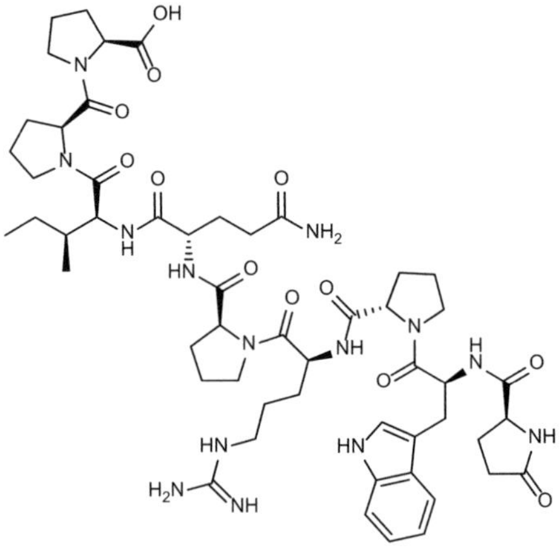

[Structure of Teprotide]

Teprotide

Occurrence:
- Not a natural substance

Effect / Indication:
- ACE inhibitors

Possible medical significance
- Antihypertensives

The world of natural substances is very colorful and we have already learned about many natural substances from plants and microorganisms in this book. The fact that we also learn from snakes and take their venoms, which they inject into prey or enemy with a bite, as a model for new drugs, is rather unusual. The structure of these venoms is also unusual, because unlike the rather small molecules, we are dealing here with a peptide that is significantly larger than all substances presented so far. If we continue to deal with the exciting snake venoms, the question still needs to be clarified what a peptide is. A first thought is certainly that it could be related to a protein and that is correct. Peptides are small proteins consisting of no more than 50 linked amino acids. This is not a uniform definition, because some researchers see the end of the line already at 20 amino acids and therefore distinguish between peptides with a length of less than 20 amino acids and polypeptides with a length of more than 20 amino acids. We are not interested in this in detail here, because it is about peptides as toxins.

In the living cell, proteins and peptides are built up somewhat differently. The large classical proteins like albumin, interferon, insulin or receptor proteins are built up with the help of a biochemical synthesis machine, the ribosome, which is supplied just in time with activated amino acids that are linked together in a certain order according to a genetic blueprint. For the peptides, there is a second biosynthesis pathway, which is rather referred to as an assembly line. Just as Henry Ford assembled parts into a Ford T at various stations using an assembly line, there is also a molecular assembly line in the cell. This assembly line is biochemically a large enzyme complex with various stations and from one station to the next, the peptide is passed on and extended by a defined amino acid. Once the length specified in the blueprint is reached, the peptide is cleaved off and possibly stored. This biosynthesis differs from the biosynthesis of complex natural substances. It is often found in microorganisms that are not capable of performing complex biosyntheses like in plants.

A question in my circle of acquaintances always confirms to me that snakes are not popular cuddly toys. The snake is considered dangerous, is feared and often pursued with great hatred. This rather negative attitude can be found as early as the dawn of mankind. With the snake in paradise, which tempts Eve to hand Adam the apple of knowledge, the snake has more than a dubious reputation among Christians. However, snakes are unjustly given a bad reputation, as they are shy animals and avoid contact with humans. Very few snakes are real venomous snakes and use their venom or, in the case of constrictor snakes, their bodies to hunt and defend themselves. There have been no deaths from snakes in Germany in the last 50 years. Nevertheless, a bite from the native and harmless adder still unjustly triggers panic reactions.

The bite of a viper is not enough to kill a human. For a body weight of 75 kg, a lethal dose of 480 milligrams is necessary. A snake injects about 40 milligrams with its bite. You can quickly calculate that more than ten snakes would have to bite, and that is not a real danger. The situation is quite different in the tropics and in Australia. Here, it is the unintentional and very dangerous bites of snakes. They feel threatened by humans when they are startled by field work or by children playing. Snakes enter huts at night to hunt mice, rats, and other rodents. People believe that snakes are after them or their children. The snake is pursued, panics, and defends itself with a bite. Estimates range from 100,000 to 400,000 deaths per year worldwide. In addition, there are 1.3 million bite victims with a high number of unreported cases.[326, 327] This is a significant number, but the true number of snake bites with sometimes fatal outcomes is estimated to be much higher.

The term venomous snake suggests that it is a separate genus or family, which is not the case is. Vipers, rattlesnakes or venomous cobras can be classified into different families like fish or monkeys. What they have in common is that they produce a venom that is stored in the teeth. There are about 600 venomous snakes on earth and all have a more or less poisonous venom cocktail in their venomous teeth, which can be dangerous to us humans. In some snakes, the venom is found in the fang and other teeth in the snake's dentition are rather atrophied, but they have reserve teeth and smaller teeth, which can be arranged in up to six rows. Venomous snakes usually swallow their prey whole and digest it over a longer period of time. In snakes, the jaw is not fixed like in humans, the upper and lower jaws can be separated to widen the mouth and swallow larger animals like rats and mice at once.

The widespread assumption that the venomous teeth are filled with venom does not apply to all venomous snakes. Some colubrids use their smooth teeth for catching and the close contact of the oral mucosa with the prey is enough for the venom to be passively absorbed by the prey. This is not very effective, as the snake has to hold its prey for a longer time. Other colubrids therefore have grooved venomous teeth, which have a groove. The venom flows from the Duvernoy's glands in the upper jaw into the wound. True tubular teeth, which function like injection needles, are found in venomous colubrids like cobras, mambas, kraits and coral snakes. These teeth, up to 30 mm long, have a mucous membrane fold at the base, which is opened by muscle pressure when biting, so that the venom can flow from the gland into the prey. The teeth can break off during defensive or hunting attacks, but are replaced within a few days. In rattlesnakes, it is common for them to replace their fangs every six weeks, always leaving one in the jaw, you never know.

What are snake venoms chemically speaking? As with plant extracts, the venom cocktail is a mixture of various peptides, proteins and some smaller substances like histamine, which are supposed to inhibit blood clotting and keep the wound open by swelling. In fresh condition, snake venom is a viscous, milky-white to yellowish liquid. The water content can be between 50 and 90%. If you remove the water, the venom can be stored in the refrigerator for years. The dry powder consists of almost 90% proteins. The typical snake venom does not exist, each species has its own composition and factors such as habitat, diet, age or conditions in snake farms have an influence. Younger snakes have a significantly more poisonous venom cocktail than older ones and even freshly hatched snakes are poisonous. What the snake venom is supposed to do is very interesting. Of course, it should be poisonous and kill the prey immediately. Who wants to slither after a prey without legs and then lose it. It also helps immensely if the prey is already pre-digested by the bite. Therefore, digestive enzymes are injected into the cocktail. This makes the prey digestible and allows it to decompose in its own digestive tract. These digestive secretions are therefore highly concentrated in the venom and make the drama of a snake bite: the skin dissolves or bursts, there are extreme swellings or massive hemorrhages.

Brazil is a country known for its many species of snakes due to its tropical location and enormous biodiversity. As humans continue to penetrate deeper into the tropical rainforests, they expose themselves to an increased risk of being bitten. This explains the high number of snake bites in this country. Sergio Henrique Ferreira (1934–2016) focused on the venom of the Jararaca lancehead *Bothrops jararaca* in his doctoral thesis and was tasked by his boss to find out what the venom mechanism is. What was the problem? A massive drop in blood pressure was known to occur in many bitten patients. Sergio Henrique Ferreira completed his doctoral thesis and found a peptide that he held responsible for the blood pressure-lowering effect. As is often the case with dissertations, one must find an end after three or four years, even if many questions remain unanswered. Ferreira knew the peptide, but did not understand its biochemical effect. It was only during his postdoc stay with John Vane in London, whom we have already met in the chapter on aspirin®, and who was dealing with pain and prostaglandins, that he was able to clarify the mechanism of action.

David W. Cushman and Miguel A. Ondetti (1930–2014, Fig. 8) were working at Squibb & Son, now Bristol-Myer-Squibb (BMS), in England

Fig. 8 Miguel A. Ondetti (1930–2014) (public domain, wikipedia)2

on a new physiological control circuit that promised new blood pressure reducers. It was the renin-angiotensin system, which regulates blood pressure in humans. While it was not yet known at the time that aldosterone, an important mineralocorticoid, also plays a major role, this control circuit was later expanded to the renin-angiotensin-aldosterone system. In summary, this control circuit can be explained as various hormones and enzymes controlling the body's volume balance. An important enzyme is the angiotensin converting enzyme (ACE), which converts an inactive protein from the kidney into an active form, leading to a narrowing of the blood vessels and indirectly to an increase in blood pressure. This ACE can also break down bradykinin, another important protein responsible for the dilation of blood vessels. This massive drop in blood pressure in humans is caused by the snake venom of the lancehead *Bothrops jararaca*. In his doctoral thesis in 1965, Sergio Henrique Ferreira found a pentapeptide in the venom cocktail, which he named BPF. He knew it had to act via bradykinin and named it Bradykinin Potentiation Factor (BPF).

Upon his arrival in London, Ferreira was not yet familiar with the renin-angiotensin system, but he brought along a vial of freeze-dried BPF, because you never know. His boss, John Vane, whom we got to know as the discoverer of cyclooxygenases as the site of action of acetylsalicylic acid in the chapter on aspirin®, was preoccupied with the renin-angiotensin system. In the 1960s, he was also dealing with high blood pressure, was a scientific advisor at E. R. Squibb & Son, and regularly met with Vice Presidents Charles G. Smith and Arnold Welsh, the boss of the two scientists David Cushman (1939–2000), Miguel Ondetti was. Cushman was a biochemist and Ondetti a recognized peptide chemist. They all rather reluctantly dealt with the ACE, which they isolated from the lungs of dogs and for which they were now looking for a good inhibitor. The state of knowledge at that time was that ACE and the renin-angiotensin system only played a role in critical, rapidly escalating episodes of high blood pressure. But Vane convinced them that ACE inhibitors could also play a role in normal high blood pressure. If this were the case, Squibb could make a big leap in the world of cardiovascular diseases. Leapfrog? Smith must have laughed when Vane used this comparison. But Smith countered that research in the field of the renin-angiotensin cycle had been going on for 30 years and the leading cardiologists had a very different opinion than he did. But Smith wanted to give BPF a chance and consulted twelve leading cardiologists worldwide, only John Laragh from Columbia University Medical School was enthusiastic about this BPF research. It was not the doubt that it would not work with the peptide from the Brazilian snake. Smith was much more concerned with the question of how a peptide that cannot be administered orally to the patient should be active in the body. Smith had several problems. No one would inject a peptide daily against high blood pressure. The synthesis of the peptide was unusually expensive. A kilogram would cost Squibb a million dollars, the therapy would not be competitive and he had to explain to the board why the company was risking its hard-earned money so riskily. Only Vane and Laragh believed in the peptide throughout the company. But Vane was to be proven right and his clear arguments convinced the board to take the risk.

Smith gave the green light, and Vane suggested to the team at Squibb to meet with his postdoctoral fellow Ferreira, who had the snake venom from his homeland here in London and was ready to collaborate with them. The pentapeptide BPF was tested and showed an excellent effect on ACE in the lungs of dogs. Research on the renin-angiotensin system woke from its slumber, as BPF was so potent that everyone immediately recognized the value of an ACE inhibitor. With Ferreira, further pentapeptides were sought, BPF was given the new name BPP (Bradykinin Potentiating Pentapeptide), as there were now other, even longer peptides that could significantly lower blood pressure in experimental animals. The next step was to test the blood pressure-lowering effect in humans. Laragh recruited 17 volunteers with essential hypertension and showed that blood pressure significantly normalized in 14 of them. This also finally convinced the Squibb board, which released funds for research and development.

All these beautiful new peptides were very active, but had a serious disadvantage: They did not survive the gastrointestinal passage because they were broken down by enzymes after oral intake. Only one peptide, Teprotide with four proline amino acids at its end, was superior. But the oral bioavailability was not satisfactory and the high price for the synthesis meant the end for this nonapeptide. Ferreira and Ondetti, both South Americans, were looking for a solution. The Teprotide molecule was simply too large to make it more orally tolerable through chemical changes. Ondetti's group searched unsuccessfully for 2,000 analogous nonapeptides until 1973. After a failed clinical trial, he was instructed to abandon the project and devote more time to antibiotic research. But like many scientists, he could not resist and continued to work on the project in secret, ignoring his boss's instructions. He, a working-class child from Buenos Aires who had to finance his own studies, was a fighter. Giving up was not an option for him. He had to find a new approach and the emerging field of computer chemistry offered an alternative. Ondetti pursued the approach of rational design, which was just in vogue. Rational design meant for him the targeted search for structures and their adaptation to the target structure of the nonapeptide. In a time without great computing power, this was a remarkable achievement. Today, the rational design of active substances is based on supercomputers and millions of structures are generated, tested, discarded or improved in silico in one day. Ondetti, who possibly had a construction kit to assemble structures three-dimensionally, must have been enormously creative, because just a year later he made the breakthrough—Captopril was born.[328]

On a rainy Wednesday morning, March 13, 1974, Ondetti and Cushman met to discuss their results and discuss the current literature. On the table was a paper by Byers and Wolfenden, which was already a year old. A short summary appeared on a small card in the series "Drugs in Prospect", which informed researchers about new chemical substances that could become interesting new active substances. Ondetti picked up the card and read: *"L-Benzylsuccinic acid—a strong inhibitor of carboxypeptidase A"*. What did this have to do with their project? The binding of amino acids in the active center of enzymes was known, but what was the deal with L-Benzylsuccinic acid? Byers and Wolfenden discussed whether an acidic group for the Carboxypeptidase A was responsible. A good question, also found Ondetti and Cushman. They wondered if there were parallels between ACE and Carboxypeptidase A? Was ACE perhaps also an exopeptidase, which nibbles proteins from the end instead of splitting them right in the middle? If that were the case, the researchers suspected from their degradation studies on angiotensin, then it would be two amino acids in the active center, the throat of the ACE, that. If there were further parallels to carboxypeptidases, they could also assume that as in the Carboxypeptidase A a zinc atom in the active center initiates the reaction. What would then be the ideal candidate? Definitely a proline, because they already knew that from the BPP5a, and they should try it with a succinic acid, as Byers and Wolfenden had suggested. From these considerations, it was decided to synthesize the Succinyl-L-Proline, which was an analog of the dipeptide glycine-proline.

The succinic acid-L-proline was a bitter disappointment, as it was 30,000 times weaker than the later Captopril, as the new ACE inhibitor was to be called. Were they right with their rational design idea? Yes, because the succinic acid-L-proline showed a selective inhibition of ACE. In pharmacology, a distinction is made between potency and selectivity. A substance can inhibit many different enzymes very strongly. An inhibitor can be weak, but very selectively inhibit only one enzyme. Exactly these are the substances that pharmacologists are looking for. Selective inhibition is important, chemists can try to increase the potency later by modifying the lead structure. That's exactly what Cushman and Ondetti did. They added a methyl group to the succinic acid-proline and tested the D-2-methyl derivative of succinic acid-L-proline. It was not only orally bioavailable, but also 15 times more potent than succinyl-proline. On March 31, 1975, the first orally bioavailable ACE inhibitor was found. Over the next two years, the potency was improved. The real breakthrough came when the acid in the molecule was replaced by a sulfur group. The effect increased 2000-fold compared to succinyl-proline. Based on these new sulfur compounds, both developed another 60 structures, among them was Captopril. It fit perfectly into the throat of the ACE and prevented a dipeptide from being bitten off.[329]

Captopril is not immediately recognizable as a chemical structure consisting of at least one amino acid. However, upon closer inspection, it becomes clear that proline is hidden in the structure. This makes sense, as BPP5a has a sequence of tryptophan-alanine and proline, which was Ondetti's creative lead structure. For him, it was the simplest structure he had ever found in his life as a chemist, but it fit so well that he wrote almost touchingly about the journey of discovery in his publication "History of the Design of Captopril and Related Inhibitors of Angiotensin Converting Enzymes".[329] For medicinal chemistry, it was the first active ingredient developed using a computer model of the BPP5a structure. Captopril marked the beginning of a technological development in which computers play an increasingly important role in the discovery of drugs. Today, modeling programs are indispensable in research, and every new active ingredient undergoes numerous in-silico studies, as we call these new research works in reference to in-vitro or in-vivo studies. What started as a leapfrog has become a quantum leap.

Paclitaxel—an Academic Success Story

Brief profile – Paclitaxel
Structural formula:

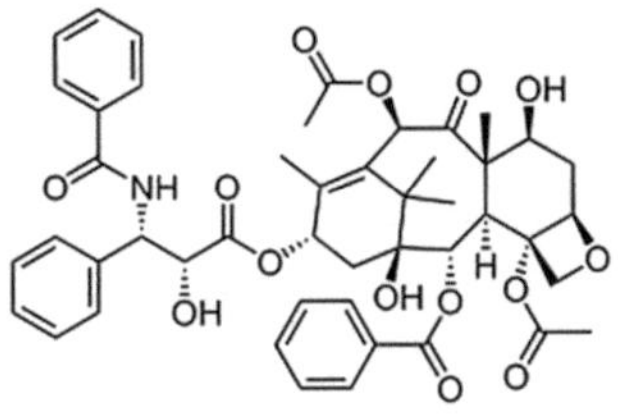

Natural substance group:
- Diterpene
- Taxane

Derived drugs

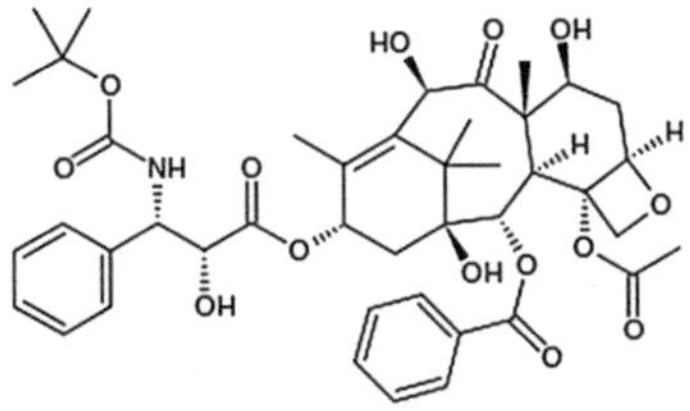

Docetaxel

Occurrence:
- Natural substance (paclitaxel)
- Semisynthetic natural substance (docetaxel)

Effect / Indication:
- Cytostatic drug

A nation like the United States has never been content with mediocre projects. Whether it was John F. Kennedy's ambitious program to send the first man to the moon, or Richard Nixon's declaration of the war on drugs in 1972. Already a year earlier, he had declared another war, the war on cancer. *"I will propose the provision of an additional 100 million US dollars to start an intensive campaign to find a cure for cancer, and I will later request further funds as they become available,"* he declared in his annual State of the Nation address at the end of January 1971. The goal, according to Andrew von Eschenbach (*1941), who accompanied this program as director of the American National Cancer Institute (NCI) for some time, should be to make cancer curable in all areas of medicine by 2015 and to lose its terror. A gigantic program was launched and of course plants and natural substances were of great importance. All over the world, a large number of plants and microorganisms were collected, cultivated or directly extracted. Up to 35,000 plants have been tested to date as part of the program using a very standardized procedure for their possible effect against cancer cells. This number may seem small when you consider that there are an estimated 370,000 higher plants on earth. However, many of these plants only occur in small numbers in ecological niches. Their collection is therefore very demanding and sometimes a test can endanger the stock if a minimum quantity is required for extraction.

It is often stated in magazines that the tropical rainforest is the treasure chest of green gold. However, it is a fallacy to believe that most medicinal plant species can be found in the tropical rainforests of Africa or Brazil. While it is true that Brazil, along with Indonesia and Central Africa, is considered the country with the greatest biological diversity, many interesting hotspots of diversity are located in arid areas or in mountainous regions. The next time you visit the Table Mountain in Cape Town, you will probably find more plant species there than in the Netherlands. The Himalayas and some mountain ranges in the Middle East are also highly interesting for botanists. Many of the medically interesting plants do not come from the rainforests, but from steppes and dry areas. There, in the truest sense of the word, the plant is fighting a battle for survival against bacteria, insects and predators. The substances formed by the plants are more like chemical warfare agents, with which they keep their enemies at bay. The ecological pressure here is particularly high. Plants produce very strong defensive substances that have proven themselves in the course of evolution and that are coincidentally also useful for us.[330]

A plant substance that is supposed to provide protection is contained in the Pacific yew, which was tested along with other plants in the war against cancer. The botanist Arthur Barcley was very active and collected many species during this time. He went into the forests of California and took a piece of the bark of the yew *Taxus brevifolia*, which he initially sent to the Wisconsin Alumni Research Foundation, which conducted the first very successful tests. The success spread, and the NCI became curious and asked for the extract and the bark. On September 8, 1964, Arthur Barcley sent 15 kilograms of the bark of the same tree, in front of which he stood in 1962 in the Gifford Pinchot National Forest in the state of Washington, to Mansukh C. Wani (1925–2020, Fig. 9) and Monroe E. Wall (1916–2002, Fig. 9) at the newly founded Research Triangle Institute (TRI) in North Carolina. They extracted the bark once again and in their experiments the extract showed an excellent and very selective effect against cervical cancer cells. However, the quantity was too small and it was not until two years later that Wani succeeded in isolating a crystal, which he named Taxol (Fig. 10). He did not know the structure, but he knew that it must be a *"beautiful ring"* that contained alcohol groups, and so the first name was created, which was later changed to Paclitaxel due to trademark law. The two researchers recognized the pharmacological activity and unfortunately found out in a literature search that the effect was already known and published in 1964, which is why no patent protection could be applied for anymore. Anyone could work with the plant, which did not arouse the interest of the pharmaceutical industry. This negative attitude of the industry led to the first academic and state-funded development program for a drug that ever existed.[331,332]

Fig. 9 Monroe E. Wall (1916–2002, l) and Mansukh Wani (1925–2020, r) (Permission Research Triangle Institute, RTI)

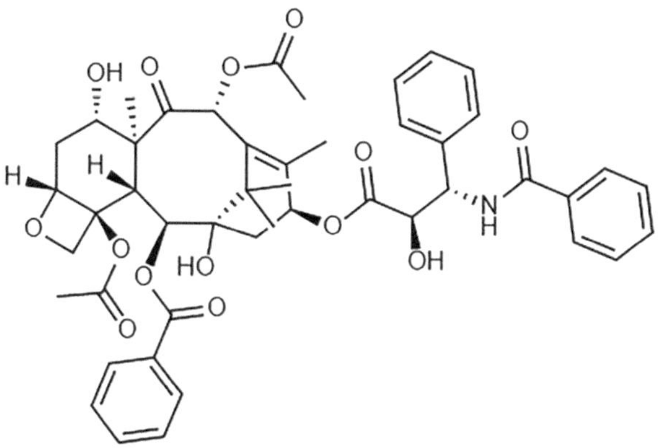

Fig. 10 Paclitaxel (work of the author)

Monroe E. Wall was a natural substance researcher all his life and began his career in 1941 in the US Department of Agriculture, where he dealt with rubber as a new type of gum. The USA was at war, the Germans were also looking for new types of rubber, and Brazil was not a reliable supplier. After the war, he was involved in the discovery of plant steroids and enriched the then popular cortisone research. However, this was to change in 1957 when Jonathan Hartwell from the National Cancer Institute visited him and was very interested in Wall's plant extracts from cortisone research. Wall sent Hartwell 1000 extracts, one of which caught his particular attention. *Camptotheca acuminata* was the name of the plant that showed a strong anti-cancer effect. In 1958, Wall isolated the active ingredient Camptothecin and tried to elucidate its structure. This became his new passion. It was the royal discipline of natural substance researchers to isolate the active ingredient from small amounts of an extract. He became a master of his craft and until his death the expert for natural substance elucidation.

Paclitaxel presented him with a special task, because the substance is only present in traces in the yew bark. Because the tree grows so slowly, a hundred-year-old tree must be felled and the bark peeled off for one gram of the substance. This trace search brought Wall and the then young Wani together. Wani, a specialist for light-active natural substances, had completed his chemistry studies in India and had come to Indiana for his doctorate. Although fluorescence can be seen very well, the fluorescent substances are only present in tiny amounts. The quinine in tonic water may be a very good example here. To isolate these few crumbs from an extract, one needs a special touch and therefore Wall offered him the work on the fluorescent Camptothecin.

The two chemists not only isolated this substance, but also Paclitaxel in 1971 (Fig. 10). Wall once remarked that Wani had *"magic hands,"* because whatever he had in his hand, he succeeded. But Paclitaxel was to remain a challenge, as nothing seemed to work. Wall and Wani struggled with the structural elucidation and Wall suggested abandoning the project, which Wani did not accept. He recalls: *"We worked on the structure in 1965 and 1966, but it was very difficult to determine and we were not successful. At some point in 1967, Monroe called me into his office to discuss it. We decided that we had so many other things to do, maybe we should just forget about Taxol. I said I didn't want to give up."* The compromise was that he should work on Paclitaxel on weekends or when there was downtime in the lab. It was once again a coincidence that led to the correct structural elucidation. Let's let Wani tell us from his memories: *"I was very excited when I read about a new technique that large molecules like this were broken down into simpler parts, each part was examined individually and then the whole structure was determined."* Wani had Paclitaxel in high purity, dissolved it in ethanol with sodium in 1970 and wanted to put the mixture in the refrigerator overnight, but he forgot his experiment. Only after ten days did he remember that he still had something important to do and he remembered his Paclitaxel mixed with sodium. He no longer remembered the approach and was annoyed about the waste of this expensive natural substance. To his surprise, it turned out that a decomposition into two parts had taken place, which could be measured well with the newly developed mass spectrometry and with Wall's help. The structure was so complex and the elucidation so difficult that Wani still needed until 1971 to decipher it. Finally, he published a version in which a side chain was in the wrong position, as it turned out later. Look at the chemical compound in the profile and consider how difficult it is still today to locate the stereochemistry and the position of all the many substituents.

The willingness of the NCI and the state funders to continue the project did not last long. Because a quantity calculation showed: To treat all those diagnosed with ovarian or cervical cancer in just one year, 120 kg would have to be isolated from 360,000 yews, which means that all Pacific yews in the USA would have to be felled. If there are other good alternatives, should such a project be funded? No, the NCI decided against the project. The money and the permission to fell at least a few trees only came later, when the mechanism of action was clarified and an intelligent alternative to production was found. Susan Horwitz in New York spectacularly discovered that the mechanism of action is astonishingly different from known cancer drugs. When a cell divides, and it doesn't matter whether it's a normal or a degenerate cancer cell, the chromosomes, the carriers of genetic information, are doubled and the pairs—to put it simply—are pulled apart on a string. These strings are microtubules-protein threads that collectively form the spindle apparatus and distribute the chromosomes safely to the old and the newly forming cell. This spindle apparatus must be imagined as being built from many protein threads that look like two cones whose bases lie on top of each other. The old drugs acted on the formation of these protein threads and prevented the assembly, but the new Paclitaxel caused an innumerable number of threads that were uncoordinated and thus made the spindle apparatus inoperable. The cell could no longer divide. This completely new effect convinced the reviewers and they released the money for further research.

The second very serious problem could be solved in an intelligent way. How can Paclitaxel be produced in sufficient quantities, as the worldwide demand would be gigantic at 300 to 500 kg per year. For ecological reasons, it would be a tragedy if all yews were felled. As an Indian colleague reported to me at a conference in Deharun, a medium-sized Indian city on the edge of the Himalayas on the Chinese border, in 2019, this is unfortunately what happened in India. There used to be many yews there. But the greed of the traders and loggers was too great, so the stock is now dangerously low. A German company had taken on this problem and looked for a sustainable alternative. This lay in the green needles, which keep growing back. From them, it 10-Deacetylbaccatin III, or 10-DAB, could be obtained, which became a semi-synthetic natural substance in a chemical step that replaced Paclitaxel. The precursor was found and extracted in the green branches of many yews as well as the widely spread English Yew (*Taxus baccata,* Fig. 11). The English yew grows eight to ten years or even faster and seemed to be an ideal raw material source.

This led ingenious Belgians to the idea of founding a start-up company and offering many garden owners the free cutting of their yew hedges. They took the branches as garden waste and extracted 10-DAB from them. Meanwhile, 10-DAB is no longer only obtained from the green needles, but biotechnologically from plant cell cultures of the Chinese Yew *(Taxus chinensis)* in a pharmaceutical company near Bremen. Through chemical modification of 10-DAB in several steps, the active ingredient Docetaxel, which is marketed by Sanofi-Aventis under the name Taxotere®, was created. Docetaxel was modified by chemists to be significantly more effective than paclitaxel. A nice side effect is the increased water solubility, which made administration to patients more pleasant.[333]

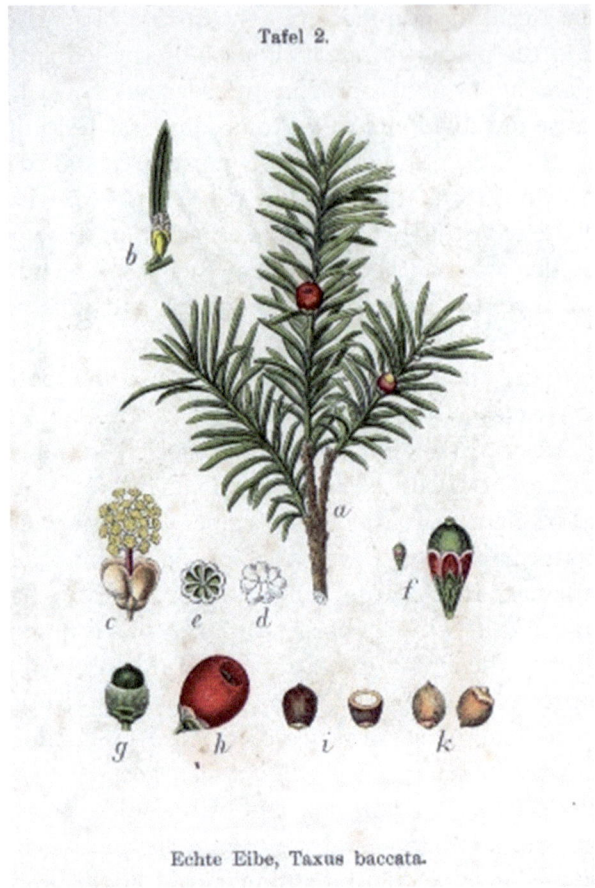

Fig. 11 *Taxus baccata* (public domain, wikipedia)1

With Genetic Engineering for the Protection of Forests

One of the most exciting developments in natural product chemistry from discovery to modern biotechnology is the story of paclitaxel. Associated with this research is biotechnology and genetic engineering, which are now so advanced that researchers understand which genes are crucial for biosynthesis in the plant and what the sequence of the known four bases adenine, thymine, cytosine, and guanine is as a gene sequence. With this knowledge, genes can be artificially assembled and introduced into microorganisms in synthetic biology. This way, an originally plant-based natural substance can be produced in yeasts or fecal bacteria. Of course, this is not trivial, as many genes are not or poorly read in the new host. These genes encode enzymes that convert metabolites into an end product in downstream biochemical reactions. Most natural substances, like paclitaxel, are so-called secondary metabolites, whose biosynthesis is coupled to the basic primary metabolism, which is almost identical in all living beings on earth. This is similar to our computers, which with an operating system like Windows perform all basic primary operations such as uploading, controlling the printer, and internet connection as a matter of course. If we want to start secondary tasks such as opening Word, surfing the internet, or playing a video game, this program must be loaded and communicate and interact with the Windows operating system. But our secondary substances are something special, and a baker's yeast might wonder why on earth it should produce this substance when it would much rather produce ethanol. After all, evolution never gave it the job of producing morphine or other secondary substances. Ideally, the yeast accepts the new genes, converts them into enzymes, and delivers its metabolites from the primary metabolism, which are taken as substrates for the new secondary metabolic pathway.

But there is a problem, because the same gene that, for example, digests a protein, is not identical between a yeast, a plant, or humans. Although we learned in school that the same four bases are found in all species, their arrangement in the gene does not make reading and translating easy. If we take up the example with the computer again, then it is comparable to a video game that runs well under Windows, but not on an Apple Mac. It is the same with a bacterium or a yeast cell, which have a similar biochemical operating system, but there are many parameters that need to be set species-specifically when a plant metabolism is to be introduced into a microorganism. This is exactly what various working groups are trying to do for paclitaxel, but more than 20 genes or enzymes from very different metabolic pathways make it a real challenge to produce paclitaxel biotechnologically from sugar as a cheap starting material. Unfortunately, no working group has yet succeeded in producing paclitaxel in microorganisms. It is too difficult to control the concert of all genes in the correct order. If we stick with music, then it is the off-key notes that hurt us unmelodically in the ear, which describe the current state of research squeaking.[334]

Tetrodotoxin—a Devilish Delight for the Palate

Brief profile – Tetrodotoxin

Structural formula:

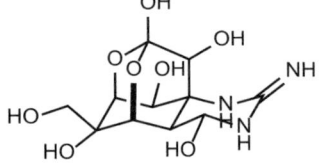

Natural substance group:
- Alkaloid

Occurrence:
- Pufferfish
- Hedgehog fisht
- Crayfish
- Snails
- Starfish
- Blue-ringed octopus

Effect / Indication:
- Poison
- Blockade of sodium channels in neurons

Possible medical significance:
- Pain relief for cancer patients

James Cook (1728–1779) left London and sailed on his second voyage of discovery with the HMS Resolution from 1772 to 1775 through the South Atlantic and South Pacific, in order to find the supposed southern continent on behalf of the Admiralty and His Majesty. Unfortunately, in vain, he found a few lonely islands, which he named South Georgia and the Sandwich Islands, but otherwise it was the cold sea and ice that he identified. On board were two Germans, Johann Reinhold Forster (1729–1798) and his son Georg(1954–1794), who participated in the second South Sea voyage for Joseph Banks, as Banks had too high demands for accommodation and collection of all possible finds. The conversion of the HMS Resolution with an additional upper deck would have too severely restricted the seaworthiness, as trial runs had shown. Joseph Banks declined to follow James Cook a second time under these circumstances. On July 13, 1772, father and son Forster, who was wanted in the Mainz Republic as a revolutionary with a warrant, set sail under the command of James Cook. The task of the two Germans was to draw and scientifically record the plants, animals, minerals, and indigenous peoples they encountered. Georg Forster was a gifted draftsman who fulfilled this task very conscientiously. He put everything on paper that did not disappear into a tree, into the water, or behind a stone. Over 600 colored drawings have survived from his South Sea voyage. Georg was always under steam and drew late into the night. September 8, 1774, off New Caledonia east of Australia was a particularly long and busy day. Georg was still drawing the fish that the cook was impatiently waiting to put in the pan. Georg insisted that science take precedence over the crew's hunger, and they had to bow to the passionately committed German, including the captain. It took a long time and it got late. Cook and the officers were very hungry and he tasted the roe and liver with the officers, which Forster allowed them because he did not want to draw them. The night was restless and in the middle of the night Cook awoke with very strange symptoms of poisoning, which we know well today. Let's read what is in his logbook: *"At three or four o'clock in the morning we were afflicted by an unusual weakness in all our limbs, which was accompanied by numbness or a feeling that one might experience if one first puts hands and feet into the fire after they have almost frozen; I had almost completely lost the feeling, nor could I distinguish heavy and light objects from each other: A pot full of water and a feather weighed equally heavy in my hand. Each of us took an emetic and then some sweets, which provided great relief. The pig of one, which had eaten the intestines, was found dead, the dogs had been thrown the head by the servants and what had wandered from our table; but soon it made the dogs sick, and they vomited everything out and were thus not heavily affected. When the natives came on board in the morning and saw the cut open fish, they immediately made us understand that it was not intended for consumption at all, showing the greatest terror of it."*

What had happened? Cook and his officers had eaten a fish whose liver and bile ducts are colonized by bacteria that synthesize a highly toxic natural substance. Just half a milligram of this tetrodotoxin can kill a human. Tetrodotoxin is not a poison that the fish itself produces, but a metabolic product of marine bacteria that live in the gall or liver of the animals.

Another fish that has the same bacteria in its organs and is very poisonous is the Japanese pufferfish. However, the pufferfish owes its fame not to Cook, but to the fascination of Japanese cuisine, which brings the fish to the table as a delicacy. If you have ever been to Japan, you have certainly visited Tokyo or Osaka. Perhaps you have wondered why some restaurants have an inflated pufferfish hanging? Here, the pufferfish Fuku of the genus Takifugu is served to you as a Japanese delicacy. Because the cook has a license for professional gutting. He dissects the Tiger Fugu *(Takifugu vermicularis)* and carefully removes the ovaries, intestine, roe, liver, and depending on the species also the skin. Only the muscle meat is non-toxic and is eaten raw. Therefore, the fish must be fresh and is transported alive. Unappetizing is that the fish's mouth is sewn shut so that no one gets injured by the fish's sharp teeth during transport.[335, 336]

In the pufferfish live bacteria of the genera *Pseudomonas, Vibrio* and *Aliivibrio,* which, like most representatives of their genus, possess the special ability to produce this strong poison. If this poison is so strong, one might think that it should be sufficient for the protection of the small fish, but the fish has another trick to defend itself in case of an attack. It swallows water, which makes it big and round. Attackers cannot swallow it because of its size and its spines. Try biting into a football, you will immediately understand why this defense is clever. But a little nibbling and sucking is possible. Especially young male dolphins like to hunt the pufferfish and absorb the tetrodotoxin from the sucked skin, which excites them like a drug. They get high on the pufferfish, grab the smaller fish, chew them gently and pass them on like the bong when smoking weed.

Do you like zombies too? These undead, these seemingly dead, who were brought back to life without a soul and wander aimlessly on this planet. No, but what does a zombie have to do with tetrodotoxin? Tetrodotoxin seems not to have revealed all its secrets yet and even in Caribbean folk medicine, tetrodotoxin may provide an explanation for the transformation of people into zombies. This sounds very dramatic, but the American ethnologist Wade Davis (*1953) traveled through Haiti in 1982 and researched the Voodoo religion and the alleged zombies, those undead who are willless and wander somewhere between the realms of the living and the dead. Wade Davis noticed that the zombies shown to him were severely paralyzed. They resembled those suffering from a fugu fish poisoning. He took samples of the so-called zombie powder, which was given to the victims, back to the USA and analyzed the composition. It turned out that the mixtures consisted of pufferfish of the species *Diodon hystrix, D. holacanthus* and

Sphoeroides testudineus, the toad *Bufo marinus* and the tree frog *Osteopilus dominicensis*. The cocktail was rounded off with the psychoactive plants *Albizzia lebbeck, Datura* spp. and *Strophanthus gratus* with its heart-active steroids.[92] This mixture leads to physical and mental paralysis. What happens to the zombies is bizarre and part of the death cult. The prepared zombie is placed in a coffin and buried alive for a few days, only to be dug up again in a large ceremony. These zombies are fed with sweet potatoes, sugarcane and Concombe Zombi, an extract of the Indian thorn apple *(Datura metel),*. Louis Ozimba was a volunteer who underwent zombification and his accounts are astonishing when he tells how he felt completely paralyzed in the coffin while fully conscious. It feels, he reports, as if ants were running over his skin and biting him. He heard his sister crying and lamenting his death. He also did not know how long he was buried, but the zombie makers, as they called themselves, left him in his coffin underground for three days. The reason for the short burial was not the diminishing effect of the zombie cocktail, but the lack of oxygen, which would have ended his life.[337]

You may wonder whether such a dangerous natural substance from the death fish of Japanese haute cuisine, which kills in tiny amounts of two milligrams, is suitable to help people and cure diseases? Two milligrams, an amount smaller than that of a salt crystal? Yes, that's what researchers wanted to find out and took on this challenge. The initial question was: How does this highly potent poison work? Once ingested, it spreads throughout the body. It binds to the sodium channels of two nerves that send and receive a signal, and it blocks the transmission. The nerve is paralyzed, the information can no longer flow. Numbness, dizziness, and coordination disorders are typical symptoms. The poison is deadly because it stops breathing. In non-lethal doses, the situation is different and can drag on for hours. The symptoms remain similar, and if you survive the next 24 hours, there is a chance that you will not die. Wade Davis sent Leon Roisin an extract from his Haiti expedition to the New York Psychiatric Institute. There, an unanalyzed extract in a concentration of 5 micrograms per 100 mg body weight was applied to the shaved abdominal skin of a monkey. This immediately led to severe swelling. The same amount injected into the abdominal cavity led to catalepsy, meaning the monkey could no longer move while fully conscious. The pulse beat much faster and the monkey no longer felt pain. Nevertheless, Ana Campos-Rìos and her colleagues wondered what benefit such a toxic substance could have. Thus, the paralytic effect was examined to see if it could be used in epileptics or other muscle cramps. Indeterminate abdominal pain like allodynia, which can be triggered by poorly localized pain, is another interesting indication. Patients may feel pain from non-painful stimuli, which can be unbearable for them. These exaggerated pain sensations are unpleasant and significantly reduce the quality of life.

Recent research results show the potential for treating neuropathies and indeterminate pain in cancer patients. Neuropathies will play an even greater role in the future as the number of patients with diabetic neuropathies will drastically increase. Neuropathies are characterized by indeterminate and chronic pain that millions of people suffer from daily. Previous therapies relied on Aspirin® or Ibuprofen, and in particularly dramatic cases, on opiates. A Spanish research group showed that tetrodotoxin in a dose of less than 30 micrograms in five studies could help patients with chronic pain above average. This is all the more interesting as tetrodotoxin is not known as a direct painkiller. In humans, it is more the case that they know they have pain, but ignore it. These are new approaches in modern pain therapy, which suggest an as yet unknown future for tetrodotoxin and its chemical modifications.[338,339]

Glossary

Acetyl group Functional group in chemistry that is derived from a carboxylic acid.

Acid, organic Chemical compound with a carboxyl group that transfers protons to another reaction partner. Functional group: R-COOH.

Active ingredient Chemical substance with a biological effect.

Alcohol Group of organic molecules that carry at least one hydroxyl group (OH).

Aldehyde Chemical compound that is an alcohol from which water (H_2O) has been removed. Functional group: R-CHO. Aldehydes are very reactive in natural product chemistry.

Alkaloid A group of natural chemicals that contains at least one nitrogen atom, has a strong effect on humans, but is not an antibiotic.

Amine Chemical compound that contains at least one nitrogen as derivatives of ammonia. Cyclic amines are considered heterocycles and are very commonly found in alkaloids.

Anesthesia State of insensitivity in pharmacology during surgical procedures, which is generally also referred to as numbness.

Base A molecule that can absorb hydrogen ions when dissolved in water. Bases produce an alkaline pH above 7.

Cancer A collective term for diseases characterized by uncontrolled growth of cells and the seeding of metastases into healthy tissue.

Chromatography A technique in chemistry used for the separation of a mixture of substances using a stationary phase (a solid that does not dissolve in the mobile phase) and a mobile phase (liquid, gas).

Condensation Chemical reaction in which two substances are combined with the elimination of water. For example, an ester from an alcohol and an acid.

Constitution In a molecule, the quantity, type, and position between the atoms are described.

Drug substance An active ingredient that has been approved by an authority for use in humans or animals.
Effect Biological effect of a chemical substance in a biological living system from the cell to the organism.
Efficacy Effect of a drug in humans.
Ester Chemical group of substances that have been formed in a chemical reaction between an alcohol and an acid, with the elimination of water. Example: Acetylsalicylic acid, Heroin.
Estrogen Group of female sex hormones. Newer term is estrogens.
Ethin A chemical gaseous substance, also known as acetylene.
Fat Chemical group of substances in which glycerol is esterified with three fatty acids. Fats can be solid (butter) or liquid (olive oil) at room temperature.
Fermentation Microbiological conversion of a substrate by cells or enzymes in biology and biotechnology.
Glucocorticoids Chemically modified steroid hormones that have an effect on glucose metabolism.
Hydrolysis Splitting of a chemical compound, in which water is consumed.
Inflammation Body's own response to harmful stimuli, leading to warmth, redness, swelling, pain.
Ketone Chemical compound that has a carbonyl group as a non-terminal functional group (R-CO-R). Ketones are very reactive.
Metabolism see Metabolism.
Metabolism Chemical transformations in the cell of substances such as food. These chemical conversions are also called biochemical reactions, which metabolize sugar and fats into energy, heat, carbon dioxide, and metabolic waste products in the body.
Molecule Group of atoms that are connected with each other.
Molecular Weight Mass of a molecule.
Natural substance Chemical compounds that are formed by an organism to fulfill a biological function.
Neurotransmitter Signal molecule that is released at the synapse of a nerve cell to transmit an electrical signal.
Oncogene An unusually activated gene that can turn a cell into a tumor cell.
Opiate Alkaloids of opium, which as natural or semi-synthetic substances have an effect on opiate receptors in humans.
Opioid Natural or synthetic substances that have an effect on opioid receptors in humans.
Oxidation Chemical reaction in which electrons are given off and oxygen is very often involved.
Pharmacognosy Science in pharmacy and is the study of plant, fungal, and animal drugs, medicines, and toxins.
Pharmacology Sub-discipline in medicine that deals with the effect and efficacy of chemical substances and drugs in the human and animal body.

Pharmacy Science of the production, testing of drugs.

pH value Chemical term for the acidity and alkalinity of a liquid. A pH value of 7 is neutral like distilled water, values less than 7 are acids and those greater are bases.

Phytochemistry Analytical science that chemically investigates the occurrence of constituents in plants.

Plant drug Dried whole plant or parts such as leaves, roots from it. The name is derived from the Old German word *gedrogt*, which means dried.

Progestogen Group of pregnancy hormones that serve the initiation and maintenance of pregnancies.

Reduction Chemical reaction in which electrons are gained and the oxidation state is reduced.

Synthesis, chemical Chemical term for the production of a chemical substance from at least two chemical starting materials.

Terpene Largest known group of natural substances, predominantly composed of carbon followed by oxygen. The general structure consists of isoprene units.

Testosterone Chemical group of male sex hormones.

References

1. Goerig, M. & Schulte am Esch, J. Friedrich Wilhelm Adam Sertürner – dem Entdecker des Morphins zum 150. Todestag. *Anasthesiologie Intensivmedizin Notfallmedizin Schmerztherapie* **26,** 492–498 (1991).
2. Wink, M. A Short History of Alkaloids. *Alkaloids* 11–44 (1998) https://doi.org/10.1007/978-1-4757-2905-4_2.
3. Almagro, B. R. A brief history of medical libraries. *Ariz Med* **42,** 176–178 (1985).
4. Ho, V. K. Y. The 'Problem' of Opium Smoking in Canton. in *Understanding Canton* 95–155 (Oxford University PressOxford, 2005). https://doi.org/10.1093/0199282714.003.0004.
5. Aragón-Poce, F. *et al.* History of opium. *Int Congr Ser* **1242,** 19–21 (2002).
6. Matthias Seefelder. Die Hinterlist der weißen Teufel. in *Opium – Eine Kulturgeschichte* 113–141 (1996).
7. Vries, P. *Zur politischen Ökonomie des Tees. Was uns Tee über die englische und chinesische Wirtschaft der Frühen Neuzeit sagen kann, 161 pages.* https://www.researchgate.net/publication/282151810 (2009).
8. Beckett, G. H. & Alder Wright, C. R. IV. – Action of the organic acids and their anhydrides on the natural alkaloids. Part II. Butyryl and benzoyl derivatives of morphine and codeine. *J Chem Soc* **28,** 15–26 (1875).
9. Stockmann, R. & Dott, D. B. Report On The Pharmacology Of Morphine And Its Derivatives on. *The British Medical Journal* 189–192 (1890).
10. Dreser, H. Pharmakologisches über einige Morphinderivate. *Therapeutische Monatshefte* **9,** 509–512 (1898).
11. Harnack, E. Über die Giftigkeit des Heroins (Diacatlymorphins). *Münchner Medizinische Wochenschrift* **46,** 881–884 (1899).
12. Floret, N. Klinische Versuche über die Wirkung und Anwendung des Heroins. *Therapeutische Monatshefte* **12,** 182–187 (1899).

13. de Ridder, M. *Heroin: Vom Arzneimittel zur Droge.* (Campus Verlag, 2000).
14. Strube, G. Mitteilung über therapeutische Versuche mit Heroin. *Berliner Klinische Wochenschrift* **23**, 993–996 (1898).
15. Hughes, J. Hans Kosterlitz (1903–96). *Nature 1996 384:6608* **384**, 418–418 (1996).
16. Kolbe, L. & Fins, J. J. The Birth of Naloxone: An Intellectual History of an Ambivalent Opioid. *Cambridge Quarterly of Healthcare Ethics* **30**, 637–650 (2021).
17. Keefe, P. R. *Imperium der Schmerzen.* (Carl Hanser Verlag, 2022).
18. OxyContin: The History of OxyContin. https://www.michaelshouse.com/oxycontin-rehab/history-of-oxycontin/.
19. Ohler, N. *Der totale Rausch: Drogen im Dritten Reich:* (KiWi-Taschenbuch, 2017).
20. Nicholas, J. S. Ross Granville Harrison, 1870–1959. *Yale J Biol Med* **32**, 407–412 (1960).
21. Rasmussen, N. On Speed. *On Speed* (2022) https://doi.org/10.18574/NYU/9780814777350.001.0001.
22. Dayer, L. E. *et al.* A recent history of opioid use in the US: Three decades of change. *Subst Use Misuse* **54**, 331–339 (2019).
23. Basch-Ritter, R. 1942-. *Die Weltumsegelung der Novara 1857–1859: Österreich auf allen Meeren.* (Akademische Druck- u. Verlagsanstalt, 2008).
24. Jacobs, M. The Life of Sigmund Freud. In *Sigmund Freud* 1–32 (SAGE Publications Ltd, 2012). https://doi.org/10.4135/9781446215128.n1.
25. Freud, S. 1856–1939 *et al.* Warten in Ruhe und Ergebung, Warten in Kampf und Erregung Januar 1884–September 1884.
26. Selbstversuche: Forschung unter Lebensgefahr. https://www.aerzteblatt.de/archiv/80003/Selbstversuche-Forschung-unter-Lebensgefahr.
27. Forschung: Gibt es heute noch Selbstversuche in der Wissenschaft? https://www.berliner-zeitung.de/zukunft-technologie/forschung-gibt-es-heute-noch-selbstversuche-in-der-wissenschaft-li.18475.
28. Selbstversuche von Medizinern | Apotheken-Umschau. https://www.apotheken-umschau.de/therapie/selbstversuche-von-medizinern-720835.html.
29. Dr Christian Kopetzki, U.-P. & Maria Huber, M. *Exposé des Dissertationsvorhabens 'Medizinische Selbstversuche'.* http://www.nzzfolio.ch/www/d80bd71b-b264-4db4-afd0-277884b93470/showarticle/af92a1b6-2f62- (2013).
30. Goerig, M., Bacon, D. & Van Zundert, A. Carl koller, cocaine, and local anesthesia: Some less known and forgotten facts. *Regional Anesthesia and Pain Medicine* vol. 37 318–324 Preprint at https://doi.org/10.1097/AAP.0b013e31825051f3 (2012).
31. Becker-Koller, H. Carl Koller and Cocaine. *Psychoanal Quart* **32**, 309–379 (1963).
32. Freud, S. Ueber Coca. *Zentralblatt Ges Ther* **2**, 289–314 (1884).

33. M Amm, K. H. „Coca-Koller" und seine Freunde. Zum 140. Geburtstag des jüdisch-wienerischen Trios: Carl Koller (1857–1944), Sigmund Lustgarten (1857–1911) und Sigmund Freud (1856–1939). *Wien Klin Wochenschr* **109,** 170–175 (1997).
34. Breathnach, C. S. Biographical sketches – 45. Koller. *Ir Med J* **77,** 335 (1984).
35. Leonard, M. Carl Koller: Mankind's greatest benefactor? The story of local anesthesia. *Journal of Dental Research* vol. 77 535–538 Preprint at https://doi.org/10.1177/00220345980770040501 (1998).
36. 150 Geburtstag: Eduard Ritsert und das Anaesthesin | PZ – Pharmazeutische Zeitung. https://www.pharmazeutische-zeitung.de/ausgabe-452009/eduard-ritsert-und-das-anaesthesin/.
37. Gordh, T., Gordh, T. E. & Lindqvist, K. Lidocaine: The origin of a modern local anesthetic. *Anesthesiology* **113,** 1433–1437 (2010).
38. Wildsmith, J. A. W. & Jansson, J. R. From cocaine to lidocaine: Great progress with a tragic ending. *European Journal of Anaesthesiology* vol. 32 143–146 Preprint at https://doi.org/10.1097/EJA.0000000000000168 (2015).
39. Kuhnert, N. 100 Jahre Aspirin. *Pharm Unserer Zeit* **29,** 32–39 (2000).
40. Desborough, M. J. R. & Keeling, D. M. The aspirin story – from willow to wonder drug. *Br J Haematol* **177,** 674–683 (2017).
41. Marson, P. & Pasero, G. *Il contributo italiano alla storia dei salicilati The Italian contributions to the history of salicylates. Reumatismo* vol. 58 (2006).
42. Gerhardt, C. *Précis de chimie organique.* vol. 2 (Fortin, Masson et Cie , 1844).
43. HS. Wunderpille Aspirin. *Geschichte* **6,** 58–59 (2003).
44. Vaupel, E. Arthur Eichengrün – Tribute to a forgotten chemist, entrepreneur, and German Jew. *Angewandte Chemie – International Edition* **44,** 3344–3355 (2005).
45. Eichengrün, A. Angebliche Curpfuscherei seitens der chemischen Industrie. Eine Abwehr. *Zeitschrift für Angewandte Chemie* **13,** 55–60 (1900).
46. Die neuen Arzneimittel im zweiten Semester 1898 _ Enhanced Reader.
47. Eichengrün, A. Mittheilungen aus dem Vereine deutscher Chemiker. 12. Überproduction an neuen Arzneimitteln. *Zeitschrift für angewandte Chemie* **39,** 892–897 (1898).
48. Meyer, H. H. Heinrich Dreser. *Archiv für experimentelle Pathologie und Pharmakologie 1925 106:3* **106,** I–VII (1925).
49. Dreser, H. Pharmakologisches über Aspirin (Acetylsalicylsäure). *Archiv für die gesamte Physiologie des Menschen und der Tiere 1899 76:5* **76,** 306–318 (1899).
50. Sneader, W. The discovery of aspirin: a reappraisal. *BMJ : British Medical Journal* **321,** 1591 (2000).
51. Feilitzsch von, H. Dr. Hugo Schweitzer and the Great Phenol Plot – The Secret War Council. http://felixsommerfeld.com/news/mexican-revolution-blog/2015/6/7/dr-hugo-schweitzer-and-the-great-phenol-plot (2015).
52. Brownlee, G. Obituary: Harry Collier. *Br J Addict* **79,** 233–233 (1984).
53. Costello, J. Obituary For Professor Priscilla Piper. 320–322 (1995).

54. Rainsford, K. D. Fifty years since the discovery of ibuprofen. *Inflammopharmacology* **19,** 293–297 (2011).
55. A brief history of ibuprofen – The Pharmaceutical Journal. https://pharmaceutical-journal.com/article/infographics/a-brief-history-of-ibuprofen.
56. A brief history of ibuprofen. *Pharmaceutical Journal* (2017) https://doi.org/10.1211/PJ.2017.20203273.
57. Stewart. An interview with Stewart Adams. (2011) https://doi.org/10.1016/j.tips.2011.10.007.
58. Nelson, E., Rolian, C., Cashmore, L. & Shultz, S. Digit ratios predict polygyny in early apes, Ardipithecus, Neanderthals and early modern humans but not in Australopithecus. *Proceedings of the Royal Society B: Biological Sciences* **278,** 1556–1563 (2011).
59. Ahmad, D. L. Opium smoking, anti-Chinese attitudes, and the American medical community, 1850–1890. *American Nineteenth Century History* **1,** 53–68 (2008).
60. Bage, P. *Woman Rebel Margaret Sanger Story Book.* (2013).
61. Chaudhury, R. R. The quest for a herbal contraceptive. *Natl Med J India* **6,** 199–201 (1993).
62. Hodenbaden – Eine Retrospektive – 20 Minuten. https://www.20min.ch/story/hodenbaden-eine-retrospektive-161205538865.
63. Shefi, S., Tarapore, P. E., Walsh, T. J., Croughan, M. & Turek, P. J. 50 Wet Heat as a Reversible Gonadotoxin Clinical. *Urology International Braz J Urol* **33,** 50–57.
64. Lauritzen, C. & Göretzlehner, G. Die kinderreichste Frau Deutschlands. *Journal für Fertilität und Reproduktion* **9,** 22–33 (1999).
65. Shahar, S. *Kindheit im Mittelalter.* (Rowohlt, 1993).
66. N.N. Mrs Sanger has never quit fight. *Arizona Daily Star* 9–9 (1961).
67. Wallace, W. The 1957 Mike Wallace Interview with Margaret Sanger. (1957).
68. Köstering, S. 'Etwas Besseres als das Kondom'. Ludwig Haberlandt und die Idee der Pille. 113–126 Preprint at (1996).
69. Lesky, E. Haberlandt-Urvater der Ovulationshemmung. *Der Prakt Arzt* **10,** 1567 (1975).
70. Bandhauer, K. Ludwig Haberlandt – der erste Entdecker der 'Pille'. *Mitt Österr Ges Wissenschaftsgesch* **24,** 111–127 (2004).
70. Noé, A. C. Gottlieb Haberlandt. *Plant Physiol* **9,** 850–855 (1934).
72. Haberlandt, L. Über hormonale Sterilisierung weiblicher Tiere. *Pflüger's Archiv für die gesamte Physiologie des Menschen und der Tiere 1927 216:1* **216,** 525–533 (1927).
73. Haberlandt, L. *Die hormonale Sterilisierung.* (1931).
74. Dimroth, K. *Das Portrait: Adolf Windaus 1876–1959.*
75. Biografie,AdolfButenandt.https://www.sammlungen.hu-berlin.de/objekte/-/9200/.
76. Karlson, P. Adolf Butenandt (1903–1995). *Nature 1995 373:6516* **373,** 660–660 (1995).

77. Schieder, W. & Trunk, Achim. *Adolf Butenandt und die Kaiser-Wilhelm-Gesellschaft : Wissenschaft, Industrie und Politik im Dritten Reich.* (Wallstein, 2004).
78. Simpson, E. & Santen, R. J. Celebrating 75 years of oestradiol. *J Mol Endocrinol* **55,** 1–20 (2015).
79. Quinkert, G. Hans Herloff Inhoffen in His Times (1906–1992). *European J Org Chem* **2004,** 3727–3748 (2004).
80. Inhoffen, H. H. Der Weg vom Cholesterin zum Follikelhormon Oestradiol. *Angewandte Chemie* **59,** 207–212 (1947).
81. Inhoffen, H. H. & Hohlweg, W. Neue per os-wirksame weibliche Keimdrüsenhormon-Derivate: 17-Aethinyl-oestradiol und Pregnen-in-on-3-ol-17. *Naturwissenschaften* **26,** 96 (1938).
82. Ebeling, H. *Schwarze Chronik einer Weltstadt.* vol. 1 (Ernst Kabel Verlag, 1980).
83. N.N. Cilip Nr. 58 – Chinesenverfolgung im Nationalsozialismus. https://archiv.cilip.de/alt/ausgabe/58/china.htm (2022).
84. Miao, Q. & Yang, D. Huang-Minlon (Huang Ming-Long) – A Pioneer of Organic Chemistry in China. *Progress in Chemistry* **24,** 1229 (2012).
85. Blümer, G.-P., Collin, G. & Höke, H. Tar and Pitch. in *Ullmann's Encyclopedia of Industrial Chemistry* (Wiley-VCH Verlag GmbH & Co. KGaA, 2011). https://doi.org/10.1002/14356007.a26_091.pub2.
86. Curzons, A. D., Constable, D. J. & Cunningham, V. L. Pharmaceutical Compounds LCA Case Studies LCA Case Studies Cradle-to-Gate Life Cycle Inventory and Assessment of Pharmaceutical Compounds. https://doi.org/10.1065/lca2003.11.141.
87. Serini, A. & Straßberger, L. Verfahren zur Darstellung von tertiaeren Carbinolen der Cyclopentanopolyhydrophenanthrenreihe. (1935).
88. Elsey, H. *Edward Curtis Franklin _ Enhanced Reader.pdf.* https://www.research-collection.ethz.ch/mapping/eserv/eth:22023/eth-22023-02.pdf (1991).
89. Lehmann, P. A., Bolivar, A. & Quintero, R. Russell E. Marker. Pioneer of the Mexican steroid industry. *J Chem Educ* **50,** 195–199 (1973).
90. Lehmann F., P. A. Early history of steroid chemistry in Mexico: the story of three remarkable men. *Steroids* **57,** 403–408 (1992).
91. N.N. The marker ‚degradation' and creation of the mexican steroid industry 1938–1945. *An International Historic Chemical landmark* 1–8 Preprint at (1999).
92. Guide to the Russell Earl Marker Papers, 1919–1984 601. https://doi.org/10.4/THEMES/REDMOND/JQUERY-UI.CSS.
93. Rudo, A., Siehl, H. U., Zeller, K. P., Berger, S. & Sicker, D. Diosgenin aus Yams als Hormonvorstufe. *Chemie in unserer Zeit* **49,** 372–384 (2015).
94. The Morning After: The Jewish Connection of the Birth Control Pill Invention – Museum of the Jewish People. https://www.anumuseum.org.il/blog-items/the-morning-after-the-jewish-connection-of-the-birth-control-pill-invention/.

95. Kendall, E. C. The development of cortisone as a therapeutic agent. *Nobel Prize Award Lecture* 1–19 Preprint at (1951).
96. Djerassi, C. Steroid research at Syntex: 'the Pill' and cortisone. *Steroids* **57**, 631–641 (1992).
97. Merkel, W. Maximilian Ehrenstein. 1899–1968. *Helv Chim Acta* **52**, 2178–2182 (1969).
98. Scheele, B. C. *et al.* Amphibian fungal panzootic causes catastrophic and ongoing loss of biodiversity. *Science (1979)* **363**, 1459–1463 (2019).
99. Greenberg, D. A. & Palen, W. J. A deadly amphibian disease goes global. *Science (1979)* **363**, 1386–1388 (2019).
100. Streller, S. & Roth, K. 50 Jahre Pille in Deutschland: Über die Heldentaten der Hormonsucher. *Chemie in Unserer Zeit* **45**, 270–291 (2011).
101. Staupe, Gisela., Vieth, Lisa. & Deutsches Hygiene-Museum Dresden. Die Pille: von der Lust und von der Liebe. 239 (1996).
102. Schwarz, S., Onken, D. & Schubert, A. The steroid story of Jenapharm: From the late 1940s to the early 1970s. *Steroids* **64**, 439–445 (1999).
103. van de Laar, A. Kastration. in *Schnitt! Die ganze Geschichte der Chirurgie erzählt in 28 Operationen* 230–244 (Droemer Verlag, 2014).
104. Butenandt, A. & Hanisch, G. Über Testosteron. Umwandlung des Dehydroandrosterons in Androstendiol und Testosteron; ein Weg zur Darstellung des Testosterons aus Cholesterin. *Hoppe Seylers Z Physiol Chem* **237**, 89–97 (1935).
105. Zondek, B. Æstrogenic Hormone in the Urine of the Stallion. *Nature 1934 133:3361* **133**, 494–494 (1934).
106. Butenandt, A. & Tscherningr, K. Über Androsteron, ein krystallisiertes männliches Sexualhormon. I. Isolierung und Reindarstellung aus Männerharn. *Hoppe Seylers Z Physiol Chem* **229**, 167–184 (1934).
107. Butenandt, A. & Tscherning, K. Über Androsteron. II. Seine chemische Charakterisierung. *Hoppe Seylers Z Physiol Chem* **229**, 185–191 (1934).
108. Ruzicka, L. In the Borderland between Bioorganic Chemistry and Biochemistry. https://doi.org/10.1146/annurev.bi.42.070173.000245 **42**, 1–20 (2003).
109. Oudshoorn, N. Laqueur en Organon. *Gewina* **22**, 1–11 (1999).
110. Oudshoorn, N. Laqueur and Organon. The university laboratory and the pharmaceutical industry in the Netherlands. *Gewina* **22**, 12–22 (1999).
111. Verlong, J. *75 Jahre Organon.* (N.V. Organon, 1998).
112. Waites, G. M., Wang, C. & Griffin, P. D. Gossypol. *Int J Androl* **21**, 313–316 (1998).
113. Liu, G. Z., Lyle, C. K. & Cao, J. Trial of gossypol as a male contraceptive. 9–16 Preprint at (1985).
114. Coutinho, E. M. Gossypol: a contraceptive for men. *Contraception* **65**, 259–263 (2002).

115. Joshi, V. R. & Poojary, V. B. Tadeus Reichstein. *Journal of Association of Physicians of India* **62,** 645–646 (2014).
116. Kendall, E. C. The Isolation of Thyroxine and Cortisone: the Work of Edward C. Kendall. *Journal of Biological Chemistry* **277,** 21–22 (1919).
117. Bettendorf, G. Tausk, Marius. *Zur Geschichte der Endokrinologie und Reproduktionsmedizin* 583–584 (1995) https://doi.org/10.1007/978-3-642-79152-9_235.
118. Reichstein, T. Über Bestandteile der Nebennieren-Rinde VI. Trennungsmethoden, sowie Isolierung der Substanzen F. a. H und J. *Helv Chim Acta* **19,** 1107–1126 (1936).
119. Wheeler, O. H. The Girard reagents. *J Chem Educ* **45,** 435 (1968).
120. Thorn, G. W. The Adrenal Cortex. I: Historical Aspexcts. *Bull Johns Hopkins Hosp* **123,** 49–64 (1968).
121. Reichstein, T. Über Cortin, das Hormon der Nebennierenrinde. I. Mitteilung. *Helv Chim Acta* **19,** 29–63 (1936).
122. Shaw, J. E. Alejandro Zaffaroni (1923–2014). *Nature 2014 508:7495* **508,** 187–187 (2014).
123. Vitamin D für die Massen | Vandenhoeck & Ruprecht Verlage. https://www.vandenhoeck-ruprecht-verlage.com/vitamin-d.
124. DeLuca, H. F. Historical Overview of Vitamin D. in *Vitamin D: Fourth Edition* vol. 1 1–12 (Elsevier Inc., 2017).
125. Huldschinsky, K. Heilung von Rachitis durch künstliche Höhensonne. *DMW – Deutsche Medizinische Wochenschrift* **45,** 712–713 (1919).
126. McCollum, E. V, Simmonds, N., Becker, J. E. & Shipley, P. G. Studies on the experimental rickets; and experimental demonstration of the existence of a vitamin which promotes calcium deposition. *J Biology and Chemistry* **53,** 293–312 (1922).
127. Hess, A. F., Weinstock, M. & Helman, F. D. THE ANTIRACHITIC VALUE OF IRRADIATED PHYTOSTEROL AND CHOLESTEROL. I. *Journal of Biological Chemistry* **63,** 305–308 (1925).
128. Holick, M. F. McCollum award lecture, 1994: Vitamin D – New horizons for the 21st century. *American Journal of Clinical Nutrition* **60,** 619–630 (1994).
129. Wolf, G. The discovery of vitamin D: the contribution of Adolf Windaus. *J Nutr* **134,** 1299–1302 (2004).
130. Pohl, R. Zum optischen Nachweis eines Vitamines. *Naturwissenschaften 1927 15:20* **15,** 433–438 (1927).
131. Wolf, G. The discovery of vitamin D: the contribution of Adolf Windaus. *J Nutr* **134,** 1299–1302 (2004).
132. Sigmungd Otto Orsenhelm, 1871–1955. *Biographical Memoirs of Fellows of the Royal Society* **2,** 256–267 (1956).
133. Wolf, G. The Discovery of Vitamin D: The Contribution of Adolf Windaus. *J Nutr* **134,** 1299–1302 (2004).
134. Haas, J. *Vigantol.* (Wissenschaftliche Verlagsgesellschaft, 2007).

135. VELLUZ, L. & AMIARD, G. [The condition of the equilibrium between precalciferol and calciferol]. *C R Hebd Seances Acad Sci* **253,** 603–606 (1961).
136. VELLUZ, L., AMIARD, G. & GOFFINET, B. [Photochemical processing in pre-calciferol tachysterol]. *C R Hebd Seances Acad Sci* **240,** 2156–2157 (1955).
137. Holick, M. F. *et al.* Photosynthesis of previtamin D3 in human skin and the physiologic consequences. *Science* **210,** 203–205 (1980).
138. Friedlieb Ferdinand Runge: Coffein aus Goethes Kaffeebohnen | PZ – Pharmazeutische Zeitung. https://www.pharmazeutische-zeitung.de/ausgabe-122017/coffein-aus-goethes-kaffeebohnen/.
139. Clark, T. *Starbucked: a double tall tale of caffeine, commerce & culture.* (Sceptre, 2008).
140. Jatinangor Rasmikayati, in E., Saefudin, B. R., Wardhana, M. Y., Tawali, A. B. & Laga, A. Luwak coffee in vitro fermentation : literature review. *IOP Conf Ser Earth Environ Sci* **230,** 012096 (2019).
141. Avallone, S., Guyot, B., Brillouet, J. M., Olguin, E. & Guiraud, J. P. Microbiological and biochemical study of coffee fermentation. *Current Microbiology Journal* **42,** 252–25 (2001).
142. Hyde Salter, B. On Some Points in the Treatment and Clinical History of Asthma. *Edinb Med J* **4,** 1109 (1859).
143. Diamant, Z., Diderik Boot, J. & Christian Virchow, J. Summing up 100 years of asthma. *Respir Med* **101,** 378–388 (2007).
144. Kossel, A. Über eine neue Base aus dem Pflanzenreich. *Berichte der deutschen chemischen Gesellschaft* **21,** 2164–2167 (1888).
145. Schultz, W. W., Van Andel, P., Sabelis, I. & Mooyaart, E. Magnetic resonance imaging of male and female genitals during coitus and female sexual arousal. https://doi.org/10.1136/bmj.319.7225.1596.
146. Osterloh, I. How I discovered Viagra. *27.04.2015* https://cosmosmagazine.com/science/biology/how-i-discovered-viagra/ (2015).
147. Klotz, L. How (not) to communicate new scientific information: a memoir of the famous Brindley lecture. *BJU Int* **96,** 956–957 (2005).
148. Ma, N., Zhang, Z., Liao, F., Jiang, T. & Tu, Y. The birth of artemisinin. *Pharmacol Ther* **216,** (2020).
149. Erster Medizin-Nobelpreis für China: Forschen für Mao. https://www.tagesspiegel.de/wissen/forschen-fur-mao-3682654.html.
150. Veale, L. An historical geography of the Nilgiri cinchona plantations, 1860–1900. (2010).
151. Brockway, Lucile. *Science and colonial expansion: the role of the British Royal Botanic Gardens.* (Yale University Press, 2002).
152. Roersch Van Der Hoogte, A. & Pieters, T. Quinine, Malaria, and the Cinchona Bureau: Marketing Practices and Knowledge Circulation in a Dutch Transoceanic Cinchona–Quinine Enterprise (1920s–30s). *J Hist Med Allied Sci* **71,** 197 (2016).

153. Goss, A. The floracrats: state-sponsored science and the failure of the Enlightenment in Indonesia. 256 (2011).
154. Goss, A. Building the world's supply of quinine: Dutch colonialism and the origins of a global pharmaceutical industry. *Endeavour* **38,** 8–18 (2014).
155. A Handbook of Cinchona Culture. Die Chinarinden in Pharmakognostischer Hinsicht dargestellt. *Nature 1883 27:691* **27,** 287–289 (1883).
156. Rawe, S. L. & McDonnell, C. The cinchona alkaloids and the aminoquinolines. *Antimalarial Agents: Design and Mechanism of Action* 65–98 (2020) https://doi.org/10.1016/B978-0-08-101210-9.00003-2.
157. Greenwood, D. *Antimicrobial Drugx, Chronicle of a Twentieth Century Medical Triumph.* (Oxford University Press, 2008).
158. Pedeliento, G., Pinchera, V. & Andreini, D. Gin: a marketplace icon. *Consumption Markets and Culture* **25,** 91–101 (2022).
159. Walker, K. (Researcher) & Nesbitt, M. *Just the tonic: a natural history of tonic water.* (Royal Botanic Gardens, 2019).
160. Zaffiri, L., Gardner, J. & Toledo-Pereyra, L. H. History of antibiotics. From salvarsan to cephalosporins. *Journal of Investigative Surgery* **25,** 67–77 (2012).
161. Norn, S., Permin, H., Kruse, E. & Kruse, P. R. [On the history of vitamin K, dicoumarol and warfarin]. *Dan Medicinhist Arbog* **42,** 99–119 (2014).
162. Stahmann, M. A., Huebner, C. F. & Link, K. P. STUDIES ON THE HEMORRHAGIC SWEET CLOVER DISEASE V. IDENTIFICATION AND SYNTHESIS OF THE HEMORRHAGIC AGENT *. *J Biol Chem* **138,** 513–527 (1941).
163. Macdonald, C. & Soll, J. Shark conservation risks associated with the use of shark liver oil in SARS-CoV-2 vaccine development. https://doi.org/10.1101/2020.10.14.338053.
164. Tsujimoto, M. A highly unsaturated hydrocarbon in shark liver oil. *Ind Eng Chem* **8,** 889–896 (1916).
165. Kar, S., Sanderson, H., Roy, K., Benfenati, E. & Leszczynski, J. Green Chemistry in the Synthesis of Pharmaceuticals. *Chem Rev* **122,** 3637–3710 (2022).
166. Jiménez-González, C., Curzons, A. D., Constable, D. J. C. & Cunningham, V. L. Cradle-to-gate life cycle inventory and assessment of pharmaceutical compounds. *The International Journal of Life Cycle Assessment 2004 9:2* **9,** 114–121 (2004).
167. Farina, V. & Brown, J. D. Tamiflu: The Supply Problem. *Angewandte Chemie International Edition* **45,** 7330–7334 (2006).
168. Johanns, E. S. D. *et al.* Een epidemie van epileptische aanvallen na drinken van kruidenthee | NTvG. *Ned. Tijdschr Gennesk* **146,** 813–816 (2002).
169. Biessels, G. J., F.H., V. & Leijten, F. S. S. Epileptische aanval na drinken van Sterrenmixthee: intoxicatie met Japanse steranijs. *Ned Tijdschr Geneeskd* **146,** 808–811 (2002).

170. WINTER, A. The Making of 'Truth Serum'. *Bull Hist Med* **79,** 500–533 (2005).
171. Gomila Muñiz, I., Puiguriguer Ferrando, J. & Quesada Redondo, L. Primera confirmación en España del uso de la burundanga en una sumisión química atendida en urgencias. *El País* (2016) https://doi.org/10.1016/j.medcli.2016.06.025.
172. Müller, J. Witch ointments and aphrodisiacs. A contribution to the cultural history of nightshade plants. *Gesnerus* **55,** 205–220 (1998).
173. N.N. *Ordnung des Bräuens.* (Stadtrat Landshut, Bayern, 1486).
174. Taberaemontanus, J. T. *Kräuterbuch.* (1664).
175. Ansati, P. Die Therapiegeschichte der Depression und die Einführung der antidepressiven medikamentösen Therapie in der BRD im Zeitraum von 1945–1970. (Medizinische Hochschule Hannover, 2013).
176. Jaffe, R. J., Novakovic, V. & Peselow, E. D. Scopolamine as an antidepressant: A systematic review. *Clin Neuropharmacol* **36,** 24–26 (2013).
177. Pastore, M. N., Kalia, Y. N., Horstmann, M. & Roberts, M. S. Transdermal patches: history, development and pharmacology. *Br J Pharmacol* **172,** 2179–2209 (2015).
178. Rolf Banholzer, S. (DE) *et al.* Methods for the synthesis of tiotropium bromide. (2012).
179. Wachtel, H., Kattenbeck, S., Dunne, S. & Disse, B. The Respimat® Development Story: Patient-Centered Innovation. *Pulm Ther* **3,** 19–30 (2017).
180. Shah, Z. *et al.* Podophyllotoxin: History, Recent Advances and Future Prospects. *Biomolecules* **11,** 603 (2021).
181. Coran, A. Y. Encyclopedia of Polymer Science and Engineering. in (John Wiley & Sons, 1989).
182. Jünger, W. *Kautschuk, vom Gummibaum zur Retorte.* (Goldmann, 1952).
183. Watson, T. Lectures on the principles and practice of physics. *Medical Times and Gazette* **2,** 349–354 (1845).
184. Semmelweiss, I. Hoechst wichtiger Erfahrungen über die Aetiologie der in Gebäranstalten epidemischen Puerperalfieber. *Gesellschaft der Aerzte Wien* **4,** 242–244 (1847).
185. Proskauer, C. Development and Use of the Rubber Glove in Surgery and Gynecology. *J Hist Med Allied Sci* **XIII,** 373–381 (1958).
186. Spirling, L. I. & Daniels, I. R. William Stewart Halsted – surgeon extraordinaire: a story of 'drugs, gloves and romance'. *J R Soc Promot Health* **122,** 122–124 (2002).
187. Imber, G. *Genius on the edge: the bizarre story of William Stewart Halsted.* (Kaplan Publishing, 2011).
188. Manteuffel, Z. von, W. Gummihandschuhe in der chirurgischen Praxis. *Zentralbl Chir* **24,** 223–556 (1897).

189. Halsted, W. S. The training of the surgeon. *Johns Hopkins Hospital Bulletin* **162,** 1–24 (1904).
190. Friedrich, C. Fritz Hofmann: Die Synthese des Kautschuks | PZ – Pharmazeutische Zeitung. https://www.pharmazeutische-zeitung.de/ausgabe-432016/fritz-hofmann-die-synthese-des-kautschuks/.
191. N.N. *Schreiben des Reichsgesundheitsrates.* (1908).
192. Hendrik Christian Voß. Die Darstellung der Syphilis in literarischen Werken um 1900 | Enhanced Reader. (Universität Lübeck, 2004).
193. Kissel, T. Ignaz Semmelweis: Der Retter von der traurigen Gestalt – Spektrum der Wissenschaft. *Der Retter von der traurigen Gestalt* https://www.spektrum.de/news/ignaz-semmelweis-der-retter-von-der-traurigen-gestalt/1574034 (2018).
194. Sepkowitz, K. A. One hundred years of Salvarsan. *N Engl J Med* **365,** 291–3 (2011).
195. Vernon, G. Syphilis and Salvarsan. *Br J Gen Pract* **69,** 246 (2019).
196. Srikala, Y., Jyothirmayee, S., Sharadha, V. & Uma, M. R. Serpenditious drug discovery in the field of pharmacy.
197. Baenninger, Alex. Good chemistry : the life and legacy of valium inventor Leo Sternbach. 168 (2004).
198. Roy, J. Lifestyle drugs, statins, and COX-2 drugs. in *An Introduction to Pharmaceutical Sciences* (ed. Roy, J.) 297–325 (Elsevier, 2011). https://doi.org/10.1533/9781908818041.297.
199. Epochen/Krankheiten: Konstellationen von Literatur und Pathologie. 287 (2006).
200. Anil, S. Oppenheim – Die Wiege der chemisch-pharmazeutischen Chinin-Industrie in Deutschland – regionalgeschichte.net. *Die Wiege der Pharma-Industrie stand in Oppenheim. Friedrich Koch und die erste deutsche Chininfabrik.* **22,** 2–21 (2000).
201. Slater, L. *Von Menschen und Ratten.* (Beltz, 2015).
202. Moll, A. *Ärztliche Ethik, Die Pflichten des Arztes in allen Beziehungen seiner Thätigkeit.* (Verlag von Ferdinand von Enke, 1902).
203. Fox, J. G. & Bennett, B. T. Laboratory Animal Medicine: Historical Perspectives. in *Laboratory Animal Medicine: Third Edition* 1–21 (Academic Press, 2015). https://doi.org/10.1016/B978-0-12-409527-4.00001-8.
204. Franco, N. H. Animal Experiments in Biomedical Research: A Historical Perspective. *Animals (Basel)* **3,** 238–73 (2013).
205. Duclaux, E. *Pasteur; the history of a mind.* (1920).
206. Darwin, C. *On the Origin of Species.* (John Murray, 1859).
207. Bisby, M. A. Animal rights. *CMAJ* **146,** 1897–1899; discussion 1899 (1992).
208. Cramer, M. Animal rights/liberation philosophy. *FASEB J* **6,** 2489; author reply 2490-1 (1992).
209. Ammann, K. Menschen, Mäuse und Fliegen. Eine wissenssoziologische Analyse der Transformation von Organismen in epistemische Objekte. in

Objekte, Differenzen und Konjunkturen. Experimentalsysteme im historischen Kontext (eds. Hagner, M., Rheinsberger, H. & Währig-Schmidt, B.) 259–289 (Akademie Verlag, 1994).

210. Logan, C. A. Commercial Rodents in America: Standard Animals, Model Animals, and Biological Diversity. *Brain Behav Evol* **93,** 70–81 (2019).
211. Taylor, F. W. *Shop Management*. (Harper & Brothers Publishers, 1911).
212. Clause, B. T. The Wistar rat as a right choice: Establishing mammalian standards and the ideal of a standardized mammal. *J Hist Biol* **26,** 329–349 (1993).
213. Wistar, C. Caspar Wistar established the first successful glass manufacturing business in North America. https://www.immigrantentrepreneurship.org/entries/caspar-wistar/.
214. Hendrickson, R. *More Cunning Than Man: A Social History of Rats and Men*. (Dorset Press, 1983).
215. Caspar Wistar. https://www.immigrantentrepreneurship.org/entries/caspar-wistar/.
216. Donaldson, H. *The Rat*. (The Wistar Institute for Anatomy and Biology).
217. Potter, M. History of the BALB/c Family BT – The BALB/c Mouse: Genetics and Immunology. 1–5 (1985).
218. The BALB/c Mouse. **122,** (1985).
219. Verband forschender Arzneimittelhersteller. *Tierversuche und Tierschutz in der Pharmaindustrie*. https://www.vfa.de/de/arzneimittel-forschung/so-funktioniert-pharmaforschung/tierversuche-und-tierschutz-in-der-pharmaindustrie.html/_5-was-hat-neue-tierschutzgesetz-gebracht.
220. Lindl, T., Välkef, M. & Unterfranken, R. Von. Tierversuche in der biomedizinischen Forschung Lindl_2005.pdf. 143–151 (2005).
221. Hanahan, D., Wagner, E. F. & Palmiter, R. D. The origins of oncomice: a history of the first transgenic mice genetically engineered to develop cancer. *Genes Dev* **21,** 2258–2270 (2007).
222. World Intellectual Property Organization. Bioethics and Patent Law: The Case of the Oncomouse. *WIPO Magazine* (2006).
223. Robins, R. Inventing Oncomice: making natural animal, research tool and invention cohere. *Genom Soc Policy* **4,** (2008).
224. Brinster, R. L. *et al.* Transgenic mice harboring SV40 t-antigen genes develop characteristic brain tumors. *Cell* **37,** 367–379 (1984).
225. Knabe, J. Thalidomid, eine unendliche Geschichte. *Pharm Unserer Zeit* **27,** 66–67 (1998).
226. Viertel, B. Reproduktionstoxikologie: Gesunde Kinder dank Ratten und Kaninchen. *Biologie in Unserer Zeit* **48,** 318–323 (2018).
227. Abraham, K., Wöhrlin, F., Lindtner, O., Heinemeyer, G. & Lampen, A. Toxicology and risk assessment of coumarin: Focus on human data. *Mol Nutr Food Res* **54,** 228–239 (2010).
228. Kobert, R. Pharmakologisches Coreferat über Erteilung von ärztlichen Gutachten über neu erfundene Arzneimittel. In *72. Tagung der Gesellschaft Deutscher Naturforscher und Ärzte* 38–45 (1900).

229. Kaur, R., Sidhu, P. & Singh, S. What failed BIA 10–2474 Phase I clinical trial? Global speculations and recommendations for future Phase I trials. *J Pharmacol Pharmacother* **7,** 120 (2016).
230. Grabar, E. Arzneimitteltest: Warum musste Guillaume Molinet sterben? | ZEIT ONLINE. *Zeit-Online* 1–2 (2016).
231. Skoot, R. *The Immortal Life of Henrietta Lacks. Oprah.Com* (Cown Publishing, 2010).
232. National Institutes of Health. Significant Research Advances Enabled by HeLa Cells. https://osp.od.nih.gov/scientific-sharing/hela-cells-timeline/#2010s (2021).
233. Jedrzejczak-Silicka, M. History of Cell Culture. *undefined* (2017) https://doi.org/10.5772/66905.
234. Taylor, M. W. A History of Cell Culture. in *Viruses and Man: A History of Interactions* 41–52 (Springer International Publishing, 2014). https://doi.org/10.1007/978-3-319-07758-1_3.
235. Nicholas, J. Ross Granville Harrison, 1870–1959. (1961).
236. Witkowski, J. A. Alexis Carrel and the Mysticism of Tissue Culture. *Med Hist* **23,** 279–296 (1979).
237. Rous, P. & Jones, F. S. A METHOD FOR OBTAINING SUSPENSIONS OF LIVING CELLS FROM THE FIXED TISSUES, AND FOR THE PLATING OUT OF INDIVIDUAL CELLS. *J Exp Med* **23,** 549–556 (1916).
238. Kumar, P. & Murphy, F. A. Francis Peyton Rous. *Emerg Infect Dis* **19,** 661–663 (2013).
239. Rodriguez, E. & Wrangham, R. Zoopharmacognosy: The Use of Medicinal Plants by Animals. *Phytochemical Potential of Tropical Plants* 89–105 (1993) https://doi.org/10.1007/978-1-4899-1783-6_4.
240. Engel, C. Zoopharmacognosy. in *Veterinary Herbal Medicine* 7–15 (Elsevier, 2007). https://doi.org/10.1016/B978-0-323-02998-8.50006-8.
241. Suárez-Rodríguez, M., López-Rull, I. & Garcia, C. M. Í. Incorporation of cigarette butts into nests reduces nest ectoparasite load in urban birds: new ingredients for an old recipe? *Biol Lett* **9,** (2013).
242. Busch, W. *Das große farbige Wilhelm Busch Album mit über 1.600 farbigen Illustrationen.* (Bassermann Verlag, 2016).
243. Diamond, J. M. Dirty eating for healthy living. *Nature* **400,** 120–121 (1999).
244. Dini, J. W. Eating Dirt. *Emerg Infect Dis* **9,** 1016 (2003).
245. Efferth, T. *et al.* Biopiracy of natural products and good bioprospecting practice. *Phytomedicine* **23,** 166–173 (2016).
246. Imran, Y., Wijekoon, N., Gonawala, L., Chiang, Y. C. & De Silva, K. R. D. Biopiracy: Abolish Corporate Hijacking of Indigenous Medicinal Entities. *The Scientific World Journal* **2021,** (2021).
247. Wullweber, J. *Das grüne Gold der Gene – Globale Konflikte und Biopiraterie.* (Verlag Westfälisches Dampfboot, 2004).

248. Drayton, R. *Nature's government, Science, Imperial Britain and the 'Improvement' of the World.* (Yale University Press).
249. Zernick, B. Die Botanische Zentralstelle für die deutschen Kolonien. in *Kolonialmetropole Berlin* (eds. Heyden van der, U. & Zeller, J.) 107–111 (Berlin Edition, 2002).
250. Lack, W. Botanische Handbücher für die deutschen Kolonien in Afrika. in *Kolonialmetropole Berlin* (eds. Heyden van der, U. & Zeller, J.) 112–114 (Berlin Edition, 2002).
251. Bernhard Zepernick. Die Botanische Zentralstelle für die deutschen Kolonien. in *Kolonialmetropole Berlin: Eine Spurensuche* (eds. van der Heyden, U. & Zeller, J.) 107–111 (Berlin Edition, 2002).
252. Little, B. Die tödliche Mode des 19. Jahrhunderts. https://www.nationalgeographic.de/geschichte-und-kultur/2018/10/die-toedliche-mode-des-19-jahrhunderts (2018).
253. Baker, D. D., Chu, M., Oza, U. & Rajgarhia, V. The value of natural products to future pharmaceutical discovery. *Natural Product Reports* vol. 24 1225–1244 Preprint at https://doi.org/10.1039/b602241n (2007).
254. ETC Group. Playing God in the Galapagos. https://www.etcgroup.org/content/playing-god-galapagos.
255. US Department of Commerce Patent and Trandemark Office. *NOTICE OF INTENT TO ISSUE REEXAMINATION CERTIFICATE, STATEMENT OF REASONS FOR PATENTABILITY AND/OR CONFIRMATION.* (2001).
256. Kolodziej, H. & Kayser, O. Pelargonium | PELARGONIUM SIDOIDES DC. NEUESTE ERKENNTNISSE ZUM VERSTANDNIS DES PHYTOTHERAPUETIKUMS UMCKALOABO. *Zeitschrift fur Phytotherapie* (1998).
257. Taylor, P., Maalim, S. & Coleman, S. The strange story of umckaloabo. *Pharmaceutical Journal* **275,** 790–792 (2005).
258. Ernst, E. Stevens' cure for tuberculosis. *J R Soc Med* **95,** 575 (2002).
259. Charles, H. Medical evidence given in the consumption cure libel action: Stevens v The British Medical Association. Contributors. **44,** (1912).
260. N.N. Medico Legal. Stevens's Consumption Cure. *British Medical Journal* 578 (1910).
261. N.N. Medico-Legal. Stevens vs British Medical Association. *British Medical Journal* 1170 (2012).
262. N.N. The Life Everlasting Plant. 3 (1914).
263. Green, J. 052: Dr Arthur Bennett and „Umckaloabo", 1914–1915 – Jeffrey Green. Historian. https://jeffreygreen.co.uk/052-dr-arthur-bennett-and-umckaloabo-1914-1915/ (2019).
264. N.N. The Report of the Select Committee on Patent Medicines. *The Lancet* **184,** 710–712 (1914).
265. Siegel, R. K. & Brodie, M. Alcohol self-administration by elephants. *Bulletin of the Psychonomic Society* **22,** 49–52 (1984).

266. Müller, J. L. Love potions and the ointment of witches: Historical aspects of the nightshade alkaloids. *Clin Toxicol* **36,** 617–627 (1998).
267. Gesellschaft & Religion – Perus Koka-Kultur bringt den Himmel ein Stück näher – Kultur – SRF. https://www.srf.ch/kultur/gesellschaft-religion/perus-koka-kultur-bringt-den-himmel-ein-stueck-naeher.
268. MCGONIGAL, K. Your Brain on Meditation. *Yoga Journal* (2015).
269. Jungaberle, A. Yoga, Tee, LSD. in *Yoga, Tee und LSD* (Schattauer Verlag, 2022).
270. Lee, M. R. The history of Ephedra (ma-huang). *J R Coll Physicians Edinb* **41,** 78–84 (2011).
271. N.N. Neues aus Japan – Nagayoshi Nagai (1845–1929). *Neues aus Japan* https://www.de.emb-japan.go.jp/NaJ/NaJ1210/nagai.html (2012).
272. Kim, H.-E. *Doctors of empire: medical and cultural encounters between Imperial Germany and Meiji Japan.* (University of Toronto Press, 1976).
273. Hartmann, R. *Japanische Studenten an der Berliner Universität 1870–1914.* (2000).
274. Chen, K. K. A pharmacognostic and chemical study of ma huang (Ephedra vulgaris var. helvetica). 1925. *J Am Pharm Assoc (2003)* **52,** 406–412 (2012).
275. Chen, K. K. Ephedrine and related substances. in (ed. Foundation, W.) 1–144 (1930).
276. Stein, S. W. & Thiel, C. G. The History of Therapeutic Aerosols: A Chronological Review. *J Aerosol Med Pulm Drug Deliv* **30,** 20–41 (2017).
277. Yamashima, T. Adrenaline/Epinephrine Hunters: Past, Present, and Future at 1900. *Emerg Med Investig* **4,** 145 (2017).
278. Yamashima, T. Jokichi Takamine (1854–1922), The Samurai Chemist, and His Work on Adrenalin. https://doi.org/10.1177/096777200301100211 **11,** 95–102 (2016).
279. Mieda, R. *et al.* Comparison of four documents describing adrenaline purification, and the work of three important scientists, Keizo Uenaka, Nagai Nagayoshi and Jokichi Takamine. *J Anesth Hist* **6,** 42–48 (2020).
280. Billington, C. K., Penn, R. B. & Hall, I. P. β2 Agonists. in *Handbook of Experimental Pharmacology* vol. 237 23–40 (2016).
281. Barnes, P. J. Drugs for asthma. *Br J Pharmacol* **147,** 297–303 (2006).
282. Gouzoulis-Mayfrank, E. & Daumann, J. *Amphetamine, Ecstasy und Designerdrogen.*
283. Edeleano, L. Über einige Derivate der Phenylmethacrylsäure und der Phenylisobuttersäure. *Berichte der deutschen chemischen Gesellschaft* **20,** 616–622 (1887).
284. *Die Zeiten verfliegen wie im Rausch.* (J. G. Cotta'sche Buchhandlung, 2019).
285. Kupfer, A. Die künstlichen Paradiese: Rausch und Realität seit der Romantik: ein Handbuch. 778 (1996).
286. Knust, C. Drogenmissbrauch: Koks für die Kunst | ZEIT ONLINE. *Zeit Magazin* 4 (2009).

287. David J. Garrow. *Rising Star: The Making of Barack Obama*. (William Morrow, 2017).
288. Bock, C. Rio Reiser: Drogen machten 'den König' kaputt. https://www.stern.de/kultur/musik/rio-reiser-drogen-machten--den-koenig--kaputt-3596344.html (2006).
289. Trimborn, J. 1971–2012. Ein Tag ist ein Jahr ist ein Leben Rainer Werner Fassbinder; die Biographie.
290. Finney, N. S. & Siegel, J. S. In Memoriam: Albert Hofmann (1906–2008). *Chimia (Aarau)* **62,** 444 (2008).
291. Hofmann, A. Historical view on ergot alkaloids. *Pharmacology* **16,** 1–11 (1978).
292. Hofmann, A. *LSD mein Sorgenkind, Die Entdeckung einer Wunderdroge*. (Klett-Cotta, 1979).
293. Passie, T., Halpern, J. H., Stichtenoth, D. O., Emrich, H. M. & Hintzen, A. The pharmacology of lysergic acid diethylamide: A review. *CNS Neuroscience and Therapeutics* Preprint at https://doi.org/10.1111/j.1755-5949.2008.00059.x (2008).
294. Steiner, U. & Leistner, E. Ergot Alkaloids and their Hallucinogenic Potential in Morning Glories. *Planta Medica* vol. 84 751–758 Preprint at https://doi.org/10.1055/a-0577-8049 (2018).
295. Jiang, H.-E. *et al.* A new insight into Cannabis sativa (Cannabaceae) utilization from 2500-year-old Yanghai Tombs, Xinjiang, China. *J Ethnopharmacol* **108,** 414–422 (2006).
296. Mechoulam, R. & Hanuš, L. A historical overview of chemical research on cannabinoids. *Chemistry and Physics of Lipids* vol. 108 1–13 Preprint at https://doi.org/10.1016/S0009-3084(00)00184-5 (2000).
297. Shahbazi, F., Grandi, V., Banerjee, A. & Trant, J. F. Cannabinoids and Cannabinoid Receptors: The Story so Far. *iScience* vol. 23 Preprint at https://doi.org/10.1016/j.isci.2020.101301 (2020).
298. Graham, E. S., Ashton, J. C. & Glass, M. Cannabinoid receptors: A brief history and 'what's hot'. *Frontiers in Bioscience* **14,** 944–957 (2009).
299. Crunkhorn, S. Drug design: Cannabinoid receptor structure revealed. *Nature reviews. Drug discovery* vol. 15 822 Preprint at https://doi.org/10.1038/nrd.2016.242 (2016).
300. Maa, E. & Figi, P. The case for medical marijuana in epilepsy. *Epilepsia* **55,** 783–786 (2014).
301. Russo, E. B. Cannabis and epilepsy: An ancient treatment returns to the fore. *Epilepsy and Behavior* vol. 70 292–297 Preprint at https://doi.org/10.1016/j.yebeh.2016.09.040 (2017).
302. Friedman, D. & Sirven, J. I. Historical perspective on the medical use of cannabis for epilepsy: Ancient times to the 1980s. *Epilepsy and Behavior* vol. 70 298–301 Preprint at https://doi.org/10.1016/j.yebeh.2016.11.033 (2017).

303. Lattanzi, S. *et al.* Efficacy and Safety of Cannabidiol in Epilepsy: A Systematic Review and Meta-Analysis. *Drugs* vol. 78 1791–1804 Preprint at https://doi.org/10.1007/s40265-018-0992-5 (2018).
304. Tomida, I., Perlwee, R. G. & Azuara-Blanco, A. Cannabinoids and glaucoma. *British Journal of Ophthalmology* Preprint at https://doi.org/https://doi.org/10.1136/bjo.2003.032250 (2004).
305. Mosaed, S. *Cannabis, Glaucoma and Intraocular Pressure.* https://www.reviewofophthalmology.com/article/cannabis-glaucoma-and-intraocular-pressure (2022).
306. Green, K. Marijuana smoking vs cannabinoids for glaucoma therapy. *Archives of Ophthalmology* **116,** 1433–1437 (1998).
307. Jarho, P., Pate, D. W., Brenneisen, R. & Järvinen, T. Hydroxypropyl-β-cyclodextrin and its combination with hydroxypropyl- methylcellulose increases aqueous solubility of Δ9-tetrahydrocannabinol. *Life Sci* (1998) https://doi.org/10.1016/s0024-3205(98)00528-1.
308. Hari, J. Anslinger Politico. *The Hunting of Billie Holiday* https://www.politico.com/magazine/story/2015/01/drug-war-the-hunting-of-billie-holiday-114298/ (2015).
309. Kinder, D. C. Bureaucratic Cold Warrior: Harry J. Anslinger and Illicit Narcotics Traffic. *Pac Hist Rev* **50,** 169–191 (1981).
310. McWilliams, J. C. *The protectors: Harry J. Anslinger and the Federal Bureau of Narcotics, 1930–1962.* (University of Delaware Press, 1949).
311. Asakawa, Y., Hashimoto, T., Takikawa, K., Tori, M. & Ogawa, S. Prenyl bibenzyls from the liverworts Radula perrottetii and Radula complanata. *Phytochemistry* (1991) https://doi.org/10.1016/0031-9422(91)84130-K.
312. Hussain, T. *et al.* Demystifying the liverwort Radula marginata, a critical review on its taxonomy, genetics, cannabinoid phytochemistry and pharmacology. *Phytochemistry Reviews* Preprint at https://doi.org/10.1007/s11101-019-09638-8 (2019).
313. Crocq, M. A. History of cannabis and the endocannabinoid system. *Dialogues Clin Neurosci* **22,** 223–228 (2020).
314. Russo, E. B. Beyond Cannabis: Plants and the Endocannabinoid System. *Trends in Pharmacological Sciences* Preprint at https://doi.org/https://doi.org/10.1016/j.tips.2016.04.005 (2016).
315. Fattore, L., Melis, M., Fadda, P., Pistis, M. & Fratta, W. The endocannabinoid system and nondrug rewarding behaviours. *Experimental Neurology* vol. 224 23–36 Preprint at https://doi.org/10.1016/j.expneurol.2010.03.020 (2010).
316. Daly, J. W. *et al.* Alkaloids from frog skin: The discovery of epibatidine and the potential for developing novel non-opioid analgesics. *Natural Product Reports* vol. 17 131–135 Preprint at https://doi.org/10.1039/a900728h (2000).
317. Evans, D. A., Scheidt, K. A. & Downey, C. W. Synthesis of (−)-epibatidine. *Org Lett* **3,** 3009–3012 (2001).

318. Miranda C., M., Navarrete T., L. & Zúñiga N., G. Niccolo Paganini: Medical aspects of his Life and work. *Rev Med Chil* **136,** 930–936 (2008).
319. van Dongen, P. W. J. & de Groot, A. N. J. A. History of ergot alkaloids from ergotism to ergometrine. *European Journal of Obstetrics and Gynecology and Reproductive Biology* **60,** 109–116 (1995).
320. Moir, J. C. Ergot: From 'St. Anthony's Fire' to the isolation of its active principle, ergometrine (ergonovine). *Am J Obstet Gynecol* **120,** 291–296 (1974).
320. Küster, H. *Am Anfang war das Korn. Am Anfang war das Korn* (Verlag C.H.BECK oHG, 2012). https://doi.org/10.17104/9783406652189.
322. Lee, M. R. The history of ergot of rye (Claviceps purpurea) II: 1900–1940. *J R Coll Physicians Edinb* **39,** 365–369 (2009).
323. Jacobs, W. A. & Craig, L. C. The ergot alkaloids XI. Isomeric Dihydrolysergic acids and the structure of Lysergic acid, 227–238 (1936).
324. Uhle, F. C. & Jacobs, W. A. THE ERGOT ALKALOIDS. XX. THE SYNTHESIS OF DIHYDRO-dZ-LYSERGIC ACID. A NEW SYNTHESIS OF 3-SUBSTITUTED QUINOLINES. **52,** 50 (2022).
325. Eadie, M. J. Ergot of rye – The first specific for migraine. *Journal of Clinical Neuroscience* **11,** 4–7 (2004).
326. Gutiérrez, J. M. *et al.* Snakebite envenoming. *Nat Rev Dis Primers* **3,** 17063 (2017).
327. Kasturiratne, A. *et al.* The Global Burden of Snakebite: A Literature Analysis and Modelling Based on Regional Estimates of Envenoming and Deaths. *PLoS Med* **5,** e218 (2008).
328. Smith, C. G. & Vane, J. R. The Discovery of Captopril. *The FASEB Journal* **17,** 788–789 (2003).
329. Cushman, D. W. & Ondetti, M. A. History of the design of captopril and related inhibitors of angiotensin converting enzyme. *Hypertension* vol. 17 589–592 Preprint at https://doi.org/10.1161/01.HYP.17.4.589 (1991).
330. The Discovery of Camptothecin and Taxol® (2003).
331. Howat, S. *et al.* Paclitaxel: Biosynthesis, production and future prospects. *New Biotechnology* vol. 31 242–245 Preprint at https://doi.org/10.1016/j.nbt.2014.02.010 (2014).
332. Renneberg, R. Biotech History: Yew trees, paclitaxel synthesis and fungi. *Biotechnol J* **2,** 1207–1209 (2007).
333. Onrubia, M. *et al.* Bioprocessing of plant in vitro systems for the mass production of pharmaceutically important metabolites: paclitaxel and its derivatives. *Curr Med Chem* **20,** 880–91 (2013).
334. Bailey, J. E. Toward a science of metabolic engineering. *Science (1979)* **252,** 1668–1675 (1991).
335. Narahashi, T. Tetrodotoxin – A brief history. *Proc Jpn Acad Ser B Phys Biol Sci* **84,** 147–154 (2008).
336. Bucciarelli, G. M. *et al.* From Poison to Promise: The Evolution of Tetrodotoxin and Its Potential as a Therapeutic. *Toxins (Basel)* **13,** (2021).

337. Kao, C. Y. & Yasumoto, T. Tetrodotoxin in ‚zombie powder'. *Toxicon* **28**, 129–132 (1990).
338. Campos-Ríos, A., Rueda-Ruzafa, L., Herrera-Pérez, S., Rivas-Ramírez, P. & Lamas, J. A. Tetrodotoxin: A new strategy to treat visceral pain? *Toxins* vol. 13 Preprint at https://doi.org/10.3390/toxins13070496 (2021).
339. González-Cano, R. *et al.* Tetrodotoxin, a potential drug for neuropathic and cancer pain relief? *Toxins* vol. 13 Preprint at https://doi.org/10.3390/toxins13070483 (2021).

Further Reading

Survey Literature

Wolf-Dieter Müller-Jahncke, Christoph Friedrich, Ulrich Meyer (2005) Arzneimittelgeschichte. Wissenschaftliche Verlagsgesellschaft mbH Stuttgart [Obwohl ein Fachbuch ist es eine gut lesbare Reise durch die Geschichte der Arzneistoffe und ihrer Anwendungen.]

Hansjörg Küster (2012) Am Anfang war das Korn. Verlag C.H. BECK oHG [Ein populärwissenschaftliches Fachbuch über die Getreidesorten, die unsere Kultur beeinflussten.]

Otto Krätz (1999) 7000 Jahre Chemie, Nikol Verlag, Hamburg [Ein schöner und informativer Bildband über die Anfänge der Chemie vom alten Orient bis in die Moderne der Chemieunternehmen heute.]

343. Gisela Graichen (2004) Heilwissen versunkener Kulturen, 3. Auflage, Econ Verlag, München [Ein optisch ansprechendes Buch mit vielen Fotos aus den wichtigen Kulturen der Welt und ihren Traditionen in der Ethnomedizin.]

Amphetamine, Adrenaline and Beta-Blockers

Oliver Sacks (1989) Bewußtseinsdämmerung, VEB Deutscher Verlag der Wissenschaften, Berlin [Sacks ist der große Psychiater mit einem fundierten biochemischen Wissen, der besondere Fälle seiner Praxis beschriebt und neurochemisch zu erklären versucht. Insbesondere L-DOPA steht im Mittelpunkt und war wahrscheinlich die Grundlage für den Kinofilm „Zeit des Erwachens".]

Aspirin

Alfred A. Knopf (1993) The Aspirin Wars: Money, Medicine and 100 Years of Rampant Competition, Harvard Business Review Press, New York [Lesenswertes Buch für alle, die sich für den Wirtschaftskrimi hinter Aspirin interessieren.]

Helga Vollmer (1997) Aspirin, ein Jahrhundertmittel macht Karriere, Wilhelm Heyne Verlag, München [Interessantes Buch zum 100. Geburtstag des Aspirin über seine Geschichten von A wie Anwendung bis Z wie Zahnschmerzen.]

Diarmuid Jeffreys (2004) Aspirin, the remarkable story of a wonder drug, Bloomsbury Publishing, London [Das umfassendste Buch in englischer Sprache zu Aspirin mit vielen Highlights aus der Wissenschaft.]

Heribert von Feilitzsch (2015) The Secret War on the United States in 1915, Eigenverlag [Das Buch zeigt die Bedeutung des Phenol und des Aspirin während des Ersten Weltkriegs in den USA und ihrer Haltung gegenüber dem Deutschen Reich auf.]

Biopiracy

Tyler Whittle (1970) The Plant Hunters, Chilton Book Company, Philadelphia, New York, London [Für Leser, die erfahren möchten, wie die großen Botaniker Pflanzen ihrer Reisen nach Europa brachten und wie diese als Zierpflanzen ihren Platz in unseren Gärten fanden.]

Florian Freistetter, Helmut Jungwirth (2021) Eine Geschichte der Welt in 1000 Mikroorganismen, Carl Hanser Verlag München [Es geht in diesem Buch um Mikroorganismen und nicht um pflanzliche Naturstoffe, aber beide Welten kennen sich und können nicht ohne einander. Die beiden Autoren bieten interessante Anekdoten aus dem Reich der Mikrobiologie.]

Renate Basch-Ritter (2008) Akademische Druck- und Verlagsgesellschaft, Graz [Wenige wissen es, dass Österreich eine Seenation war. Wie James Cook mit der Endeavour, fuhren österreichische Wissenschaftler auf der Novara um die Welt und versuchten sie zu verstehen.]

Quinine

Femme S. Gastra (2001) De Geschiedenis van de VOC, Walburg Pers, Zutphen [Ein spannender Bilderband über das mächtigste Wirtschaftsunternehmen, das es in der Geschichte gab. Texte sind auf Niederländisch geschrieben.]

Henry Hobhouse (2001) Sechs Pflanzen verändern die Welt, Klett-Cotta, Stuttgart, Kapitel Chinarinde, S. 19–66, 4. Auflage

Kim Walker und Mark Nesbitt (2019) Just the Tonic, A natural history of tonic water. Royal Botanic Garden, Kew [Ein wunderbar illustriertes Buch, das die

Geschichte der Chinarinde bis zur Entdeckung und Veredlung von Gin und Tonic Water beschreibt. Texte sind auf Niederländisch geschrieben.]

Mark Honigsbaum (2001) The Fever Trail, Pan Books, London [Die Geschichte des Chins und der Chinarinde aus der Sicht zweier Wissenschaftler, die als Biopiraten von Kew Garden in die Anden geschickt wurden.]

Victoria Finlay (2016) Das Geheimnis der Farben, List Taschenbuch, Berlin [Eine amüsante Kulturgeschichte der Farben wie das Mauvein. Mit viel persönlichem Einsatz beschreibt die Autorin die Geschichte und die Bedeutung der Farben und Färbetechniken für die Menschheit.]

Drugs

Norman Ohler (2015) 3. Auflage. Der totale Rausch. Drogen im Dritten Reich. Kiepenheuer & Witsch [Zusammenfassung der Anwendung und Konsequenzen des Methamphetamin in der Weimarer Republik und dem Dritten Reich.]

Hofmann, Albert (2001) LSD – mein Sorgenkind. Klett-Cotta Verlag [Sehr unterhaltsame autobiografische Beschreibung seiner Forschung zu LSD, Psilocybin und der Zauberwinde.]

Exzellente Seite zur Geschichte der Inhalation. http://www.inhalatorium.com/

Poisons

Wie finde ich heraus, ob der Giftfrosch giftig isst? https://www.youtube.com/watch?v=Ryu5Yu0nQuU

Cindy Engel (2004) Wild Health, Gesundheit aus der Wildnis, animal learn Verlag, Bernau, 3. Auflage 2016 [Ein interessanter Einblick in die Verhaltensbiologie von Tieren, die Pflanzen und Mineralien zur Selbstmedikation nutzen.]

Rubber and Medical Latex

Gerald Imber (2011) Genius on the edge: The bizarre double life of Dr. William Stewart Halsted. Kaplan Publishing, New York [Die Biografie eines exzentrischen Arztes mit all seinen Tiefen und Höhen.]

Wolfgang Jünger (1952) Kautschuk, vom Gummibaum zur Retorte, Wilhelm Goldmann Verlag, München, Wien [Das Buch ist alt, aber es ist wahrscheinlich die letzte vollständige Zusammenfassung der technischen Geschichte des Kautschuks.]

Stefan Winkle (2021) Geschichte der Seuchen. Anaconda Verlag, München [Sehr interessante Beschreibung der Syphilis (S. 516–617), aber auch andere Infektionskrankheiten wie der Malaria (S. 702–181) und Tuberkulose (S. 83–152).]

Karl Aloys Schenzinger (1953) bei I.G. Farben. Wilhelm Andermann Verlag, München, Wien [Die Erzählung der Kautschukproduktion vom Samen bis zur Gummiherstellung in Europa.]

Caffeine and Viagra

Mary Banks, Christine McFadden (2016) Coffee, Lorenz Books, Anness Publishing, London [Wunderbares Buch in englischer Sprache, in dem die Geschichte, Kultur und Biologie des Kaffeetrinkens geschildert wird.]

Cocaine

Sigmund Freud und Martha Bernays (2015) Warten in Ruhe und Ergebung, Warten in Kampf und Erregung. Die Brautbriefe Band 3. Herausgegeben von: Gerhard Fichtner, Ilse Grubrich-Simitis, Albrecht Hirschmüller. Verlag S. Fischer, Frankfurt a.M. [Privater Briefwechsel, der sehr persönliche Einsichten in die Entdeckung des Kokains als Lokalanästhetikum zeigt.]

Henry Hobhouse (2001) Vierte Auflage, Kokastrauch, S. 277–350, Klett-Cotta, Stuttgart [Eine historische Beschreibung der Anwendung des Kokastrauches und sein Einfluss auf unsere westliche Kultur.]

Thomas Köhler (2019) Die Zeiten verfliegen wie im Rausch. J.G. Cotta'sche Buchhandlung, Stuttgart [Eine kurze kulturelle, pharmakologische und pharmazeutische Beschreibung der meist genutzten Drogen in unserer Gesellschaft.]

Modern Diseases and Randomness in Science

Vera Buck (2015) Runa, Limes Verlag, München [Spannendes Buch über eine junge Frau in der Salpeterie, die vermeintlich unter Hysterie leidet und behandelt werden soll. Das Buch gibt einen tiefen Einblick in die psychiatrischen Diagnosen, Behandlungen und das Leben in dem größten psychiatrischen Krankenhaus Europas seiner Zeit.]

Modern Drug Research

Gerhard Klebe (2009) Wirkstoffdesign: Entwurf und Wirkung von Arzneistoffen. Kapitel 1 Grundlagen der Arzneimittelforschung. Spektrum Akademischer Verlag, Heidelberg. 2. Auflage [Wenn auch etwas älter, aber das Standardwerk, um moderne Wirkstoffforschung zu verstehen.]

Tonja Koeppel und Leo H. Sternbach (1986) Transkript eines Interviews mit Tonja Koeppel bei Hoffmann-La Roche Inc. Nutley, New Jersey am 12.03.1986.

Philadelphia, PA. The Beckman Center for the History of Chemistry. [Interview mit persönlichen Ansichten zu L.S: Leben und die Entwicklung der Benzodiazepine.]

Peter Slater (2005) Von Ratten und Mensch, die berühmtesten Experimente in der Psychologie. Belz Weinheim und Basel. [Das Wissenschaftsbuch des Jahres 2005, das die Psychologen und die Ideen hinter den Experimenten beschreibt.]

Enrique Ravina (2010) The Evolution of Drug Discovery. Wiley VCH, Weinheim [Ein englisches Fachbuch, das aber einfach die gesamte Geschichte der modernen Wirkstoffentwicklung erklärt.]

Morphine and Other Opiates

Michael de Ridder (2000) Heroin: Vom Arzneimittel zur Droge, Campus Verlag, Frankfurt [Das Buch beschriebt sehr detailliert, wie die Bayer AG aus dem Morphin eines der mächtigsten Schmerzmittel machte und bis zum Verbot verkaufte.]

Stephen R. Platt (2019) Imperial Twilight: The Opium War and the End of China's Last Golden Age, Vintage Books, New York [Englisches Buch, das sich detailliert über das Verhältnis von England und China auslässt oder wie China aus Profitgründen abhängig gemacht wurde.]

Patrick Radden Keefe (2022) Imperium der Schmerzen, Carl Hanser Verlag, München [Ein Buch, das man nicht mehr aus der Hand legt. Es beschreibt die Geschichte und das Wirken der Familie Purdue auf dem Gebiet der Benzodiazepine und Opiate. Vor dem Hintergrund der US-Opiatkrise ist es sehr aktuell.]

Pharmaceutical Companies

Goldschmidt, Saskia (2014) Die Glücksfabrik, dtv Verlagsgesellschaft, München [Die niederländische Übersetzung des Buches De Hormoonfabriek, die die Gründung und den Aufbau der Firma Organon am Beispiel des Testosteron beschreibt.]

Steroids and the Pill

Wolfgang Fröbius (1989) Ein Siegeszug mit Hindernissen, Schering Aktiengesellschaft, Berlin [Lesenswertes Heft der Schering AG, die die Steroidentwicklung aus industrieller Sicht beschreibt.]

Interview von Margaret Sanger (1957) auf YouTube: https://www.youtube.com/watch?v=q4pwJas4En0

Peter Bagge (2013) Woman Rebel, the Margaret Sanger Story. Drawn & Quartely [Ein englischer Comic, der die politische Arbeit von Margaret Sanger beschreibt.]

Shahar Shulamith (1993) Kindheit im Mittelalter. Rowohlt Taschenbücher. [Ein spannendes Buch über Sexualität, Kindererziehung und familiäre Strukturen im Mittelalter. Es ist aber wenig informativ über die Geschichte der Arzneipflanzen.]

Wolfgang Frobenius (1989) Ein Siegeszug mit Hindernissen. Schering Aktiengesellschaft [Ein sehr umfassendes Werk über die Geschichte des Ethinylestradiol aus dem Hauptlaboratorium der ehemaligen Schering AG.]

Gisela Staupe, Lisa Vieth (1996) Die Pille, Rowohlt Verlag Berlin [Das Buch wurde aus Anlass der Ausstellung „Die Pille. Von der Lust und der Liebe" im Deutschen Hygiene-Museum im Jahr 1996 veröffentlicht.]

Animal Experiments and Tissue Cultures

Richard David Precht (2016) Tiere denken. Goldmann Verlag München. [Ein tiefgründiges Buch über das Verhältnis Mensch zu Tier, das in einem Kapitel die ethische Frage der Tierversuche bespricht.]

Rebecca Skloot (2012) Die Unsterblichkeit der Henrietta Lacks: Die Geschichte der HeLa-Zellen. Goldmann Verlag [Eine berührende Lebensgeschichte der Henrietta Lacks und eine gute geschriebene Geschichte über den Siegeszug der HeLa-Zelle.]

Vitamin D

Jochen Haas (2007) Vigantol. Wissenschaftliche Verlagsgesellschaft Stuttgart [Sehr guter geschichtlicher Überblick über Adolf Windaus und die Entdeckung des Vitamin D.]

Peter Karlson (1990) Adolf Butenandt: Biochemiker, Hormonchemiker, Wissenschaftliche Verlagsgesellschaft, Stuttgart [Sehr umfassende Biografie über Adolf Butenandt.]

Other Finds

Der alternative Nobelpreis, den keiner will, aber alle zum Nachdenken anregt. https://improbable.com/

MIX
Papier aus verantwortungsvollen Quellen
Paper from responsible sources
FSC® C105338

If you have any concerns about our products,
you can contact us on
ProductSafety@springernature.com

In case Publisher is established outside the EU,
the EU authorized representative is:
**Springer Nature Customer Service Center GmbH
Europaplatz 3, 69115 Heidelberg, Germany**

Printed by Libri Plureos GmbH
in Hamburg, Germany